LAND RECLAMATION – EXTENDING THE BOUNDARIES

PROCEEDINGS OF THE SEVENTH INTERNATIONAL CONFERENCE OF THE INTERNATIONAL AFFILIATION OF LAND RECLAMATIONISTS RUNCORN/UNITED KINGDOM/13–16 MAY 2003

Land Reclamation
Extending the Boundaries

Edited by

Heather M. Moore & Howard R. Fox
Department of Geography, University of Derby, UK

Scott Elliott
Leyden Kirby Associates, UK

A.A. BALKEMA PUBLISHERS   LISSE / ABINGDON / EXTON (PA) / TOKYO

COVER-PHOTO:
Taf Bargoed Valley, Merthyr Tydfil, Wales, UK. Three former collieries extending to over 60 ha reclaimed for a community park, climbing centre and minewater discharge treatment works. Photograph courtesy of Welsh Development Agency and Groundwork.

Published by: A.A. Balkema, a member of Swets & Zeitlinger Publishers
www.balkema.nl and www.szp.swets.nl

ISBN 90 5809 562 2

Land Reclamation – Moore, Fox & Elliott (eds)
© 2003 Swets & Zeitlinger, Lisse, ISBN 90 5809 562 2

Table of Contents

Legislation and Land Reclamation

Addressing Difficult Sites

BSSS/IPSS – Soils and Land Reclamation

Increasing Biodiversity

Reclamation Research: Achievements and Challenges

Soil Remediation

Soil Restoration

Case Studies

Innovative Techniques

Chinese Case Studies

Preface

Major programmes of land reclamation, aimed at providing new uses for previously developed land, were pursued in the United Kingdom and in other industrialised countries around the world throughout the latter part of the 20th Century. Such programmes have formed a critical first step in the regeneration process and have been largely driven by public sector funding and management.

Over recent years, there has been a noticeable shift in emphasis of public sector funding for regeneration away from the large scale physical changes which reclamation has sought to achieve. Yet, at the same time, problem sites still remain in significant numbers and there has been a politically driven emphasis on the reuse of previously developed land (so called 'brownfield sites') for new development. The consequence has been the necessity to involve a wider range of stakeholders – from public, private and voluntary sectors – in the successful delivery of reclamation and regeneration.

Sustainability of both the process and the outcomes has been an ever present guiding principle for all stakeholders. This theme provided the focus for the last International Land Reclamation Conference held in the UK (Nottingham 1998). The messages from that event – notably the essential need for a thorough analysis of the land resource and for multidisciplinary working – remain as fundamental principles of good practice. The additional themes emerging in the 21st Century encompass the need for robust and innovative approaches to the assessment and delivery process.

The papers published in these proceedings address a range of reclamation issues and capture experience from the continents of Europe, Africa, and America as well as the UK. Administrative, technical and legal perspectives are considered and a wide range of case studies are presented; with the objective of sharing knowledge and experience for the benefit of all those involved in defining and delivering project outcomes.

The papers were presented at Land Reclamation 2003, the seventh in a series of conferences organised under the banner of the International Affiliation of Land Reclamationists (IALR). Previous conferences in the series were held in USA, Australia, Canada and China. Organisations in these countries provide the core membership of IALR. The conference was jointly organised by the National Land Reclamation Panel (NLRP) and the British Land Reclamation Society (BLRS), in association with the Institute of Professional Soil Scientists and CL:AIRE.

The National Land Reclamation Panel has existed for over 30 years as a representative body of Local Authorities throughout the UK active in land reclamation, urban regeneration and restoration of mineral workings. Its principle aims are to disseminate information, provide unbiased technical advice to Local Authorities and promote research into reclamation of derelict and contaminated land.

The British Land Reclamation Society is a multidisciplinary organisation with members drawn from both public and private sector and acts as a forum for discussion on land reclamation and regeneration and for the promotion of best practice and research.

CL:AIRE is a public/private partnership, incorporated in 1999. Its purpose is to demonstrate and research new technologies for the sustainable and cost-effective remediation of contaminated land. Included within the technical remit are innovative methods for site characterisation and monitoring.

The Institute of Professional Soil Scientists (IPSS), founded in 1991, is a professional body which aims to promote and enhance the status of soil science and allied disciplines. It prescribes the professional standards accepted by its members and strives continually to advance their competence encompassing scientific/technical expertise. Many of its members are actively involved in land reclamation research and consultancy.

Land Reclamation – Moore, Fox & Elliott (eds)
© 2003 Swets & Zeitlinger, Lisse, ISBN 90 5809 562 2

Conference Organising Committee

Co-chairman of the Organising Committee: Howard Fox (*University of Derby*) and
Dave Morris (*Dudley Metropolitan Borough Council*)

For the British Land Reclamation Panel
Scott Elliott *Leyden Kirby Associates*
Heather Moore *University of Derby*
Malcolm Reeve *Land Research Associates*
Steve Smith *Welsh Development Agency*
Elizabeth Simmons *Robert Long Consultancy Ltd*

For the National Land Reclamation Panel
John Ellison *Halton Borough Council*
Alison Hesselberth *Newcastle-upon-Tyne City Council*
Fiona Stagg *Stoke-on-Trent City Council*

For the British Society of Soil Science
Scott Young *University of Nottingham*

For CL:AIRE
Paul Beck

The Organising Committee are grateful for the support of the following sponsoring agencies and companies:
Welsh Development Agency, English Partnerships, Scottish Enterprise, Northwest Development Agency,
Forestry Commission, United Utilities, Peel Holdings, Halton Borough Council, Environment Agency.

Editors:
Heather Moore
Department of Geography, University of Derby
Howard Fox
Department of Geography, University of Derby
Scott Elliott
Leyden Kirby Associates

Catalysing Regeneration

Land Reclamation – Moore, Fox & Elliott (eds)
© *2003 Swets & Zeitlinger, Lisse, ISBN 90 5809 562 2*

The municipal engineer in regeneration and the "Brownfield Debate"

P. Kirby & S. Elliott
Leyden Kirby Associates, Bury, Lancashire, UK

ABSTRACT: This paper charts the growth of concern in society about the effects of urban deprivation and the mechanisms put into place to deal with this problem. The paper goes on to outline how this concern has combined with increasing "environmental awareness" to demand a response by engineers, particularly in relation to the art of the re-using and recycling despoiled or "brown" land. Consideration is given to the manner in which engineers have needed to expand their awareness to deal with the various environmental issues. The Legislative and policy issues driving the "brownfield" remediation sector are considered, as are a typical range of technical problems and some suggestions as to their solution.

1   INTRODUCTION

Municipal Engineering as a distinct profession has long been merged into the more laterally organised profession of Civil Engineering. However, the design and provision of new and improved municipal schemes still exercises many Engineers and attracts increasing amounts of financial investment. The engineer engaged in such work has always been required to produce his or her work in conformity with the requirements of society. Society now requires that such work makes the best possible use of existing resources and places the least possible demands upon the environment.

This paper looks at how Municipal Engineers have needed to adjust their activities particularly in the need to become more involved in the field of "increasing the value of land" in relation to the reuse of previously used "Brown Land".

2   THE CHANGING ROLE OF THE ENGINEER

2.1   *Introduction*

The vast majority of municipal reclamation projects have fallen within the province of the engineer with regard to their detail design and construction. The engineer has needed to absorb and implement the effects of environmental legislation, in implementing these projects, which are still required to improve our standard of life and safety.

Much that is really useful and valuable about the control of engineering projects is inherently embedded in individual experience combined with scientific judgment. To be competent in the management of projects one needs a mixture of formal knowledge, skills and behaviors.

Engineers are therefore expected to gain from experience of best appropriate practice, which will, if applied, give the best chance of successful project outcome.

It is in the application of the more recent need to also obtain the "best environmental performance" from a project that the engineer now needs to add to their traditional skills.

2.2   *"Traditional" engineering project management*

Traditionally engineers have been expected to participate at many stages, during the currency of a particular Project, in order to achieve its delivery both to time and to the most appropriate cost.

2.2.1   *Pre-project assembly and procurement*

In advance of project implementation, the engineer is often required exercise skill and experience in "setting up" the project. This activity will be in accordance with the client's general procurement procedures.

Familiarity with procurement procedures and an understanding of achievable objective setting will be needed, together with experience in resourcing and the need to construct robust communication links will be required. These will include:

– the provision of procurement advice, in relation to assembly of strategic project objectives and project scoping;
– assistance with selection process for project contractors;

- advice upon the definition of project related objectives;
- advice upon project quality plan compilation and production;
- advice upon appropriate project procurement method and appropriate conditions of contract.
- advice upon required resources in relation to the supervision of a project; and
- implementation of lines of communication.

2.2.2 The engineer as project "champion"

Traditionally, the engineer has been expected to assume the role as project leader or champion. The engineer's responsibilities in this role are to:

- achieve, by the effective demonstration of experience and capability, an acceptance of the engineer in the position as team leader; and
- clearly define and communicate the client's requirements and expectations of a project and link these to the overall objectives of the project.

The achievement of an acknowledged position of "leadership" is perhaps the most difficult to define in concrete terms. It is a position achieved by a combination of experience, maturity, demonstration of personal integrity and application of interpersonal skills. A commitment to the client's strategic and project specific objectives will also form a crucial part of this role.

2.2.3 Essential participant in a project team

A major advantage in facilitating the management of any project will be an appreciation and knowledge of the modus operandi, technical factors, philosophy and driving forces applying to other organizations that form part of the project team.

The engineer requires an appreciation of the roles of the respective organizations. For example, it is important to understand how contractors, who carry the responsibility for project construction, operate. Conventionally this has been within the framework of formal specifications and conditions of contract. The crucial requirement of any contracting organization is in delivering a project in a manner that also provides a profit to the contractor.

In addition, there are also the designers, who are responsible for providing the technical design of the projects. The design of a particular scheme may be the responsibility of the same organisation as that responsible for its implementation. However, the role of designer and project engineer are quite different and an appreciation of the need for the non-partial role of the project engineer is crucial to a project's performance.

2.2.4 Programme/contract progress monitoring

The engineer is also required to possess a familiarity with all types of contract delivery systems, including programming techniques, particularly an understanding of contract critical path identification. The need to appreciate the effect of resource allocation upon the achievement of project milestones is also important as is the ability to distinguish between "real" and "perceived" delays and the techniques available to negate the effect of delays and keep the project to programme.

Experience in project cash flow and project finance requirements is essential and this experience should encompass:

- an understanding of the relationship between contract programmes and actual progress;
- an understanding of the governance of critical activities upon progress;
- the effects of delays, particularly unforeseen circumstances;
- the effects on project financial performance of delay and disruption; and
- consequences of delays and procedures in mitigation.

3 THE INTEGRATION OF ENVIRONMENTAL RESPONSIBILITIES INTO PROJECTS

3.1 Introduction

The days of a project being considered successful by relating it purely to the achievement of technical objectives are no longer tenable. As a consequence, to the responsibilities outlined above, the engineer will now require the commitment to impart the concept of "environmental responsibility" in relation to the reclamation project.

This will require to be achieved by a combination of personal vision and acceptance aligned with the objectives and a thorough understanding of the "whole life environmental costs" of a project related to the current technical and legislative guidance relating to environmental protection.

3.2 Knowledge of legislation, technical and legal guidance and the regulators

In order for the engineer to fulfill his or her enhanced role, as environmental guardian is required to achieve an appreciation and familiarity with the principles of environmental legislation, particularly the Environmental Protection Act 1990 and accompanying statutory guidance.

An appreciation of the technical aspects of qualitative and quantitative modeling systems used to conduct risk appraisal particularly related to contamination issues is also considered essential.

Equally important is an understanding of:

- health and safety legislation;
- waste management regulations; and
- construction, design and management regulations.

The existence and legitimate interests of other responsible organizations, including Local Authority Planning and Environmental Health Departments is important as is an awareness of aspects of the project that may be politically sensitive or require public relation skills.

4 BROWNFIELD DEVELOPMENT

The factors that have combined to produce the vast amount of brownfield land in the UK are a consequence of the decline in traditional heavy industries in and adjacent to our major conurbations.

This had the effect, in addition to widespread unemployment, of leaving vast tracts of derelict and unused land that were formally the location of "old" industries. These industries were usually based on potentially contaminating activities, such as heavy engineering, metal production, chemical production and refining, mining, etc.

Although these sites had very obvious problems, they were usually situated close to population centres and unless recovered and re-used would simply remain as a continuing "sore" and in some cases an actual physical danger. Consideration of their beneficial re-use also would have the added advantage of relieving pressure on "green" land.

At the same time appreciation of the effects of environmental pollution had become to be better understood, usually, regrettably, as a result of a series of massive disasters such as Seveso in Italy 1976, Bhopal in India 1984. Mercury ingestion from contaminated fish in Japan, asbestos related problems and the Love Canal site in the USA also increased the awareness of the general public to contamination issues.

It is important at the same time not to lose sight of the purpose which underpins the UK Government's Regeneration Programme. This specifically mentions "needs and priorities associated with poverty and deprivation .... (including) long term and youth unemployment, low skills levels, un-competitive industry, poor health and education, a rundown physical environment, benefit dependency, ... lone parents .... loss of social cohesion, ethnic minority disadvantage and high levels of crime and drug misuse". The conversion of despoiled land to productive use must therefore be seen as an important component of this social objective, but perhaps not an end in itself?

5 CURRENT LEGISLATIVE FRAMEWORK

5.1 *UK legislation*

The UK contaminated land regime has been developing since 1989 and has recently been finalised in the form of Part IIA of the Environmental Protection Act 1990 (DETR 2000).

The UK legislation has been framed with the intention of not following the US route, in particular to avoid the pursuit of "deep pockets", such as the banks and financial institutions, who may have an interest in a particular piece of land.

However, the principle of "polluter pays", retrospective liability, without time limit and liability under certain circumstances for subsequent occupiers is contained within the regulations. The UK regime thus remains quite severe.

The primary responsibility for the identification and institution of remediation of contaminated land, considered likely to hold the "significant possibility of causing significant harm" to the environment, is placed with Local Authorities. Under certain circumstances, Local Authorities are able to force remediation where sites are considered to be causing serious harm to the environment.

The regime introduces and confirms the Source/Pathway/Target principles to the issue of the requirement to deal with a contaminated site. The practical aspects of this risk based approach are discussed later.

It is necessary before the regulators require clean up, that a demonstration has been made that there is a pollutant linkage between a contaminant (Source) and a Target. Thus, if a contaminant cannot "escape" or is causing no demonstrable harm, the regulators cannot require clean-up. However, as may be appreciated, the problems of demonstration of either the existence or lack of a linkage (Pathway) are likely to be technically complex and expensive.

The legislation also defines the responsibility of persons with an interest in a particular site. This paper will not go into great detail on this subject however, it is interesting to note that under certain circumstances, an owner or occupier of a site may be liable for clean-up even if they did not cause the pollution in the first place.

5.2 *UK planning policy*

National Planning policy has now been aligned with the objective of directing new development to "Brownfield" sites, including sites that may be contaminated.

5.2.1 *Planning policy guidance note 23 (DoE 1994)*
"The principle of sustainable development means that, where practicable, brownfield sites, including those affected by contamination, should be recycled into new uses and the pressures thereby reduced for the greenfield sites to be converted to urban industrial or commercial uses. Such recycling can also provide an opportunity to deal with the threats posed by contamination to help the environment." (DoE 1994).

This statement encourages the dual objective of recycling contaminated land and reducing the pressure upon "greenfield" sites. Thus contamination remediation will be encouraged and the protection of sensitive green sites ensured.

The Government's policy set out in PPG 23, towards tackling land contamination is that the works, if any, required to be undertaken for any contaminated site should deal with any unacceptable risks to health or the environment, taking into account its actual or intended use. This is known as the "suitable for use" approach, the aims of which are stated to be:

– to deal with actual or perceived threats to health, safety or the environment;
– to keep or bring back such land into beneficial use; and
– to minimise avoidable pressures on Greenfield sites.

The proposed use of a contaminated site following development will therefore determine the extent to which remediation is required. For example, if housing is proposed on a contaminated site, then it is likely that a higher level of remediation will be required than if the proposed use was for an industrial purpose or a recreational open-space area.

5.2.2 Planning policy guidance note 3 (DETR 2000)

The Government's approach to housing is set out in this revised PPG3, which states that local planning authorities should: "Provide sufficient housing land but give priority to the re-use of previously developed land, bringing empty homes back into use and promoting the conversion of existing buildings within urban areas, in preference to the development of Greenfield sites." (DETR 2000).

The guidance goes on to state that the Government's national target is that 60% of additional housing should be provided on previously developed land or through conversions. The Government expects this national target to be underpinned by regional targets set in regional planning guidance and local planning authorities are told that they should each adopt their own land recycling targets.

The PPG also seeks to apply a progressive approach to the release of housing land. In other words, the Government does not expect Greenfield sites to be released for development until other options, including the use of previously developed sites within the urban area, have been considered. It is stated that where applications do not meet the criteria of the sequential approach, local planning authority should reject development proposals for Greenfield sites where "there is a realistic unrealised potential to develop on suitable previously developed urban sites".

The guidance clearly demonstrates that the Government is determined to encourage the re-use of previously developed land. Although not all previously developed land will be contaminated, it is inevitable that a certain percentage of such sites will be contaminated, some seriously.

6 CURRENT STANDARDS IN RECLAMATION

6.1 Background

Prior to the 1990s, an engineer's involvement in the reclamation sector placed great reliance upon a mechanistic approach to solving the problem. Sites were investigated in a more or less methodical manner, for the presence of what was known as a "Red List" contaminants. The contaminants were identified by reference to testing laboratories and compared with a series of values determined by a government sponsored committee known as The Interdepartmental Committee for the Reclamation of Contaminated Land (ICRCL 1987). Where gaps in the ICRCL standards existed, these were filled by reference either to Continental or US Standards.

The ICRCL standards were intended to be used with the benefit of appropriate professional judgement, but generally have tended to be used in an unintended manner, as absolute standards. The materials identified as being above ICRCL "Action" levels, were generally excavated and transported to either licensed external tips or engineered repositories. In addition this procedure generally led to "clean" material being imported to replace that discarded. This is an extremely simplistic description of what was a pragmatic and predictable solution to a perceived problem.

What has been realised, however, is that the continuation of this procedure would ultimately fill every available licensed facility in the country with material which would ultimately present a problem of control and containment to future generations. The implementation of the Landfill Tax, although usually not applied to reclamation activity, has had the effect of making repository space much more expensive for all "through the gate materials".

The industry, over the last five or so years, has now moved to a Risk Based Approach, together with the introduction of more sophisticated in-situ remediation techniques.

6.2 Risk based approach

Described very simply, the Risk Based Approach is a procedure, which sets out to determine whether a particular contaminant can, by a credible mechanism, cause harm to a given target. A wide range of targets are defined which include the wider, environment, eco-systems, living organisms (including humans) and property.

As may be seen, the existence of a contaminant, no matter how virulent, can be perfectly "safe" if there is a lack of a route, or pathway, to a target. The Risk Based Approach is, therefore, largely the consideration of such linkages, and remediation techniques may be considered as the determination of how such linkages may be severed.

Risk Assessment may be a qualitative demonstration of lack of a credible linkage or, usually in cases where relatively mobile contaminants are concerned, a quantitative or mathematical model, which could, for example, calculate the rate of spread of a contaminant plume.

The UK approach to managing contaminated land is risk based. Risk management principles underlie the legislative requirements of Part IIA of the Environmental Protection Act and the "suitable for use" approach used in other contexts such as planning and development control (DETR, EA & NHBC 2000).

The approach is founded on the concept of the contaminant-pathway-receptor relationship, or pollutant linkage. Each site therefore, must be subject to an individual risk assessment with the sensitivity of the proposed use critical to its outcome. The nature of the pollutant linkages will depend on the proposed end-use, the design and layout of the proposed development, whether or not it has gardens or landscaped areas and what materials and precautions are to be used in its construction.

In terms of risk to human health, inhabitants of private residential housing with private gardens are at a far greater risk from contaminant hazards than inhabitants of flats or houses with managed gardens or than people working in a proposed business or industrial development. That risk is also influenced by factors such as the nature of the contamination, its location, the quantities and concentrations present, the physical and chemical properties of the soil and the bioavailability of the contaminant (DETR, EA & NHBC 2000).

The first step to assessing the potential contamination issues associated with a brownfield site is to identify the hazard. Desk studies are a fundamental part of this process and involve the reviewing of historical maps, aerial photos, geology and hydrogeology maps and trips to local library archives. This information can then be used to create a conceptual model, which highlights the potential sources, potential pathways and potential receptors associated with a particular site. To complete this model properly the end-use of the site needs to be known.

If a hazard has been identified then the next stages of the process involve the risk estimation and the risk evaluation of that hazard. The risk estimation will involve some form of ground investigation, which will be undertaken to collect sufficient data to allow estimation of the risks that hazardous substance may pose to defined receptors under defined conditions of exposure. The risk evaluation is the process where all the information for the site is reviewed to decide whether the estimated risks are unacceptable.

Up until the 14th March 2002, guideline values for contaminants encountered on sites in the UK were based on the target threshold and action values given in the Interdepartmental Committee on the Reclamation of Contaminated Land report 59/83 (ICRCL 1987). However, after months of waiting, local authorities can now use new guideline values developed to be used as a generic assessment criteria in risk estimation for the protection of human health. These are based on the recently released Environment Agency reports CLR7 (DEFRA & EA 2002a), CLR8 (DEFRA & EA 2002b), CLR9 (DEFRA & EA 2002c) and CLR 10 (DEFRA & EA 2002d) in conjunction with CLR Tox 1-10 (DEFRA & EA 2002e) and CLR GV 1-10 (DEFRA & EA 2002f).

The new guideline values are based on the Contaminated Land Exposure (CLEA) Model, which evaluates the exposure of humans via different pathways that can occur for a range of different land-uses. Guideline values are then set for each contaminant in the soil for each land-use.

The risk assessment procedures also need to assess risk to other potential receptors such as controlled waters (rivers, streams and groundwaters) and ecosystems. The Environment Agency has published tools such as the "Methodology for the Derivation of Remedial Targets for Soil and Groundwater to Protect Water Resources" (EA 1999) which provides a framework for considering the risks to water quality.

6.3 Problems to be considered

An example of the potential range of problems, due to contamination from previous uses can be highlighted using the example of a site with an historic use as a steelworks.

The acidity or alkalinity of the soil can range from low pH (acid) to high pH (alkaline), both of which can give problems. Acid soils can affect building materials and increase the solubility of metals and other contaminants. Highly alkaline soils can also attack some structural materials. Although alkaline conditions can reduce the mobility of some contaminants, they can under some circumstances increase the solubility of some metals. Both extremes of pH are potentially harmful to humans and the aquatic environment and may poison plants (phytotoxic).

Residual metallic compounds from steel-making could include: Arsenic, Cadmium, Chromium, Lead, Mercury, Selenium, Manganese, Copper, Zinc and Nickel. Whilst some of these metals may be tolerable or even essential to human beings at "trace" levels, some have no known beneficial function and can accumulate in the body, to ill effect. A good example is the association between high lead levels and reduced intelligence, in children. Copper, Nickel and Zinc are of particular concern due to their tendency to accumulate in plants.

Sulphates, if acid soluble, are phytotoxic, or if water soluble are aggressive to concrete. Phenols, which are a product of coal carbonisation, can permeate "plastic" water pipes and taint water supplies, whereas chlorides can attack buried metal pipes. Cyanides, are common, arising from coal carbonisation and blast furnace gases and can range in nature from complex salts to "free" cyanides. Their potential for harm varies considerably. Ammoniacal Nitrogen is another likely contaminant deriving from steel manufacture. Sulphates, Phenols, Cyanides and Ammoniacal Nitrogen are harmful to the aquatic environment.

The other likely classes of contamination are oils, tars and elemental sulphur. These are determined by the relation to the proportion in a soil that is soluble in toluene, known as Toluene Extractable Matter (TEM). Some TEM's, particularly, mineral oils and fuels oils are of concern by virtue of their danger to human health and potential to pollute water resources.

A further class of materials associated with coal carbonisation are the Aromatic Hydrocarbons and PCBs. The latter are generally associated with transformer oils. These materials are of considerable concern due to their toxicity, persistence and carcinogenic (cancer causing) properties and in the case of PCB's, their ability to accumulate in the food chain.

As may be seen from the forgoing, the number of problems associated with the redevelopment of this type of site is complex, before the engineering starts. This is by no means exaggerated and has been confirmed in the authors' experience.

Besides aggressive contamination, steelworks are invariably characterised by slag deposits. These slags in some areas can be extremely unstable and can expand and present severe problems for any buildings or infrastructure.

6.4 Higher technology remediation solutions

The escalating cost and environmentally negative aspects of the traditional "dig and dump" techniques, are increasingly leading to the development and expansion of alternative technologies.

These can be assembled in generic groups as:

– Thermal methods
– Physical methods
– Chemical treatments
– Biological methods and
– Stabilisation and solidification.

It is perhaps not the purpose of this paper to exhaustively discuss the full range of available techniques, which are very well described in a number of publications, in particular The CIRIA Special Publication No 107 Remedial Treatment for Contaminated Land (Harris et al. 1995). However, one needs to compare what may be seen as, simplistically, the "cheapest" solution, with the "whole life" environmental cost of a particular remedial alternative. For example the cost of un-sustainability in terms of the loss of repository capacity, the environmental disbenefits of the required transport to tips, and cost and transport of replacement "clean" material, needs to be weighed against what may initially be seen as the more "expensive" alternative technologies. One perhaps needs to consider that it is beyond argument that all repositories and barriers must eventually fail, no matter how well engineered.

7 SUMMARY

This paper attempts to draw together an account of one profession's need to amend and improve its knowledge base and operational procedures in response to the changing concerns of society.

The arena of brownfield land reclamation demonstrates the need for engineers to appreciate the advice and expertise of the many specialists involved in the sector. At the same time, the engineer needs to appreciate the technical and legislative factors that control the safe re-use of despoiled land. The environmental specialists and regulators also need to be aware of the responsibilities of the engineer as project manager and the various, sometimes competing pressures that the engineer as project manager needs to reconcile.

However, the principle that the best possible use must be made of the most appropriate resources, in order to preserve the environment and prevent, as far as possible, any further degradation is one that is demanded by society and wholeheartedly endorsed by the engineering profession.

REFERENCES

DoE. 1994. *Planning Policy Guidance Note No. 23: Planning and Pollution Control*. London: HMSO.
DETR. 2000a. *Environmental Protection Act 1990: Part IIA. Contaminated Land Circular 02/2000*. Norwich: HMSO.
DETR. 2000b. *Revision of Planning Policy Guidance Note No. 3: Housing*. London: HMSO.
DETR, EA, and NHBC. 2000. *Guidance for the development of housing on contaminated land*. Environment Agency. R&D Publication 66.
DEFRA and EA. 2002a. *Assessment of risks to human health from land contamination: An overview of development of Soil Guideline Values and related research. Report CLR7*. Swindon: WRc plc.

DEFRA and EA. 2002b. *Priority contaminants report. Report CLR8.* Swindon: WRc plc.

DEFRA and EA. 2002c. *Contaminants in soil: Collation of toxicological data and intake values for humans. Report CLR9.*, Swindon: WRc plc.

DEFRA and EA. 2002d. *Contaminated Land Exposure Assessment Model (CLEA): Technical Basis and Algroithms. Report CLR10.* Swindon: WRc plc.

DEFRA and EA. 2002e. *Toxicological reports for individual soil contaminants, reports CLR9 TOX1-10.*, Swindon: WRc plc.

DEFRA and EA. 2002f. *Soil Guideline Value reports for individual soil contaminants, reports CLR10 GV1-10.* Swindon: WRc plc.

EA. 1999. *Methodology for the Derivation of Remedial Targets for Soil and Groundwater to Protect Water Resources, R&D Publication 20.:* Bristol: Environment Agency.

Harris, M.R., Herbert, S.M., and Smith, M.A. 1995. *Remedial Treatment for Contaminated Land: Volume VII Special Publication 107.* London: CIRIA

ICRCL. 1987. *Guidance on the assessment of redevelopment of contaminated land 2nd* Edition. *ICRCL Paper 59/83.* DoE, London: HMSO.

Land Reclamation – Moore, Fox & Elliott (eds)
© 2003 Swets & Zeitlinger, Lisse, ISBN 90 5809 562 2

Barriers to the effective implementation of the Part IIA contaminated land regime and recommendations for progress

Bill Baker
Independent Land Contamination Consultant

John Handley
School of Planning and Landscape Architecture, University of Manchester

Colin Hughes
University of Manchester Environment Centre, UK

1 INTRODUCTION

It is now three years since Part IIA of the Environmental Protection Act 1990 (Part IIA) which deals with the regulation of contaminated land came into force in England in April 2000. The regime introduced a radical risk-based approach to managing the widespread problems of contamination often associated with the large areas of old industrial land. It provides an effective framework for progress in identifying those sites which are causing or threatening harm to the environment, and in encouraging the confident regeneration of "Brownfield" sites.

This paper reviews the progress achieved in the first three years and presents an analysis of the barriers which exist to the more effective implementation of the regime and the universal acceptance and application of the principles of risk assessment and management on which the new regime is based. It presents an analysis of the variation in the response and policies of the local authorities and the Environment Agency to the new regime and its objectives. Special attention is paid to the effectiveness of the investigation and remediation of land contamination under Part IIA, as funded by the Supplementary Credit Approval (SCA) Contaminated Land funding programme administered by Department for Environment, Food and Rural Affairs (Defra) with the technical support of the Agency.

(NB part of Defra was included in the previous Department of the Environment, Transport and the Regions (DETR) prior to 2001).

Recommendations are made with the aim of reducing some of the most serious barriers to enable the considerable benefits of the regime to be realised.

2 DEVELOPMENT OF LEGISLATION

In the UK there was an early recognition in the 1960s/70s of the growing problem of vast tracts of land being abandoned by failing heavy traditional industries, and the potential hazards of associated contamination. Government grants were made available for reclamation and the Interdepartmental Committee for the Redevelopment of Contaminated Land (ICRCL) was formed to develop technical guidance for the effective management of the most common types of contaminated site. This guidance has not proved to be entirely satisfactory in its widely varying application.

The need for a scientifically sound risk-based approach was recognised, and led to the development of Part IIA of the Environmental Protection Act of 1990, introduced by section 57 of the Environment Act 1995. This part of the 1990 Act was eventually implemented in April 2000 in accordance with the Statutory Guidance (DETR Circular 02/2000), published after extensive public consultation.

The Part IIA regime requires local authorities with regulatory duties to inspect the land in their areas and identify those sites which fall within the statutory definition of "contaminated land", where contamination is causing significant harm or where there is a possibility of such harm to a range of sensitive receptors.

Part IIA receptors include (most importantly) human beings, ecological systems, livestock and crops, buildings and controlled waters. The regime incorporates the new definition of "contaminated land", based on risk assessment, and the principles of "polluter pays", reasonableness and fairness, cost benefit, sustainability and openness and transparency in its application. There is then a duty to secure remediation of the land to prevent the harm, ideally carried out by the "appropriate person(s)" (the polluters) preferably by voluntary agreement.

However, there is a formal procedure for consultation with "appropriate persons" to effect satisfactory remediation in line with agreed remediation statements, or the issuing of remediation notices where there is dispute, with the regulator to carry out remediation if this is ignored, with the costs being recharged. All regulatory actions are to be recorded on a public register.

The Environment Agency's (the Agency) role is one of support and guidance and the regulation of "Special" sites as defined by regulation, according to the nature of the contaminants, the context of contamination and the ownership of the site (MoD). Although the legislation is based upon the current condition of land, its major implications are for the application of risk-based principles to the redevelopment of all land affected by contamination in the processes of planning and regeneration, by which it is expected that land posing unacceptable risks will be remediated. This is confirmed by the pending revision of much supporting guidance and associated regulation which is being published for public consultation in 2003.

3   PROGRESS OF THE LEGISLATION

Three years after implementation the rate of apparent progress with the new regulatory system is disappointing and described by some critics as being at a "snail's pace"! There are many reasons for this situation and for such judgements to be made, and these will be reviewed and analysed in this paper.

Various estimates of the extent of the problem have been made, ranging from 50,000 to 100,000 potentially contaminated sites, affecting a total area of around 300,000 hectares (DETR 02/2000). In this context therefore the number of sites identified as being "contaminated" three years after implementation i.e. 49, including 13 Special sites (as at January 30th 2003) seems very small indeed.

4   LOCAL AUTHORITY REGULATORS

The local authorities are the primary regulators under Part IIA, and the main drivers for the acceptance and application of the risk based best practice on which the new regime is based. The performance of local authorities in this area is of course highly variable due to a wide range of influences. These include political and financial priorities, technical expertise, experience, understanding and awareness, external pressures on the planning process, and the local profile for environmental and regeneration policies.

The first requirement placed upon them was the development of their individual strategy for the inspection of the land in their area. This was to have been completed within the first 15 months following implementation. By 2002 most of the strategies had been satisfactorily completed and published, but the next stage of systematic inspection has not been progressed effectively by many authorities. There are many barriers to progress in this field which is a relatively new area of responsibility and technical consideration for many of the smaller authorities.

a) Resources are always an issue in local government finance, but although an increase in Standard Spending Assessment (SSA) was provided by central government to each authority to deal with contaminated land matters, this was not ring fenced. In many cases inadequate, or no provision was made in the council's budget for these purposes.

Capital funding to finance the costs of investigation and remediation of contaminated sites, is available by the provision of the Supplementary Credit Approval (SCA) fund by Defra, with technical assessment carried out by the Agency. Sites whose management is financed in this way may be "orphan", owned by the council, or the subject of regulatory action, or in need of urgent action. The uptake of these funds which are now exclusively for the support of Part IIA activities (or those arising from the requirements of section 161 of the Water Resources Act 1991 administered by the Agency) has not been high, as reflected by the low numbers of determined "contaminated" sites, or "special" sites.

b) Technical Expertise and Experience in this field was recognised to be limited in all areas of responsibility, and a working party involving the Local Government Association (LGA), the Agency, DETR and the Chartered Institute of Environmental Health (CIEH) was formed in 1996 to deal with this problem. An agreement was brokered through the Joint Training Initiative, whereby the Agency would organise and deliver a joint training programme for local authority and Agency staff to cover all the major areas of technical concern. This programme was pursued in the years before implementation, but it was not until Autumn of 2001 that procedural training courses were arranged for the local authorities based upon a special Local Authority Guide to the Application of Part IIA.

Since then a new initiative has been agreed whereby the Agency's training role will be more limited, with the CIEH responsible for organisation of specialised technical training for its members working for the local authorities in this field.

After these delays, and with the progression of such a programme, it is envisaged that there will be a significant improvement in performance, with increased experience of implementation and relevant training.

c) The introduction of a complex new regulatory regime requires procedural and technical guidance to enable the regulators to undertake their new duties in an effective and consistent way. The programme for the development and publication of such guidance for the Part IIA regime, and associated supportive training has been consistently late and delayed until after the date of implementation. This has been a significant barrier to the progress of the regime, and the reasons for this will be discussed along with an analysis of the Agency's role and responsibilities for these matters.

d) The general attitude of local authorities towards the public determination of a site as "contaminated" under the Part IIA regime is naturally one of extreme caution, due to the serious implications which are likely to arise in the minds of the local community, site owners and environmental groups.

Unless such information on the nature of the contamination and the means of its remediation and control is carefully managed it is recognised that there is a likelihood of blight, stigma and extremes of fear and psychological harm affecting all with an interest in such a site.

For these reasons there is a widespread policy being adopted which promotes the voluntary remediation and management of sites in an initially informal way to avoid adverse publicity, and to prevent the extremes of misunderstanding which may prejudice any reasonable environmental action especially on highly sensitive sites.

The following two situations are presented to illustrate their dilemma with the prescriptive approach to regulation which is implicit in the contaminated land regulations:

Example Case study 1: An old badly managed landfill site in the North has a large quantity of asbestos fibre waste deposited near to the surface on a poorly vegetated wooded slope which is bounded at its lower end by a road, a primary school and an established area of Victorian residential working class development. There are problems of structural slope stability, the regular exposure of fibre from minor slippage and the natural failure of birch trees. There is a project designed to remove the fibre and dispose of it to an alternative landfill site, but this is extremely expensive and planned to take place over an extended period. The local community is disadvantaged by the constant fear of the hazards arising from airborne fibres, which are seldom detected by monitoring

exercises, and the refusal of local building societies to consider loans on nearby housing.

Example Case study 2: Local authorities are approaching owners of contaminated sites to inform them of the fact that they are minded to designate their site as "contaminated". They then suggest that they may wish to accelerate their plans for redevelopment, or undertake some remediation which would enable them to avoid determination and entry of the site on to a public register, and any consequential blight or local community concern etc.

5 ENVIRONMENT AGENCY BARRIERS

5.1 *Early setbacks*

When the Agency was first formed in 1996, it strongly reflected the culture and policies of its major constituent parts, – the National Rivers Authority (NRA) predominantly, the waste regulation functions of the local authorities, and Her Majesty's Inspectorate for Pollution (HMIP) for the regulation of industry.

In these early days, priority was given to the management of those primary regulatory regimes which operated under established legislative authority. A much lower priority was given to the development of the new contaminated land regime, and the resignation from the Agency of key staff involved in its original formulation within the DoE/DETR further reduced the impetus for its progress towards implementation. This led to inevitable delays in the development of the supporting procedural and technical guidance which was essential for the introduction of a complex new regime.

5.2 *Delays in guidance development and publication*

Ideally such supporting guidance should have been available to enable the regulators to adequately prepare for implementation by April 2000. In the event, the three most important guidance documents were delayed as indicated:

"Model Procedures for the Management of Contaminated Land" – CLR11 – due for publication mid 2003.

"Local Authority Guide to the Application of Part IIA of the Environmental Protection Act 1990" – published by the Agency, September 2001.

"Contaminated Land Exposure Assessment (CLEA) Model" – CLR7,8,9 and 10 – published March 2002.

5.3 *Agency staffing*

New funding from the (then) DETR made available to the Agency from 1998 for contaminated land issues was invested in the recruitment and provision of contaminated land officers in the Areas (26), Regions (8) and within the National Groundwater and Contaminated Land Centre (NGWCLC). However, their deployment has recently been reorganised under the new "Better Regulation In The Environment" (BRITE) project.

5.4 *Limited consultation*

Good working relationships have been established between these Agency's officers and the local authorities in most areas. However the national policies with respect to the scope of the Agency's guidance and consultation on contaminated land issues, which are derived from an interpretation of the Agency's Part IIA duties and powers, are significantly restricted to the following circumstances:-

1. Part IIA consultation with the local authorities on – "inspection strategies", the identification of contaminated land as "special" in accordance with the "special site" regulations, and the pollution of controlled waters;
2. Statutory consultation concerning planning and development proposals restricted to issues of the protection of groundwater and water resources, development on or close to landfill sites, and flood defence, with an express reluctance to provide advice and support related to the essential decisions of risk assessment for human health in the most sensitive context of housing development.

5.5 *The importance of training*

The decision of the Agency to reduce the priority for the joint training of local authority contaminated land staff implicit in the early Joint Training Initiative, and to concentrate its resources on the training of the Agency's own regulatory staff, has had a significant effect on the rate of progress and understanding of the regulatory regime. More important, perhaps, is the likely adverse effect on the progress towards the acceptance and promotion of the effective application of the culture of risk-based best practice by the local authority regulators. They are the main drivers for the improvement in the standards of investigation, assessment and remediation which it is necessary to obtain from the developers and their consultants in the processes of planning, redevelopment and regeneration. Although there are many of the larger unitary, metropolitan authorities, with their high priority urban regeneration policies, who are able to demand these higher standards, general levels of competence within the wider contaminated land community and the regeneration industry it serves are often low.

New arrangements for a comprehensive training programme for the local authority regulatory staff are now being planned, supported by Defra/the Agency with the CIEH taking a leading role in the development and delivery of technical training courses.

5.6 *Effective published guidance*

Despite these criticisms of the Agency's policies and achievements there is a formidable volume of valuable guidance which has been developed to support the wider world of industry. This has come largely from the work of the NGWCLC and the Scotland and Northern Ireland Forum for Environmental Research (SNIFFER) with the support of Defra, but its impact is severely limited its low level of promotion and publicity. Details of Agency guidance on contaminated land can be found and is summarised in the progress report "Dealing with Contaminated Land in England" published in September 2002, and the guidance of greatest significance can be found on the Agency's web site.

This covers the protection of water resources, systems for the identification of groundwater vulnerability zones, CLEA, alternative risk assessment tools, the management of the communication and understanding of contaminated land risks, and the "Guidance for the Safe Development of Housing on Land Affected by Contamination". This latter publication which was jointly developed by the National House Building Council (NHBC) and the NGWCLC within the Agency, provides a user-friendly application of the "Model Procedures" principles to the most risk sensitive process of residential development.

6 DEFRA SCA FUND PROJECTS – BARRIERS

The Supplementary Credit Approval (SCA) contaminated land fund is a key factor in the progress of the new regime. It exists to provide capital support for the local authorities (and the Agency) to enable the regulators to deal with the high proportion of potential and identified "contaminated" sites that require management through Part IIA of the Environmental Protection Act 1990.

These sites may be "orphan" sites, where there is no owner or commercial interest which can be identified as having responsibility for the contamination and its remediation. Eligible sites can also be sites owned by the local authority, or sites being managed by the local authority or the Agency in the context of firm regulatory action, where the "hardship" provision applies, or where there is a need for emergency action.

Research work carried out at the University of Manchester Environment Centre (UMEC) has examined barriers to the effective investigation and remediation of sites under the Defra SCA contaminated land funding programme (Robinson 2003). The focus of the research is a questionnaire survey undertaken in the Summer of 2002, where local authorities in receipt of SCA contaminated land funding were asked to comment on the significance of "barriers" to the investigation and remediation of land contamination under the SCA contaminated land funding programme. In addition, the Agency's own SCA funded programme of land contamination investigation and remediation was monitored for a period of 18 months to establish what, if any, barriers existed.

For the purposes of this research, the barriers were grouped into the following categories:

"Institutional" – barriers that stem from the internal working or functions of bodies that seek to regulate land affected by contamination, or from the *interaction* of such organisations;
"Regulatory & Legislative" – barriers that are imposed by *government authorities and agencies* through specific statutes, regulations, policies and programmes;
"Technical" – barriers associated with the technical aspect of site investigation and remediation projects, including management tools available (e.g. for risk assessment);
"Financial & economic" – barriers associated with financing investigation and remediation projects, including economic aspects of investigation and remediation.

Questions on the level of understanding of the SCA contaminated land funding programme revealed that although the majority of local authority respondents were aware of the aims and background to the programme, their confidence in their own management of the technical and financial aspects of projects was less assured.

The respondents generally confirmed the variable nature of experience in SCA project-type work. This supports the conclusions drawn by Parkinson *et al.* (2000), where it was confirmed that there was a wide variation across the local authorities in England with respect to their preparedness for the new regime. Further additional work in isolating the different types of local authority responding to the questionnaire survey would most likely have revealed more detail of where the variations in expertise existed (for example, rural district, urban metropolitan & county councils).

The final section of the questionnaire looked at overcoming the barriers to effective management of contaminated land SCA projects asking respondents to rank (in order of significance) the different barriers experienced by them in the consideration of sites for project funding and their progression, once approval

was gained. There was then an opportunity for respondents to make recommendations to overcome those barriers (if any). The results from this section demonstrate that there were common themes in the perception of the existence and significance of barriers, and also in recommendations to overcome them. For the consideration of sites for funding, the most significant barriers recorded by the respondents were *Institutional* and *Financial & economic*. This was also reflected in the recommendations made by the respondents in overcoming the barriers.

For barriers to the progression of projects within the programme (once funding was assured) *Institutional* and *Financial & economic* barriers were again significant, but *Technical* barriers were also presented as being important. Additionally, regulatory and legislative barriers were conspicuous when applying environmental legislation, (for example, waste management licensing) to remediation project works.

6.1 *Recommendations for dealing with observed barriers*

The recommendations focussed on *Institutional* and *Financial & economic* barriers, with particular reference to the administration of funding (provision through grant finance rather than additional borrowing), and the timing of applications and their assessment. Recommendations were made for strategic and operational requirements (Table 1).

Table 1. Recommendations.

Organisational recommendations

- Review of existing environmental legislation applicable to site investigation & remedial works is required. A single licence applicable to this work is needed
- Provision of SCA funding, particularly to local authorities should be simplified. Grant aid provision to LAs could be considered
- Technical training for all regulators should continue, focussing on identified requirements of the practitioners

Operational recommendations

- Dedicated project management staff resources should be in place for each project for the duration of works. Their skills should reflect the multifunctional nature of land contamination work
- Technical assistance for both Agency and LA project managers should be available through specialist support teams provided at the outset of the works
- Community involvement throughout the duration of works should be encouraged, including in selection of remedial method

7 NEW INITIATIVES – REASONS TO BE CHEERFUL

7.1 *Practical benefits of the new regime*

As part of the original concept of the Part IIA regime it was always assumed that the formal regulatory process would only be necessary for those sites where some regulatory pressure was necessary to achieve remediation. It was thought that by far the greater proportion of sites affected by contamination would be dealt with through the processes of planning and redevelopment (DETR 02/2000). Thus where a site was being considered as a potential "contaminated" site, a favoured option on the part of the owner may involve accelerating their plans for redevelopment. This would thereby effect a satisfactory remediation, avoiding the inclusion of the site on a public register as being, or having been determined as "contaminated" with all the perceived disadvantages of potential blight, stigma and adverse reputation which would flow from such a declaration.

This, and the preference of many local authorities for pursuing an initial informal approach to the owners of potentially contaminated sites, probably provides the most likely major reason for the apparent low level of recorded progress of new regime, as evidenced by the small number of sites formally identified by the regulators.

It can thus be argued that the implementation of the new regime is having the intended overall beneficial environmental effect, by encouraging an increase in the number of sites affected by contamination being remediated.

Once a site affected by contamination is being considered for redevelopment, however, it is important that the risk-based principles of assessment and risk management which underpin the Part IIA regime are applied to the development process, and that the local authority regulators demand that the necessary standards implicit in such action are sensibly and effectively applied. The benefits to all interested parties in the acceptance and application of the risk-based culture are enormous in terms of the much higher levels of confidence which can be gained for the developer, funder/investor, environmental regulator, and most importantly the end user and the local community.

7.2 *Revision of guidance and regulation*

In order to ensure that these standards and related benefits are achieved with all new developments, adherence to the risk based approach is being demanded in two new documents which have been drafted by the Office of the Deputy Prime Minister (ODPM). These documents are scheduled for publication for public consultation early in 2003. They are a revised Planning guidance document and a revised Building Regulations Part C, both dealing with contamination.

The proposed Planning Guidance is expected to confirm the requirement for all sites to be assessed for the potential presence of contamination, and best practice principles applied to the proposed remediation and redevelopment.

The proposed amended Building Regulations Part C extends the scope of the regulation from merely the footprint of the building to that part of the site which is occupied by the users or residents, i.e in most cases the entire site. It requires that the risk-based approach to contamination be adopted and makes extensive reference to published guidance and particularly the NHBC/Agency "Guidance for the Safe Development of Housing on Land Affected by Contamination" and other Agency, BRE and CIRIA guidance documents.

7.3 *Compatibility with other regulatory regimes*

Since the establishment of the Environment Agency, there has been concern expressed by developers and others working within the construction industry about the perceived incompatibility of the other regulatory regimes, and especially waste regulation, and the inconsistency of its application to the process of planning and development. This involves the definition of marginally contaminated soil and fill as waste and the consequent requirement for waste mangement licensing when such material is handled or treated in the course of groundworks prior to the construction process.

The system was also seen to discourage the application of new more sustainable treatment of contamination by innovative process based methods which are advocated by government. Following pressure from the new technology vendors the Agency introduced the "Mobile Plant Licence" and supporting guidance relating to its use. This has served the new technology sector of the industry well, but the difficulties concerning the non-treatment approaches to remediation remain.

In response to the recognition of this problem by the Urban Task Force (1999), a working party has produced recommendations for a "Single Regeneration Licence". This proposes that a single licence dealing with all the necessary regulatory issues involved with site specific development proposals should be administered by the Environment Agency (in England and Wales). It is designed to ensure that an integrated approach is developed towards all the areas of interest to the Agency, with the Local authority planning authority retaining regulatory responsibility for standards of remediation relating to human health. It is anticipated that details of this new development will be subject to consultation mid – 2003.

7.4 Competence of consultants and standards of site assessment

It was also recognised by the Urban Task Force (1999) that there were serious problems associated with the inconsistent standards of assessment of sites in the process of sale, purchase and development as carried out by consultants, and that this led to the risk of inadequate valuation and unsatisfactory remediation.

The highly variable quality of consultancy services in the contaminated land field is a significant cause of concern especially in the context of local authority commissioning of their own technical consultants and their assessment of development proposals often supported by inadequate investigation, site assessment and risk management.

A working party was established to develop the concept of the "Land Condition Record". This is a standard reporting format that provides a structured summary of existing information concerning the condition of a site, and identifies gaps in information, inadequacy, caveats and assumptions made, with the aim of ensuring a more consistent assessment of the current condition of sites under consideration.

A system has now been established for this purpose supported by an accreditation process whereby consultants covering a wide range of different disciplines associated with land contamination issues can demonstrate their competence in this field and in the application of recognised risk-based best practice through examination and interview. This leads to the confirmation of a successful consultant as a Specialist in Land Contamination – SiLC (www.silc.org.uk/).

To date there are only a limited number of such qualified consultants, but the system should provide in the future a higher level of confidence to the developer, landowner, funder, and regulator in the competence of the consultants they seek to commission.

8 SUMMARY RECOMMENDATIONS

The progress of the Part IIA contaminated land regime and associated barriers and issues over the first three years has been reviewed, along with some important initiatives supporting the promotion of the risk-based best practice approach to redevelopment. Arising from this analysis, there are some well identified areas where there are opportunities through changes in policies and the redirection of priorities and resources which would be most likely to lead to greater progress in this field. The following recommendations are therefore made in this context.

8.1 Resources, expertise and training

- There is a case for reassessment of central government fund allocation with some consideration for a degree of "ring fenced" finance to the local authorities to ensure at least a minimal level of activity in this area.
- The progress made by the Agency in the provision of reasonably qualified and experienced staff being deployed to support contaminated land regulation is impressive, and is designed to provide national regional co-ordinated cover.
- There is a strong case for the provision of a greater level of technical support and assistance by the Agency, especially to those local authorities who have the greater need for basic initial assistance. It would need a radical change in the Agency's general policies on consultation, and some modest additional financial support from Defra for this initiative.
- The importance of training as a vital and essential requirement is clear. There are prospects for a new agreement between Defra/the Agency and the CIEH for commitment to a co-ordinated comprehensive on-going training programme for the local authority regulators. Feedback from training events has highlighted a serious problem concerning a generally poor level of co-operation between the EHOs dealing with the technical aspects of contamination and the planners, in the management of the planning and redevelopment process. Technical support is often sought only when it is too late to enforce acceptable standards, and there is an urgent need for an awareness raising campaign on a national scale with the co-operation of the appropriate professional bodies. Economic development managers, engineers and lawyers, also need some similar initiative in order to encourage significant improvements in awareness and performance. This should then lead to an increased demand for higher standards from all involved and an associated greater degree of confidence in the effectiveness and sustainability of development and regeneration projects.

The need for training is widespread and applies not just to the regulators but to all professionals working within the regeneration and construction industry. There are some notable consultants and academic institutions active in this area, mostly with their own market oriented agendas, but there is a strong case for some form of national co-ordinated training authority to manage training in this field, and promote consistency.

8.2 Agency policies

- The potential influence of the Agency in the promotion of best practice and in its facilitation in the context of redevelopment is considerable. Current policies are very firmly against any action outside the restrictive areas of statutory consultation concerning the protection of groundwater, water resources, development on or near to landfill sites and flood defence. The Agency has a clear concern about the implied liability of the provision of advice which is applied in the context of other regulatory regimes managed by the local authorities. However there is an opportunity to build constructive partnerships not only with the local authorities, but with the Regional Development Agencies. They have relatively large regeneration budgets, a significant proportion of which will be spent on remediation of land contamination, and closer technical liaison with the agency could help to ensure that best practice is followed and regulatory difficulties avoided by early consultation.

- Throughout the development and construction industry there is an expectation for the Agency to act as the national technical authority on contaminated land matters, to which all with particularly difficult problems can turn, whether as regulators developers or consultants. Although it is recognised that there would be significant cost implication associated with the adoption of this role, the widespread benefits which would flow from a change in policy along such lines would be enormous.
- The local authorities' most urgent need is for greater support particularly with the application of the new risk assessment tools, CLEA, SNIFFER etc to decision making in both the context of Part IIA and planning and redevelopment, and relaxation of current policies merely in this area would pay large dividends.

REFERENCES

DETR, *Environmental Protection Act 1990: Part IIA Contaminated Land*, Circular 02/2000, 2000, Norwich, England: The Stationary Office.

Environment Agency 2002 Fact Sheet No. FS-06, *Fact sheet for the Contaminated Land Exposure Assessment (CLEA) 2002 model*. September 2002, Bristol.

Parkinson, R., Newman, A. & Nathanail, P. 2000 Contaminated Land: State of Local Authority Preparedness. *Land Contamination & Reclamation* 8 (2): 133–144.

Robinson, S. 2003 SCA Questionnaire (unpublished research) UMEC.

Environment Agency Guidance:
Dealing with Contaminated Land in England, 2002.
Local Authority Guide to the Application of Part IIA of the Environmental Protection Act 1990, 2001.
Guidance on the Safe Development of Housing on Land Affected by Contamination, EA/NHBC, 2000.
Communicating Understanding of Contaminated Land Risks, EA/SNIFFER, 1999.

Land Reclamation – Moore, Fox & Elliott (eds)
© 2003 Swets & Zeitlinger, Lisse, ISBN 90 5809 562 2

Millennium Coastal Park, Llanelli – Wales' largest coastal regeneration scheme

Phillip Holmes

Welsh Development Agency, South West Wales Division, Llys-y-Ddraig, Swansea

ABSTRACT: Llanelli and its surrounding communities grew up around the thriving development of heavy industry along the town's coastline. The gradual decline in these traditional heavy industries left the scars of a 200 year legacy of industrial dereliction. Since 1980, the WDA and the Local Authority joint venture partnership has driven forward the regeneration of the Llanelli coastline, securing investments of £60 m to transform over 800 ha of former industrial land. The centrepiece of the regeneration, the *Millennium Coastal Park*, includes 22 km of revitalised coastline, stretching from the Loughor estuary in the East to Pembrey County Park in the West. This ambitious project attracted funding from a host of sponsors, including WDA, the local authority, Millennium Commission, the Welsh Assembly Government and the European Commission. Sustainable reuse of materials has provided cost effective solutions to environmental problems, delivering both a sensitive and imaginative design, transforming 520 ha of industrial wasteland into a haven for wildlife and a magnet for visitors. The Millennium Coastal Park has transformed the image of Llanelli and is destined to become a visitor attraction of international status.

1 INTRODUCTION

This paper describes a bold and visionary solution to the severe problem of economic decline and dereliction left in the wake of the heavy industry closures in South Wales. In particular, the industries of coal, iron & steel, which having brought prosperity to the Llanelli area, left behind a post-industrial wilderness and a legacy of derelict, neglected and unsightly land. A partnership approach has since changed the fortunes of the area, transforming the 22 km coastal belt of derelict wasteland into an important environmental and economic asset to the region. Few areas in Britain have undergone such a stunning transformation, where the Millennium Coastal Park is an exemplar in high quality, originality and appropriate design, together with sustainable reclamation techniques. The £30 m Millennium Coastal Park project has "given the coast back to the people" and opened a prosperous new chapter for a town left in economic decline in the wake of the Industrial Revolution.

2 A PROUD INDUSTRIAL HERITAGE

Llanelli and its surrounding communities grew up around the thriving development of heavy industry along the town's coastline. For many decades the tall chimneys of steel, copper and tin mills dominated the skyline and gave the town its alternative name "tinopolis". In close proximity to the local abundance of coal, mining operations and coal exportation in the Llanelli area date as far back as the 1500's. Networks of interlinking canals were soon established to transport materials from nearby pits to the harbours for export.

As the copper, iron and coal industries in the town grew and prospered, so did the port activity, with the first dock, "Carmarthenshire Dock" being built in 1795. This dock was instrumental in the development of the trade of Llanelli, and is the oldest surviving dock in Wales. As production levels further increased, a number of other docks were constructed in the area, with construction materials often including molten slags from nearby works used to reclaim the land and to act as a sea defence.

Iron making at Llanelli can be traced back to 1866 when the Old Castle Iron and Tinplate Company was formed. However, it was the discovery of the open hearth steel making process which signified the end of iron forges, and saw steel come into general use given its more economical and superior metallurgical properties in tinplate manufacture. A number of steel and tinplate works thrived in the area, although of

Figure 1.   The former Duport Steelworks.

Figure 2.   Sandy Water Park (The former Duport Steelworks site).

particular importance was the former Duport Steelworks, which saw nearly a century of steel making on the site before closure in 1981, Figure 1. At the peak of its industrial might, Llanelli produced around 50% of the world's tinplate, and interestingly, accounted for a significant proportion of the world's saucepan production. This gave rise to Llanelli's famous Welsh theme song, "Sospan Fach", which translated, means "little saucepan". Sadly, the Corus Trostre Tinplate Works is the only operational example of Llanelli's tinplate producing legacy in the area today.

3   NEW BEGINNINGS

The revitalisation of the Llanelli coastline really began in earnest in 1980 with the reclamation of the former MOD Royal Ordnance Factory at Pembrey. This major reclamation scheme was the first of many schemes carried out in the area by the Welsh Development Agency and the Local Authority as part of a rolling programme of land reclamation. In partnership with the then Llanelli Borough Council, the former munitions factory that was established by Alfred Nobel, and which produced TNT and Tetryl explosives up to the mid 1960's, was acquired and reclaimed to create the Pembrey Country Park. The Pembrey Country Park now extends to 200 ha of forestry and 7 miles of "Blue Flag" beaches.

The next major reclamation scheme commenced in 1986 and involved the regeneration of the former Duport Steelworks, Llanelli. After a bitter struggle, the Duport Steelworks closed in 1981 with a loss of 1100 jobs, dealing a severe blow to the economy of the whole area. Although the initial reaction at the time was to seek out new owners to take over the site, the effort was with no avail. A realisation that the declining heavy industries that had once served the area and its communities so well, would not be returning, forced Llanelli Borough Council to rethink their regeneration policies. As a result, imagination and foresight saw the 50 ha derelict eyesore transformed into Sandy Water Park, a parkland of new rolling landscape with a 6 ha lake and sites prepared for housing, leisure and commercial uses (Figure 2). The works were carried out using plant hire, rather than a civil engineering contractor and involved earth moving in the order of two million tonnes of material. The works cost £2.5 m and took almost three years to complete.

Following the success of the Duport reclamation scheme, the first of a number of joint ventures between the WDA and the Local Authority were formed. This ensured continued investment with reclamation of a number of key sites and infrastructure serving the sites in advance of development. The most significant of these derelict sites included the Carmarthen Bay Power Station, R.T.B Mills (formerly Burry works), North Dock, and the levelling of 0.5 miles of slag banks up to 15 m in height. Other schemes included Llanelli Gasworks, a number of railway sidings and locomotive sheds, foundries, scrap yards, and several colliery sites.

4   CELEBRATING THE NEW MILLENNIUM

In May 1995 the first plans were drawn up by the Llanelli Coast Joint Venture to create the Millennium Coastal Park, a symbol of Llanelli's emergence from social and economic decline. The key elements of the proposals included:

– securing the coastline for public access in perpetuity;
– creating a leisure resource of national status;
– creating and conserving wetland habitats of international importance;
– helping to revitalise the economy through investment opportunities, particularly in leisure and tourism;

- demonstrating sustainable development and environmental care;
- interpretation of local culture and history and
- links with communities.

The development programme for delivering the Coastal Park was phased over a number years, culminating in summer 2000 when the national Eisteddfod, Wales' largest cultural event celebrating Welsh music, poetry, literature and arts was scheduled to be hosted at Llanelli. The event would attract over 160,000 visitors, signifying the arrival of the Coastal Park, providing the first ever purpose built Eisteddfod venue, "Eisteddfod Fields". The application for the whole project was made to the Millennium Commission in Autumn 1995.

On Christmas Eve 1995, the Millennium Commission approved the application for 50% funding for the £27.5 m project, with match funding coming from the local authority, the WDA and a number of other public and private organisations. A dedicated project team responsible for delivery of the project was quickly assembled, with detailed design works being carried out by engineering consultants. The main elements of the 22 km Coastal Park, which stretches from Loughor Bridge in the East to Pembrey Country Park in the West included:

- *Pembrey Country Park* – 200 ha of open parkland, forest and the "Blue Flag" Cefn Sidan beach;
- *Burry Port Harbour* – revitalised docks with marina and 8 ha of mixed use development sites;
- *Burry Port Community Woodland* – mixed woodland planting with a campsite, fishing and nature lakes, a giant earth sculpture and a land bridge connection over the main Swansea to Fishguard railway line;
- *Pwll Greenway* – coarse fishing lakes and a new sports/visitor pavilion, footpath, cycleway and land train routes with picnic areas;
- *Eisteddfod Fields* – main festival events area.
- *Millennium Park Centre* – beach and seafront park, visitor facilities, look-out tower, cycle hire point and 2nd land bridge connection over railway;
- *North Dock* – impounded dock with water training school, bridge and road access, seafront promenade, Sustrans cycleway, 11 ha of mixed use development sites;
- *Machynys Peninsula* – links to Machynys ponds SSSI, coastal defence improvements;
- *Golf Course* – 18 hole championship golf course and residential development sites, coarse fishing lake, birdhide and view point;
- *The Swannery* – major 60 ha extension to existing wildfowl centre, Europe's largest wetland habitat and Black Poplar woodland, fishing lakes, cycleway route, visitor centre and educational facilities;
- *Loughor Greenway* – eastern gateway to the park, coastal path improvements to Loughor bridge and

- *SUSTRANS* – a cycle route running the whole length of the park and linking into the National cycleway network.

5 GREEN RE-ENGINEERING OF THE LLANELLI COASTLINE

From first inception through to completion, sustainability has been a key consideration on the Millennium Coastal Park project (Figure 3). The concepts of "green-engineering" and "sustainable development" have not been taken as mere token descriptions but as living principles in every aspect of the design and delivery of the scheme. From the very outset, the project focused on the re-use of materials from within and around the Park. This sustainable ethos was behind the recycling of silt, sand, ash and slag, much of which resulted from the Park's industrial heritage.

The reclamation of the former *Carmarthen Bay Power Station* area of the site is an exemplar in sustainable reclamation and re-use of materials. During the working life of the former power station at Burry Port, thousands of tonnes of pulverised fuel ash (PFA) was generated, which was deposited in a marshland area adjacent to the site. Following comprehensive risk-based chemical and geotechnical analysis of the ash deposits, supported by the findings of the vegetative growth trials, it was agreed that the PFA was acceptable as an engineering material for use within the Park. It had a number of uses:

- The ash was used as a top soil to cap-off contaminated land on former industrial sites, in particular, the Burry Port community woodlands.
- PFA, lime and ordinary Portland cement was used as a base for sections of the footpaths and cycle ways.
- PFA "soil" was blended with digested sludge from the local sewerage works to provide a cost-effective fertiliser substitute and a medium for tree growth.
- Some 120,000 m^3 of PFA was used as a lightweight fill in the construction of two giant land bridges

Figure 3.   The Millennium Coastal Park.

Figure 4. One of the two land bridges over the Swansea to Fishguard main line.

Figure 5. The revitalised North Dock, Llanelli.

that spanned the main Fishguard to Paddington railway line (Figure 4). The bridges carry part of the Millennium Coastal Park continuous cycle way and won the 1999 George Gibby Major Project Award, Wales' top civil engineering award. Construction of the three-pin arched bridges was one of the most complex civil engineering schemes in the park. Both bridges are made of pre-cast reinforced concrete and sections weighing up to 20 tonnes were lowered into place over the railway line between midnight on Saturday and 10 a.m. Sunday morning for a series of weekends. The haul road used to transport the excavated PFA latter became the foundation for the National Cycle Network route between Llanelli and Burry Port.
- A giant earth sculpture was formed using 160,000 m^3 of PFA. The PFA was excavated from nearby deposits forming leisure lakes in the park.
- The PFA was used to raise levels on a marshy area of the Park (Dyfatty marsh) to create a 10 ha development site.

Elsewhere within the Park, crushed materials from demolished buildings in Pembrey industrial park at the Park's western end were used as "general fill" for the cycleway link between Burry Port and Pembrey – this tackled the litter pollution as well as creating the foundation for the cycleway. Another 30,000 tonnes of crushed material from a local landfill site were used as general fill and drainage material along the Park's £1.5 m promenade and in the Park centre.

The regeneration of the former North Dock, built to export millions of tonnes of tinplate, presented a further opportunity to demonstrate sustainable use of materials (Figure 5). In delivering the "Millennium Quays" maritime centre, the silt dredged from the harbour was tested using vegetative growth trials and successfully used as a cover material for contaminated

areas on adjacent land and on other areas of the Park as a growth medium for landscaped areas. The dredging of the Burry Port Harbour also offered a source of material for re-use on the Park, where 60,000 m^3 of silt was spread over contaminated areas of the former Carmarthen Bay Power Station. Further use of natural materials saw 50,000 m^3 of sand excavated from within the Park to create fishing lakes, utilising the sand deposits to form a surface of the fairways at the Machynys Golf Course.

6   A PARTNERING APPROACH WITH MODERN FORMS OF CONTRACT

The "partnering" concept was embraced on the project from day one, especially given the limited budget available and where cost-effective engineering solutions were deemed essential. A total of eight consultancy firms were employed to design various parts of the project, all using A.C.E conditions of contract. During the construction of the park, approximately thirty works contracts were awarded, utilising many different contract forms. Of particular success was the use of the Engineering Construction Contract. Option C of the contract was found to provide sufficient flexibility, whilst providing the client partnership with a high degree of cost certainty and most importantly, positively encouraging a proactive value engineering approach by the contractor. At the time, this was considered quite a bold approach, since this was the first experience of using the ECC form of contract by the client.

A particular success of the ECC form of contract was the land bridges contract. Using a design and build procurement method, delivery of the scheme whilst encompassing the important principles of obtaining high quality in the shortest time possible yet still achieving best value for money was achieved.

A positive site management approach contributed to a continuous flow of management ideas and design alternatives, ultimately contributing to a saving of 13% against an original target cost of £3 m and ensuring project completion to programme (Edmunds & Hogan 2000). The Seafront Promenade element of the project, an 855 m long 6 m wide wave shaped coastal pathway also benefited from the ECC Option C contract. Tight financial control, without sacrificing quality, resulted in 12% savings on the target cost.

7 CATALYSING ECONOMIC REGENERATION

The former industrial landscape of the Llanelli coastline is now on the threshold of becoming one of the premier commercial development opportunities in Wales and a leading tourist destination. This radical transformation, which has been driven forward by the Llanelli Coast Joint venture, is one of the largest environmental enhancement projects in Britain, covering 22 km of Llanelli coastline and over 800 ha of land. The main achievements include:

- *Land Reclamation*
  - In excess of 400 ha of land reclaimed
  - 50 ha of extensive landscaping
  - £18 m of public sector investment
- *Transport*
  - New road infrastructure, including a Coastal Link road
  - £6.5 m of public investment
- *Residential*
  - Over 400 new homes on a range of sites for all market sectors
  - £26 m private sector investment
- *Recreation & Leisure*
  - A Country Park
  - £2.5 m Sandy Water Park
  - £6 m coastal protection works
  - £1 m private sector development
- *Millennium Coastal Park*
  - £14 m Millennium Commission funding, £16 m public/private sector match funding

- North Dock 10 ha mixed use development site at the heart of the Coastal Regeneration Project
- *Industrial*
  - 18 ha Delta Lakes Business Park, suitable for B1, B2 and B8 classes of use
  - 15 ha of development land at Burry Port
  - £1.5 m public sector investment.
  - Removal of overhead electricity cables and 11 pylons along 4 km of coastal belt, re-routing the electrical cables below ground, removing visual blight on development land.

The level of investment attracted by the Joint Venture has been achieved by creating a visionary development strategy, which has been both robust and flexible. An appropriate development framework has established a strong urban structure that has been dynamic enough to respond to changes in market need, resource availability and new opportunities. By setting the highest standards of design and development, the legacy of industrial dereliction along the Llanelli coastline has now been transformed into the spectacular Millennium Coastal Park. The area has become a high quality and desirable place to live, work and visit, and has provided the catalyst for a bright new future for Llanelli and its surrounding communities.

ACKNOWLEDGEMENTS

Martyn Williams, Project Engineer, Carmarthenshire County Council.
Mike Edmunds, Land Bridges Project Manager, Arup, Cardiff.

REFERENCES

Edmunds M & Hogan (2000) Milllennium Coastal Park: Llanelli Land Bridges.

Land Reclamation – Moore, Fox & Elliott (eds)
© 2003 Swets & Zeitlinger, Lisse, ISBN 90 5809 562 2

Reclamation and regeneration in the new economies of Eastern Europe

Gwyn Griffiths
Land Reclamation Director, Welsh Development Agency

Hana Chlebna
Deputy Director, Czechinvest

ABSTRACT: The needs for a multifunctional approach and for the sharing of international experience have been notable and growing mantras during the forty year life of the evolving art or science of land reclamation. They have however generally been pursued between practitioners and Nations at loosely similar positions on the learning curve. The rapidly changing political situation and the imminent accession of Eastern Block States to the European Community demand a rapid catching up process. This process applies partly to techniques and processes, but more importantly to Political and Community expectations and the creation of management and funding processes. Whilst these have to be evolved, owned and managed locally, close liaison with experience and lessons gained elsewhere should speed up and simplify that learning process and avoid the repetition of errors made previously.

It is sometimes forgotten that the Czech Republic has a long and distinguished industrial history. This is not solely based upon the exploitation of large scale reserves of brown and hard coal, coupled with Iron Steel. The Czech Republic, which is approximately four times as large as Wales, and has a population of 10 m, has a tradition of high quality precision engineering, for instance in the field of armaments, automotive engineering and heavy manufacturing and fabrication, along with the production of fine glassware and chemicals.

At independence, the major State owned industries including coal, steel and power generation were privatised, but the entry of these industries into an ever more competitive market place with increasing world commercial pressure and heightened environmental standards has been followed by large workforce reductions and closures.

In the face of local unemployment levels of approximately 20%, Czechinvest, the Czech Government Inwards Investment Agency has achieved rapid and impressive success in attracting high quality Foreign Direct Investment. However, this has achieved primarily on greenfield sites because of the need to respond quickly to the rising unemployment in the absence of the necessary time, structure, mechanism or funding to reclaim the derelict sites from which the unemployment has arisen.

However, despite the impressive achievements in attracting new employment, Czechinvest and its Czech partners have recognised that they are faced with an urgent need not only to create new employment, but to start the extensive and lengthy process of comprehensive regeneration of those communities affected by the quite rapid loss of the older basic industries.

Czechinvest has tapped the experiences of those countries and organisations, particularly in Western Europe, who have experience of dealing with the effects of rapid industrial decline. Their initial links focussed particularly closely upon job creation and the attraction of Foreign Direct Investment. They have developed particularly close links with Scotland and Wales and have tapped experience and expertise in relevant areas.

Issues which they have started to address to move the process of regeneration forward are:

– Political acceptance of the existence and likely scale of the Regeneration process to be addressed.
– Political buy-in to the need and social and economic justification for comprehensive regeneration.
– Examination of possible administrative and financing frameworks for delivery of regeneration.
– Technical guidance regarding project assessment and prioritisation, procurement, design strategy, design standards, project delivery and the ongoing development and site development management process.

This process has been progressed by means of a number of Czechinvest sponsored Phare Programme funded studies commencing in the year 2000 with the preparation of a short Land Reclamation Strategy for the Czech Republic which was written by Eres Consultants. This strategy reviewed the nature of the problems facing the Czech Republic and made outline recommendations as to the manner in which the task could tackled, taking account of experiences and progress made elsewhere in Europe.

In presenting this project to the Ministry of Industry, the Consultants were at pains to point out the impacts and benefits that would accrue, albeit at significant initial cost, over perhaps fifty years.

This study was followed by the now complete Vitkovice Demonstration Project. This major project focussed upon the closed "heavy end" of the privatised Vitkovice Steelworks in Ostrava. Ostrava is the heavily industrialised third City of the Czech Republic, with a population of 300,000.

The purpose of this project, carried out by NEI-Ecorys has been to devise a reclamation and re-use strategy for the Lower Vitkovice site, as a pilot for future projects throughout the Czech Republic. It has also provided advice upon the establishment of a Brownfield Regeneration Unit, provided training for Czech politicians and public servants and advised upon project appraisal and funding mechanisms.

Whilst it was accepted that the Lower Vitkovice site was a complex choice as the focus of such a study, it has proved to be extremely valuable, containing as it does such features as continued partial occupation and use of buildings, the ongoing need for rail links, heritage listing of the 16 ha Blast Furnace, Coking Plant complex and adjoining coal mine coupled with high levels of contamination along with the usual challenges of providing new infrastructure and integrating the regenerated site and its proposed uses into adjoining communities.

Completion of this study has seen the quite sudden emergence of an encouraging consensus across a complex group of stakeholders including a newly elected local mix of political parties. This may prove to be the most important outcome of the project.

Czechinvest is now seeking Phare Programme funding to build upon this newly gained momentum by carrying out a survey to identify and prioritise the rumored 6500 derelict sites, and to detail the reclamation requirements of five of those sites for early implementation.

It is then hoped that these projects will be able to benefit from a cocktail of local and European Funds to set in train a long, challenging but exciting process of regeneration, supported centrally but managed locally, leaning where necessary upon the good experience and hard and bitter lessons learned elsewhere throughout Europe and the world.

Ms Chlebna focuses upon the background, needs and aspirations of the Czech Republic whilst Gwyn Griffiths deals with particular challenges and opportunities.

Land Reclamation – Moore, Fox & Elliott (eds)
© 2003 Swets & Zeitlinger, Lisse, ISBN 90 5809 562 2

Multifunctionality and scale in post-industrial land regeneration

C. Ling & J. Handley
School of Planning and Landscape, University of Manchester

J. Rodwell
Unit of Vegetation Science, University of Lancaster

ABSTRACT: Post-industrial regeneration in the UK has been dominated by strategies to redevelop sites for industrial, commercial and housing end uses; however mainstream strategies for the regeneration of derelict land have not made significant inroads into the total stock. The strategy is limited in its potential to tackle the problem in those areas that have the greatest extent of dereliction and associated social problems. Moreover, it ignores the real potential for brownfield sites for biodiversity and nature conservation. This paper explores a holistic, multifunctional approach to regeneration. All landscapes are inherently multifunctional even though regeneration of post-industrial landscapes does not take this into account and focuses on specific economic, social or environmental goals resulting in a non-optimised regeneration solution. Any methodology that aims to improve the sustainability of regeneration should be valid for the integration of processes operating at multiple spatial scales, and allow holistic approaches for all projects regardless of scale within the context of strategic regional planning. The concept of multifunctionality appears to answer some of the questions asked of regeneration. The concept uses a holistic approach and looks at the landscape not as a pattern of land uses but as a dynamic system providing a variety of functions that work together in the landscape in an integrated way.

1 INTRODUCTION

Derelict and vacant land formed by post-industrial economic restructuring in urban and urban fringe locations, blights local communities and reduces the development potential and quality of life in affected areas (Handley 1996). This legacy is the result of the change from an economy dominated by the production of goods from raw materials, with industry located near the constituent raw materials, to an economy where the location of industry is primarily determined by the attractiveness of environment and location of the relevant workforce (e.g. Clark et al. 1993). This further enforces the lack of investment of redevelopment in areas badly affected by dereliction.

Regeneration of this land in the UK has been dominated by strategies to redevelop sites for industrial, commercial and housing end uses (National Audit Office 1988, Department of the Environment, Transport and Regions 1998). Mainstream strategies for the regeneration of derelict land have, however, not made significant inroads into the total stock, and for reasons such as poor location, low land values and technical problems, a substantial portfolio of damaged and neglected land persists (Handley et al. 1998).

Despite the continued and persistent presence of derelict land in many areas of the UK there is still a strong commitment to "hard" redevelopment of brownfield sites. The Urban Task Force report (1999), from which national policy is developing, has been criticised for being over reliant on the redevelopment of brownfield land for housing when this strategy has been shown to be limited in its potential to tackle the problem in those areas which have the greatest extent of dereliction and associated social problems. Regeneration projects often focus on specific economic, social or environmental goals. This narrow focus may compromise some of the potential benefits inherent in any given landscape by overlooking functions within it (Handley 1996, 2001).

The Urban Task Force report also ignores the real potential of brownfield sites for biodiversity and nature conservation (Frith 2001). The problem faced by planners is to ensure that solutions found, and options chosen, in the pursuit of regeneration take full account of the potential within the post-industrial landscape for economic, environmental and social benefits. Sustainable development requires an approach to the regeneration of post-industrial landscapes in which all three factors are given proper emphasis, and the regeneration of post-industrial landscapes in a sustainable way is a key requirement for the achieving of potential in regions where such landscapes exist. A holistic approach to regeneration exploring the full potential of a landscape would address these issues.

The focus of this paper is to assess the potential for the concept of multifunctional landscapes (e.g. Brandt et al. 2000) as a suitable conceptual framework for post-industrial regeneration and how such a concept may be incorporated into the planning and implementation of regeneration at different scales.

2 MULTIFUNCTIONALITY AND REGENERATION

According to Brandt and Vejre (in press) there are three types of functionality within any landscape.

1. A spatial combination of different functions related to separate land units (spatial segregation).
2. A number of different functions devoted to the same land unit, but separated in time typically in certain cycles (time segregation).
3. An integration of different functions within the same or overlapping land unit, at the same or overlapping time (spatial integration or "real multifunctionality").

Multifunctionality is now seen to offer great potential in the post-productivist agricultural landscape but its possible application to other types of landscape, including the post-industrial landscape, has received limited attention (Maier and Shobayashi 2001).

All landscapes are inherently multifunctional (Fry 2001); however the regeneration of post-industrial landscapes does not, at least explicitly, take this into account. Spatial segregation reflects the existing philosophy of regeneration projects. Here design focused master plans for mixed used development emphasise the quality of the built environment and the public realm with limited regard for the wider environment (Urban Task Force 1999 and DETR 2000a). Criticism levelled at the Urban White Paper suggests that regeneration policy lacks a multifunctional philosophy, and overlooks the potential for natural capital inherent in post-industrial landscapes.

But as Frith (2001) reports "*Representations by English Nature, the Wildlife Trusts and others during the Paper's preparation process, on the value of biodiversity and natural features in contributing to the quality of the urban realm, have been almost completely ignored. The Urban White Paper reflects perhaps a stubborn myopia within DETR as to the role of nature in our towns and cities – not only as habitats for wildlife, but its environmental functions and contribution to the health and well-being of people.*"

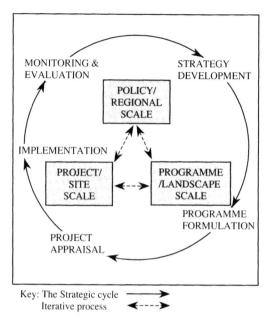

Key: The Strategic cycle ⟶
Iterative process ◀----▶

Figure 1. An iterative and cyclical view of the strategic approach to regeneration and the relevance of scale. Adapted from Johnson et al (1992).

However the OECD's Committee for Agriculture suggests that the application of a multifunctional approach is often problematic due to different processes or functions operating at different scales (Maier and Shobayashi 2001). As Forman (1995) states "*We are left with the* paradox of management. *One can more likely cause an effect at a fine scale, whereas success is more likely to be achieved at a broad scale.*"

In reality the development of strategic regeneration takes place in a way that reflects these issues of scale. This view is illustrated in Figure 1.

It follows, therefore, that any assessment framework should be able to work at any scale so as to adequately inform decision making from the regional/policy making scale, through programmes focusing on landscapes, to the site focused project scale. The methodology therefore needs to be able to compare functions that operate at different scales, recognising that there will be a degree of contradiction and mismatch between these scales.

This paper explores the relevance of the multifunctional approach at different scales in the context of land reclamation strategies.

3 FUNCTIONS WITHIN THE LANDSCAPE

Multifunctionality, as interpreted by Brandt et al. (2000), concerns the integration of land uses in a

Table 1. Elements of multifunctionality. (after: Brandt et al. 2000).

Function	Description
Ecological	An area for living
Economic	An area for production
Socio-cultural	An area for recreation and identification
Historical	An area for settlement and identification
Aesthetic	An area for experiences

landscape made of assorted functions and processes into a cohesive sustainable whole. These functions are summarised in Table 1.

The approach considers the landscape not as discrete utilitarian packets but as a network of processes and functions. Multifunctionality should be less limited by scale than spatial segregation which is dependent on the size of the land-use units within the landscape. Working with the ecological, social and environmental functions within the landscape, not with land-uses, will best deliver the holistic and integrated decisions that result in more sustainable landscapes (e.g. Rodwell and Skelcher 2002).

4 EXAMPLE OF MULTIFUNCTIONALITY AT DIFFERENT SPATIAL SCALES

We are currently exploring whether multifunctionality can be applied as a principle at all scales of regeneration. We illustrate this here with an example of this application in general terms for a post-industrial region of England. The aim is not to fully describe the extent of functionality within the region but to illustrate that multifunctionality can exist in all landscapes and at all scales.

4.1 Regional scale

Britain's post-industrial landscape is a legacy of geological resources. Much industry developed around areas such as coalfields, ore resources and other mineral deposits. An examination of geological maps showing these areas (British Geological Survey 1996 and 1999) therefore points to regions where the post-industrial landscape has a definable boundary and character resulting from these underlying geological patterns. In addition, geological units affect the landscape character of an area through the impact on soils, topography and vegetation and the influence these attributes have on the cultural and economic activity within the area.

Examples of a post-industrial landscape at the regional scale can be drawn from areas with a concentration of natural resource that has resulted in

Figure 2. The East Pennine coalfield and English Nature Natural Area (British Geological Survey 2002 and English Nature 2002).

extensive extraction and associated industry over many years since the industrial revolution. One such example is the Pennine coalfield (British Geological Survey 1999) as illustrated in Figure 2. The recognition of this region as an English Nature Natural Area (English Nature 1998) suggests immediately that this region can be identified as being multifunctional in addition to it having an identifiable industrial geology.

At the regional scale therefore the Pennine coalfield could be described as being multifunctional in the following way.

4.1.1 Historical functionality
The major historical features of this region are the impacts of industry on the development of landscape character and the pattern of human settlement. Consequently the presence and development of landmark industrial buildings such as mill chimneys, coal mines (ranging from mediaeval bell pits to modern opencast mines), associated transport networks and industrial towns in the landscape has, through time, shaped the pattern of population and strongly

contributes to the identity of the region (Countryside Agency 1999).

4.1.2 *Ecological functionality*

The region has an ecology that consists of both semi-natural landscapes and habitats that have developed in relation to interactions of climate, geology and soils, alongside newly formed and distinct plant and animal communities on areas previously despoiled by various industrial activities. Ecological targets for the region include the reduction in the loss of hedgerows and the creation of wetlands and heathlands as well as protecting and providing new habitat for animals such as Water vole, Brown hare, European otter, Pipistrelle bat and Grey partridge (English Nature 1998).

4.1.3 *Economic functionality*

General industrial restructuring is taking place in the region. With activity changing from heavy industrial and mining, to an economy with a greater amount of technology and service based industries. Policy is geared towards a shift from supporting industrial sectors across the region, historically based on the natural resource availability, to developing clusters of technologically based industrial developments in the Advanced Engineering and Metals, Food and Drink (including agriculture), Digital Industries, Bioscience and Chemicals sectors. There is a requirement to stimulate new business start-ups and attract more investment. These policy measures will need to be enhanced by improving skills and training in the region and improving the environment by regeneration and by protecting and enhancing natural assets (Yorkshire Forward 2001)

4.1.4 *Socio-cultural functionality*

The region is socially diverse with council wards from the top 5% to the bottom 5% of deprived wards according to the Indices of Multiple Deprivation for 2000 (Department of the Environment, Transport and Regions 2000b). Therefore social aspects of regeneration will need to be concentrated on those areas where deprivation is greatest, and perhaps encourage a degree of integration between the two communities.

4.1.5 *Aesthetic functionality*

The main aesthetic features in the landscape of this region have, until very recently, been related to the industrial heritage, specifically: mill chimneys, mines, iron and steel works, spoil heaps, and other features. The impact of these on landscape character is significant and the industrial landscape tends to dominate other, fragmented, semi-natural and agricultural features in the majority of the landscapes within the region (Countryside Agency 1999). This impact is not necessarily welcome, and in the post-industrial period has been very considerably changed.

4.2 *Landscape scale*

Within a post-industrial region there are likely to be concentrations of industrial decline forming intensively post-industrial landscapes characterised by the visual and environmental impacts of economic restructuring. The identification of these landscapes can be observed from satellite imagery, as indicated by the concentration of bare, abandoned ground in such surveys as the Land Cover Map of Great Britain 1990 (CEH 2002). In the region described above a post-industrial landscape that can be identified in this way is the Dearne Valley, east of Barnsley in South Yorkshire.

Figure 3 illustrates the location of the Dearne Valley within the Natural Area (the total area of the valley extends east of the Natural Area but is not included in this study as it has a different geological and landscape character). The points distributed over

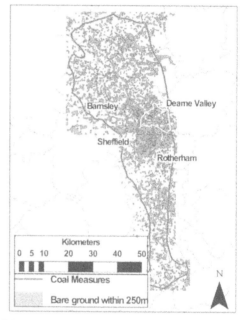

Figure 3. The identification of a post-industrial landscape within a region. The grey regions represent those areas with bare ground within 250 m. The concentration of bare ground in the central part of the natural area is coincident with the urban areas Sheffield, Rotherham and Barnsley. Also part of this area with a concentration of bare ground is the Dearne Valley (ringed). This area is different in character to the other three; in that it is predominately not urban – the bare ground being the result of coal related activity and therefore indicative of a post-industrial landscape. The displayed data illustrates the distribution of inland bare ground as plotted by the Land Cover Map of Great Britain 1990.

map show the presence of "inland bare ground" ording to the classification of the Land Cover Map 0. Bare ground includes, amongst other types of d cover, derelict land where regeneration has not en place. These data have therefore been used to ntify landscapes, combined with site visits and map dy, which have significant amounts of derelict land. Looking at the elements of functionality at this le shows a different set of characteristics to that at regional scale.

.1 *Historical functionality*

storically the Dearne Valley has been an agricul- al-industrial landscape with the history of mining other industries stretching back certainly to the h century and probably to the Roman occupation he area (Hey 1979 and Hill 2001):

'There is evidence that coal was mined sporadi- ly during the Roman occupation. This was probably ng a narrow arc along the outpost from Rotherham the west of Barnsley and onwards to Wakefield' ll 2001).

This inherited rural-industrial landscape is therefore najor part of the character and culture of the area.

.2 *Ecological functionality*

e ecological fabric of the Dearne Valley landscape determined by the river's track through the sand- ne/shale cuestas. The area is dominated by indus- l dereliction and associated land reclamation jects though this has provided many incidental and tored habitats for a variety of species. Water vole live in some of the artificial water- irses. Artificial wetlands have been created on sub- ence flashes caused by mining. Natural succession aking place on many derelict sites, such as along ndoned railways and spoil heaps, with the appear- e of many floral and faunal species, some of ich are endangered or threatened. For example wn hares, lapwings, skylarks (Rotherham Urban ldlife Group 2000).

.3 *Economic functionality*

til 1980 the economy of the Dearne Valley was ngly dominated by the coal industry. Today the coal ustry does not exist as a generator of economic pro- ctivity; during the 1980s and 1990s 11,000 jobs ectly attributable to the coal industry were lost in the arne Valley as a whole (Owen 1992). New types of ployment are gradually taking its place, particularly l centres (Booth 2001). Indeed it is estimated that re may be more jobs created by 2005 on former coal ne sites than were lost to the mine closure in previ- s decades although most of these are of lower status d pay than the previous employment opportunities wen 1992, Booth 2001). However, unemployment

and poverty is still high so further economic develop- ment is still required.

4.2.4 *Socio-cultural functionality*

The main challenge to the socio-cultural environment in the area is skills and training, with the majority of the workforce trained in disappearing manual jobs, and not the skilled and office jobs promoted by the Regional Development Agency Yorkshire Forward. The wards that are present within the Dearne Valley area are almost all characterised by high scores on the deprivation index with poor health and high unem- ployment etc (DETR 2000b).

4.2.5 *Aesthetic functionality*

The landscape of the Dearne Valley is aesthetically characterised by the juxtaposition of agricultural and industrial landscape features and a complex pattern of settlements. Alongside softer features such as farm- land, woodlands, and seasonally flooded meadows – "ings" and mining linked subsidence flashes are large areas of derelict land and spoil heaps, now mostly regraded (South Yorkshire Forest 2000). The priori- ties for the visual landscape are therefore seen to be to create a more aesthetically pleasing landscape by working to restore visual quality in the derelict areas whilst retaining the local identity derived from those parts of the industrial heritage that have formed the landscape over many centuries.

4.3 *Site scale*

Although the regional and landscape scales may be appropriate scales for building sustainability and functionality into the regeneration process, in reality it is at the site scale decisions are made regarding development on a practical basis.

Within the Dearne Valley landscape are a number of sites that have been developed as part of the wider

Figure 4.   Manvers West (Rotherham MBC 2002).

regeneration process in the valley. One of these is the site at Manvers West, a 113 hectare site between Wath upon Dearne and Bolton upon Dearne, on the river itself. The site was previously used as part of the Manvers complex of collieries and included some of the largest railway marshalling yards in Europe. The site is currently under redevelopment but displays a number of other existing functions that connect it to the wider landscape.

4.3.1 Historical functionality

The modern era of coal mining associated with the site commenced in the 1870s with the first coal bought to the surface in 1879. The mine was called Wath Main Colliery, it merged with the Manvers complex to become Mavers Complex West in 1986 and finally closed in 1988 (Hill 2001). Over 100 hundred years of activity has left a significant historical legacy in the culture and development of the local area. Prior to mining the site was agricultural, bisected by a railway line. 1:10,560 County Series maps from 1855 refer to the area as "Great Moor" and the fields are served by "Moor Lane" (Landmark Information Group Ltd and Ordnance Survey 2002).

4.3.2 Ecological functionality

The site of Manvers West forms part of the flood plain of the Dearne River. The site of Manvers West itself has some wetland habitat in existence in the form of a lake and a stream. There has been both natural regeneration, since the mine closed, with birch scrub invading, and reclamation with landscaping and tree planting. Brown hares, kestrels and partridge have moved on the site and areas of reed and reed-mace have begun to develop by the water features (Rotherham Urban Wildlife Group 2000).

4.3.3 Economic functionality

Manvers West is situated between two settlements, Wath upon Dearne and Bolton upon Dearne. Both these settlements are categorised by the high unemployment and low average income associated with loss of traditional employment such as that previously located on the Manvers West site. The rest of the Manvers Colliery Complex has now been redeveloped for a variety of business uses including manufacturing, engineering, distribution and customer service centres. The local authority is proposing to extend this investment and development onto Manvers West but also including recreation activities in the form of a golf course in the northern part of the site along the river (Rotherham MBC 2002).

4.3.4 Socio-cultural functionality

Due to the exclusion of people from the site by the extent of dereliction and the closure of the mines it may be thought that there would not be much in the way of socio-cultural activity associated with the site. However, the site is actually used for both informal and formal recreation, with the development of a golf course on the northern edge. In addition the Trans Pennine Trail – a long distance foot and cycle path – runs along the boundary of the site, connecting many parts of industrial heritage in the north of England from Southport on the west coast to Hornsea on the east.

4.3.5 Aesthetic functionality

Immediately after the abandonment of coal mining on the site the aesthetics of the place were marred by the industrial detritus found there. Since that time there has been a combination of natural regeneration and landscape reclamation that has greened the site and resulted in a covering of scrub and immature woodland. The aesthetics are marred by the lack of management resulting in fly tipping and erosion on the banks of the stream that crosses the site. The site forms an open space between the settlements of Bolton upon Dearne in the north, Wath upon Dearne in the south and new industrial development replacing coal mining to the east.

5 SIGNIFICANCE FOR REGENERATION

Any methodology that aims to improve the sustainability of regeneration should be valid for the integration of processes operating at multiple spatial scales, and allow holistic approaches for all projects regardless of scale within the context of a strategic regional planning. The temporal dimension is also very significant – working with processes to generate sustainable landscapes that can develop over time to respond to changing needs and demands rather than imposing solutions only appropriate in the context of a short project timescale.

We contend that the creation of sustainable landscapes means moving away from the development of large areas of land in which the needs of only one single function are addressed, with others overlooked. This also includes many "mixed-use" developments where, although from a human perspective a site may have many "functions" from a holistic perspective it may only be optimised for two – economic and socio-cultural being the most likely.

Forman (1995) has argued that planning for sustainability is most appropriate at the regional and landscape scales, and this requires an approach that focuses on the holistic landscape rather than a single issue. In addition as has been demonstrated there is a great deal of multifunctionality at the site scale and therefore individual sites need to be examined as being part of the landscape and region, and integrated into it through the planning process.

Looking at current regeneration proposals in the areas study shows that this pattern persists. In the regional context the single environmental target proposed by Yorkshire Forward is concerned with Greenhouse Gases, this amongst about two-dozen other actual or proposed targets (Yorkshire Forward 2002). While other landscape related aspects of multifunctionality may not be considered to be in Yorkshire Forward's sphere of interest, the creation of a new sustainable will require holistic thinking (e.g. Hermann and Osinski 1999, Naveh 2000).

At the landscape and site scales the situation is rather different. In the Dearne Valley there is much work being undertaken for all parts of the functional spectrum from the development of historical features in the landscape by the local authorities in the area to physical and social regeneration by a variety of organisations with social, economic and ecological agendas. However there appears to be a tendency to parcel up the land, distributing it between the various regeneration organisations. Thus, while regeneration for a variety of objectives is occurring, there is no strategic thinking as to how the landscape itself is functioning as a whole or how each individual site contributes to the processes and functions within this landscape.

What is required is a methodology that allows decisions for planning and regeneration to incorporate multifunctionality, thereby working with the landscape rather than just in the landscape. This should result in decisions that optimise the functionality of the landscape and thereby deliver a more sustainable and cost effective regeneration solution.

The concept of "real" multifunctionality (Brandt and Vejre, in press) appears to answer some of the questions asked of regeneration, serving the requirement of sustainable landscape planning in the post-industrial environment. The concept uses a holistic approach to the landscape but also looks at it not as a pattern of land uses but as a dynamic system providing a variety of functions that work together in that landscape in an integrated way.

ACKNOWLEDGEMENTS

This work is part of a joint ESRC/NERC research studentship.

REFERENCES

Booth, C. 2001. *Call Centre Development in the Dearne Valley Enterprise Zone, Yorkshire and Humberside: A UK Case Study for the NWMA INTERREG IIc SPECTRE Project*. Sheffield: Sheffield Hallam University Centre for Economic and Social Research.

Brandt, J., Tress, B. & Tress, G. (eds). 2000. *Multifunctional Landscapes: Interdisciplinary approaches to landscape research and management – Conference material for the conference on "multifunctional landscapes". Centre for Landscape Research, Roskilde, October 18–21, 2000.* Roskilde: Centre for Landscape Research.

Brandt, J. & Vejre, H. in press. *Multifunctional Landscape – motives, concepts and perspectives.*

British Geological Survey. 1996. *Industrial Mineral Resources Map of Britain.* D.E. Highley, G.R. Chapman, G. Warrington and D.G. Cameron compilers. Bristol: NERC and BGS.

British Geological Survey. 1999. *Coal Resources Map of Britain.* G.R. Chapman compiler. Bristol: NERC and the Coal Authority.

British Geological Survey. 2002. *Coal Resources Map of Britain [CD-ROM].* Nottingham: British Geological Survey.

Centre of Ecology and Hydrology. 2002. Landcover Map of Great Britain 1990. http://www.ceh.ac.uk/data/lcm/.

Clark, M., Burall, P. & Roberts, P. 1993. A Sustainable Economy, in A. Blowers, (ed.) *Planning for a sustainable environment: A report by the Town and Country Planning Association.* London: Earthscan.

Countryside Agency. 1999. *Countryside Character Initiative: Yorkshire and Humber.* Cheltenham: Countryside Agency.

Department of the Environment, Transport and the Regions. 1998. *Planning for the communities of the future.* London: DETR.

Department of the Environment, Transport and the Regions. 2000a. *Our Towns and Cities: Delivering an Urban Renaissance.* London: DETR.

Department of the Environment, Transport and the Regions. 2000b. *Indices of Deprivation.* London: DETR.

English Nature. 1998. *Natural Areas: Nature Conservation in Context.* Peterborough: English Nature.

English Nature. 2002. *Natural Area Boundary Data.* Peterborough: English Nature.

Forman, R. 1995. *Land Mosaics: The Ecology of Landscape and Regions.* Cambridge: Cambridge University Press.

Frith, M. 2001. Brown before green – biodiversity and the urban white paper, *Ecos* 22 (1): 6.

Fry, G.L.A. 2001. Multifunctional landscapes – towards transdisciplinary research. *Landscape and urban planning* 57: 159–168.

Handley, J. 1996. *The Post-industrial Landscape: A Groundwork Status Report.* Birmingham: The Groundwork Foundation.

Handley, J. 2001. Derelict and despoiled land – problems and potential, in C. Miller (ed.). *Planning and Environmental Protection.* Oxford: Hart.

Handley, J., Griffiths, E.J., Hill, S.L. & Howe, J.M. 1998. Land restoration using an ecologically informed and participative approach, in H.R. Fox, H.M. Moore & A.D. McIntosh. *Land Reclamation: Achieving Sustainable Benefits.* Rotterdam: Balkema.

Hermann, S. & Osinski, E. 1999. Planning sustainable land use in rural areas at different spatial levels using GIS and modelling tools. *Landscape and Urban Planning* 46: 93–101.

Hey, D. 1979. *The Making of South Yorkshire.* Ashbourne: South Yorkshire Metropolitan District Council.

Hill, A. 2001. *The South Yorkshire Coalfield: A History and Development.* Stroud: Tempus Publishing.

Johnson, D., Martin, S., Pearce, G. & Simmons, S. 1992. *The Strategic Approach to Derelict Land Reclamation.* London: Department of the Environment, Directorate of Planning Services.

Landmark Information Group and Ordnance Survey. 2002. *1:10,560 County Series: Yorkshire.* Exeter: Landmark Information Group.

Maier, L. & Shobayashi, M. 2001 *Multifunctionality: Towards an Analytical Framework.* Paris: OECD.

National Audit Office. 1988. *The Department of the Environment Derelict Land Grant.* London: HMSO.

Naveh, Z. 2000. What is holistic landscape ecology? A conceptual introduction. *Landscape and Urban Planning* 50: 7–26.

Owen, G. 1992. Regenerating the Deanre Valley. *The Yorkshire and Humber Regional Review* 2 (1): 11.

Rodwell, J.S. & Skelcher, G. 2002. *The ecoscapes and plant communities of Cheshire.* Lancaster: Lancaster University Unit of Vegetation Science.

Rotherham Metropolitan Borough Council. 2002. *Rotherham Buniness Information.* Rotherham: Rotherham MBC.

Rotherham Urban Wildlife Group. 2000. *The Dearne Valley.* Rotherham: The Rotherham Urban Wildlife Group.

South Yorkshire Forest. 2000. *South Yorkshire Forest Plan 2000.* Sheffield: South Yorkshire Forest.

Urban Task Force. 1999. *Towards an urban renaissance.* London: DETR.

Yorkshire Forward. 2001. *Regional Action Plan for the Yorkshire and Humber Economy.* Leeds: Yorkshire Forward.

Land Reclamation – Moore, Fox & Elliott (eds)
© *2003 Swets & Zeitlinger, Lisse, ISBN 90 5809 562 2*

Reclamation or redevelopment: the Oasis Principle

S.D. Steffens
Oasis Mine Reclamation Consultants, Evergreen, Colorado, USA

ABSTRACT: Reclamation of lands disturbed by mining and quarrying is often implemented with minimalist goals. When final reclamation commences, the income revenues for the site are gone and low-cost rehabilitation of the lands is necessary. Reclamation using the "Oasis Principle" can provide redeveloped areas next to areas of ecologically-based reclamation. Like an oasis in the desert, the contrast between these areas can enhance the value of both. Attributes and intrinsic values of the site are explored to identify alternative uses that are economically feasible, socially beneficial, and perpetually viable. Such uses can be integrated with conventional range, forest, and wildlife reclamation to maximize the benefit to the mine operator while still complying with national, provincial, or state regulations. A method for determining feasible post-mining uses for a mine-site is presented, including assessment of a site's attributes, screening and ranking of potential uses, and economic comparison with conventional reclamation methods.

1 INTRODUCTION

When a mine has reached the end of its useful life and the ore is exhausted, a new phase, reclamation, begins with its associated expenses and variety of issues. By that time sales revenues are also exhausted, and the cost of reclamation may be an expense with no offsetting income. Prudent mine operators will have planned ahead to implement the required reclamation measures efficiently and at minimal cost. Some may have investigated alternative post-mining uses and incorporated those into the plan. In the state of Colorado, USA, reclamation laws allow the mine operator to incorporate uses other than conventional reclamation measures. Conventional measures may include range, forest, and wildlife enhancement, hereafter referred to as RFW reclamation. However, alternative uses are rarely included in the initial (official) Reclamation Plan and many times are not considered until too late in the mine's life. Some of the reasons for avoiding consideration of alternative uses in the initial plan are the following:

- In Colorado, alternative uses must be approved by the regulating agency as an integral part of the Plan. It may be premature to commit to post-mining uses other than conventional reclamation when a mine is first developed.
- For a mine with a lengthy production life, there may be many changes in the regional, local, and site socio-economic picture as well as technological advancements during the life of the mine. Those changes could vastly alter the feasibility of a post-mining use, which justifies delaying consideration of alternative post-mining uses to a later date.
- Planning for redevelopment, as compared to RFW reclamation, may require selection of a successor or contractor to implement the construction, manage the site, and in some cases provide expensive long-term care and maintenance. Again, premature planning could prove to be fruitless and a wasted effort.

For all of these reasons it is easiest to plan and implement a simple RFW reclamation plan. Commitment to such a plan throughout the life of the mine is relatively safe and certain. Performance warranties and bonds for reclamation need only reflect increased costs for construction and the occasional technical revision to the permitted disturbance.

While RFW reclamation is reasonably safe and simple, it may not yield the best, or any, return on the investment of labor and materials to reclaim the site. Most RFW measures also won't take advantage of valuable resources that are often associated with a working mine, such as the electric power-grid, water systems, heating fuel systems, etc. Alternative uses that provide a revenue stream after the mining is complete can capitalize on these resources and should be considered.

A method for evaluating a wide range of alternative uses, quickly and comprehensively, provides flexibility in determining the best post-mining use at the latest date possible, and with the greatest certainty.

The objective of the "Oasis Principle" is to find areas or centers of value within the site that, when developed, will complement the surrounding areas that are reclaimed by conventional means. Alternative uses within these areas may provide revenues that not only pay for their development but offset some of the costs of reclamation and long-term care (if such long-term care is required).

The selection of feasible alternative uses must address the following factors:

– Does the mine-site have the basic infrastructure and resources to support the alternative use?
– Is the alternative use compatible with the local geography and culture and will it be compatible with the surrounding areas treated with ecologically-based (RFW) reclamation?
– Are there regional or local constraints that would limit or exclude implementation of the alternative use?
– Is there a market or demand for the alternative use?
– Is the cash-flow/rate-of-return for the alternative use worth the effort and is it more favorable than conventional RFW reclamation?

The following is a procedure for assessment of alternative uses that will answer these key questions. The objective of the process is to find an optimum arrangement that will minimize mine-closure costs, help to provide a seamless transition after mine closure for support of the local economy, integrate with ecologically-based reclamation, and be self-sufficient and sustainable. A creative component of the procedure encourages investigation of non-standard as well as common uses.

1.1 Definitions

Certain terms are used in the remainder of this presentation that have specific meanings, as follows:

Attribute – A resource such as power, water, or gas that is required for a particular land-use.
Intra-site – Attributes at a specific location within a site, as compared to merely available somewhere at the site.
RFW reclamation – Ecologically based reclamation measures for range, forest, and wildlife enhancement.
RLS value – The value of a potential use at the mine-site from the perspective of Regional and Local needs or restrictions and attributes that are available at the Site.
Value center – A location within the mine-site where one or more attributes, necessary for a particular use, are available and exploitable.

2 EVALUATION AND SELECTION OF ALTERNATIVE USES

The objectives of an integrated redevelopment/reclamation plan are the following:

– Find the most productive Value centers within the site and the best use of them.
– Redevelop those Value centers while reclaiming the remainder of the site with conventional RFW enhancements.
– Begin planning as early as possible, but late enough in the mine-life that the plan will reflect modern technologies and land-uses that are marketable.
– Integrate redevelopment, where possible, with the mining plan to maximize the use of on-site resources and attributes.
– Select only redevelopment options that provide a benefit, direct or indirect, to the property when compared with conventional RFW reclamation.

2.1 General procedure

This method for evaluating alternative uses includes three main phases. The first phase is to determine what uses are potentially feasible based on their apparent value. The second is an economic comparison with conventional RFW reclamation to decide what portion of the site should be redeveloped instead of reclaimed. The final phase is to determine the best organization for reclamation and redevelopment. These phases are illustrated on Figure 1.

Phase one of the evaluation actually has three secondary steps, as shown in Figure 1. The first step (Step A) is to employ a site-specific model that can be used to test and rank the feasibility of a large number of potential uses. Many uses will be eliminated because of prohibitive or prejudicial reasons relevant to the Regional or Local setting. The remaining uses are ranked for their Site value, with secondary modifiers related to their Regional or Local feasibility and restrictions. The result of this step is a list of ten alternative uses that are worthy of further consideration at the Site.

The second step (Step B) is to narrow the field of potential uses by ranking or eliminating uses based upon mine-site availability of required infrastructure.

This step introduces the concept of "Value centers" where specific resources such as power, water, advantageous topography, and other attributes exist. In this step the top-ten candidates are also ranked by their compatibility with surrounding areas where RFW reclamation will be employed.

The third step (Step C) of the first phase is to perform preliminary estimates of probable construction costs and forecasts of long-term revenue for each of the final candidates for redevelopment. The present

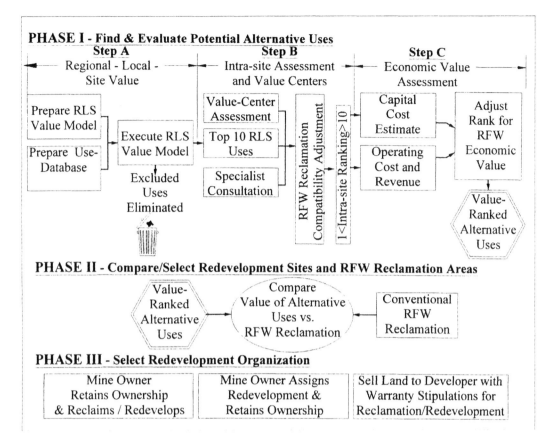

PHASE I - Find & Evaluate Potential Alternative Uses

Step A	Step B	Step C
Regional - Local - Site Value	Intra-site Assessment and Value Centers	Economic Value Assessment

Figure 1. Oasis value assessment method, flow diagram.

worth of each use is calculated and compared to the projected expense of RFW reclamation at that location. The ten uses are ranked again according to their relative economic value.

When a final list of candidates has been characterized and evaluated, each or all of those uses are compared in Phase II to the expected site-wide cost/benefit of conventional RFW (Range-Forest-Wildlife) reclamation. One redevelopment use may stand out as the best option, or several uses at different value centers, or possibly RFW reclamation is most appropriate throughout the entire site.

The final phase is to select an organizational method for redevelopment.

One element of the process that is considered vital is the use and availability of special expertise consultation. Examples may be consultants that are knowledgeable and experienced in the development of sporting facilities, or of theme parks, or of specialized industries and commerce. When a site is first assessed for potential uses (Phase I, Step A) these consultants would provide suggestions of creative, non-standard uses that have

potential. During the economic evaluation (Phase I, Step C) they would provide order-of-magnitude estimates of cost and opinions regarding the feasibility of special uses. The same consultants could help coordinate the final design of the project. In all three cases they would be relied on to provide special insight and efficiency in their area of expertise.

3 PHASE I – ASSESSMENT AND SELECTION OF ALTERNATIVE USES

3.1 *Step A -Regional, Local, and Site (RLS) value assessment.*

The initial selection of potential uses is illustrated in Figure 2. The objective of this phase is to prepare a short-list of alternatives that will meet the following requirements.

– The use is compatible with the Region, including its geography, climate, culture, and national or provincial laws.

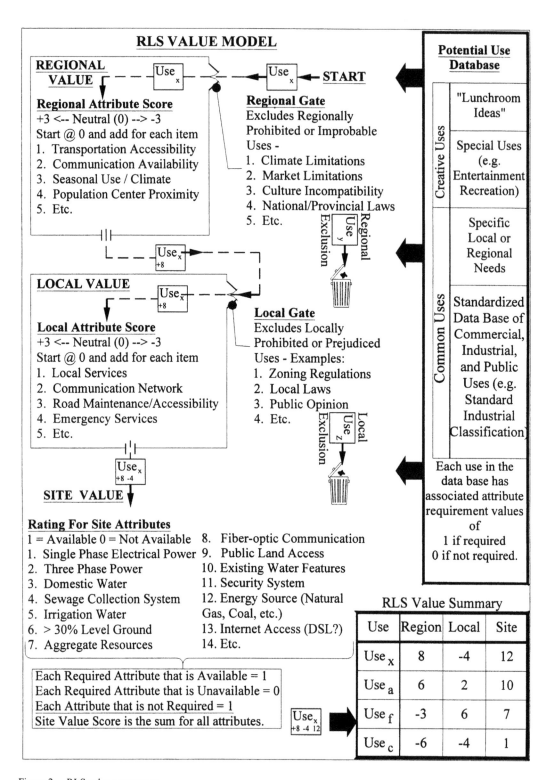

Figure 2. RLS value assessment.

– The use is compatible with the Locale and is not prohibited or excluded by local (county or municipal) regulations.
– The use, in some cases, fills local needs such as schools, service facilities, or special-use facilities.
– The use can be supported by the Site resources such as utilities, topography, etc.

The RLS value assessment includes two components – a Potential-use Database with required attributes for each use and a RLS decision model that can test the database to filter out all but potentially feasible alternatives.

3.1.1 Potential-use Database
The Potential-use Database should be as comprehensive as possible and should include a creative component of non-standard uses as well as common ones. Two possible databases for common uses could be the *North American Industry Classification System – United States, 2002 (NAICS)* or the *United Nations Standard Products and Services Code (UNSPSC)*. The NAICS has replaced the *Standard Industrial Classification (SIC) Index*. Both codes include most common industrial and commercial enterprises. They are used by marketing professionals as a tool for determining potential markets and could be easily adapted for assessing the attributes of potential uses.

A second type of common use is specific regional or local needs for public services (e.g. schools and public service facilities). Although public services may also be a regional or local value attribute, they are treated as a use. In this manner, a lacking or inadequate public service facility can be assessed and compared on an equal basis with private uses.

Creative uses may be unconventional or extraordinary businesses and enterprises that would not normally be considered. Such ideas are the result of "thinking outside the box." For example, aquaculture facilities, theme entertainment, or botanical gardens are not normally associated with mines. However, an extraordinary botanical display is the Butchart Gardens in Victoria, British Columbia, Canada, a site that was formerly a limestone quarry. The No. 31 (Coal) Mine in Lynch, Kentucky, USA, is being converted to a mining museum with tours in an effort to partially offset the decimating economic effects that resulted from the mine's closure. An underground mine tunnel in Colorado once was used as a tree nursery, growing seedlings for reclamation by artificial light in an otherwise perfect climate. These are three examples of non-standard, or special uses.

One source for providing ideas for such special uses is the network of consultants described previously. A second source may be internal to the operating mine's staff. There can be a surprising variety and depth of insightful ideas generated by the "lunchroom banter" of mine and other industrial workers. Formalizing brainstorm sessions about "what can we do with this mine when it's closed" may be a valuable resource.

The required site attributes or resources for each potential use in the Potential-use Database are quantified for later comparison with the attributes that are actually available at the site. A computerized database, such as Microsoft Access, is a convenient tool for assigning and quantifying attributes for a potential use.

3.1.2 RLS assessment model
The second component of the Regional-Local-Site (RLS) value assessment is a model that will test each potential use and determine whether it is plausible and worthy of further consideration. As illustrated by Figure 2, each use is subjected to five tests for either scoring or elimination.

1. A use must first pass through a "Regional Gate." This gate eliminates uses that are infeasible for fundamental reasons related to the region surrounding the mine-site. The reason for exclusion may be related to climate, market, culture, geographic limitations, or national/provincial regulations. For example, a commercial fishing or ski area venture in a desert, or a large concert hall located far from a population center, would not merit further consideration.
2. Regional attributes are then scored by comparing the use requirements to the availability of the resource on a regional scale. For example, the regional value for transportation would be high if the site is close to an interstate highway (or autobahn), or low if access to the site is remote and time-consuming. Other attributes could include climate statistics (e.g. degree-days per year), proximity to population centers, and proximity to necessary regional markets. If a particular use does not depend on regional values it is scored neutral. If the regional value is an advantage/disadvantage to the use it is scored positive or negative, respectively. The sum of the regional values is finally assigned. In the example shown on Figure 2, Use_x has a regional value of $+8$.
3. The use then passes through a "Local Gate" before it is considered further. The Local Gate would include limitations of the immediate locale. These might include planning and zoning restrictions (such as building codes), prohibited uses (such as speedways near residential areas), or even a vociferous public opinion against such use.
4. The use is next evaluated with respect to the availability of local services such as its requirement for roads, emergency services, communications (e.g. the availability of cellular service or DSL internet), and other such infrastructure. As with the regional attribute score (item 2 above), the use is assigned

a local attribute score based on the availability of required local services. If they are not important then it is a neutral score, etc. In the example, Use $_x$ has a local value of −4, which indicates that local resources and support may be lacking.

5 The final rating for a use is the Site Value, which is a measure of whether the required attribute resources for the use are available at the site. Examples of site attributes are electrical power and other utilities, topographic characteristics, construction materials such as sand and gravel, energy sources, and water resources. Each attribute is scored 1 if it is necessary and available or if the attribute is not needed. If a required attribute is NOT available it is scored zero. The sum of the individual scores is the Site Value. The Site Value for Use$_x$ in the example is 12.

The Regional, Local, and Site Values are tabulated and sorted as shown on Figure 2 to provide the short-list of the top ten candidates for redevelopment. That list is then used in the next step for intra-site evaluation.

3.2 *Step B – intra-site evaluation and value centers*

The RLS value assessment (Step A) determines whether the required attributes are available somewhere at site while the Intra-site evaluation determines whether those attributes are in reasonable and feasible proximity within the site. In this manner, potential uses that may not be feasible to develop will be rated much lower than others where all or most of the resources are readily available. The assessment is based on the concept of "Value centers."

3.2.1 *Value centers*

Value centers are the areas within a site where the availability of resources, or attributes, is maximized. For example, for a use requiring common utilities such as domestic water or sewer, the value center would be the area where those services are either available or nearby. The value center for an agricultural use may be areas with rich soils and water, the point being that value centers may be different for each use. Figure 3 illustrates a method for determining, for each potential use, whether there is a value center that will meet all or most of the requirements for the use. The figure also shows how the top-ten uses are re-ranked for their compatibility with the site and economics.

Utilities, geographic features (water, landforms, and topography), construction resources (clay, aggregates, timber), water resources, and other specialty resources are shown on Figure 3a, the map of a generic mine-site. In simple cases it will be readily apparent where the different focal points for redevelopment (value centers) are present. In larger sites with a complex system of utilities, the value centers may

be more numerous and difficult to identify. A simple rating system helps to sort out the alternatives.

3.2.2 *Attribute rating*

Each potential use derived in the RLS evaluation has been assigned a rank based on that evaluation, from one (lowest) to ten (highest). The rank is next devalued by the number of attribute requirements that are NOT available within a value center. For example, if a use requires three phase power, domestic water, and sanitary sewer but one of those attributes is missing then the RLS rank is devalued by one point. If another location within the mine-site has all three attributes, the rank is restored. In this manner the uses derived in the RLS assessment are reordered according to the specific availability of resources within the site boundaries.

3.2.3 *RFW compatibility*

The use ranking is next adjusted according to the compatibility of each use with the conventional RFW reclamation in the surrounding areas. Each ranking is modified according to whether it blends with ecologically-based reclamation plans. A high rating reflects that the use is consistent with enhancement of range and forest in surrounding areas and a corresponding low rating indicates incompatibility. A scale of −3 to +3 is suggested for most sites. However, if the RFW reclamation areas are particularly sensitive to redevelopment then a magnified scale can be used to provide better definition and weight to ecosystem compatibility.

3.3 *Step C – economic evaluation*

In this third step of Phase I the projected economics of each use in the short-list is addressed and the use ranking is modified again to reflect its value compared to RFW reclamation of the same area. A preliminary, order-of-magnitude cost estimate is prepared for the construction or implementation costs of each use in the list. A projection of future income and operation/ maintenance (O/M) cost is also determined, characterized as an annual annuity of net revenue. A present worth of the use is then determined, calculated by conventional cash flow/rate-of-return procedures.

The objective of this final step in the Intra-site assessment is to modify the rank of each use according to its economic value compared to the cost of conventional reclamation for the same footprint area. (The objective of this step is NOT to determine the feasibility of redevelopment. That occurs in Phase II.) If the economics of a potential use show that it will be a less desirable investment than RFW reclamation, i.e. it will be more expensive in the long-term, that use should be excluded from further consideration. The possible exception is "local need" uses such as

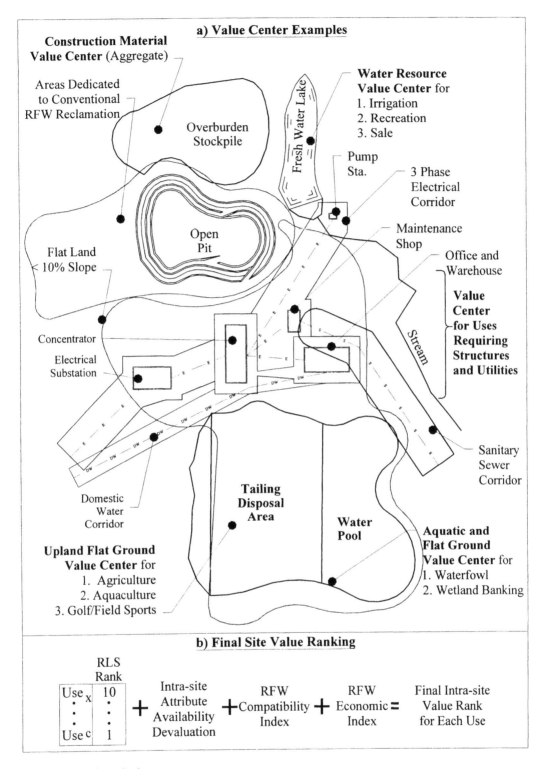

a) Value Center Examples

Construction Material Value Center (Aggregate)

Areas Dedicated to Conventional RFW Reclamation

Overburden Stockpile

Water Resource Value Center for
1. Irrigation
2. Recreation
3. Sale

Fresh Water Lake

Pump Sta.

3 Phase Electrical Corridor

Open Pit

Maintenance Shop

Office and Warehouse

Flat Land < 10% Slope

Value Center for Uses Requiring Structures and Utilities

Concentrator

Stream

Electrical Substation

Sanitary Sewer Corridor

Domestic Water Corridor

Tailing Disposal Area

Water Pool

Upland Flat Ground Value Center for
1. Agriculture
2. Aquaculture
3. Golf/Field Sports

Aquatic and Flat Ground Value Center for
1. Waterfowl
2. Wetland Banking

b) Final Site Value Ranking

RLS Rank		
Use $_x$	10	
.	.	
.	.	
.	.	
Use ^{c}	1	

$\begin{array}{c} \text{RLS} \\ \text{Rank} \\ \boxed{\begin{array}{cc} \text{Use}_x & 10 \\ \cdot & \cdot \\ \cdot & \cdot \\ \text{Use}^c & 1 \end{array}} \end{array}$ + Intra-site Attribute Availability Devaluation + RFW Compatibility Index + RFW Economic Index = Final Intra-site Value Rank for Each Use

Figure 3. Intra-site evaluation.

41

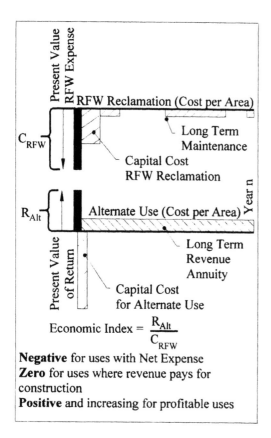

$$\text{Economic Index} = \frac{R_{Alt}}{C_{RFW}}$$

Negative for uses with Net Expense
Zero for uses where revenue pays for construction
Positive and increasing for profitable uses

Figure 4. Calculation of economic index.

schools and public service facilities. There may be other negotiable incentives, administrable by the local government that will offset the negative worth of those uses and make them justified. Examples of such incentives may be tax credits, subsidies, or preferential consideration on other local projects.

Figure 4 illustrates the method for determining the economic index that is applied to the use ranking. Each cash-flow/rate-of-return bar shows the construction cost as a present value, a long-term annuity of either expense or revenue, and the present worth of the alternative. For RFW reclamation the present worth will be the present-value of the expense (C_{RTW}). For use alternatives the present worth (R_{Alt}) may be either an expense or a net income, either of which is shown in units of present value. Each value is expressed as cost per unit area of the footprint of the use.

The present worth of each potential use is then divided by the present value of the expense for RFW reclamation to arrive at the economic index. For example, if a potential use has a negative present worth (i.e. net expense) it would have a negative economic index to reflect the fact that it will be an

overall expense but not as expensive as RFW reclamation. A use for which the long-term revenues will exactly offset the capital cost will have a zero economic index. Uses that will provide positive present worth will have increasingly positive economic indices to reflect the fact that the alternative use will pay for itself as well as offsetting the expense of reclaiming surrounding lands. These indices are then added (or subtracted) from the top-ten ranks. The system can be modified as needed to make the economic effect more or less robust and important.

The final result is that the rank of each candidate from the RLS screening will have been modified to reflect the availability of required resources, to reflect its compatibility with RFW reclamation in the surrounding area, and to reflect its potential for offsetting the expenses of reclamation in other areas.

4 PHASE II – COMPARISON WITH CONVENTIONAL RFW RECLAMATION

In this phase the economics and benefits of implementing the entire, composite plan of redevelopment and reclamation is evaluated. The total cost for developing the alternative uses is compared with the total savings that will accrue by not treating those areas with RFW reclamation measures. Other non-economic benefits of the alternative uses are compared, as well. This procedure can be as simple or as in-depth as required for the site being evaluated. However, the comparison should be based on the following considerations:

– Present worth cost comparison.
– Size of redevelopment (i.e. the footprint of each use within the site) and the relative portion compared to the total site area.
– Organization options for implementing the project.
– Long-term suitability and stability of alternative uses.

In the simplest form of economic comparison, the projected total cost for RFW reclamation is reduced by the amount attributable to the footprint area that may be committed to the alternative uses. The present worth of the alternative uses is then added to provide a composite reclamation cost for the project with both reclamation and redevelopment features. These costs can be compared to the RFW reclamation cost to determine the benefits of the redevelopment option and whether they are sufficient to justify implementation.

5 PHASE III –

When, or if, the redevelopment option has been accepted, the organizational method for implementation of the plan should be considered. The need for long-term care and maintenance is an important consideration in this selection. If, for example, the mine has environmental problems that will require a treatment system, the mine owner may need to retain long-term ownership and close management of the site. Alternately, if the site is expected to be self-sustaining after reclamation is completed, the owner may want to sell the property and assign the responsibility. An intermediate arrangement might be assignment of the reclamation/redevelopment to a managing agent while retaining ownership.

Determining the best management arrangement can be a very complex matter and is beyond the scope of this paper. However, it should be considered when evaluating the economics of alternative uses because any assignee will expect to receive a portion of the benefit from redevelopment. The magnitude of this commission may affect the justification for the chosen redevelopment option.

6 EXAMPLES AND APPLICATIONS OF THE OASIS PRINCIPLE

Oasis Mine Reclamation Consultants is developing the programs and databases for applying the procedures and methodology described above. The Oasis Principle and these methods have not been formally tested with a detailed assessment. However, there are several mine-sites in Colorado where the concepts are applicable. Four of these mine-sites have provided the basis for formulating the principle.

All four sites (three metal mines and one quarry) have several years of expected production life before closure-reclamation will commence. The three metal mines are the Henderson and Climax molybdenum mines owned by Phelps Dodge and the Cresson gold mine owned by Anglo Gold Ltd. Lafarge North America, Inc operates the specification aggregate quarry. Each of the four sites has distinctly different regional and local values and site-specific attributes, described below.

6.1 Climax Mine and Eagle Park Reservoir

The Climax Mine is located at the Continental Divide at an elevation of about 11,000 feet (3350 m) and about two hours travel time from Denver, Colorado. The 14,000 acre (5660 ha) facility includes several large industrial buildings; two water storage reservoirs; the mine production and stockpile area; and large expanses of tailings that have been covered by rock from the mine as the first step of reclamation.

The key regional attribute limitations for the site are transportation (remote), climate (severe and cold), and population accessibility (remote). Strong positive

regional values include its location at the headwaters of three major river systems (for water supply) and its close proximity to four major ski areas (for recreation).

Local values and limitations are somewhat mixed because the mine is within three county jurisdictions. The regulations, codes, and master plans of the three counties are quite different.

Intra-site attributes include a wide availability of utilities, water resource structures, flat land (albeit mine tailings), steel buildings, and access to public land. These attributes define the value centers that include industrial, recreation, and even residential uses (to provide housing for nearby ski-area workers).

One of the great successes of mine redevelopment in Colorado has been the rejuvenation of the Eagle Park Reservoir at the Climax Mine. The reservoir was formerly an impoundment of molybdenum oxide tailings. Thousands of tons of tailing have been relocated and the facilities have been converted for use as a water storage reservoir. Eagle Park Reservoir is now an outstanding fishery and a water supply for making artificial snow at the world-famous Vail Ski Resort. Sale of the water resources paid most of the expense for moving the tailings and rejuvenating the reservoir.

6.2 Henderson Mine and Mill

The Henderson Mine and Mill facilities also straddle the Continental Divide and three river basins. As with the Climax Mine, water resources and winter sports recreation are positive regional attributes, while transportation and population accessibility are limitations. The 12,500-acre (5070 ha) facility is similar to the Climax Mine and includes industrial buildings, a tailing impoundment, and storage yards.

The mine and mill sites are, however, quite different from each other with respect to local values because they are in two counties with different local issues and goals. The mine site is in a relatively pro-development county that is hindered by lack of land suitable for building. Seventy percent of the county is federally owned and restricted. The flat areas at the mine and the availability of utilities provide high local and site values and define the value centers.

The mill site, on the other hand, is in a remote area surrounded by federal forestlands. Local government has discouraged all post-mining uses except conventional RFW reclamation. However, the availability of utilities and flat terrain provides a wide variety of value centers. Development of water resources has the highest value potential, for the same reason that Eagle Park Reservoir has been successful. The mill site has existing reservoirs that could be converted for domestic and recreation use and it is ideally located to divert water among three river basins. The 900-acre (400 ha) tailing impoundment has good potential for uses that require relatively flat terrain. Consolidation and settling of the tailing may be a hindrance to such uses, however.

6.3 Cresson Mine

The Cresson Mine is located within an hour's travel time of Colorado Springs (a large population center). The 5800-acre (2360 ha) site is in the historic Cripple Creek Mining District. Casino gambling is a major tourist attraction to the area.

Positive regional values include transportation accessibility to Colorado Springs and the site's close proximity to that population center.

The local commerce and government regulations are centered on casino gambling and its associated services. The best post-mining value of the mine-site may be uses that capitalize on the historic-mining and gambling, i.e. tourist trade.

Utilities, buildings, and aggregate resources are the predominant site attributes that also define value centers. The Valley Leach Facility (when completed) will be a vast area of relatively flat terrain. Creative post-mining uses that can be built on that terrain but that are not sensitive to differential settling of the soils would define additional value centers.

6.4 Lafarge specification aggregate quarry

This 253-acre (102 ha) quarry is located on the west edge of the Denver Metropolitan area. Regional, local, and site values are high because of its location. The quarry is nestled between Interstate Highway 70 (the primary transportation route into the mountains), public recreation parks, and a small amusement park that has operated with marginal success for fifty years.

Oasis Mine Reclamation Consultants performed a detailed assessment of the quarry in 2000 as a private research project. The methods described in this paper are based on the result of that evaluation. The conclusions of that study were that the quarry had good potential for the following post-mining uses:

- An expanded world-class theme entertainment park.
- An education and research center integrated with the theme park.
- A transportation and parking center for mass transit into the mountains.

7 CONCLUSIONS

Redevelopment of mine-sites can be effectively integrated with ecologically-based reclamation by focusing on value centers within the site and choosing new uses that are compatible with the surrounding range and forest. The development of these "Oases" can result in cost savings, revenue generation, and a more diverse reclamation result.

Land Reclamation – Moore, Fox & Elliott (eds)
© *2003 Swets & Zeitlinger, Lisse, ISBN 90 5809 562 2*

The market for environmental goods and services related to regeneration in the North West

J.P. Palmer
Sinclair Knight Merz, Manchester, UK

ABSTRACT: Envirolink North West is the North West Development Agency's environmental goods and services cluster champion. In order to assist in promoting the North West's environmental businesses Envirolink commissioned Sinclair Knight Merz to carry out an assessment of the market for environmental goods and services related to regeneration in the North West. The current total market for regeneration is estimated at £466 million per annum and is set to increase at 7% per annum. About 5% of this money will be spent on environmental goods and services. The sector is fragmented and diverse with over 500 companies providing services related to regeneration. Some elements are strong such as environmental consultancy with good experience in regeneration and remediation. There is also a strong regional science base. However links between industry and research and higher education are not as strong as they could be. Principal drivers of remediation method are the implications of Part IIA regulation, landfill directorate, CLEA guidance etc. all of which will require innovation on the part of providers of goods and services related to remediation. Current mechanisms for encouraging innovation are not well developed. This paper will discuss ways of improving environmental goods and services provision in the North West and encouraging innovation to meet the challenges posed by changing regeneration needs and focus.

1   INTRODUCTION

This paper considers the results of a project undertaken to provide an analysis of markets for environmental products and services in relation to regeneration of the North West of England. The work was commissioned by Envirolink North West, the environmental goods and services sector champion in the North west, on behalf of the North West Development Agency's (NWDA).

The NWDA has set out its thinking on regeneration in a consultative document "Regeneration Prospectus"(NWDA 2001a). In the consultation document the NWDA states: "*The challenge faced by the NWDA has been how to give greater weight to the economic elements of regeneration in a way that is relevant to the needs and opportunities of the North West while ensuring that a comprehensive approach is taken to dealing with areas of deprivation.*" It goes on to describe in some detail how the agency sees regeneration playing a part in each of the four themes of the regional strategy:

– Investing in business and ideas;
– Investing in people and communities;
– Investing in infrastructure;
– Investing in image and environment.

The provision of relevant, sustainable and quality environmental goods and services will be key to the success of initiatives in all of these four themes. The Regeneration Prospectus also fully accepts the recommendations of a review of land reclamation completed in 2001 which recommends the trebling of the resources available for land reclamation in the region (NWDA 2001b).

It is against this backcloth that the assessment of the market for environmental goods and services has to be set – looking at both the existing players and situation and the future impacts of policies and innovations.

The market is driven by those who plan, contract or fund such services and includes:

– Agencies such as NWDA, Urban Regeneration Companies (URC), English Partnerships, European Investment Bank, Regional Selective Assistance etc
– Private sector developers
– Public sector – local government, Highways Agency etc.

In addition, there are indirect opportunities that can arise from regeneration programmes, new environmental regulation, regeneration policies and funding. Some of these might be in existing markets, others in emerging markets and include:

- Spin-off from innovation and new technology;
- Revised strategies for rural development and tourism;
- Transportation policies;
- "Greening" of run-down areas and investment in arts, heritage and recreation to make them more attractive to inward investors and somewhere for people to live.

Emerging legislation and regulation such as the aggregates, landfill and carbon taxes and accelerated tax relief for remediation announced in the 2002 budget will also affect the market for these services. Economic instruments will heavily influence the built environment and tax relief of 150% of remediation costs in particular will make brownfield development more attractive. Landfill and aggregates taxes will encourage recycling and innovative technology. Renewable energy is a rapidly developing sector that will play a part in regeneration in the region. Simultaneously, the Land Reclamation strategy outlined by NWDA (NWDA 2001b) will increase "soft end" schemes such as community forests, renewable energy crops and greening the landscape. It is therefore expected that the balance of goods and services provision will change in response to these initiatives.

The focus in this project was on environmental goods and services which use environmental skills rather than services such as architecture, buildings conservation and the social sciences which can lead to environmental improvements in the wider sense.

The North West covers almost 14,000 square kilometres from Cheshire in the south to the northern border with Scotland. The region is a mix of areas of outstanding beauty and protected landscapes with areas of industrial dereliction, a consequence of its long industrial history. 25% of England's derelict land and one third of the country's poorest quality rivers are in the North West. Agriculture accounts for 80% of land use and there are more than 400 important habitat and wildlife areas.

The region has been in the process of regeneration over decades. During this period, the focus of regeneration has changed from one of discrete and often small-scale reclamation for economic redevelopment of brownfield land to a region-wide strategic approach that encompasses economic regeneration and sustainability. The definition of regeneration for the purpose of this study is therefore wide ranging and includes:

- The reclamation of brownfield land for development
- Transport initiatives intended to improve access and infrastructure

- Improvements from "greening" the region to encourage tourism, increase the region's attractiveness as somewhere to live and work and regenerate land unsuitable for economic redevelopment
- New environmental technology that is intended to provide employment in place of traditional employment activities that have been displaced. This could include the development of energy crops as part of a re-focus of energy policy towards renewables.
- Clean-up of water resources including rivers, canals and the Mersey Basin.

2 METHODS

Information was gathered by questionnaires to relevant organisations, follow-up meetings with all the key sector groups and stakeholders. Significant emphasis was also placed on concurrent focussed interviews with key players.

The method and tasks used to meet objectives were:

- A review of existing data and information provided by strategy reports, annual and company reports and website information
- Circulation of questionnaires to buyers and suppliers of environmental products and services
- Telephone and "face to face" interviews.

Interviews were held with representatives of the NWDA, Mersey Basin Campaign, East Lancs Partnership, Sustainability North West, the North West Chemicals Industry, English Partnerships, Royal Institute of Chartered Surveyors (RICS), Groundwork, Centre for Sustainable Urban Futures (SURF), several of the local authorities and some of the major developers in the region.

3 THE MARKET

3.1 The UK

Recent European Union spend on all environmental products and services is $98 billion and this is expected to double by 2007. The UK proportion of this has been estimated from information provided in the EU Eco report for JEMU (Table 1) (JEMU 2002).

3.2 The North West region

A previous study for Envirolink North West into the environmental technology sector concluded that there are more than 700 companies employing 24,000 people and with £1.3 billion in annual sales (Enviros 2000). Of these, 90% are small or medium enterprises (SMEs) of which 62% offer consulting services and 12% analytical services. 30% of their customers are

Table 1. The UK proportion of European Union spend on environmental goods and services.

Activity	% of total market	Value*	Regeneration related
Waste management	40	4.54	✓
Waste water treatment	39	4.43	
Air pollution control	7	0.86	
Consulting and engineering services	7	0.86	✓
Contaminated land and remediation	3	0.43	✓
Environmental monitoring and instrumentation	2	0.32	✓
Energy management	1	0.22	✓
Other	1	0.11	
Total		11.77	

*Values shown are US $ billions.

Table 2. Funds for regeneration (£ million unless otherwise specified).

Source	Value of funds nationally[1]	Value of funds regionally[2]
EU structural funds 2001–2005	2126.4	£1.6 billion
English cities fund	100 million for assisted areas nationally and anticipated to leverage in up to £1 billion.	
Single regeneration budget 2001–2005[3]	7654.9	
Regional (NWDA) to 2002–2006		95
Private developers 2001–2005	1134.4	100
New deal for communities 2001–2005	614.5	75
Lottery 2001–2005	359.9	35
Other 2001–2005[4]	1946.9	186
Coalfield fund	385 million nationally. Projects and priorities in North West still in development.	
Total	14,322	2091[5]

[1] Data taken from MSI 2000, English Partnerships and NWDA.
[2] Figures provided for EU structural funds, others calculated based on annual MSI figures 1996–2000 (MSI 2000).
[3] Replaced by Single Financial Framework that will include SRB, NDC and other funds.
[4] Other includes Millenium Commission, City Challenge, CNT, NOF (New Opportunities Fund) and Groundwork.
[5] MSI predictions for total expenditure in the region in the period 2001–2005 are £2338.5 million (MSI 2000).

in the construction industry and are related to regeneration. The findings of the study correlate broadly with the JEMU findings in terms of sub sector breakdown (Table 1). However there appears to be a significantly higher percentage of contaminated land work in the regional study. This may be explained by the greater quantity of derelict land in the region, which is disproportionate in comparison with the national average, together with an accelerated reclamation programme.

3.3 Funding sources

The principal sources of funding are interlinked and the relationships are not always clear. There are a number of regeneration initiatives for which funding

is either already available or is actively being sought. Table 2 shows the principal sources of funds for and investment in regeneration.

The majority of funds are focussed on the major urban areas of Manchester and Liverpool but also include the activities of the Regional Rural Recovery Plan and the East Lancs Partnership. These are under the auspices of the NWDA. In August 2000, the NWDA announced a funding package of £1 billion for 20 regeneration partnership schemes across the region (NWDA 2001c). This was under Round 6 of the Single Regeneration Budget which was subsequently disbanded by the government and replaced by the Single Financial Framework. The NWDA strategy, produced in consultation with regional partners, is now expected to be implemented through Local

47

Strategic Partnerships (LSPs) between public and private sectors.

Land reclamation is an important element of economic regeneration in the region. There have been a number of studies and reports that have influenced strategy for regeneration. The two most significant are "*Reclaim the North West*" (NWDA 2001d) and "*Land Reclamation to further economic development and regeneration*" (NWDA 2002). Both indicated a need to move away from historic "*small scale and sometimes disconnected reclamation schemes*" to strategic reclamation incorporating a greater degree of soft end use reclamation, such as that of the Forestry Commission's Newlands project which aims to turn derelict land into woodland.

The NWDA identified reclamation costs of between £25,000–£50,000 per hectare for soft end uses, costs significantly below those for commercial, industrial or residential development. The NWDA intends to quadruple the level of investment in soft-end use reclamation as well as investing in remediation for economic regeneration. NWDA considers that between £1 billion and £2 billion of investment will be needed to eliminate derelict, underused, neglected and contaminated land in the region. An early estimate indicated that there are at least 11,000 hectares of affected land in the region and more derelict land being created is outstripping the rate of reclamation. Accurate figures on the quantity of land have yet to be compiled and some estimates exceed this estimate significantly.

The NWDA has made a number of recommendations in its report on land reclamation including re-allocation of land use in Regional Planning Guidance, Structure plans and Unitary Development Plans where there is over provision of some land allocations that might impose unrealistic or impractical land use. There is a "mismatch" of land use which in some cases in based on anticipated demands for industrial floor space that is now out of date. Plans should provide a framework for more green space and Regional parks. Landscapes created under the reclamation programme will require trusts for maintenance.

The Agency set out its regional strategy objectives in a Regeneration Prospectus published in March 2002. In terms of physical development the aim is to ensure that government targets of 60%, as a minimum, of new housing to be built on brownfield land or through adaptive re-use and refurbishment of buildings. By 2002 brownfield land in the region should be reclaimed at a rate of over 1,100 hectares per annum, thus reclaiming 5% of the total by 2004 and 17% by 2010.

3.4 *Environmental goods and services proportion of market*

The market for environmental goods and services is calculated to be approximately £210 million per annum increasing to £250 million per annum by 2005. This is based on a proportion of the available funding rather than the amount of money needed for total regeneration. An average of 5% of regeneration funds is considered to be applicable to environmental goods and services based on the following assumptions:

– Regeneration includes activities over and above the reclamation and redevelopment of land including for example, industry initiatives in the nuclear, chemical and other sectors to improve land quality and reduce impacts and new routes for reducing or disposing of wastes as a consequence of new regulations.
– Although environmental costs can form a very high proportion of regeneration costs where land is heavily contaminated. For most sites the bulk of the resources are spent on traditional engineering activities such as earthworks and drainage.

4 PRODUCTS AND SERVICES

The study identified 712 companies in the region providing environmental products and services in relation to regeneration. These include environmental consultants, laboratories, site investigation and demolition companies, landscape contractors and companies engaged in waste management and energy services.

4.1 *Analysis of questionnaires – suppliers*

"Suppliers" include site investigation, demolition companies and suppliers of products but does not include environmental consultants who are analysed separately. Suppliers were questioned regarding the nature and value of services, the proportion and value of work related to regeneration, number of staff employed, their main customers, type of service and primary influences for customer selection.

The regeneration market generally represents a low proportion (typically less than 40%) of total market share for these suppliers. However 17% of companies responding reported that work in regeneration activities

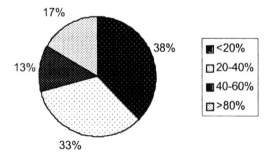

Figure 1.   Proportion of work related to regeneration.

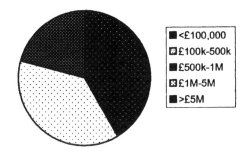

Figure 2. Annual value of regeneration work.

accounted for more than 80% of the workload. Site investigation and asbestos removal companies reported highest share of workload whereas solar energy companies typically spend less than 40%, often less than 20% on regeneration related work.

The approximate annual value of the regeneration market for a supplier in the North West region is generally between £100,000 and £500,000 and typically less than £1 million. Services for groundworks, site investigations and asbestos removal commanded highest values. Insulation, pollution control and solar power were reported of lowest values but identified as areas of growth.

The number of staff employed by a supplier for work in relation to regeneration is generally between 1 to 10 and typically less than 50. 30% of respondents employed more than 10 employees on regeneration work – mainly in waste management, site investigations and asbestos removal. Higher staffing levels were found in companies offering services in waste management, insulation materials, site investigation, asbestos removal and groundworks.

Suppliers generally consider that cost and quality are of equal importance to customers purchasing goods and/or engaging services in relation to regeneration in the North West region. Services in asbestos removal, pipes and fittings and site investigations were reportedly selected purely on the basis of cost.

Several suppliers reported timing and speed of service as an influencing factor.

The suppliers principal customers for work in relation to regeneration are contractors/civil engineers and local authorities. Other significant customers include environmental consultants and architects.

Services in ecology and land management are thought to be increasing but (ENDS 2002), are characterised by low fee rates. The emphasis of NWDA strategy on soft-end uses for site regeneration, the establishment of new woodlands and bio-diversity parks will increase the demand for these and landscape design and contracting services significantly.

New products and services are primarily channelled through consulting firms, industry and the private sector. In the order of 60% of suppliers considered that their products and/or services had not changed significantly in the last two to five years, in response to market needs relating to regeneration. However the emphasis on sustainability is now being reflected in new developments. Major developers now include eco ratings or BREEAM assessments for new build and innovation, particularly in energy, will affect products and services in the sector. The responses received suggest either that access to these markets is not being met locally or that the level of awareness among local suppliers is low.

Some suppliers reported a change in services due to increase in refurbishment as opposed to new build and the availability of new piling techniques and new insulation materials.

Some suppliers are proposing new products and/or services in respect of regeneration in the North West region. These mainly relate to waste management and pollution control probably driven by regulation. Further regulation and incentives for household energy efficiency announced for review in the 2002 budget will also stimulate new products and services. Encouraging innovation in this sector may be a key to building clusters of leading skills in the North West.

4.2 Analysis of questionnaires – environmental consultants

Consultants were questioned regarding the value of the regeneration market, numbers employed, customers, type of sub-contracted services and influences of sub-contractor selection.

The proportion of work in relation to regeneration varies across the sector. It appears that for a small number of larger consultancies, work in regeneration represents a significant part of the workload whereas smaller companies have a more diverse range of work.

The annual value of regeneration work is generally less than £500,000 per annum even for larger consultancies. More than half of those questioned reported annual values of £100,000 or less and only 11% exceeded £500,000 per annum. Phase I and II investigations dominate this work. Services in respect of waste management in regeneration are expected to increase. Those in strategic studies and planning reflect the current state of change in regeneration.

Private sector clients dominate due to the past focus on economic regeneration. The private sector will continue in importance, although with a change in the type of development. Mixed end use, travel development areas, commercial development and residential housing is expected to increase whilst industrial development is in decline. However a change in focus with a greater emphasis on "greening" activities and soft-end regeneration will increase the volume of work from the public sector.

SMEs predominate in the sector with the majority of consultancies employing small teams on regeneration activities supported by sub-contracted specialist services. 87% of the consultancies employ between 10 and fifty staff on regeneration related activities.

Environmental consultants sub-contract services for laboratory analysis, ecological assessments, geotechnical investigations, landscape services and asbestos surveys. In general, contracts are let locally with the exception of some aspects of ecology and archaeology that are contracted outside the region. Consultants were generally satisfied by the cost and quality of service, rating quality as important as cost. In some cases quality was the dominant criteria.

The value of sub-contracted services varies. 50% of contracts had a value of £5000 or less. Of the remainder, only 3% were above £500,000. In these cases the costs of the sub-contracted services were of the same order as the fees to the consultants.

5 MARKET TRENDS AND INFLUENCES

Markets for urban regeneration in the north west are mature with expenditure increasing at a significant rate and Liverpool Vision and Manchester in particular attracting substantial public and private sector investment. Regeneration in coastal and rural areas will focus on renewable energy, new crops and tourism.

It is clear from the market assessment findings that there will be considerable growth in the activities associated with regeneration over the next few years. An essential part of any regeneration process is remediation of the land which is being regenerated either for hard or soft end-use. Remediation will be a principal driver for environmental goods and services. Both the assessment of land remediation needs and the carrying out of remediation are areas still being driven by new regulation. The standards to which remediation has to be achieved are being challenged not only by regulators and funders but also by government and EU policy. These influences are, for example, driving towards more use of brownfield sites on the one hand, and waste minimisation and hence in-situ remediation rather than landfill on the other. For example landfill tax which has been rising at the rate of £1/tonne per year will increase by £3/tonne per year after 2002 in order to reach the UK's targets under the Landfill Directive. The long term target is £35/tonne landfill tax.

The challenge is not only on technical grounds:

- The insurance industry has been grappling with smarter ways to minimise liability for pollution on brownfield sites;
- Landowners and developers are becoming much more aware of the consequences of site being

perceived as having long-term environmental liabilities associated with them;
- There are significant barriers to on-site remediation because of the way that the waste management licensing system is being applied.

None of these features are specific to the North West region however the North West will be one of the principal regions in which these challenges have to be met because of its industrial legacy and its regeneration targets. It is also a region short of suitable landfill sites. Where there is challenge there is the opportunity for innovation. Innovation will be needed both to meet the technical challenges and in the way that businesses respond in management and marketing.

5.1 *How fit is the North West's environmental goods and services sector to meet these challenges?*

Scrutiny of the links between regeneration players shows that there are differences in the number and types of link that the players have into regeneration activities. Environmental consultants have the most links and relate to virtually all players in service provision. Importantly they link into all parts of the development/regeneration promotion network in both public and private sectors. Nevertheless environmental consultants are not as far up the "food chain" as advisers such as planning consultants and architects. Other potential players are much further away from mainstream regeneration activity and are commissioned by environmental consultants and contractors. This concept of distance is illustrated in Figure 3.

Important points are:

- Some suppliers of materials and equipment are relatively close e.g. laboratories that specialise in

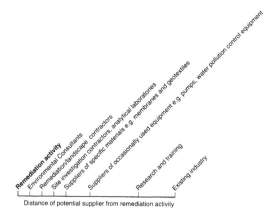

Figure 3.

contaminated land analysis and suppliers of geo-textiles and membranes. Their representatives can talk knowledgeably about their products to remediation advisers.
- Some suppliers are some distance from the remediation activity and may not be familiar with it.
- There appear to be few links between the higher education establishments and remediation in the North West. There are some exceptions on the training side but there appears to be little remediation led research. There is a strong science base in the North West and opportunities using this to bring innovation to remediation problems should be addressed. The proposed NW Centre of Excellence should address this.
- Existing industry may have sites which need remediating and this will fall within the normal development cycle or as a response to a pollution incident. The North West Chemicals Initiative are being proactive in assessing contaminated land associated with the chemicals industry in the North West. However more importantly, existing industry may have technology which could be applied to remediation problems in the North West and this expertise should be drawn into the regeneration/remediation capability in the North West.

The North West has a strong and experienced remediation consultancy base. However there are few North West based specialist remediation contractors. This is for a number of reasons:

- There is little home-grown innovative remediation contracting in the UK. The UK is behind some European countries and the USA in developing remediation technology largely because the regulatory constraints on "dig and dump" began to be enforced later in the UK. The result is that most innovative remediation methods are licensed in from overseas as the market demands. Specialist remediation contractors therefore tend to operate nationally;
- The remediation companies that have arisen in the UK tend to have arisen from University research (e.g. Scotland and South Wales). There has been apparently little research in fields that lead to new remediation methods in the North West. This does not mean that there is no relevant research at present or research teams that could not be focussed on remediation topics;
- There have been (and still are) regulatory constraints and/or caution with respect to some new techniques in the UK.

The environmental goods and services sector is therefore rather fragmented in its ability to deliver the full range of services needed in the North West and is not well prepared to meet the innovation challenges that regulation and policy are mounting.

5.2 *Opportunities for innovation promotion*

There is no doubt that on-site remediation will increase as government policy bites and there are considerable opportunities for developing expertise in the North West which cannot only be used in the North West but can be exported elsewhere. Much future remediation may well be based around "process" remediation techniques which are those which use some equipment to remediate the site either in-situ or ex-situ, on-site or off-site. Choosing where the North West should focus in trying to initiate innovation in this area is a function of a number of factors:

- The nature of the principal remediation problems in the North West – there are likely to be a few sites with severe problems (e.g. acid tar pits). The technology developed will limit application in any one region but could be applied nationally or internationally. There will be many sites with less severe but common problems which would benefit from a new approach and where the approach could have wide application in the North West and could also be applied nationally;
- The existing relevant science base;
- The capability within SMEs for all or part of the solution;
- The capability in heavy industry for all or part of the solution.

A way forward would be to:

- Gather together the principal stakeholders who can provide information concerning the points above;
- Agree on the focus innovation should take and the sites to which it is relevant;
- Check against initiatives in other parts of the country;
- Carry out a SWOT and sensitivity analysis of the approach;
- Canvas for funds/support;
- Agree a programme;
- Implement.

Principal aims would be to:

- Initiate relevant research;
- Establish pilot remediation in the North West;
- Apply the technology.

6 ANALYSIS OF STRENGTHS, WEAKNESSES, OPPORTUNITIES AND THREATS IN THE SECTOR

The SWOT analysis given below indicates a number of issues to be addressed if the region's environmental sector is to capitalise on the opportunities offered by regeneration work.

Table 3.   Actions to promote the North West's environmental goods and services sector's activities in regeneration.

Principal finding	Action	Mechanism
1. There will be considerable growth in regeneration-led environmental activities over the next few years and opportunities for companies in the north west to exploit these.	Publicise survey findings.	Through Envirolink presentation and website and through presentation at conferences and seminars.
2. There are constraints for companies in the region to access markets because of long and complex supply chains for services.	Help develop relationships between buyers, suppliers and intermediaries.	Continue "meet the buyer" events focussing on regeneration. Look to promoting regional flagship events.
3. Assessment of land needs and carrying out regeneration activities will continue to be driven by new regulation.	Publicise developments and forthcoming regulation.	Continue existing regulation updates and consider workshop on waste management, the Landfill Tax and new opportunities in both waste management and regeneration. Establish training courses for professionals in the region on new regulation and initiatives.
4. There are barriers to site remediation and adoption of innovation because of the way waste management system is being applied.	Publicise situation and support lobbying for change.	Through information update and closer links with Environmental Industry Commission (EIC) lobbying programme. Support single remediation permit initiative.
5. NWDA is tasked with setting up a Centre of Excellence for regeneration in the region.	Support NWDA in setting up a Centre of Excellence for regeneration in the region.	Interface between NWDA and stakeholders.
6. There is a need to encourage links between Higher Education Institutes (HEI), industry and environmental practitioners.	Through support for Centre of Excellence.	
7. There is a need to encourage higher standards in remediation in order to improve regeneration programmes and also allow companies in the region to exploit the market.	Encourage registration in SiLC (Specialists in Land Condition) scheme.	Lobby and continued interface between sector, NWDA and other major players.
8. Improve links with bodies such as Royal Institution of Chartered Surveyor (RICS), National Federation of Demolition Contractors, regional Institute of Waste Management, Institution of Civil Engineers, Environmental Industries Commission.	Publicise work and activities of Envirolink in relation to regeneration. Consider workshop for relevant regional associations.	Continue developing links such as that already established with the chemical sector through NWCI.
9. Lack of innovation and North West derived solutions.	Promote communication and innovation.	Stakeholder meetings to determine remediation innovation, linking with HEI, consultants etc.

Strengths
- NWDA support for the development of clusters in the region.
- Preference for use of local companies.
- Some leading edge technology.
- Some multi-disciplinary set ups.
- A strong consulting sector.
- Strong science base in HEI.

Weaknesses
- Fragmented and diverse sector with a large number of small companies not necessarily focussed on regeneration.
- Long supply chains and reliant on others to access work.
- Weak links to innovation.
- Lack of marketing and sales skills by many companies.
- Inability to retain graduates and skilled labour in some parts of the region.
- Limited networking particularly by smaller companies.
- Competition among parts of the region and local authorities to access funding.
- Need for stronger links between industry and HEIs.

Opportunities
- Emerging regulation will drive the sector and encourage higher standards and fees and use of new technology.
- Opportunity for enhanced links with trade associations to enhance Envirolink membership and thereby access to support for the sector.
- Support for the establishment of a centre of excellence for regeneration in the region.
- Develop relationships between links in the service supply chain.

Threats
- Continued emphasis on costs of services rather than quality and innovation in both public and private sector a deterrent to development.

7 CONCLUSIONS

The principal findings and recommendations are given in Table 3 and are intended to:
- encourage innovation
- mitigate constraints to site remediation activities
- improve links between industry, education, suppliers of environmental products, local government and professionals working in regeneration
- encourage high standards and allow the sector to exploit opportunities within the region and elsewhere
- raise awareness of opportunities within the sector
- support the NWDA in the establishment of a regional Centre of Excellence for regeneration.

REFERENCES

ENDS, (2002), Environmental Consultancy Market Analysis, UK2002/03.

Enviros, (2000), *Environmental Technologies and Services in the North West of England: Sector Mapping, Outline Strategy and Plan.*

JEMU, (2002), *Global Environmental Markets and the UK Environmental Industry: Opportunities to 2010.*

MSI Database (2000) *Urban regeneration UK.*

North West Development Agency, (2001a), *Regeneration Prospectus Consultation Draft.*

North West Development Agency, (2001b), *Land Reclamation.*

North West Development Agency, (2001c), *Annual Report 2000–2001.*

North West Development Agency, (2001d), *Reclaim the Northwest.*

North West Development Agency, (2002), *Regeneration Prospectus.*

Creating multi-functional landscapes – a more enlightened approach to land reclamation

Richard Cass
Senior Partner Cass Associates, Liverpool, Great Britain

ABSTRACT: Derelict industrial land provides opportunities for the creation of new landscapes on a scale which are rare in a crowded, heavily developed country such as Great Britain. Too often, opportunities are lost because of a lack of imagination, a preoccupation with technical issues, and inflexible planning policies. Another difficulty is the separation of 'hard' from 'soft' after-uses in the planning of schemes. This can be caused by restrictive planning policies, by a lack of awareness of what is possible, and by economic and funding pressures. This paper uses practical examples to illustrate what can be achieved when these difficulties are tackled. It demonstrates the benefits of taking a creative, holistic approach to land reclamation as a key component of effective urban regeneration. It shows how, by combining 'hard' and 'soft' after-uses landscapes can be created which are multi-functional, and socially and economically sustainable.

1 INTRODUCTION

As the first industrialised country in the world, Great Britain has long experience in facing the problems associated with the decline and closure of traditional industries. At a large, regional scale, these started with the coalfields, in the 1960s, mainly in the north of England, but have since extended throughout virtually all industrial regions. As well as coalfields, the industries affected have spread to include steel, ship-building, petro-chemical, and textiles. Over the past 20 years, with the globalisation of the economy, all traditional manufacturing has been affected. Reductions in military expenditure have also lead to large areas of land becoming redundant.

A common problem in all of these areas is the legacy of large scale dereliction and contamination, and poor environmental conditions, together with widespread social problems of unemployment, poor housing, education and health. New investment and the jobs, housing and other social facilities which follow it, is attracted to high quality locations which do not suffer from these problems.

Although it does not solve all of the problems, transforming the quality of the environment is an essential part of regenerating these difficult areas. This can be a very difficult and expensive process, and it takes a long time to reverse the decline. However, the existence of difficult problems, including large areas of land, can create opportunities, provided there is the skill, imagination and determination to grasp them.

Great Britain is a very long way from solving all of these problems. The amount of derelict land has remained roughly constant for many years, as more industries close and new dereliction replaces land which has been reclaimed. Sometimes, land which has been reclaimed once falls back into dereliction, as no positive use has been found for it, or the techniques used were not successful.

Very often, an 'engineered' approach is taken, where technical problems such as ground conditions, contamination, or road access are addressed, but the environment is still poor, so that no-one wants to live, work or invest there. This, is a very big problem in many regions of the world, including Great Britain. Beauty and bio-diversity are not easy to put into a financial appraisal, and yet they have a profound effect on quality of life and long-term economic success.

2 BASIC PRINCIPLES

During 30 years experience of land reclamation, three basic principles have been developed which can be applied to the regeneration of large, derelict sites. They are:

- Work with the site
- Work with nature
- Work with the community

In addition to these, there is a need to address in an integrated, 'holistic' way, a range of often complex technical and environmental issues, and not let the process be dominated, or possibly obstructed, by single issues or narrow interests. Normally, radical change is required if problems are to be resolved, and this demands a creative, broadly based approach.

3 ASSET CREATION

Another important principle is to identify how the greatest asset value can be created from land which is derelict. Sometimes this may be a social or environmental asset rather than a financial one, but very often it is the case that the greatest asset value is created when all three are brought together. In recent years, this has come to be defined as 'sustainable development', although it is all too rare for the principles of sustainable development to be backed up in practice. This often requires a 'hybrid' approach to regeneration, whereby financial assets are created alongside social and environmental ones. This makes it possible for the private sector to play an effective role in the regeneration process, but it is normally essential for the public sector to invest initially in reducing the risks and creating conditions where market forces can operate. This may be in the form of direct grant, or in the form of initial investment which may be repaid over time, depending on actual financial performance.

Another important principle which has been learnt from Britain's experience is the need for the public and private sectors to work in partnership to achieve effective regeneration. Neither side can do it on their own. In the examples given below, various models for this have been developed.

4 EXAMPLES OF SUCCESSFUL REGENERATION

This paper describes four examples carried out by the author where large scale dereliction has created opportunities for economic, social and environmental regeneration.

4.1 *Silksworth Colliery, Sunderland (1973–6) (65 ha; colliery site)*

This was a large colliery site in the south-west part of the city, on a prominent site surrounded by large areas of housing and close to a major approach road to the city centre (see Figure 1). It had an area of 65 ha and presented a range of severe problems. These included a burning, unstable tip, pollution of air and water, local flooding, large areas of derelict buildings, extensive contamination and a dangerous, unsightly environment. An initial feasibility and masterplanning study concluded that the best opportunity for effective regeneration was to create a town park on the site. Large scale earth works required to eliminate the burning shale created an opportunity for an entirely new landform. The masterplan took advantage of this to include an artificial ski-slope, an athletics arena, playing fields, lakes for recreation and nature conservation and extensive informal recreation, all within a strong new landscape framework.

Technically, the earthworks process was highly complex. All burning material had to be excavated, re-spread and compacted. Boulder clay was excavated from the site to cover the shale and prevent air and water having access to it, preventing re-ignition and acid drainage. Two lakes were formed as part of this process. Natural limestone and burnt, red shale were excavated from other parts of the site and used as cover and surfacing, and demolition materials were crushed and spread beneath the sports pitches as a drainage layer. In all, around 2.5 million tonnes of material were moved.

No material either left the site, or was brought to it. The reclamation process was completed in 1976 and over the next ten years a wide range of facilities were developed on the site. The site is one of the most successful new urban parks in Britain and has made a significant contribution to the regeneration of the city.

4.2 *Riverside, Liverpool (1980–6) (100 ha; petro-chemical, docks and landfill sites)*

At the time it was carried out, between October 1981 and March 1983, the Riverside Reclamation Project was described as the most difficult and technically demanding example of industrial dereliction to have been tackled in Europe. It was certainly the fastest. Initially the site for the UK's first (and so far only) International Garden Festival in 1984, the reclamation programme created an opportunity for a permanent, high quality landscape as an integral part of the regeneration process.

In the Merseyside Development Corporation's regeneration strategy, the reclamation and infrastructure was originally programmed to be complete by 1984 – i.e. in two and a half years. The decision to hold the Garden Festival meant that this period was reduced by a year. It also meant that the quality of the landscape had to be very high and suited to the demands of an event which were largely unknown at that time.

4.2.1 *The role of the festival*
The Festival's primary role was urban regeneration. It would mean that this large area of severe dereliction would be reclaimed quickly and to a high standard. It would bring about a dramatic improvement to the environment along the waterfront in the south of the city, release large amounts of land for development

Before reclamation, Silksworth Colliery, Sunderland, 1973 © Cass Associates

After reclamation, Silksworth Colliery, Sunderland, 2002 © Cass Associates

Figure 1.

and improve access to the city centre and the South Docks. Uniquely it would attract large numbers of people to a previously derelict part of Liverpool, promoting the city, boosting confidence and encouraging investment.

4.2.2 *The problems of the site*
The Riverside site contained a wide range of problems:

- a silt filled dock.
- two derelict jetties and associated river walls.
- three large petro-chemical depots, heavily contaminated, containing a wide range of structures and widespread, uncontrolled tipping.
- a large domestic refuse tip, producing large amounts of landfill gas.
- uncontrolled tipping of often contaminated materials.
- poor and unpredictable ground conditions.

In addition to the problems of industrial dereliction, the site was largely flat and featureless, was severely exposed to winds and salt spray and was almost entirely devoid of vegetation. It stretched for almost three kilometres along the river and had a total area of around 100 hectares (see Figure 2).

4.2.3 *The reclamation programme*
Some basic facts and figures will illustrate what had to be done during the reclamation programme to transform the site.

- 400,000 tonnes of silt were removed from the dock.
- 1 million tonnes of sand were dredged from the river for use as clean, structural fill.
- 0.5 million tonnes of unsuitable and contaminated fill were removed from development areas and used to create a new landform over the domestic refuse.
- 0.4 million tonnes of domestic refuse were excavated to create the site for the Festival Hall, and redeposited to form "the Big Hill."
- 0.2 million tonnes of boulder clay were excavated from one of the petro-chemical sites to provide a water and gas-proof seal over the domestic refuse.
- 150,000 tonnes of topsoil were imported from Runcorn New Town.
- 250,000 trees and shrubs were selected and purchased in advance from nurseries throughout the UK.
- An on-site nursery was set up to handle all delivering, holding and supply of plant material.
- 5 km of irrigation water main were installed.
- A landfill gas protection and extraction system was researched, designed and installed (the first time in the world that an actively gassing landfill site was used for intensive public use); since 1984 up to a megawatt of electrical power has been produced from the gas each year; no evidence of trees being damaged by gas has been detected.
- 3 km of new road were constructed, together with all services.

The landscape design was based on the close integration of landform and planting. Shelter was created by a highly sculptured new landform along the river edge. This led onto a series of hills and valleys, culminating in a large hill, 40 metres high. This landform created a complex pattern of enclosure, within which a wide range of features was provided such as gardens, water features and play areas.

4.2.4 *Successful regeneration*
The success of the Riverside project in regenerating the area is unquestioned. Evidence of this success includes the following facts and figures.

- It is the only area of Liverpool to have experienced population growth over the past 10 years; most other parts of the city have declined, a few have remained static.
- Following the success of the Garden Festival, Liverpool has experienced very considerable growth in tourism.
- A wide range of developers have invested in the area, mainly in new housing, meeting the needs of a wide section of the community.
- Local shops, public transport and other services benefit from the increased population.
- A fast, convenient and very attractive new road has been provided through the South Docks to the city centre; this has removed through traffic from densely populated inner city neighbourhoods.
- Large numbers of people live in attractive new housing close to the city centre, increasing the use of existing services and reducing the need for greenfield sites on the edge of the city, which are costly in infrastructure and environmental terms.
- Around £100 million of private sector investment has been attracted to the site.
- The environment along 3 km of the Mersey has been dramatically improved, replacing severe dereliction with an attractive promenade, set against a green and beautiful landscape.

4.3 *Bold Moss, St Helens (1995 – present) (300 ha; colliery and power station sites)*

In 1994 Cass Associates was commissioned by National Power, the landowner, to carry out a development feasibility study into the future of a large, redundant power station site on the south-east edge of the town (see Figure 3). The site was part of a 300 ha area of dereliction including a colliery and two large tips, on either side of the Liverpool to Manchester railway line. The area had been identified as a

Before reclamation, Riverside, Liverpool, 1982   © Cass Associates

After reclamation, Riverside, Liverpool, 1989   © Cass Associates

Figure 2.

59

Before reclamation. Bold Moss. St Helens. 1995 © Cass Associates

After reclamation. Bold Moss. St Helens. 2001 © Cass Associates

Figure 3.

strategically important part of the town's regeneration programme.

The site presented a range of problems, including extensive contamination, severe underground obstructions, poor drainage and a degraded environment. Several small watercourses crossed the site, mostly in culverts. The site has small areas of landscape and ecological value, including young woodland, heath and wetland.

After careful examination of the planning, technical, environmental and social issues, a 'hybrid' scheme was proposed for the power station site, including new housing set within a new greenspace framework. This provided an attractive setting for the new community and created strong links to the surrounding areas.

The masterplan for the site is based on a strong landscape framework consisting of three main elements. Along the northern part of the site is a new lake, with existing and new woodland and grassland creating a large new informal park. Running down the centre of the site is a formal avenue, linking the park with the open countryside and creating a focus to an otherwise flat, featureless landscape. This is punctuated at the northern end by a conical hill, formed from surplus material, symbolising the history of mining on the site. At the southern end is a 'village pond,' providing a setting for the new housing, a collector for site drainage and flood storage. A medical centre, shop and pub/restaurant overlook the pond.

Around the perimeter of the new housing linear parks are created, with recreated streams running through them. Landform and planting shelter and enclose the development and screen adjacent industry. Around 25 ha of new housing is being developed, with eventually around 1500 people living on the site. The project is very popular and is making a significant contribution to the local community. The total project cost is around £7 million and has been carried out by the landowner and a private developer. A substantial grant was provided by the Regional Development Agency during the reclamation stages, but this has been re-paid due to the land values being higher than originally anticipated.

4.4 *Royal Ordnance Factory, Lancashire (1995 – present) (300 ha; munitions manufacture)*

As a result of radical restructuring in the defence industry, munitions manufacturer activities at the Royal Ordnance site in Central Lancashire have been substantially reduced. This resulted in a legacy of redundant land and buildings covering an area of 320 hectares, one of the largest derelict sites in North-West England (see Figure 4).

The planning framework for the re-use of this major site was established following detailed discussions over a five year period between British Aerospace (now BAE Systems) as the land owner and the three local authorities with a direct interest (Chorley and South Ribble Borough Councils and Lancashire County Council). The relevant statutory plans reflect the spirit of partnership between the site owner and the local authorities and designate the site as suitable for mixed use development based on the theme of an urban village. The ultimate scheme is firmly based on the balanced mixed-use approach of an urban village.

The mix of landuses and the structure of the urban village has been derived from an integrated approach to planning, development and environmental issues. There has been a particular emphasis on understanding the complex relationship between the remediation strategy and the development strategy. The BAE Systems Environmental Services Group has worked closely with Cass Associates to ensure that the process of land remediation is not progressed in isolation, but is very much an integral element of the sustainable re-use of derelict land.

A key feature of the remediation strategy is a 'confidence scrape' exercise which will overcome the small risk posed by the possibility of explosive articles lying at, or near the surface level. The exercise allows cleared areas to be present for future development but it generates a significant amount of material for disposal. In the interests of sustainability, this material needs to be retained and re-used within the site. It will be used to form the landscape framework which is an integral part of the scheme. Furthermore, because the depth of the 'scrape' is a function of the ultimate landuse, and because the process is destructive in nature, it has been necessary to weigh the requirement for a robust approach to remediation against environmental and development parameters. The outcome is a balanced approach which accounts for a wide range of relevant factors and is not unduly dictated by the re-engineering of the site.

The masterplan for the overall development is derived from a comprehensive and integrated approach which draws together a wide range of key issues – landuse and remediation have been considered in parallel with transportation, environmental and engineering matters. It is inevitable that tensions between each of the key issues will surface, but it has been the function of an overall masterplan to respond to the tensions and to arrive at a form of mixed development which is sustainable and realistic.

Detailed discussions took place with the Regional Development Agency over the possibility of grant support. Eventually, it was decided that market demand for the site should create values which would fund the reclamation and infrastructure of the works and the site was marketed on this basis. This method was successful and a developer was selected in 2000, with reclamation starting soon afterwards. In 2000 the

Before reclamation, Royal Ordnance Factory, Lancashire, 1996 © Cass Associates

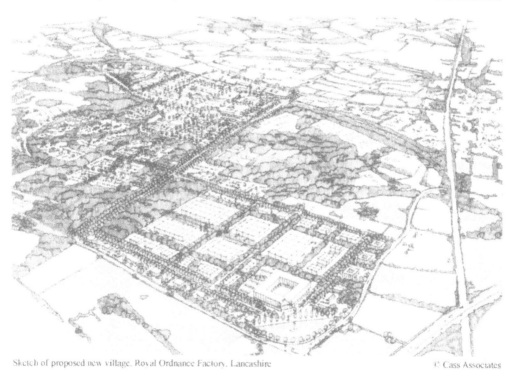

Sketch of proposed new village, Royal Ordnance Factory, Lancashire © Cass Associates

Figure 4.

project won a regional award for planning achievement from the Royal Town Planning Institute.

5 CONCLUSION

The reclamation of derelict and contaminated land is a complex process which is informed by the experience and resources of a range of professional groups. Land remediation, however, should not be viewed as an end in itself. It is essential to form close links between remediation, landuse and environmental considerations to ensure that the recycling of land is cost-effective and sustainable.

Land owners and local authorities need to work in partnership to secure the benefits of recycled land. The partnership needs to be open and constructive. Landowners should be given the opportunity to generate value from their investment in remediation whilst the local authority needs to be imaginative and pro-active in its attitude to landuse.

The resources going into creating landscapes for people are very small in comparison to those going into engineering infrastructure and decontamination. Much greater recognition should be given to the role that successful, high quality landscapes play in effective regeneration of derelict land. The economic benefits of this approach are irrefutable.

These four examples demonstrate a range of approaches to how large, difficult regeneration projects, can be tackled. They illustrate the need for radical thinking to the regeneration process and the importance of linking economic, social and environmental issues closely together in order to achieve practical, effective and above all sustainable solutions to complex problems. They also demonstrate the need for long term commitment if these difficult issues are to be tackled effectively. They provide clear evidence that there is a range of ways in which the public and private sectors can work effectively together in dealing with large scale, difficult regeneration projects.

In all cases, landscape planning and landscape design had a powerful influence not just on the final development, but on the techniques that were used to solve the severe technical problems which each site presented.

Land Reclamation – Moore, Fox & Elliott (eds)
© 2003 Swets & Zeitlinger, Lisse, ISBN 90 5809 562 2

Land reclamation and industrial heritage in Cornwall: catalysing regeneration

C.L. Wilson
South West of England Regional Development Agency

R.P. Sainsbury
Cornwall County Council

ABSTRACT: Cornwall contains more recorded derelict land than any other county in England – Wickens, Rumfitt & Willis (1993). However, much of this is associated with past metalliferous mining activity. It is also currently proposed for inscription by UNESCO as a World Heritage Site, which inevitably affects much of the land reclamation activity in the county, posing both a challenge and an opportunity. The paper reviews the positive and negative aspects using project examples. Finally, it will outline the positive contribution these works are making to the local economy, compared with more conventional land reclamation for "hard end" use/direct employment space.

1   BACKGROUND

The Cornubian orefield of Cornwall and West Devon contains an extraordinary suite of minerals capable of producing antimony, arsenic, cobalt, copper, lead, manganese, silver, tin, tungsten and uranium. Based on this and Cornish ingenuity, the area became the greatest producer of copper and tin in the world during the eighteenth and nineteenth centuries. However, the threat of competition and fact that the largest and richest reserves lay within the granite and metamorphosed sedimentary rocks many fathoms below the surface, required remarkable advances in hard-rock mining and engineering technologies.

During the 18th and 19th centuries these advances transformed the landscape, economy and society of the region, placing it at the very forefront of the Industrial Revolution. This lasted until the mid 1880's, when there was a dramatic fall in the market price of copper, due to new discoveries in South Australia and North America. Also at this time, the working levels were getting deeper and the grade of ore was getting lower in many of the mines, with the result that a large majority of the Cornish mines were abandoned by the end of the nineteenth century.

Today however, distinctive physical reminders still persist within the Cornish landscape, including imposing engine houses, industrial harbours, tramways, foundries, fuse factories, mining towns and villages, numerous non-conformist chapels, the impressive houses and gardens of the mineral lords, the modest smallholdings of the ordinary miners, technical schools, miners' institutes and geological collections established for the aspiring student.

2   DERELICT LAND GRANT

Cornwall's mining past has also left the county with a legacy of derelict land, the area of which, in 1993, was greater than that of any other county in England. This was reflected in a survey carried out by the Department of the Environment – Wickens, Rumfitt & Willis (1993), in which the South West region had the second largest area of derelict land in the country (in excess of 5,500 hectares). Within this, the largest area of land was that previously used for mineral extraction (in excess of 3,500 hectares), virtually all of which was located in Cornwall and West Devon.

Another significant difference evident in the report, was that in both Cornwall and the rest of the region, the largest concentration of derelict land was in rural areas, whereas in other regions it was predominantly urban. Little of this land was thus to be found in areas normally considered suitable for redevelopment into industrial or residential use.

As a result, much of the land reclamation works in Cornwall has been aimed at creating public open space for leisure and recreation, albeit that the primary output of these projects was the reclamation of derelict land. Typical examples are the reclamation of:

– Geevor Tin Mine as a museum;
– A large derelict site at the entrance to Camborne, including the remains of North Roskear Mine, a land fill site, an old gas works and tin streaming works, to create Tuckingmill Valley Park;
– Carn Marth Quarry to form an open-air theatre.
– Boscarn Park and Wheal Jewel, reclaimed to provide a permanent site for travellers.

Most of this work was initially funded under the Derelict Land Act (1982), with grant availability at up to 100% rate of intervention because of the level of deprivation in the county.

As well as the normal shaft capping, decontamination and building consolidation works, a typical project would start with archaeological and ecological surveys, and would include archaeological and ecological watching briefs during the execution of the works.

Outputs were recorded in terms of area of land reclaimed, number of buildings consolidated and numbers of shafts made safe. Projects were generally assessed in terms of value for money per hectare of land reclaimed and per shaft capped/made safe.

3 ECONOMIC RETURN

Cornwall's mining past has also had a profound effect on the county's economy.

By the mid 1800's over a century of specialisation in mining resulted in an occupational structure dominated by the extractive industries, similar only to the coal mining area of South Wales. In 1861 the Redruth Registration District had the second highest percentage in any England or Wales of men employed directly in mining – Deacon (1998).

The distinct absence of diversification into other industries or of the ability to maintain a competitive edge in heavy engineering produced an economy over-specialised and over-dependent on mining – Payton (1992).

The closure of the mines thus resulted in a deep socio-economic depression that lasted from the late nineteenth century to the war years of the twentieth century – Payton (1992) and from which, it could be argued, it has never fully recovered.

The post-war economy has relied heavily on agriculture, fishing and tourism, with little in the way of heavy industry. Although agriculture and fishing are in the decline, tourism is still a central element of the economy, both now and in the future.

Today's its environment is recognised as one of Cornwall's greatest assets, with its unique combination of natural and built elements which makes it attractive both to visitors/tourists and as a place in which to live and work.

Land reclamation in the county therefore tends to be focused in a very different way to other parts of England, with the emphasis being very much on environmental improvement to enhance the "visitor experience", to improve recreation and leisure facilities to serve the tourist industry and to attract new businesses and inward investment.

4 CORNWALL LAND RECLAMATION STRATEGY

In 1996 English Partnerships took over the administration of the Derelict Land Grant and required the local authorities to focus more on the economic benefit gained from the land reclamation, in line with their priorities. This, together with the availability of European Regional Development Funds (ERDF) under the Objective 5b programme, which also relied on economic return and social benefit, accelerated the shift in focus of land reclamation projects towards recreation, leisure and tourism.

This, together with the need for the local authorities in the County to work together and have a common goal, resulted in the "Cornwall Land Reclamation Strategy 1997–2000" – Sainsbury (1997), produced by the local authorities as a co-ordinated and integrated approach to land reclamation across the county.

This identifies four key objectives:

– To enhance economic and employment potential;
– To preserve and enhance industrial heritage;
– To improve environmental quality;
– To create and enhance facilities for public recreation & leisure.

The availability of funding from the Objective 5b programme facilitated a move away from 100% funding through Derelict Land Grant. English Partnerships actively encouraged match-funding, which enabled the development of more holistic projects, which in turn were often able to greater benefit than the sum of the individual elements.

Between 1997 and 2001 over thirty sites were reclaimed in Cornwall with Land Reclamation funding of around £3 million per year, which in turn was used as match funding to lever in a further £5 million from ERDF, landfill tax credits, the local authorities and other sources.

At the same time there was an increasing awareness of environmental issues, which was reflected in an increase in the scope of works on many projects to include mineralogical surveys and watching briefs, and the three "B's", bat, badger and bryophyte surveys.

In order to satisfy the funding bodies, additional outputs were also recorded, to measure any associated economic value/benefit, in terms of the number of additional visitors attracted and existing visitor numbers safeguarded.

5   WORLD HERITAGE SITE BID

The Cornish Mining Industry is now considered to be of international heritage importance, in recognition of which the UK Government included it in the Tentative List of 25 sites that it wishes to see designated by UNESCO as World Heritage Sites over the following 5–10 years.

The case for World Heritage Site (WHS) status, as presented in the Tentative List – DCMS (1999), is based on the following five characteristics:

– The extraordinary suite of minerals in the Cornubian orefield, with 440 species out of a worldwide total of some 3,300.
– The crucial part played by Cornish miners, adventurers, and Cornish engineers between 1800 and 1860 in the development of steam technology and mining technology throughout the world, including the Trevithick's steam engines, the use of coal gas for lighting, Davy's safety lamp and Bickford's mining safety fuse.
– The outstanding survival of the mining landscape and its associated industrial concerns, urban development, rural settlements and miner's small holdings, great houses, parks and gardens, mineral railways and mineral ports.
– The comprehensive character of the statutory protection, conservation and long term management measures covering this wider historic mining landscape now in place and presently in progress.
– The existence still, of a well documented Cornish mining diaspora of mining sites (physically so similar to mine sites in the region) and Cornish people. This is particularly well represented today in South Australia, and the United States of America.

To qualify for inscription, the County Council has to submit a bid to UNESCO demonstrating the international significance of the proposed "site". A detailed Management Plan also has to be submitted, demonstrating how the "site" is being and will continue to be preserved and managed.

With Land Reclamation funds historically having been used so extensively in the past throughout the county on mine sites, this was naturally seen as a potential source of funds to continue this work.

6   THE CHALLENGE

6.1   *Funding constraints*

With the formation of the Regional Development Agencies came a noticeable shift in priorities. The new RDA's were required to focus even more closely on economic regeneration, with less emphasis on social works.

Applicants were encouraged to seek alternative funding sources, such the Heritage Lottery Fund (HLF) and English Heritage's Heritage Economic Regeneration Scheme (HERS), for heritage-based land reclamation projects.

However, even with these additional sources, the shortlist produced as part of the World Heritage Site bid, of engine houses, chimneys, chapels, villages etc. requiring restoration works, posed onerous demands on all available external funding, not to mention the local authorities own funds.

The RDA, Government Office for the South West and HLF asked the County Council to prioritise the list of potential projects. With all the political interests involved, this posed quite a challenge.

A scoring system was developed, based on the:

– Heritage value;
– Potential economic benefit to Cornwall plc; and
– Achievability of the scheme, eg was land ownership known.

These scores were then summated to produce an overall ranking of the projects.

6.2   *Project programming*

The evolvement of this new, quite complex cocktail of funding for projects posed a further challenge in programming the actual works.

HLF generally require a two phase project development, with quite long decision times involved at each stage. This inevitably delayed project start dates and extended the overall programme.

The Objective 1 programme of funding is only available until the end of 2006, by which time all projects are required to be completed. Three year projects commencing in 2003 thus face quite a tight timetable.

6.3   *Scope of works*

To ensure an appropriate level of statutory protection on key sites within the proposed WHS area, English Heritage brought forward the Cornish mine sites within its Monuments Protection Programme for industrial archaeology. As a result, many of the sites proposed for reclamation in Cornwall became or were likely to become scheduled. In relation to the project works, this in turn required:

– A higher level of recording;
– Tighter specification of any consolidation or safety works; and
– More consultation with English Heritage.

This altered significantly the scope of works for many projects. Archaeological surveys had to be expanded to include such items as EDM surveys of structures and buildings prior to consolidation. Additional mortar samples were required, prior to selecting a suitable mix for any pointing works.

The works in the Tamar Valley experienced particular problems. Prior to this large regeneration project, most of the building consolidation of engine houses and chimneys had been carried out in the west of the county or up on Bodmin Moor, where vegetation was generally sparse.

The Tamar Valley, by comparison, contained fairly lush woodland and many of the old mining structures were heavily covered with ivy. This made it virtually impossible to guarantee the amount of works required prior to removing this growth. However, the cost of scaffolding, say, a chimney to remove the ivy prior to the works to facilitate a more detailed survey, and then leaving it there until the works could commence, was prohibitive.

A compromise was required, which meant that the level of cost certainty at the start of the works was not as high as would normally have been expected. This, together with the higher archaeological specification for the works, resulted in an increase in the cost of between 20 and 25% of some of the building consolidation contracts.

7 THE OPPORTUNITY

Whilst the potential for WHS status seemed to increase the amount of work involved, it also presented an opportunity for Cornwall to broaden and strengthen its visitor base, attracting Cornish emigrants and their descendants from Australia, USA etc and special interest groups.

In this context, land reclamation projects in the county were developed with a slightly different focus, aimed at extending the visitor season into the shoulder months and not necessarily trying to attract more visitors, but those with a higher spend per capita.

It was also recognised that the appeal of Cornish Mining would be somewhat different to say, that of Stonehenge, Bath, or some of the other World Heritage Sites in the UK. However, the disused mineral tramways offered the opportunity to create a network of cycleways linking the mine sites, visitor centres, ports and harbours with the mining towns and villages.

The Mineral Tramways project was thus conceived, covering the central mining area of Camborne-Pool-Redruth and Gwennap-St Day, and stretching from Hayle and Portreath on the north-west coast to Devoran on the east coast. The first section of this following the Portreath Tramroad has attracted over 80,000 visitors in its first year of opening.

Hayle and Perran Foundries also offer the county the opportunity to build on its innovative and creative engineering past. This theme is reflected in the prospectus for the new Combined University for Cornwall, with its hub at Tremough campus, Penrhyn.

8 PREDICTING THE OUTCOME

Having identified these new goals, it was obviously necessary to try to:

- Predict the individual economic benefits which might accrue from each project; and
- Assess the overall economic benefit which could be achieved through gaining World Heritage Site status.

A study was commissioned in September 2002 to develop a prediction methodology and to try to quantify the answer to the second point above. This work reflected a step change in the project appraisal process carried out by both Government Office and the RDA's.

Whilst in one sense the Objective 1 programme could be seen as continuing the work of the 5b programme, there are significant changes, including much tighter appraisal process and a more thorough scrutiny of output prediction methodology and more onerous requirement for monitoring.

Even greater consideration had to be given to baseline surveys/indicators, against which any proposed economic change could be monitored. More detailed monitoring programmes are necessary.

The success of any tourism-related project is also equally dependent upon the amount of effective interpretation and marketing provided. This is also now required as an essential component of any scheme.

Thus the challenge posed by land reclamation in Cornwall no longer centres on the reclamation techniques involved, but on the economic justification of the scheme and how the outputs can be predicted and monitored.

Many of the projects centre around improving the "public product", either in terms of creating new footpaths and cycleways, toilets, public safety works or streetscape environmental improvements.

Whilst it is extremely difficult to quantify the economic benefit associated with such works, it is evident that they act in many cases as a catalyst for regeneration.

In the mining villages of Gunnislake and Calstock, the streetscape improvements revived community pride in their environment. Various businesses improved their shop fronts and the local pub opened its room to staying visitors again.

In addition to the World Heritage study referred to above, a number of attempts have been made to quantify this economic return.

Between 1999 and 2001, the National Trust commissioned a number of studies in the region, including

"Valuing Our Environment" – Tourism Associates (1999) and "The Economic Impact of the Public Product" – Atlantic Consultants (2001).

"Valuing Our Environment" identified that 78% of all trips to the South West are motivated by the conserved landscapes of the region, attracting a spend of £2,354 million and supporting 97,200 jobs.

In terms of the public product, the second study confirmed the generally accepted view that the public product is important, with the natural and historic heritage being of particular importance.

It also indicated a possible 4.3 FTE jobs supported through off-site visitor spend for every job supported directly on conservation-related public product projects.

This, however, includes linkage and multiplier effects, indirect benefits which are extremely difficult to monitor with any degree of confidence in relation to a single project in isolation, particularly where there may be a number of other tourism-related projects within the project area or district.

9 EXIT STRATEGY

Finally, and of equal importance to the above, there is a need for an effective exit strategy from any project, albeit for the:

- Subsequent management of the project post completion;
- Continued marketing to inform potential visitors of the new facilities;
- Over-arching World Heritage Site Management Plan, an essential element of the bid to UNESCO; and
- Funding of subsequent management and maintenance works, at the end of the Objective 1 Programme and when Cornish economy has hopefully recovered to the point where it no longer requires RDA funding.

10 FUTURE REDEVELOPMENT

On the wider scale, the process of preparing a World Heritage bid and the associated Management Plan has produced several valuable new studies, not only analysing the potential economic benefit that could arise from inscription, but also re-assessing our industrial and urban settlements.

It was recognised that regeneration funded through Objective 1 offered an unparalleled opportunity for contemporary contributions in urban design and architecture to the built environment of towns in Cornwall and on the Isles of Scilly, but one which could also have an equally disastrous effect on the existing historic environment.

The synergy between the historic environment and economic regeneration was recognised and strongly advocated in the "Power of Place" review carried out by English Heritage (2000). This has led to the development of a pilot study in historic characterisation of 18 key towns, identifying elements of the urban fabric that combine to create "distinctiveness" and a "sense of place".

"Characterisation is not intended to encourage or provide for imitation or pastiche: rather it offers a sound basis on which the 21st century can make its own distinct and high quality contribution to places of abiding value" – Newell (2002). Although presenting somewhat of a challenge, this is already making a valuable contribution to the development of plans for the regeneration of St Austell.

11 SUMMARY

Experience in Cornwall has shown how land reclamation for "soft" end use can bring not only environmental benefit but also positive social and economic return.

The potential of World Heritage Site status is helping Cornwall to look forward, to build on its past, using it as a firm foundation from which to progress into the twenty first century.

Finally, more recently in the historic environment of our towns and cities, the county is taking the first steps in proving that heritage conservation can be a positive driver for redevelopment, rather than the constraint it is normally considered.

REFERENCES

Atlantic Consultants. 2001. The Economic Impact of the Public Product. National Trust.

DCMS. 1999. World heritage Sites: Tentative List of the United Kingdom of Great Britain & Northern Ireland.

Deacon, B. 1998. Proto-regionalisation: The case of Cornwall, Journal of Local & Regional Studies, 18:1, 27–41.

English Heritage. 2000. Power of Place.

Payton, P. 1992. The Making of Modern Cornwall.

Newell, Kate. 2002. Cornwall & the Isles of Scilly Urban Survey: St Austell. www.historic-cornwall.org.uk.

Sainsbury, P.R. 1997. Cornwall Land Reclamation Strategy 1997–2000.

Tourism Associates. 1999. Valuing Our Environment: A Study of the Economic Impact of Conserved Landscapes & of the National Trust in Cornwall. National Trust.

Wickens, David, Rumfitt, Andrew & Willis, Ray. 1993. Survey of Derelict Land in England 1993.

Land Reclamation – Moore, Fox & Elliott (eds)
© 2003 Swets & Zeitlinger, Lisse, ISBN 90 5809 562 2

Chatterley Whitfield sustainable energy centre

A.R. Morrison
Stoke-on-Trent City Council, Stoke-on-Trent, Staffordshire, UK

ABSTRACT: The former Chatterley Whitfield colliery in Stoke-on-Trent is a Scheduled Ancient Monument. A partnership, including local residents, was formed to find a viable future that will generate sufficient income to maintain the many buildings and continue to tell the story of the site's heritage. A business plan is nearing completion showing that the commercial demonstration of sustainable energy technologies and manufacture of related equipment, with academic research and training, is an appropriate use. Remediation of ground problems is not straightforward because of the possible impact on the monument. There are issues to be resolved resulting from the decision to preserve buildings that were designed with only a limited lifespan.

1 INTRODUCTION

Few buildings remain in Britain to indicate the significance of the coal industry to the country's development over the last two hundred years. Some landforms provide reminders of the related tipping, but in many cases whole communities that owe their existence to coal mining have lost the physical links.

English Heritage scheduled the former Chatterley Whitfield colliery in Stoke-on-Trent as an Ancient Monument because it is the most complete set of colliery buildings remaining in the country (Figure 1). It therefore has the same level of statutory protection as Stonehenge. The oldest building at Chatterley Whitfield dates from 1883, so the site is not nearly as old, and the buildings were erected with a limited life expectancy, but, like Stonehenge, the designation effectively means that the complex of thirty-four buildings will be preserved in its current form for the foreseeable future.

This raises many challenging questions about how and indeed even why steel-framed buildings with single-skin brick infill panels should be retained, and there will presumably be some interesting debates to discuss such issues, but there are wider questions to consider about the future role of the whole site.

Figure 1. The Chatterley Whitfield Site.

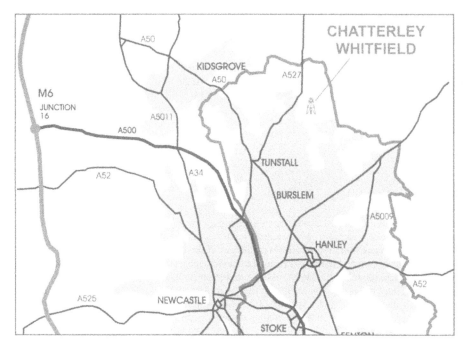

Figure 2.   Location of Chatterley Whitfield.

2   SIGNIFICANCE

The Chatterley Whitfield colliery was the largest pit in the North Staffordshire coalfield (Figure 2). Electricity was introduced underground before the start of the 20th century, and as a result, material could be brought to the surface for sorting rather than sort it underground. The resulting tip was, at one stage, the largest in Europe, and before it was remodelled and reclaimed during the 1970s, it was nearly twice its current height above the surrounding land.

In 1937, the colliery became the first in the country to produce 1 million tons of coal in a year, though it had nearly managed this before the end of the 19th century. The Pithead Baths were among the earliest in the country, provided in 1938 by the Miners Welfare Institute, many years before they were required by law. Nearly four thousand men worked at the site at its peak, though numbers steadily declined with mechanisation.

The surface of Chatterley Whitfield was closed in 1976, though coal continued to be worked underground and taken to the surface via Wolstanton Colliery about five miles away to the west.

A museum operated from the site from 1978, housing the national coal-mining collection. It was the establishment of this museum shortly after the colliery closed that prevented the wholesale demolition of buildings that happened at many other sites.

With the closure of Wolstanton colliery in 1986, the underground pumping and ventilation were switched off, so the shafts flooded. As a result, the museum could no longer take visitors underground, so a mock underground experience was built to replace this facility.

Declining visitor numbers and rising operating costs meant that the museum went into liquidation in 1993. The national collection was then transferred to Caphouse Colliery near Wakefield.

3   CHATTERLEY WHITFIELD TODAY

Chatterley Whitfield is now a group of mostly empty buildings, some interesting in their own right and others less so, ranging from the distinctive buildings housing the headgear, through the remnants of the boiler house exposing a row of ten Lancashire boilers to the elements, to anonymous sheds and 1960s buildings. All of them need improvement, and some are in a dangerous condition. The site is surrounded by open space at the edge of the North Staffordshire conurbation, and the tranquility of encroaching nature has replaced the former noise, dirt and bustle.

The site is within the North Staffordshire Green Belt, and there is a large supply of industrial land elsewhere in the city with better access to the motorway network. With its scheduling and Listed Buildings, many people saw the site as merely a set of constraints that had no future other than eventual ruin.

4 PARTNERSHIP

A major conference was organised by Stoke-on-Trent City Council in February 1999, involving a wide range of local and national figures and chaired by the Chief Executive of the British Urban Regeneration Association, to brainstorm a possible way forward. From the conclusions reached, a number of actions were taken.

Stoke-on-Trent City Council, English Heritage, the Regional Development Agency Advantage West Midlands and Joan Walley MP formed the Chatterley Whitfield Partnership. The project was given high priority by all of the partners, and both the Chairman and Chief Executive of English Heritage and the Leader and Chief Executive of the City Council attend Partnership meetings.

Early on, the Partnership established some principles for the project:

- It must be self-financing
- It must bring benefits to the surrounding communities
- Buildings should be retained and re-used rather than demolished
- The surrounding open space should be retained
- Any development must respect the industrial character of the site.

English Heritage launched its Buildings at Risk Register for 1999 at Chatterley Whitfield, using the occasion to announce its commitment to the project and contribution of over £1million.

A bid was submitted to the Heritage Lottery Fund for the staff necessary to develop and implement a detailed Action Plan, and to commission a comprehensive audit of the site. Detailed information was required, especially an accurate idea of costs involved, since previous attempts at finding a solution had foundered on unrealistically low estimates. The timetable for this work, leading to the adoption of a Masterplan for the site, proved to be optimistic, but it has been completed and is proving to be of great benefit.

A database of all interested parties was established, to assist in the process of consultation and involvement. This has been expanded repeatedly, and now consists of four hundred and fifty entries. Of these, nearly two hundred and fifty people receive minutes of the community meetings held every month. A community development strategy was prepared, to ensure that local people have a real influence on the forming of development proposals, including the establishment of the Friends of Chatterley Whitfield. This group was formally established in 2001, and is represented on the Partnership. Sir Neil Cossons, Chairman of English Heritage and a former trustee of the Chatterley Whitfield museum, agreed to be its patron.

The importance of the site was stressed in policies at different levels, including the review of the development plan. It was later included in strategies for North Staffordshire such as the Regeneration Zone and European funding strategies, and identified as a Priority Investment Area for the Single Regeneration Budget programme. Inclusion in these strategies was crucial to justify the project as a priority for funding.

The recently established Coalfields Regeneration Trust was contacted to determine its involvement, and the Chief Executive agreed to join the Partnership.

It was from the initial conference that the idea of sustainable energy first emerged. The link between energy of the past and energy of the future seemed particularly appropriate. The emphasis placed on this idea varied over time, and at some points it was submerged by other themes, but it remained in the background until resurrected later.

5 PHASE I

A programme of emergency and urgent repairs was set in motion, while the audit was undertaken. A condition survey assessed the cost of restoring the buildings to various degrees of weatherproofing and re-use, while detailed photographs and drawings were made of all of the elevations and floor plans. A topographical survey established details of the site and surrounding areas of open space.

An industrial archaeologist assessed the significance of the buildings and the many artefacts around the site, an ecologist considered the natural history and an oral history expert interviewed about two dozen former miners at Chatterley Whitfield to investigate how the site operated. These elements were then combined to form a Conservation Plan, with policies to govern how the site should be restored and re-used. For example, although many of the showers and lockers were removed from the Pithead Baths during the 1980s, there are still many left. The Conservation Plan states that a large number of remaining items is not sufficient reason to remove some of them – they should all be retained. It also states that the industrial character of the site should be respected and maintained rather than prettified.

At the same time as these surveys, a programme of community involvement was set up, deliberately adopting a slow and steady approach. Local residents were invited to one of three brainstorming sessions to generate ideas for the future use of the site. This was followed up by an event to explain the results of the sessions, and it was agreed to start holding community meetings every month to discuss progress and ensure opportunities for residents' views to be incorporated in the project.

These meeting are still held two years later, with an average attendance of between twenty and thirty people. Attendance on trips to other projects is also good. These range from the Apedale Heritage Centre, another former colliery in neighbouring Newcastle-under-Lyme, to Haig Colliery in Cumbria, New Lanark in Scotland and the Centre for Alternative Technology in Machynlleth in Wales.

This commitment and enthusiasm from residents is all the more impressive given the lack of visible progress on the site. It is partly because local residents are keen to see a currently derelict site improved for the benefit of the surrounding communities, but it is also because there is so little left of the coal mining industry and the residents want young people to know about the role the site played in the city's heritage.

Consultants prepared a masterplan, using the information from the surveys, suggesting a mixed-use development, in three phases over many years, comprising light industry, leisure, hotel, and residential uses. Renewable energy was mentioned, but only as a minor aspect. This approach, however, proved to be a blind alley, because market testing showed there was little demand for such facilities in the area.

6   PHASE II

In a concerted effort to change direction, funding was secured from the Department of Trade and Industry, Advantage West Midlands and the Single Regeneration Budget, for three studies on infrastructure improvements, the potential for renewable energy, and developing a business plan. These studies were commissioned in January 2002, and the draft final reports were produced in October that year. It was essential for all three firms of consultants to work closely together so that the recommendations fitted together and the costs incorporated in the business plan.

6.1   Sustainable energy

Consultants considered eleven different sustainable energy technologies:

- Wind
- Biomass
- Bio-oil
- Anaerobic digestion
- Combined heat and power
- Solar
- Ground sourced heat pumps
- Energy storage
- Coal bed methane/abandoned mine methane
- Municipal solid waste
- Fuel cells.

The relevance of each technology to Chatterley Whitfield was examined, including the level of resources required, the key issues, the road infrastructure required, building use and the potential appropriate scale of development.

Three scenarios for site development were developed to explore how these technologies could be utilised: an energy theme park, a sustainable energy centre and a sustainable industrial estate.

6.1.1   Energy theme park

This would involve a small-scale public demonstration of renewable energy technologies as an adjunct to the historic monument, with new clean sustainable energy sources (the future) set in the context of traditional fossil fuel extraction and use (the past). A public sector lead would most likely be required in this scenario, with uncertain economic viability.

This scenario would have a relatively low capital cost and would have little impact on the monument, but would offer only modest regeneration benefits and there would be little likelihood of private sector investment.

6.1.2   Sustainable energy demonstration centre

This scenario would be a commercial demonstration of a mix of renewable energy technologies coupled with university-led research and development, involving some new build within the site boundary. As well as demonstration, power would be produced; for example electricity would be sold to the grid, bio-oil sold commercially and space heat sold in winter to buildings on site. Part of the site would be used specifically for heritage interpretation. A regeneration company comprising a combination of public, private and academic organisations could lead in this scenario.

There would be a medium capital cost, with some impact on the monument, and grants would be required to make some renewables technologies work, but the economics would provide at the very least a break-even position and private sector investment would be feasible for some technologies. Discussions with universities show that there is considerable interest in the research aspects. This scenario is the one most likely to lead to a balanced development that generates sufficient income to maintain the site without adversely affecting the acknowledged heritage and environmental considerations.

6.1.3   Sustainable industrial estate

This scenario would involve large-scale commercial and industrial activity, requiring a significant land take in the Green Belt to accommodate development of the area around the existing core site. Electricity would be sold to grid, on site and to nearby users, and heat sold to tenants with high heat requirements.

A large land development company might be interested in leading in this scenario.

Under this option, revenue would be maximised to support the site's future and the economics may be good, achieved through economies of scale, but there would be a major impact on policy, heritage and environmental considerations that may well be insurmountable.

6.2 *Development of the chosen scenario*

Weighing up the considerations of each of the three scenarios, both the Chatterley Whitfield Partnership and the City Council adopted the second option.

The different technologies were then assessed against ten criteria:

– Fit with scenario
– Impact on historic monument
– Resource availability
– Market confidence
– Environmental impact
– Social benefits
– Capital requirement
– Grant aid requirement
– Economic potential
– Implementation timescale.

A combination of technologies and scales of development was then recommended. Some technologies, such as anaerobic digestion and biomass, are appropriate to generate revenue, while others, such as photovoltaic cells, would not be economic at this scale but their principles could be demonstrated. Wind power is still being assessed through an anemometer on top of the conical tip, where there is room for only one turbine. Energy storage, coal bed methane and thermal treatment of municipal solid waste have been discarded as options for the Chatterley Whitfield site.

6.3 *Infrastructure*

The consultants appointed to consider infrastructure improvements were commissioned to consider services, intrusive site investigations, hazardous materials in buildings, remediation strategy, movement strategy, and key outputs and costings. The minimum works required to remove existing health and safety risks were also identified.

There is some contamination of the site, with diesel being the main substance. There are also slightly elevated arsenic concentrations, and sulphate protection of concrete foundations will be required. There is a lot of made ground across the site, and there are buried concrete structures and hard standing, including large service-carrying culverts.

The main area that could accept new development is part of a tipped area, and is unsuitable for conventional foundations due to the likelihood of differential settlement. Possible alternatives include dynamic and vibro compaction, or replacing made ground with engineered fill, though the impact on existing buildings of these approaches will have to be tested.

It is usual practice when reclaiming colliery sites to remove all surface materials to the underlying natural ground and replace this with clean fill. This removes surface contamination and subsurface obstructions such as buried foundations and redundant services and pipes. At Chatterley Whitfield, the buildings are to be retained, which restricts this approach. Furthermore, there is the question of whether the existing roads and pavements are to be retained as part of the heritage value of the site, in which case the areas between the buildings must be left undisturbed as well. This, however, raises questions of health and safety and may hinder the installation of new services.

There are shallow coal measures underlying the site, and in certain defined areas there is the risk of collapse by crown hole pillar failure, the potential for spontaneous combustion and the possibility of methane emissions. The relative timing of grouting beneath foundations and the renovation of the buildings will have to be assessed.

The steep sides of the western tip will have to be re-graded, and surplus material could be filled over the flatter areas of the tip. This could potentially be designed to benefit the site drainage and the conditions for an area to demonstrate the growing of energy crops. Soil or soil forming material will need to be imported for this.

A new access road will be required, including a roundabout at its junction with the public highway, but this is considered to be relatively straightforward. The construction of this road will deal with culverts that are currently in poor condition. The long culvert under the conical tip, however, is in urgent need of a detailed condition survey.

6.4 *Business plan*

The business plan takes the conclusions of the infrastructure and sustainable energy studies, and combines them to assess the likely capital requirements, income generation, operating costs and options for an appropriate style of organisation to be established to implement the project.

At the time of writing, this work is not yet complete. However, contact has been made with many organisations including private sector firms and educational establishments. Considerable interest has been shown in the idea of a sustainable energy centre, and although the market is uncertain, it seems that there is a strong chance that the project will succeed.

Three zones have tentatively been identified: for heritage, sustainable energy, and people. The heritage area comprises buildings mainly to the sides of the complex which are too specialised in their design to have an economic re-use. These include the heapsteads and the tubhall, while the powerhouse, with its winding engine and compressors, could be a visitor centre to tell the site's story as well as the wider story of energy past, present and future. The larger sheds and buildings are suitable for the generation of renewable energy or the production of renewable energy equipment, and the remains of the boiler house could be the site of an iconic energy advice centre. The pithead baths, while retaining the showers and lockers for visitors to see, could also provide research facilities, training opportunities and start-up units.

The ground floor of the Main Office, at the entrance to the site, is currently being converted to a community office, through funding from the Heritage Lottery Fund, English Heritage and the local Single Regeneration Budget programme aimed at young people.

Young people are a crucial part of the project, but much remains to be done to encourage their involvement. No applications were submitted for the recently-advertised post of Youth Development Worker, so alternative approaches are being considered.

The community group has been active in many areas while the current studies were underway. Volunteers acted as guides and stewards for Open Days and visits by private groups, as well as providing refreshments for visitors. They have kept the site clear of litter, carried out preparatory work for the conversion of the Main Office, and cleaned the former deployment centre for use while the Main Office was being converted.

7   ENGLISH PARTNERSHIPS

In September 2002, it was announced that Chatterley Whitfield had been included in the expanded National Coalfields Programme. English Partnerships bring their own requirements to the project, but will fit in with the existing partnership arrangements. They will have to be satisfied that the conclusions drawn from the range of studies already undertaken are based on a sound analysis of the site, and that there are sustainable markets for the proposed uses. However, the need to continue with the work achieved to date, particularly with the local community, is recognised.

English Partnerships will consider the costs of all infrastructure remediation, even in a "do nothing" scenario. Details are not yet clear, but the project has already benefited from the involvement of English Partnerships by being taken more seriously in discussions with other funders.

8   NEXT STEPS

In January 2003, the Partnership will appoint architects to prepare a scheme for the whole site. This will consider the treatment of buildings and spaces, and will guide the restoration and conversion of all buildings on the site. The architects will also be asked to design emergency repairs to the Hesketh Heapstead, which is in a dangerous condition, in a way that will avoid duplication when it is fully restored.

A two-stage application will be submitted to the Heritage lottery Fund, initially for funding to design the repair and restoration of several high priority buildings, and secondly, when the full design is complete, for the funding to carry out the work. The architects will be asked to carry out the initial design work to support the first stage of the application.

An ambitious programme of community activities for 2003 has been prepared in discussion with local residents. This includes restoring the winding engine, moving archive material to better storage conditions and sorting through the many papers and films. Decorating the community office will provide a base for these activities.

An application for planning permission to change the use of the site to a sustainable energy centre will be submitted during 2003, and it is important that the issues and proposals are clearly understood by the surrounding communities. It would be all too easy for rumours and misinformation to spread, and already there have been some examples of mistaken assumptions about the project. A programme of meetings, exhibitions, Open Days, demonstrations, school projects, promotional material and trips to other projects is being put together to inform local residents and give them further opportunities to become involved in the project.

Detailed discussions with many organisations are required, both with firms which may wish to locate at the site and with potential investors.

The main step in the next twelve months, however, will be a start on site of the improvement work. This will demonstrate to all concerned that the Chatterley Whitfield project has reached a crucial stage.

9   CONCLUSIONS

The Chatterley Whitfield regeneration project is a major scheme that has taken several years to reach a point where it is taken seriously as an important development site. It will take many more years for the proposals to be fully implemented. The heritage importance of the site has made it unique, hence the willingness to tackle its problems, even though there are others that might be more easily achieved. Heritage can have the power to stimulate regeneration.

In the course of the work to date, the project has raised some challenges, not all of which have yet been answered:

- Finding revenue-generating uses for a heritage site in the Green Belt in an area with a plentiful supply of alternative development land
- Balancing revenue generation with impact on the monument and surrounding area
- Conserving buildings in effect forever when the buildings were only built with a limited lifespan and are in a poor condition
- Balancing the needs of future developers with impact on the monument
- Maintaining the involvement of local residents in a long-term project with little visible progress to date
- Demonstrating improvements to the energy efficiency of single skin buildings
- Incorporating energy efficiency and sustainable energy technologies in heritage buildings.

Finding solutions to these questions will be just one part of the next few years work at Chatterley Whitfield.

Land Reclamation – Moore, Fox & Elliott (eds)
© 2003 Swets & Zeitlinger, Lisse, ISBN 90 5809 562 2

Land reclamation in the North East: the last 30 years

Alison Hesselberth
Newcastle City Council

David Hobson
Waterman Environmental

ABSTRACT: The North East of England has a long and glorious industrial heritage. This has led to a legacy of derelict land, but also to opportunities for reuse and regeneration. Government priorities and means of funding have changed over the last 30 years. The paper examines how land reclamation has progressed from the 1960's to the present time. It also looks at what has happened to land reclaimed during the earlier times and draws conclusions about the lessons that can be learned.

1 INTRODUCTION

Land reclamation has evolved from fairly basic beginnings with limited objectives to the modern process of remediation with exacting standards and the production of land for specific uses.

The process has been driven by changing government policies and priorities and a myriad of different rules relating to the availability of financial support.

This paper seeks to illustrate the evolution of the national land reclamation programme by drawing from the experiences on Tyneside from the early years to the present time.

2 IN THE BEGINNING

Land Reclamation in the UK began in the 1960's and was initially targeted at restoring the devastation caused primarily by the coal mining industry. In the North East region this work was largely promoted by the County Councils of Northumberland and Durham supported by central Government grant aid. These early schemes largely involved visual improvements to closed colliery sites and waste heaps. In many cases this simply involved planting of woodland directly into the colliery shales with little other work. In other cases sites were regraded to soften the industrial landscape and basic drainage and fencing provided. The end uses were inevitably either public open space or agricultural grazing land.

3 TYNE AND WEAR COUNTY COUNCIL 1974–1986

3.1 *The political background*

In 1974, as a result of Local government reorganization, the metropolitan local authority areas of Tyneside and Wearside were changed. A new County Council was created whilst the District Councils were also given new powers. Land reclamation was a concurrent function, but despite the obvious difficulties of such an arrangement, by and large there was reasonable cooperation between the 6 authorities.

3.2 *The economic background*

By the 1970's the North East's traditional industries such as mining, ship building and their support industries were in dramatic decline. Unemployment was rising and much of the public housing stock was in poor condition. With the closure of collieries, shipyards and factories large areas of derelict land were being created where the owners had neither the money nor the inclination to clear up. The value of such land was generally low.

3.3 *Government policy*

Both national and local government were embarked on the twin policies of trying to save the declining industries from further contraction and also to encourage new industry, particularly if it brought inward investment to the region.

The Government's approach was to encourage and assist Local Authorities to acquire and then reclaim the derelict land that was being created. The objectives continued to be to achieve environmental improvement, but from 1974 onwards increasingly the government tried to target funds on schemes with an economic dimension.

3.4 Financing of land reclamation

During the whole of the period of the existence of Tyne and Wear County Council (1974–1986) financial support from central Government was largely provided through Local Authorities. This was mainly in the form of Local Government Derelict Land Grant, which was provided at up to 100% of costs depending on the region. The grant was very restrictive in the type of work to which it could be applied since the rules seem still to be geared to the traditional concept of site clearance, regrading and rudimentary re-vegetation which had been used in the early reclamation programmes. The rule book was a closely guarded secret at the Department of the Environment and those seeking grant became very innovative in the presentation of projects.

The grant was available on an annual non-transferable basis. This resulted in the approval of extra schemes in February each year in order to reduce underspending. This was a good time to gain approval for projects that fitted poorly into the eligibility requirements. The historic Armstrong Bridge in Jesmond was saved as a result of refurbishment carried out under Derelict Land Grant. The grant was given on the basis of a scheme for demolition of the industrial relic and reclamation of the site. The cost of refurbishment was of course slightly less than the cost of the alternative reclamation scheme.

Additional funding mechanisms were gradually introduced, the most notable being the Government's Urban Programme. This enabled more hard works such as retaining walls and roads to be included in schemes. Tyne Wear County Council had a policy of making additional funds of their own available to enhance reclamation projects.

Attempts were made to encourage the private sector to undertake land reclamation, by offering Private Sector Derelict Land Grant. Unfortunately the terms on which it was offered were unattractive. It was available at 50% of the loss made after the increase in land value was taken into account. Therefore the grant was only available if you set out to lose money. Very few commercial organizations operate on that basis.

3.5 Priorities

Over time, a number of mechanisms were introduced by the Government in an attempt to direct expenditure to projects which would produce land with a "hard" end use involving economic development. These included the introduction of Category A projects which were given priority and Category B projects which received a smaller slice of the annual grant cake. Category A schemes had to have specified end users who were committed to taking an interest in the land and developing it on completion of the scheme. This had a mixed level of success.

Examples range between the former Sunderland Airport, which contained a colliery waste heap and was prepared for Nissan and a scheme to turn derelict land to the north of Scotswood Road in Newcastle into a market garden for a group of local unemployed residents. The former was developed immediately, but the latter is still awaiting development to take place 20 years on.

At that time it was still believed that land that had been used for industry should remain in that use. As a result there was a lack of enthusiasm for schemes that created housing development on former industrial sites. One such example was approved for the former North Sands shipyard in Sunderland, but was abandoned after the abolition of the metropolitan county councils in 1986. A simple scheme of site clearance and regrading was undertaken instead to create a site which was held available for new industry. More than 10 years later the site was developed for housing.

3.6 Standards

Standards of reclamation were consistent with the current policies and technical knowledge of the time. Initially contamination was generally only recognized in obvious cases and risk assessments were rarely undertaken. The ICRCL guidelines, which were published in 1978, began to focus attention on the chemical quality of land and by 1986 the issue had become much more widely recognized. During this period colliery wastes were generally not treated as potentially contaminated and therefore subject to little testing.

Physical reclamation standards gradually improved with the move towards hard end uses and geotechnical conditions became more important.

3.7 Range of projects

A total of around 70 projects were completed by Tyne and Wear County Council with probably a similar number completed by the 5 district councils in the Tyne and Wear area over the period 1974–1986.

Projects ranged from traditional colliery reclamation schemes producing agricultural or recreational open space to reasonably sophisticated remediation projects producing land of economic value and dealing with significantly difficult physical conditions and contamination.

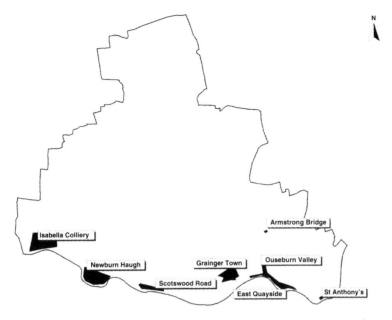

Figure 1. Map of Newcastle upon Tyne showing location of sites used as case studies.

4 CASE STUDIES

4.1 *Isabella colliery*

An area of approximately 50 ha of land on the North bank of the River Tyne between Newburn and Throckley lay derelict for many years following the cessation of mining and the closure of a railway line. The site contained a number of interesting features including a scrap yard, railway track bed, waste heaps, lagoons, shafts from which mine water continuously discharged and an area used by the water company for the disposal of liquid sludge from its settlement beds at a nearby treatment works. Natural drainage of the area had been disrupted by mining subsidence and flooding of the surrounding fields occurred during winter periods. The mine water was badly discoloured with "yellow ochre" and this was discharged via approximately 1.5 km of open water course into the River Tyne. The site was not considered to be contaminated and therefore was not subject to chemical testing.

The site was reclaimed to create what is now known as the Tyne Riverside Park at Newburn. It was financed through 100% Derelict Land Grant, additional funding from the County Council and a small contribution from the Sports Council.

For a period of about three years prior to reclamation, surplus soils were brought onto the site from various sources. These were mainly from private developments being undertaken in the region and soils primarily comprised subsoils and clays. In most cases a small charge was made for accepting the materials and this provided additional funds for the project. These operations were not considered as waste disposal and therefore were not licensed under the Control of Pollution Act 1974.

The scheme involved the creation of a new landform by regrading and covering over with the imported soils to form a growing medium. The scrapyard was cleared and visually "dirty" materials were removed off site for disposal in landfill. Mine shafts were capped, but not filled and the drainage of the area improved as far as possible by cleaning out channels and the installation of a flap valve upstream of the mine water discharge in order to prevent back flow into the areas worst affected by flooding.

The most challenging part of the scheme was the treatment of the area used for disposal of water company sludge. This had collected in a depression to a depth of about 3 m and was of a consistency similar to jelly paint. Dewatering of this material was achieved by spreading it in thin layers and allowing it to freeze during the winter. The material was also found to, be useful as a soil enhancer although, because of the high moisture content, it was extremely difficult to handle. Provision for future disposals of sludge was provided in cooperation with the Newcastle and Gateshead Water Company. The water company agreed to spray future discharges of sludge in order to spread it in a very thin layer that would be capable of drying over the winter. A large grassed area was therefore created under which a layer of gravel and system of pipes were installed. This would not have been feasible without a ready source of a large quantity of gravel. This was obtained by the excavation of a large slope into the bank of the River Tyne which comprised excellent quality river gravel. This operation was considered eligible for Derelict Land Grant because it was essential to the restoration of the site. The excavated area conveniently produced a slipway which was concreted over and this is still the largest access into the river for recreational purposes today.

The whole site was landscaped creating areas of grass and woodland with about 30% being returned to grazing land and the remainder being made available as a country park. The scheme was awarded first prize in the RICS/Times Conservation Awards – RICS (1982). It is now a well used and mature country park.

4.2 St Anthony's Tar Works

St Anthony's Tar Works was constructed in the early 1900's on the top of a large heap of ships ballast. This consisted of Thames river gravel brought in colliers returning to the north east coal fields. It was situated in the east end of Newcastle and ceased operations in 1980. The works had not been demolished when it came into County Council ownership.

The main features comprised many tanks and old boilers most containing residues from the tar distillation process, boilers and furnaces, two deep tar wells and large concrete pitch beds containing solid pitch. Broken asbestos insulation covered much of the site and buildings. The tar wells still contained liquid tars, but had also been partly filled with rubble. The upper 2–3 m of the site comprised industrial fills and ashes most of which were badly contaminated with tars and distillation products. The underlying 15 m of gravel was found to be relatively clean, except at its base. The gravel ballast had been tipped onto the original foreshore and river bank and it was discovered that there was a heavily contaminated layer approximately 2 m thick at the interface between the gravel and the original ground. This was leaching into the River Tyne which is tidal at that point and a slick was continually present around the entire frontage of the site.

The reclamation scheme involved specialist removal of the asbestos, emptying and removal of the various tanks, excavation of the tar pits and the pitch beds and demolition of all the structures. All materials were taken off site and contaminated upper layers treated by a traditional "dig and dump" approach.

Following consultation with the Northumbrian Water Authority it appeared that the contamination of the river was not of great concern at that time although it was thought that some attempt should be made to reduce the flow of hydrocarbons into the river. The deep layer of contamination was not considered accessible from the surface through 15 m of dense gravel. It was therefore decided to install an interceptor drain along the river edge which was discharged into a large multisection interceptor. This was based on a similar solution adopted in 1978 for the treatment of the Ottovale coke and tar works site at Blaydon. It was noted during the construction of these measures that the leachate was extremely aggressive, to the extent that even so called chemically resistant protective clothing seemed to disintegrate after exposure. The interception solution had a dramatic effect on improving water quality in the river, although it clearly created a long term maintenance commitment to keep it operating.

The site was reshaped to create a facility known as Walker Wheels. This was for young people to learn about car maintenance and improvement and included a track for testing and general enjoyment. It was hoped that this would divert their energies away from car theft and vandalism. After a number of years the project ceased and the site was subsequently filled over.

The river edge interceptor ceased to be effective after a number of years. It has now been replaced with a pumped extraction and treatment system. It is understood that the problems with aggressive leachate are having adverse effects on this system.

4.3 Scotswood Road

Scotswood Road is a major route into Newcastle from the West. The land alongside it had become a considerable eyesore with derelict buildings and old structures relating to various industrial uses.

A reclamation scheme was prepared to landscape most of the area in order to provide a better visual appearance on this important transport corridor. In order to achieve the necessary priority for the scheme and to ensure that grant aid would be available an end user for at least part of the site had to be identified. This would then allow the scheme to be given Category A status by the Department of the Environment. The user selected was a group of local residents who proposed to develop a market garden on the land and then supply the neighbourhood with fresh vegetables at value prices. The scheme was approved for 100% Derelict Grant aid and it was even awarded additional funds from a private trust. Sadly, although an area suitable for the market garden was created, with 300 mm of imported topsoil to provide a good growing medium, the business never started.

An additional interesting aspect of this project, which was undertaken by a labour controlled county council during the miner's strike, was the opencasting of about one third of its area to remove old mine workings. A significant quantity of coal was produced during the strike and this was delivered to a coal stocking site which was under an NUM picket.

4.4 Sunderland Airport

In 1985 Nissan Motors identified Sunderland Airport as the preferred site for their new UK factory. The site which was approximately 300 acres contained a colliery waste tip, was poorly drained and the ground levels were unsuitable for a factory of the size and nature required. A land reclamation scheme was prepared which would remove the colliery waste away from the area designated for the factory. This would then be used to form the basis of a test track. This enabled an application to be made for Derelict Land Grant as a Category A scheme. The works were carried out in 12 weeks and included earthworks in excess of 1 million cu m, temporary drainage works and a storm water storage reservoir.

5 NEWCASTLE CITY, MODERN TIMES

5.1 Tyne and Wear Development Corporation

Following the abolition of Tyne and Wear County Council, the Government, in 1987, looked to a new Development Corporation to tackle the worst areas of derelict land at that time, concentrating on the 26 miles of riverside along the rivers of Tyne and Wear. The Development Corporation was a property-based regeneration agency, with a "Master Developer" role, and designated 2375 ha as an urban development area, which included the Newcastle quayside.

In the 18th century, Newcastle had been one of the busiest ports in the country, prospering on the Tyneside coal industry, salt-making, leather and glass production industries. Shipbuilding and heavy engineering had also grown rapidly. But as these activities had declined, there was a need to bring derelict and under-used land back into effective use.

The Development Corporation set out a masterplan and community strategy to tackle the declining areas and revive the East Quayside and St Peter's riverside into thriving business areas, with high quality housing, leisure and an improved environment.

5.2 Changing priorities

From 1989, the Government started to consider a range of measures to tackle the large areas of unused and underused land that existed, including policy changes and alternative financing arrangements, to prevent land from becoming derelict.

In April 1994, Derelict Land Grant ended and English Partnerships developed the Land Reclamation Programme. In 1998, the Government set a national target of 60% of all new housing developments to be sited on brownfield sites, defined as previously used or developed land that was not currently in full use.

A significant change occurred in April 1999, with the formation of the Regional Development Agencies (RDAs). The statutory purposes of the new RDAs were to further economic development and regeneration, promote business investment and competitiveness, promote employment and to contribute to sustainable development.

Subsequently, in 1999, the RDAs adopted the land reclamation responsibilities of English Partnerships, and took over the administration of the Land Reclamation Programme, funded through their physical regeneration budgets. This resulted in a change of emphasis towards reclamation schemes with end commercial uses in view. The RDAs developed policies to target funds at commercial sites.

This was reinforced in December 2001 when the government set targets for the RDAs in four areas: development of brownfield sites, creating and safeguarding jobs; improving skills and learning opportunities; and improving the rate of start-ups. This led to a focus on land reclamation projects that could deliver in all four areas.

Land Reclamation Teams within local authorities therefore began to work more closely with their economic development colleagues and regeneration partnerships, and derelict land redevelopment within Newcastle focused on commercial and residential end-use schemes. The RDA, One NorthEast, developed a major remediation scheme at Newburn Haugh, to the west of the city, with an industrial end use. New housing developments, business parks and retail outlets were undertaken at sites to the east and west of the city centre.

In July 2002, following a review of English Partnerships that began in October 2001, the Deputy Prime Minister announced its new role as a key delivery agency in the regeneration of towns, cities and rural areas, the "urban renaissance". In particular, English Partnerships would focus on the development of strategic brownfield land, beginning with a list of prioritised sites, and would produce a national brownfield strategy.

5.3 Financing of schemes

English Partnerships took over the administration for the reclamation of derelict land from 1995 through two main programmes.

The Land Reclamation Programme provided monetary assistance for local authorities, and the Partnership Investment Programme (PIP) assisted the private sector – English Partnerships (1995). PIP was based on the principle of "gap funding" where the amount awarded was that necessary to bridge the gap between development costs and forecast end value. It enabled developers to go ahead with otherwise non-commercially viable projects to bring derelict or disused sites back into full economic use. A major interruption to this funding regime occurred in December 1999, when the European Commission announced its decision that PIP was in breach of State aid rules and the scheme was stopped. Although previously approved schemes, "PIP survivors", were allowed to complete, this decision had a significant impact on the future funding of land reclamation schemes. Despite receiving a positive decision in 2002 on new gap funding schemes within Assisted Areas, local authorities, regeneration partnerships and developers had lost the momentum that had evolved before the original decision was made. Local authorities were left with the option of using their CPO powers to assemble land for direct redevelopment, but the number of reclamation schemes reduced sharply as a result.

Traditional land reclamation schemes thus lost favour in exchange for schemes that would not only remove dereliction or contamination, but would produce economic outputs in terms of jobs, industrial/commercial space and housing, with associated benefits to the community and the environment.

In Newcastle, no land reclamation schemes took place from 1999, other than those which were included as part of a housing or economic development scheme. Large areas of derelict land alongside the river in the outer suburbs remain unused, as the cost of redevelopment makes them commercially unviable. The majority of the city's derelict sites are in areas where the local land and property market is operating at a very low level. Soft end use schemes have not been considered as they would not only fail to deliver economic outputs but would also represent revenue implications for future years.

With no specific budgets for land reclamation, funding had to be sought as part of wider schemes, assisted by funding streams such as Neighbourhood Renewal Fund or the RDA's Single Programme. RDAs responded by restricting approvals to schemes that delivered across their four key targets. However, some flexibility has recently started to emerge in other regions, with Advantage West Midlands offering 100% of the costs of reclamation to local authorities to reclaim derelict or contaminated land under a new "Land Reclamation and Environmental Improvement Programme". A high quality environment and physical framework, with minimal or no evidence of derelict or underused land, is recognised as a significant factor in most investment decisions, particularly those involving companies in the technology sector, whose aspirations are high.

In January 2000, Newcastle City Council published its Going for Growth Green Paper, which set ambitious targets for creating new jobs and housing and made the case for a regeneration strategy that would be ambitious and long-term and overcome the declining population.

The first regeneration plans for the Outer East and the West End of Newcastle were produced for consultation in June 2000 and revised plans were produced – Going for Growth (2001).

Building on the success of the riverside development during the 1980's, the Green Paper recognized the riverside as having great potential to be the focus of new regeneration for the East End, with opportunities to assemble sites for new housing development and to overcome the problems of high unemployment and underused and derelict sites. It also aims to improve the quality and accessibility of the Walker Riverside Park, including enhanced pedestrian and cycling access, provide a shipbuilding/heritage museum and to prioritise actions for removal of eyesore vacant properties in the area. The importance of utilising the landscape available in the area has been recognised. Whilst many of the regeneration plans are at design brief stage only, there is considerable interest from housing developers in the sites becoming available through demolition programmes.

One particular success story of the East End is the Ouseburn Valley, an area of previously derelict land to the east of the City Centre. Much of the land was contaminated from previous industrial uses and landfill.

Having completed a SRB2 project in 2001, redevelopment of the area as an urban village is continuing, using SRB6, to provide a mix of commercial, residential and leisure uses, building on the area's unique character and heritage. The valley has developed over the past ten years as a location particularly for small businesses, and has become increasingly popular with the creative industries.

A variety of buildings, including old, redundant warehouses, are being completely refurbished and converted to offer new, flexible office space. To date nearly 200 small units have developed and the area now contains over 250 businesses employing over 1,500 people.

In the West End, there have been similar demolition programmes to the East End, to remove housing in key areas and open up the area towards the new neighbourhood hearts. In particular, the land adja-cent to the river has been targeted for development. There are plans for a new Scotswood urban village of 3,000 high quality homes, with fine views over the Tyne.

Within the City Centre, the Grainger Town SRB 3 Project has regenerated the historic core of the city, providing a high quality environment and bringing derelict buildings back into economic use, and encouraging residents back into the City. Both the Ouseburn and Grainger Town projects have brought significant private sector investment into the area though partnership ventures.

5.4 New standards

From 1986, there was generally a raised awareness of the need to identify and manage the risks associated with contaminated land. However, contaminated land was viewed as a subset of derelict land, and its potential risk to the environment continued to be underestimated.

Having set the national target in 1998 of 60% of all new housing developments to be sited on brownfields, the issue of liability and financial risks that could deter landowners and developers from participating began to be considered. The uncertainties resulting from the lack of an agreed national set of guidelines for safe levels of contaminants were a potential barrier to achieving the target.

The concept of harm to humans, living organisms and property was eventually recognised and defined in the Environmental Protection Act 1990, and, in 1997, the provisions of the Environment Act 1995 relating to contaminated land, were implemented. Decisions on the determination of contaminated sites were based on site-specific risk assessments, which considered pathways between sources of contaminants and potential targets. Achieving appropriate land reclamation standards thus became a much more significant issue for developers and local authorities.

The Environmental Protection Act 1990 requires local authorities to identify and investigate potentially contaminated sites. Supplementary Credit Approval funding is an important source of financing for site investigations and remediation schemes on local authority land.

The funding has been used to finance several schemes at Newcastle, including the remediation of Byker City Farm, a community facility on the outskirts of Newcastle, which was the site of the Ouseburn Lead Works from 1898. Levels of lead in soils greatly exceeded ICRCL levels at 16,000 ppm. This resulted in alarming headlines in the local press, as risks were identified to young children who had visited the farm over the years, and to staff who had consumed herbs and vegetables from the farm. Exposed soils were removed to a depth of 1 m and a clay cap was provided to the site, using SCA funding in 2001.

A number of old landfill sites were also investigated and remediated using SCA funding.

Although the credit has recently been restricted to sites identified as contaminated under the legal definition, the fund is extremely useful for local authority owned sites, with a soft end use planned.

In the absence of more reliable criteria, ICRCL standards have continued to be used, although their relevance to a range of end uses was often questioned.

The production of the CLEA site assessment model – DEFRA (2002) has introduced a more scientific approach to site assessment, which should give developer's greater confidence in acquiring contaminated sites for redevelopment, and greater consistency between local authorities.

5.5 *Conclusion*

After Tyne and Wear County Council was disbanded, the main focus of land reclamation activity in Newcastle and the adjacent riverside was the urban development area designated by Tyne and Wear Development Corporation.

The changing funding regimes for local authorities was challenging enough, but the transfer of the administration of the Land Reclamation Programme to eight separate RDAs in 1999, each with their own regional strategies and priorities, introduced greater complexity. The greater awareness of risk management for potentially contaminated sites, the delay in releasing technical guidance and the sudden cancellation of gap funding in December 1999 were key factors in the demise of land reclamation by local authorities.

Given the fact that none of Newcastle's former coalfields have been included in the recent revision to the National Coalfields Programme, and none of the city's derelict sites have been prioritized by English Partnerships, it is predictable that the main use of derelict land in the area will continue to be for housing or commercial use, where it is economical to do so.

6 CASE STUDIES

6.1 *Newburn riverside industry park*

Newburn Haugh, as it was formerly known, was a heavily contaminated industrial site in an area of high unemployment 4.5 miles to the west of Newcastle on the banks of the Tyne. Originally, it was home to Lemington Glassworks, the Stella North Power Station and the Anglo Great Lakes graphite works.

The site was the focus of a major land reclamation initiative to develop the area as a new industrial park, and, at 92 ha, it is one of the country's largest land reclamation schemes, with the potential to create up to 5,000 new jobs.

It is well on course to become an exceptional 21st Century location for business. In March 2001, the first phase of the development commenced, comprising nine industrial units totalling 10 ha.

The overall scheme is a £46 million investment in reclamation and infrastructure, with £33 million funding from ONE North East, £7.1 million European Regional Development Funding and £5.2 million from the Capital Challenge Fund, secured by Newcastle City Council. Decontamination and reclamation works cost in the region of £22.5 million.

The aim is to secure jobs for local people on the park and the City Council has launched a training initiative to assist people into work. ONE North East is currently marketing the site as a development opportunity, and the RDA is also servicing the 92 ha site to enable some 180,000 sq.m. of industrial and commercial floorspace. In fact, the RDA plans to move into new office accommodation on the flagship site.

6.2 *South Benwell business development scheme*

After the war years, the heavy industry and factories, which had stretched along the banks of the Tyne and along Scotswood Road, declined. Tyne and Wear Development Corporation undertook a land reclamation scheme to provide a successful new Business Park to the south of Scotswood Road.

To the west of the Business Park, the Vickers Tank factory still operates on Scotswood Road, alongside new car showrooms and haulage distribution centres.

Land to the north of Scotswood Road occupied by housing, suffered years of decline, vandalism and social problems, resulting in demolitions during the late 1980's. However improvements began in the mid 1980's with reclamation and landscaping of the road corridor and the development of the Newcastle Business Park to the South.

To the north of Scotswood Road, a development consortium is now working in collaboration with local residents and the City Council to regenerate a large area of vacant land, including that created by clearance of parts of Buddle Road housing estate. Work started on site in January 2000, and infrastructure for the new business units is well underway.

The gross area of the site is 7.2 ha, of which 4 ha are developable in 4 business units. This development promises to bring over 1,000 jobs to the area, attracting £17 million in private investment. It is funded by £1.6 million ERDF, £1.6 million SRB2 and £4.3 million from ONE North East, to fund site remediation, infrastructure, servicing and landscaping works.

7 LESSONS TO BE LEARNED

7.1 *Retention of records*

The importance of retaining records has been repeatedly demonstrated. Sadly this is often neglected.

Very few of the files or other records originally held by Tyne and Wear County Council on reclaimed sites still exist. It is believed that, despite being archived, they were not permanently retained. Details of all the case studies in section 4 of this paper have therefore been produced essentially from the memory of the writer. Details of the materials left in place at St Anthony's tar works were known about at the time of the original reclamation, but this information was not available to the designers of the new treatment system. Records of the drainage system at the former Isabella Colliery are no longer available and site

investigations have been needed to establish the arrangements of drains for discharging mine water from remaining mine shafts into the surface water channels. A system of cut off cells and drains exist at the former Ottovale Tar works for which no records exist.

Without records, appropriate essential maintenance may not be implemented and when new work is required additional costs in undertaking site investigation are inevitable.

Authorities should ensure that records are properly indexed and securely archived in perpetuity.

7.2 *Flexibility of financing*

The funding of Land Reclamation has been subject to a variety changing rules and regulations over the years. Although these have generally been intended to focus resources on appropriate priorities they have very often found to be restrictive and in some cases leading to less than desirable results. Examples can be found in the case studies referred to previously.

The site at Scotswood Road had to be prepared for a dubious end use which never materialized. This did however enable a worthwhile environmental improvement scheme to receive priority funding.

Funding of the restoration of the historic Armstrong Bridge was judged to be ultra vires, but if the rules had been adhered to it would have had to be demolished (and at a higher cost).

A mix of funding sources were applied to the Newburn Riverside project .This has enabled a wide variety of land remediation, infrastructure and development to take place and large amounts of private sector investment has been attracted.

Rules for the funding of reclamation and regeneration projects should be as flexible as possible to allow project specific judgments to be made. Annual funding regimes create pressure to push schemes through.

7.3 *Future maintenance*

It is of paramount importance that, after completion, adequate resources are made available for ongoing maintenance. If this is not likely to be forthcoming then the scheme should be changed to create either a lesser commitment or an end use capable of funding the necessary post reclamation activities for the foreseeable future. The revenue implications for local authorities with declining council tax incomes should be carefully considered.

A good example is the former Isabella Colliery which has benefited from continued maintenance and investment to produce an extremely well used and successful facility. Conversely the St Anthony's tar works was reclaimed for a use which could not be sustained.

Projects must take into account the likely availability of future maintenance and running costs. Failure to maintain sites results in the waste of the resources invested at the time of reclamation.

7.4 Changing standards

Decisions on the standards of reclamation need to bear in mind the possibility for future changes in what is acceptable. Although standards must be appropriate, application of the bare minimum may often be a false economy. Changes in water quality standards, for instance, have meant that discharges into rivers which 20 years ago were accepted by the regulators despite being clearly undesirable, are unacceptable now. The cost of treating now is much greater than if dealt with at the time of reclamation.

A high standard should be aimed for. Costs should not be limited to the achievement of only the minimum acceptable standard.

7.5 Flexibility in future land use and realism in what can be achieved

Most sites are reclaimed with a specific end use in mind. A former ship building yard in Sunderland was reclaimed for an industrial for which there was no economic demand. Land at Scotswood was prepared for an unrealistic development which never took place. The results were the waste of resources, land remaining unused (and in danger of reverting to dereliction) and the need for further work and costs to bring the land into beneficial use.

Land should be prepared for future uses which are realistic.

7.6 Value of environmental improvements to economic regeneration

The transfer of the LRP to the RDAs introduced a greater emphasis on achieving future economic use for reclaimed sites. However, there are many examples to demonstrate that a high quality attractive environment is a major factor in encouraging inward investment. A good example is the Grainger Town area of the city, where the poor quality paving has been replaced, levering in over £100 million private sector finance. It is hoped that new funding regimes of RDAs and the new role for English Partnerships in completing exemplar schemes on their portfolio of sites, will reinforce the economic value of environmental improvement.

The importance of environmental improvements in enabling and encouraging economic investment should be fully recognized by funding organizations.

7.7 Regionalisation of development agencies

The fragmentation of land reclamation between the Regional Development Agencies has resulted in some inconsistencies between the financing and management of land reclamation in different regions, and this is further complicated by the formation of sub-regional partnerships to vet bids in some regions.

Whilst regional strategies are to be welcomed as a way forward, the sharing of ideas and experiences of land reclamation between RDAs would ensure that lessons could be learnt and good practice shared.

7.8 Local authority land reclamation teams

The knowledge and expertise of the specialist teams has been dispersed.

Local authority land reclamation teams may need to be reformed.

REFERENCES

DEFRA (2002). R & D Publication CLR10, The Contaminated Land exposure Assessment Model (CLEA). DEFRA / Environment Agency.

English Partnerships. 1995. Investment Guide. English Partnerships.

Going for Growth. 2001. Regeneration Plans Newcastle City Council.

ICRCL (Interdepartmental Committee on the Redevelopment of Contaminated Land) 1987 Guidance on the Assessment and Redevelopment of Contaminated Land.

RICS. Conservation Awards. 1983. Royal Institution of Chartered Surveyors.

Landscapes for People

Land Reclamation – Moore, Fox & Elliott (eds)
© 2003 Swets & Zeitlinger, Lisse, ISBN 90 5809 562 2

Post coal-mining landscapes: an under-appreciated resource for wildlife, people and heritage

I.D. Rotherham, F. Spode & D. Fraser
Centre for Environmental Conservation and Outdoor Leisure, Sheffield Hallam University, UK

ABSTRACT: Devastated post coal-mining landscapes are considered ripe for reclamation, with considerable funding directed to their "greening" and recovery. This work often has significant benefit – social, environmental and economic. It is increasingly apparent that post coal-mining sites may have considerable inherent interest and value (Eyre & Luff 1995, Lunn & Wild 1995). This interest may be ecology and hence nature conservation value, and local history, heritage and culture (Middleton 2000). Softened by natural, spontaneous plant and animal communities they may be attractive with local distinctiveness – important in coalfield landscapes. Potential of post-industrial sites for heathland regeneration has been noted (Rotherham 1995). The conservation interest of spontaneous communities and sensitive restoration schemes is very significant and unique to its place and time. The juxtaposed ecology and history generated opportunities not to be repeated. The potential of this value for sensitive and sustainable regeneration of these landscapes is of huge interest (Rotherham & Lunn 2000).

1 INTRODUCTION

The frequently desperate state of many British wildlife species has been presented and discussed by many over the last fifty or sixty years. Cowley *et al.* (2000) for example discuss the declines in butterflies in the UK, and Marren (2001) assesses extinctions in local floras in recent decades. Despite often poor datasets and information bases it has been clear that many wildlife habitats and their associated plants, animals and fungi have been inexorably squeezed out of an increasingly urbanised, industrialise and agri-industrialised landscape. The individual causes vary but the general trends are to do with competition for space and often the rapid abandonment of traditional land management techniques and practices on which many of these species now depend (e.g. Marren (2001). Marren suggests that one native plant species is lost from each English County each year. The trends and concerns arising from them resulted in two developments during the 1980s and then the 1990s. The first was the production of regional and city-wide Nature Conservation Strategies to guide the planning process (e.g. Bownes *et al.* 1991). These were very much in the context of, and in response to the Nature Conservancy Council's strategic vision "*Nature Conservation in Great Britain*" (1984). They were also quite clearly given impetus by the statutory tool of Planning Guidance to Local Authorities. The second was the move during the

1990s and continuing now, to develop "Biodiversity Action Plans" – at National, Regional and Local levels. The impetus for this was in part at least in the follow-up from the Rio Conference in 1992. Biodiversity Action Plans were a part of the UK Government's commitment to the Convention on Biological Diversity signed at the Rio Summit. The outputs from the work at national level have then been a series of key reports and policy statements such as "*Biodiversity: The UK Action Plan*" (1994). The precursor for this was produced in the build-up to Rio, namely "*This Common Inheritance*" (1990), a very effective summary of key aspects of the state of the nation's nature conservation and environmental resource. In 1994 "*Sustainable Development: The UK Strategy*" (Anon. 1994) was published to provide the over-arching context for these initiatives.

Much emphasis of the nature conservation aspects of these documents is focused on the need, quite rightly to halt the decline and to safeguard relict sites and relict populations. In some cases projects have specifically sought to engage government agencies such as English Nature in partnerships with NGOs such as the Wildlife Trusts, the RSPB, or Plantlife, to address specific nationally threatened plants and animals. At the same time, there have been increasing concerns regarding the potential impacts of larger scale changes such as possible global warming, and also the fallout of atmospheric pollutants such as acid rain or nitrogen depositions. The first will influence animal and plant distributions and can lead to both extinctions of native species and replacement by those from warmer climates. The latter tends to increase soil and water nutrient levels and causes more sensitive and often uncommon species to be replaced by more aggressive, generally commoner, opportunistic species. These trends have been addressed by national surveys and reports, for example: "*Measuring change in British vegetation* (Bunce *et al.* 1999), and "*Causes of change in British vegetation*" (Firbank *et al.* 2000).

These serious conservation issues and major strategic initiatives have generated responses at levels from national to regional and local. However, as is so often the case with otherwise laudable ideas, the concept of Local Biodiversity Action Planning comes with no budget attached. It is expected that Agency regional offices, along with NGOs and Local Authorities, will address these issues from existing budgets. Since the success of an LBAP must ultimately depend on the quality of information that helps develop and inform the document, and then on its uptake and use by practitioners at all relevant levels, this is a serious problem. Local Biodiversity Action Plans are now being produced for the districts of the South Yorkshire Region, although they are neither comprehensive nor consistent from district to district. Barnsley Metropolitan Borough Council for example has led a partnership approach culminating in a comprehensive plan published in 2002. In the neighbouring district of Sheffield, the LBAP is a more piecemeal and sequential publication. These along with the National BAPs do give reasonable guidance for the strategic consideration of the ecology of large-scale site restoration and reclamation across the County.

Another initiative that took shape during the 1990s and is now embedded in the landscape vision for different regions of the UK is that of "Natural Areas". These have been assessed and defined in broad terms for all parts of the country and essentially form a backdrop for more specific BAP issues (1999).

In the South Yorkshire coalfield area for example, this land holding (mostly controlled by Yorkshire Forward) now ripe for intensive restoration is potentially the single biggest contributor to the region achieving nationally recognised targets for environmental recovery such as Biodiversity Action Plan targets.

Presently this opportunity is not recognised and many sites of regional conservation value are being "reclaimed" and "restored" for supposed environmental benefit. It is argued that this is a missed opportunity for long-term recovery and sustainability. Finally it is suggested that these areas could be the centrepiece for integrated restoration schemes involving local people more fully and more genuinely, and could generate significant and sustainable economic recovery through outdoor leisure and adventure activities and through the development of initiatives such as biomass fuels for energy. These issues were discussed in terms of heaths, acidic grasslands and bogs at the National Lowland Heathland Conference in 2002 and a regional initiative was announced – "*Northern heaths and commons for people and wildlife*". This has potentially very strong links to the sensitive restoration of degraded and abused landscapes across the region. Whether or not the critical support from key agencies and drivers will be forthcoming is now the challenge.

2   THE IMPORTANCE OF POST-INDUSTRIAL SITES

It is against this background that the missed opportunities at local and regional levels become most apparent. Increasingly in recent decades the potential for nature conservation on actively and passively restored or conserved post-industrial sites has been recognised. The article "Nob End, Bolton" (Shaw & Halton 1998) highlights the potential of an abandoned chemical works now designated as a Site of Special Scientific Interest. Shaw (1998) discusses this potential of post-industrial sites in more detail.

Papers by a number of authors during the 1990s began to stress the potential importance for nature conservation of post mining sites. These include a range of different materials mined and a range of processes used. However, the sheer scale of the Yorkshire Coalfield, with the adjacent regions of Derbyshire and Nottinghamshire results in these landscapes being of extraordinary significance in the South Yorkshire region. The potential significance and conservation importance of post-industrial sites was emphasised by Rotherham (1999). One key point made was the potential stability of some of the communities and habitat-types of these areas due to the extreme environmental conditions of some post-industrial landscapes – extremes of pH, of water-logging, of drought, of physical construction of the terrain, and very importantly often very low levels of available soil nutrients. Site contamination may also be a hugely important factor in allowing stress tolerant species to persist, by eliminating the more aggressive competitors.

The recognition of these values was highlighted by papers from Middleton (2000) and then by Lunn (2001), and by Rotherham *et al.* (2000). These focused in particular on post coal-mining sites in the Yorkshire and South Yorkshire regions.

This recognition is only now being translated into positive action and many good sites have been lost. It is also clear that many extensive reclamation areas have been generated and the knowledge in terms of the science and technology for restoration is good. The ability to create large and successful water bodies is evidence by projects such as Rother Valley Country Park near Killamarsh, and Old Moor in Barnsley.

These and the accidental mining subsidence flashes have major impacts on both landscape and ecology. However, many schemes still produce extensive and often rather monotonous woodland plantings, with very poor grassland and sometimes wetlands. Opportunities to create more interesting features such as acidic grasslands, heaths and bogs are rarely taken.

3 A MISSED OPPORTUNITY

Coal mines have opened and closed across this region for many centuries. However, the major impact was over the period from the mid-1800s until the 1990s. From the late 1980s and through the 1990s the catastrophic scale of industrial collapse was apparent across South Yorkshire and into regions adjacent. As the political, environmental and social tensions pulsed across the Yorkshire Coalfield, the fate of many pits and their buildings and landscapes was sealed. Closure was often followed by decommissioning, by demolition and by tidying up, and then some form of site restoration and renewal. Some of these sites made safe and restored, form the hub of new economic regeneration. However, many sites, in whole or in part will have no active development. In effect they will help reform the green structure and fabric of the region.

It is here that the missed opportunities become most apparent. The agencies such as Yorkshire Forward are charged with the responsibility to rejuvenate the region – primarily through economic regeneration. In recognition of the severity of this task, the South Yorkshire Region was granted "Objective One" status by the EU. However, it is increasingly recognised that a high quality and sustainable environment provides the necessary backdrop for successful post-industrial economic and social regeneration. In this respect this portfolio of abandoned post-coaling sites is the most significant resource available now or in the future, to restore damaged landscapes, to conserve threatened species, and to engage local communities in their environment.

It is surprising therefore that the strategic vision, the so-called "joined up thinking" is absent from the restoration plans for these areas. The sites are considered individually and not strategically, and they are restored with no driving reference to regional "Natural Areas", and certainly not to either Local or National Biodiversity Action Plan species, habitats, or targets.

The portfolio of sites includes thirty-one major locations across the South Yorkshire County. Some of these are very large, covering many hectares. Getting further information has proved difficult although discussion with Yorkshire Forward officers has indicated that there is no strategic overview to embed reclamation into the aims and objectives of either natural areas or of Local Biodiversity Action Plans. Individual

sites have, largely through effective local lobbying had conservation restoration schemes accepted. The major colliery reclamation project at Grimethorpe for example includes heathland and passive use of natural succession as parts of the overall scheme. The Glasshoughton site has included some experimental heathland creation work.

One major ecological reclamation scheme has been the creation of the new and very large wetland centre at Old Moor in the Dearne Valley. Indeed this serves to emphasize the potential of the larger portfolio if strategically harnessed. Not only will this help meet many targets in the Barnsley LBAP, but also it has already drawn down to the region several million pounds of grant aid and will provide sustainable jobs through education, nature conservation and associated leisure and tourism. To give an indication of the potential contributions of such an approach to a local economy relevant data are given in the table below based on on-going research in South Yorkshire, including the Old Moor Nature Reserve site and others (Table 1).

It is perhaps this economic impact of sensitive and relevant restoration and reclamation that has been the vital missing link in the ecological argument. These landscapes have potential to be regenerated into thriving ecological systems, but ones that also have tremendous vibrancy for local people, and which importantly can contribute significantly to regional economic renewal. Some example information on this aspect is given in Table 1 to illustrate the point in relation to created wetlands.

Work in progress on a scoping evaluation of another restoration scheme in the area, at Potteric Carr near Doncaster, suggests that the proposed nature reserve extension will contribute around 4–500 jobs to the region over a five- to ten-year period.

This economic assessment indicates that the opportunities being missed are not merely ecological but financial too. Furthermore, the cultural resonance of these areas is also frequently overlooked.

Involvement of local communities in the full process of planning and reclamation can reap significant benefits, as evidenced by the Westwood opencast restoration scheme near Barnsley. Here the project has restored woodland areas and regenerated interesting and attractive wetlands, heaths and meadows. All were at no cost to conservation bodies and the site is self-financing in terms of its ongoing management (Rotherham & Lunn 2000).

Inspection of papers such as Middleton (2000) and Lunn (2001) adds a further twist to the current situation. Many of these post-industrial sites are already of immense nature conservation value and it can also be argued that they possess huge cultural resonance and potential, along with in some cases, important and early industrial archaeology. The post-coaling sites are often havens for National BAP species such as

95

Table 1. The contribution of water and wetland nature reserves to tourism and leisure economics.

A Summary of Economic Impacts

- Initial start up (capital investment and grant aid draw down) of a facility such as a major nature reserve and visitor centre. Short-term employment for construction of a major new site = up to 85 or more jobs.
- Running (revenue costs and generation, grant aid, commercial function and induced local impacts); anything from perhaps 10–100 direct jobs and a substantial induced impact.

The implications of the case studies

- Economic impacts from immediate and short-term (research, design, construction and establishment), to long-term leisure and tourism benefits.
- Short-term economic inputs may be relatively modest (£1,000–10,000) for a site recovery/restoration scheme to very major (£100,000–1,000,000) for reclamation and creation projects.

Direct and induced impacts

- With visitor figures from 100,000 to 600,000 day visits per year for the area's wetland sites, and site-based staff, these projects might contribute from £1m to £6m per year to the local economy through visitor spend.

A recent study by Sheffield Hallam University for Severn Trent Water showed the impacts of
Carsington Water reservoir and its associated conservation and educational facilities to be

- Nature conservation benefit of huge regional importance – provision of opportunity for many people of close contact with nature; provision of viewing opportunities (hides *etc.*) for bird watching; good interpretation; excellent access for diversity of users.
- A major positive element of the site.
- Local people regard the social and economic impacts as positive. Some point to increased trade, as well as enhanced amenity value.
- Thirty-one full-time and forty-two seasonal/part-time jobs created at the site; up to 436 jobs created in the local, rural economy as a result of an estimated annual visitor spend of over £14,125,00.
- Over one million visitors now use the site each year.

skylark, grey partridge and corn bunting, among others. Great crested newts, dragonflies, water voles and orchids may be abundant in and around ponds and marshes accidentally left on abandonment of the industry. Furthermore, important plant communities – heaths and grasslands in drier areas, and fen, marsh, and even bog in wet zones are also found and sometimes in abundance. These communities and species already contribute to the targets for local and national BAPs and yet are being accidentally lost through "restoration" and "reclamation" projects.

It is often suggested that post-industrial or derelict sites are accidental, damaged and of little value or potential. In actual fact this assessment is far from the truth. Limited in terms of nutrient levels, and often with extreme physical and sometimes chemical environments, these areas offer opportunities for species of both animals and plants adapted to survive such difficult conditions. These are often rare species. They also provide habitats for the many species of conservation significance that have been squeezed out of the wider landscape by disturbance and by nutrient fallout and other contamination or eutophication. A former colliery site at Holbrook near Killamarsh (Rotherham 1999) is essentially abandoned and unmanaged. Yet it has over twenty species of *Cladonia* lichen, and over ten species of *Sphagnum* moss. For these alone it is the most significant site in the region. And yet the Sheffield Wildlife Trust in the 1990s refused to support lobbies for the site's conservation, and it still remains

under threat. This is despite being home to over twenty species of butterfly, and numerous Red Data and BAP animals and plants – several of which are especially protected. (These include Badger (*Meles meles*), Harvest Mouse (*Micromys minutes*), Water Vole (*Arvicola terrestris*) and Great Crested Newt (*Triturus cristatus*)). Interestingly too, the site includes a coal spoil heap naturally regenerating to birch and oak woodland. The conditions here are extreme desiccation with extensive bare ground. This is now the only remaining such area in the District and is not only an interesting wildlife habitat, but also a unique piece of cultural and landscape archaeology too.

It is also assumed that these sites are inherently replaceable. They are not. Unlike many urban commons sites that have transient communities and species that are essentially able to "hop" from one "habitat island" to the next, many of these post-industrial sites represent unique time-lines of accidental and unrepeatable history. Furthermore, many also have acquired species from the wider landscape at the time of their creation and development (e.g. Rotherham 1999). In many cases these same plant and animal species have been lost from that wider environment and so could not re-colonise a new derelict site even if it were created and the conditions suitable. Some of these locations now hold examples, albeit impoverished, of the landscape at the time of the early Industrial Revolution now largely lost or degraded. This in itself is exciting. However, there is a further issue. In the present climate of intervention conservation there are suggestions to reintroduce species formerly lost (e.g. *Sheffield Biodiversity Action Plan*, 2002). For these conservation programmes to be successful it is necessary to have extensive and sustainable sites with the extreme conditions as already noted. Without this they will either fail and or require management input that will not be sustainable in the long-term. The unique histories and extreme conditions of many post-industrial sites, as well as their extensive nature make them eminently suited to the long-term survival of these animals and plants – and often without the need for reintroduction.

4   WHAT IS UNIQUE?

Quite clearly it can be argued that the evolution of post-industrial sites may give them unique context and value. This is not always the case and they are not always either of interest or sustainable. It is therefore important to be able to discriminate based on good information.

The occurrence of water and waterlogged areas may give considerably enhanced value to a site. However, for areas such as post-coaling landscapes the occurrence of extensive areas of open, bare ground may be a unique feature. Associated with extreme conditions (drought, seasonal water-logging, contamination, extreme pH *etc*) such wildlife habitat can be hugely important. Depending on aspect, bare ground can foster micro-habitat that is essential for many unusual and rare species, particularly invertebrates. Many of these cannot survive outside such areas, and these conditions become evermore rare in our increasingly eutrophic environment. Key (2000) notes the importance of bare ground for the conservation of invertebrates. This was a point noted in 1995 by Eyre & Luff.

In some sites such as Holbrook (and indeed in many post-coaling areas) illegal off-road motorcycling, in the absence of conservation management, may help maintain both bare and disturbed areas. This may enable survival of key invertebrates beyond the point of the ecological succession when they would normally be lost. In wet areas such *ad hoc* disturbance can enable the survival of large numbers of plants such as marsh orchids.

Placed in a strategic context and supported and enhanced the contribution of accidentally re-colonising and sensitively restored sites could be:

1. Significant.
2. Sustainable; (as many of these areas undergo successional change only slowly due to the extreme site conditions).
3. Low cost to create and to maintain.

In many situations it is the extreme and unproductive conditions of soil and substrate that make these areas exciting, diverse and sustainable. In some cases mild toxicity of soils and substrates again allows a diverse and interesting ecology to develop. Restoration targeted at alleviating site stresses often succeeds in undoing all the ecological positives in the name of landscape improvement and reclamation.

Unguided natural colonisation of these areas often results in attractive and relatively sustainable areas of secondary woodland, of heathland and flower-rich grasslands. With modest intervention in terms of occasional management, these attributes could largely be maintained and even enhanced.

5   POLICY DIRECTIONS

The necessary strategic guidance is probably in place to help guide and facilitate the processes of positive recovery. It is clear however that strategy and policy are only being converted into practice on a rather random basis – and this often as a result of lobbying by local champions or activists. Awareness of regional and national policy context in terms of ecology and landscape seems to be low at the cutting edge of restoration practice in the region. The necessary detailed local knowledge is also generally poor. In part this is a result of a chronic lack of the necessary and accessible information. Consultation with the main agencies involved confirms that there is no reliable database on either existing lowland heaths or acidic grasslands across the region; and certainly nothing on potential restoration and creation sites. More worryingly even after many years of arguing for resources to gather such information it appears no nearer to being done. This lack of data surely compounds and exacerbates the problems due to lack of awareness by decision makers and restoration practitioners.

At a local and regional level Local Authority and Partnership policies such as Nature Conservation Strategies and Biodiversity Action Plans become increasingly important. The *Sheffield Nature Conservation Strategy* (1991) was one of the first to designate *ALL* heather areas protected as *Sites of Scientific Interest* or their equivalent in the planning process. This policy document also highlighted the opportunities afforded by creative site restoration. More recently the Local Biodiversity Action Plans or LBAPs have been emerging. The *Barnsley Biodiversity Action Plan* (2002) for example, does begin the process of addressing some of these issues through recognition in the planning process and positive steps in encouraging appropriate habitat creation. Even here however, the targets for Lowland Heathland include identification of targets sites for restoration and creation and monitoring of development proposals, but much less is said about the associated acidic grassland. In practice dry acidic grassland and *Calluna* or *Ulex* heath are often parts of the same complex. Again there are positive statements about the need to identify suitable sites for restoration to flower-rich neutral grassland too. Importantly this landmark publication also highlights the value of post industrial and derelict land for nature conservation in the District. It also (and rather worryingly in terms of confirming the thrust of this present paper) notes restoration and development of sites as the prime cause of loss. This is followed by drainage of land, unsympathetic restoration, and poor aftercare and maintenance of sites that have been restored or landscaped to accommodate wildlife. Various actions are proposed including surveys and evaluation of development proposals. Interestingly and despite the considerable published research on the region's heaths over a fifteen-year period, the *Sheffield Biodiversity Action Plan* demonstrates no awareness of post-industrial sites and opportunities in terms of acidic grasslands and heaths. Given the lack of awareness by the nature conservation lobby it is little wonder that restoration and development practitioners are missing the opportunities.

6   WHAT IS NEEDED?

The core problem is that of a limited window of opportunity. In the South Yorkshire Region there is a tremendous resource that requires restoration, but in sympathy with the area. The derelict sites have an accidental and often incidental resonance with the region's former and potential ecology. They also have a cultural significance and depth for the local communities. Whilst the relationships may be painful due to the economic, social and indeed environmental tensions associated with heavy industry over around two hundred years, the attachments run deep. The landscape and both economic and social vitalities must be reclaimed and restored, but the long-term ecological functions require thought too. More often than not, the opportunities are being overlooked and the resonance with local ecology and community being given at most, a superficial veneer.

The chronic lack of good and accessible information, together with lack of awareness by both conservation and restoration practitioners, compounds the problems. Furthermore, despite the huge budgets for regeneration and for development, there is a crippling lack of funding for the necessary information gathering and even less for the lobbying and advice needed to turn words into actions.

It has been suggested in discussion with others such as Peter Shepherd and others involved in the Nottinghamshire Heathlands projects that there are opportunities to develop synergies and support. These might link the heaths and commons of West, East, and South Yorkshire, to projects already underway in Nottinghamshire, and even Lincolnshire and Derbyshire. Our suggestion has been a project based around the theme of "*Northern heaths and commons for people and wildlife*". This may be a way forward but if so it is an initiative that will need real support from the key players in regional regeneration. So far there is little sign of this happening (Rotherham 2002).

Most of these landscapes and their sites came about because they were important to local people. Today, heaths and commons have generally forfeited their cultural economic function. However, they are still popular recreational and amenity areas. As such they still have economic value and cultural worth. Perhaps this could provide a focus for attempts to safeguard them, and a justification (if needed) for their conservation for future generations to value and enjoy. The economic and nature conservation values and benefits are increasingly recognised, so should the local cultural attachments.

Many authors (Hetherington & Jamieson 2000) for example, stress the importance of the role of the planning process for conservation. Working effectively within the planning process for more positive conservation may be a way forward. This does necessitate information and resources in order to be so proactive. However, the idea of assessing restoration proposals and insisting where appropriate on a positive contribution to the relevant BAP targets (where they exist) is a start. However, many sites have already been "treated" or developed, and many more will have been done before any major initiative can be implemented. It remains to be seen then whether intervention even late in the process can perhaps draw down some of the hoped for benefits. Some of this intervention could be on already restored sites, perhaps if the restoration work so far were viewed as part treatment towards a longer-term objective such as heathland, marsh or bog for example. Without intervention many restored landscapes are green mounds with planted trees and "improved" grassland often of little value to wildlife and little stimulation to people. Much has been written on the potential for "gardened" landscapes to contribute to nature conservation objectives (e.g. Good 2000, or Ansell *et al.* 2001). There is also an extensive literature on grassland creation for conservation benefit (e.g. Jones 2001). As a second best aspiration it may be possible to safeguard some remaining sites with more sensitive restoration schemes, and to intervene positively in those already treated to pull them back to a more ecologically rich successional process. This would have the benefit of viewing less ecologically sound schemes as positive starting points for these richer landscapes. It could also be relatively low cost and may produce savings in terms of longer-term maintenance budgets.

7 CONCLUSIONS

The technology now exists to restore or at least reclaim most post-industrial derelict sites. However, often in the rush to reclaim there are major opportunities being missed and a unique chance to contribute to conservation targets at many levels is being lost. This argument is presented within the case study context of the South Yorkshire Region in England, but the principles are mostly transferable.

There are continuing problems of a lack of resources to provide the necessary information to better inform decision makers and practitioners. This is not a new problem but it is a continuing one.

It is now suggested that not only are there ecological opportunities but that these extend to local cultural resonance and to economic renewal also. Making links to issues of leisure and tourism and relates aspects of nature-based leisure and tourism will be increasingly important.

The potential for long-term sustainable landscapes is identified and discussed, and the need to safeguard current opportunities through more sensitive schemes, and within a limited timeframe is noted.

However, there is still the possibility to intervene in existing schemes to divert the ecological successional processes to a more valuable and more sustainable direction. It is suggested that this could be combined with site conservation in new schemes to maximise benefits within the established but underappreciated strategic context. Furthermore, this approach could be applied in a way that involves the local community and that addresses sites from the very smallest to the very largest. The desired endpoint is a landscape that gives most benefit to both local wildlife and to local people, and is generally sustainable within a context of minimal maintenance intervention. There is a vision, and we do have the technology. However, for success in realising this vision there is a need for a strategic awakening and then the appliance of science.

REFERENCES

Anon. (1990) *This Common Inheritance*. London: HMSO.

Anon. (1994) *Biodiversity; The UK Action Plan*. London: HMSO.

Anon. (1999) *Natural Areas in the Yorkshire & the Humber Region*. Peterborough: English Nature.

Anon. (2002) *Barnsley Biodiversity Action Plan*. Barnsley: Barnsley Metropolitan Borough Council.

Anon. (2002) *Sheffield Local Biodiversity Action Plan – Species Action Plans and Habitat Action Plans*. Sheffield City Council and partners, Sheffield.

Ansell, R., Baker, P. & Harris, S. (2001) The value of gardens for wildlife – lessons from mammals and herpetofauna. *British Wildlife*, 13 (2): 77–84.

Bownes, J.S., Riley, T.H., Rotherham, I.D. & Vincent, S.M. (1991) *Sheffield Nature Conservation Strategy*. Sheffield: Sheffield City Council.

Bunce, R.G.H., Smart, S.M., van de Poll, H.M., Hill, M.O., Watkins, J.W. & Scott, W.A. (1999) *Measuring change in British vegetation*. Grange-over-Sands, Cumbria: Institute of Terrestrial Ecology.

Cowley, M., Thomas, C., Thomas, J. & Warren, M. (2000) Assessing butterflies" status and decline. *British Wildlife*, 11 (4): 243–249.

Eyre, M.D. & Luff, M.L. (1995) Coleoptera on Post-industrial Land: a Conservation Problem? *Land Contamination and Reclamation*, 3 (2): 132–134.

Firbank, L.G., Smart, S.M., van de Poll, H.M., Bunce, R.G.H., Hill, M.O., Howard, D.C., Watkins, J.W. & Stark, G.J. (2000) *Causes of change in British vegetation*. Grangeover-Sands, Cumbria: Institute of Terrestrial Ecology.

Good, R. (2000) The value of gardening for wildlife What contribution does it make to conservation? *British Wildlife*, 12 (2): 77–84.

Hetherington, M. & Jamieson, D. (2000) Planning for wildlife – an insider's guide. *British Wildlife*, 12 (1), 1–6.

Jones, A. (2001) We plough the fields but what do we scatter? A look at the science and practice of grassland restoration. *British Wildlife*, 12 (4): 229–235.

Key, R. (2000) Bare ground and the conservation of invertebrates. *British Wildlife*, 11 (2): 183–191.

Lunn, J. (2001) Wildlife and mining in the Yorkshire coalfield. *British Wildlife*, 12 (5): 319–326.

Lunn, J. & Wild, M. (1995) The Wildlife Interest of Abandoned Collieries and Spoil Heaps in Yorkshire. *Land Contamination and Reclamation*, 3 (2): 135–137.

Marren, P. (2001) "What time hath stole away". Local extinctions in our native flora. *British Wildlife*, 12 (5): 305–310.

Middleton, P. (2000) The wildlife significance of a former colliery site in Yorkshire. *British Wildlife*, 11 (5): 333–339.

Rotherham, I.D. (1995) Urban Heathlands – their conservation, restoration and creation. *Land Contamination and Restoration*, 3 (2): 99–100.

Rotherham, I.D. (1999) Urban Environmental History: the importance of relict communities in urban biodiversity conservation. *Journal of Practical Ecology and Conservation*, 3 (1): 3–22.

Rotherham, I.D. (2002) *Hanging by a Thread – a brief overview of the Heaths and Commons of the north-east Midlands of England*. In: *The Proceedings of the 2002 National Lowland Heathland Conference, Nottingham*. (In press).

Rotherham, I.D., Cartwright, G. & Watts, R. (2000) Airport, Steelworks or Historic Landscape – the Sheffield Airport development as a case study of integrated site planning. *Liverpool: SER 2000 Conference, (Abstract paper)*.

Rotherham, I.D. & Lunn, J. (2000) Positive restoration in a "Green Belt" open-cast coaling site: the conservation and community benefits of a sympathetic scheme in Barnsley, South Yorkshire. *Liverpool: SER 2000 Conference, (Abstract paper)*.

Shaw, P. (1998) Conservation management of industrial wastes. *Journal of Practical Ecology and Conservation*, 2 (1): 13–18.

Shaw, P. & Halton, W. (1998) Nob End, Bolton. *British Wildlife*, 10 (1): 13–17.

Land Reclamation – Moore, Fox & Elliott (eds)
© 2003 Swets & Zeitlinger, Lisse, ISBN 90 5809 562 2

Reclamation in Gateshead: creating landscapes for people

M. Poremba
Gateshead Council, Gateshead, Tyne and Wear, England

ABSTRACT: The Borough of Gateshead has a history of intensive industrial activity. Gateshead Council has carried out a large number of land reclamation projects since the 1980's. A variety of reclamation techniques have been used and developed over the last 16 years. Current uses of reclaimed sites include industry, waste recycling, housing, sports & leisure, burials and locations for works of art.

1 INTRODUCTION

The Borough of Gateshead has an industrial background, although well over half of its land area is rural. Major industrial development was based initially on the availability of cheap coal. This gave rise to a range of industries including coke, gas, electricity, chemicals, glass, iron and steel, locomotives and other heavy engineering.

Access to the River Tyne was of great importance for the import and export of materials and so the major concentrations of activity developed close to the river. Coal mines and their associated industrial uses were, however, distributed throughout the Borough.

Much of the Gateshead bank of the River Tyne is affected by solid industrial contamination, which has resulted in poisoned ground and, in some cases, polluted leachate.

The former Gateshead County Borough and Durham County Councils became active from the earliest days of the national derelict land reclamation programme in the mid 1960's. The abolition of the Tyne and Wear County Council in 1986 resulted in a large upsurge in the level of the Borough Council's reclamation programme for the rest of the 1980's, and left the Borough Council with sole responsibility for land reclamation work within its area. It also coincided with the start of work on the site of the 1990 National Garden Festival.

2 THE NATIONAL GARDEN FESTIVAL

The National Garden Festival occupied a series of linked sites totalling approximately 81 ha, extending southwards from the River Tyne along the lower 2 kms of the Team Valley.

With the exception of Eslington Park, a former quarry and tip which had been reclaimed 25 years earlier and had been used since then as playing fields, the sites were all connected with the industrial heritage of the North East. By the late 1970's they had been abandoned by the public sector corporations and were in desperate need of reclamation. The sites included the former Redheugh Gas Works, Norwood Coke Works and Thomas Ness Tar Works which were grossly contaminated with the wastes and by-products of their respective processes.

By the early 1980's the Borough Council had begun to tackle the problem with the acquisition of most of the Redheugh Gas Works site and the implementation of the initial phases of a long term reclamation programme funded by a 100% Derelict Land Grant.

The potential attraction of the Garden Festival concept, when a further round of bids was invited by the Department of the Environment in 1983, was obvious. Priority access to Derelict Land Grant allocations meant reclamation within 5, not 25 years. With an emphasis upon this and related objectives, and upon its capacity to provide and operate a Festival of considerable attraction, the Council's bid to host the 1990 Festival was successful.

The four principal sites that were reclaimed were Redheugh Gas Works, Norwood Sidings (Dunston), Norwood Coke Works and Thomas Ness Tar Works. Each site had its own distinctive characteristics and problems (Figure 1).

Norwood Coke Works, the southernmost and largest of the Festival sites, covered 22 ha. The Coke Works closed in the early 1980's. Contaminants included

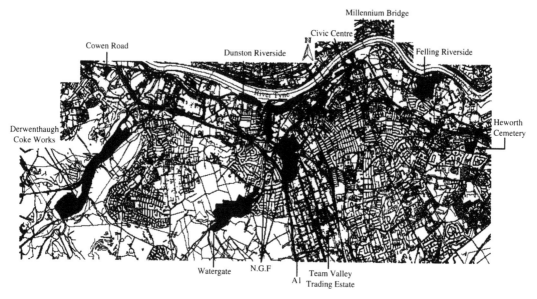

Figure 1.   Location of reclamation sites.

phenols, toluenes, lead, arsenic, sulphides and sulphates, together with combustible materials which represented a fire risk.

Housing was the Council's preferred afteruse. Ground investigations and technical analysis demonstrated that this objective could be achieved. The reclamation methods adopted provided a suitable platform for housing on a net developable area of 13 ha.

The scheme had four main components. First, substantial mounds were created around the perimeter of the site to create a landscaped framework and minimise the external noise of road and rail traffic.

Second, the coal storage areas from which combustible materials had been extracted were capped with a 300 mm blanket of crushed dolomitic stone dynamically compacted to a load bearing capacity of 80 KN suitable for standard house types.

Third, a drainage system was installed to act as the main surface water system for the eventual housing development (and for the Festival in the short term) and to deal with two ground water springs which emerged on the western boundary of the site.

Fourth, and finally, the site was "finished" with a layer of compacted clay approximately 1.5 m thick to accommodate development infrastructure etc. whilst preventing the dolomite blanket from being penetrated and its performance being impaired.

The remainder of the site was treated as a contaminated waste tip and accepted contaminated arisings from works on other parts of the Festival site. This greatly reduced the overall cost of the reclamation works but required quite stringent control measures to ensure that the area was fully reclaimed. This area

will remain permanently in Council ownership as amenity open space.

Agreement was reached prior to the Garden Festival for the sale of the land to Shepherd Homes of York, who provided a cluster of 22 show houses for use during the Festival. Shepherds subsequently sold on part of the site to Bellway Homes and both housebuilders have developed the site with a total of 321 houses now occupied.

On the Dunston site, 32 Sheltered housing units were constructed before the Festival opened and 60 "Social" housing units, (for rent), were subsequently constructed.

The former Redheugh Gasworks site is now the subject of Wimpey and the Hemmingways' development proposals, (Staiths South Bank). Works started on site in October 2002.

This area, originally reclaimed for employment use, will now be developed as one of the first "home zone" housing developments in the UK. The £70 million scheme is expected to provide 688 homes.

The total cost of reclaiming all four sites, for the Garden Festival, was £7 million – an average of some £140,000 per hectare.

The visitor total was 3.1 million, 42% of which were from outside Tyne & Wear.

Geographic visitor profile; 58% Tyne Tees area; 15% Yorkshire Borders; 6% Scotland; 8% London/ South/South West; 12% Rest of UK; 1% Overseas

Further expenditure was required in order to create a national event on the site. £41.6 million was spent, but there was an income of £37.9 million, including £22 million in sponsorship. The shortfall of £3.7 million was underwritten by Gateshead MBC. However, 1000 jobs were created in building the Festival and 1400 further jobs were provided by the event.

Since the Garden Festival other major projects have been completed.

Derwenthaugh Coke Works and Watergate Colliery, both now approaching the end of their establishment phases, were designed for different, but primarily "soft", end uses.

3 DERWENTHAUGH COKE WORKS

The Derwenthaugh Coke Works scheme covers a series of linked sites extending some 2.5 km along the lower Derwent Valley and comprising a total area of 50 ha. The land was formerly occupied by the coke works and associated chemical plant, a coke stocking area, coke and coal waste disposal areas, workshops and a large spoil heap associated with the Clockburn Drift coal mine. The reclamation process was complicated by the meandering course of the River Derwent alongside which the various uses were located. The Council bought the site from National Smokeless Fuels Ltd in 1987 following the closure and partial dismantling of the coke works. Reclama-tion work began in 1991. The reclamation scheme contained the following elements:-

- demolition of remaining structures and removal of building foundations
- excavation, screening and washing of coal and coke waste to recover reusable fuel and prevent future spontaneous combustion of the material
- renewal or diversion of various drains which crossed the site
- two new river crossings to enable contractors vehicles and plant to reach all parts of the site
- dealing with the chemical contamination which had resulted from the coke making and by-product manufacturing processes
- stabilising the slopes of the Goodshields Haugh spoil heap
- creation of the final land form and landscaping of the site for future use.

The most challenging part of the reclamation site was the area of the former coke processing and by-product manufacturing plant which had been contaminated by a number of hazardous chemicals. The Council had to develop a range of innovative physical, chemical and biological solutions to this highly complex problem. These were all included in a decontamination contract, which commenced in June 1993. Initially programmed to run for 18 months, the contract had to be extended when the contamination was discovered to be more extensive than first indicated.

The first task, on this part of the site, was to isolate the contaminated area from incoming groundwater flows and protect the adjacent River Derwent from contaminated outflows. This was achieved by means of an underground "cut-off wall". The wall was constructed using a "Vib Wall" technique developed by French civil engineering company Bachy. A large piling rig was used to drive an "H" shaped lance equipped with grout pipes into the ground to rockhead. As the lance was withdrawn a cement and bentonite grout was placed under pressure to form an "H" shaped grout column. By successively driving the lance so that the flange position coincided, a continuous cut-off wall was produced.

To address the problem of high levels of benzene vapours which existed within the contaminated ground, a Dual Vacuum Extraction Plant was commissioned. A network of extraction wells was drilled into the ground and linked by pvc pipes to two extraction plants. The vapours were "sucked out" of the ground and passed through the extraction plant for aeration treatment before finally being discharged into the atmosphere.

This process was followed by dewatering of the contaminated area and treatment of the contaminated water. This was pumped to a treatment plant where floating oil was skimmed off for offsite disposal and the water treated chemically by a series of operations to reduce the contaminants to acceptable levels. The treated water was then pumped to a holding lagoon where it was chemically analysed to check the treatment had been successful before being discharge to a soakaway.

Following dewatering it was possible to tackle the solid contamination. A permanent encapsulation area was constructed for the containment of material which could not be treated, for example tars and thixotropic chemicals. It was formed by utilising the concrete bases of the former coke ovens and constructing clay bunds lined with crushed concrete, sand and a 2 mm thick chemical resistant pvc liner.

Treatment beds were constructed for the biological treatment of other contaminated solids. This was done using a layer of chemical resistant liner sandwiched between two 500 mm layers of clay. Lined cut-off ditches were constructed to catch any rainwater which ran off the contaminated material.

The material was first screened to remove large stones, then placed on the treatment beds. Air was drawn through the stockpiled material and the material was sampled and tested to establish the nutrient requirements for the successful biodegradation of the contaminants by bacteria. The nutrients in a water soluble form were then applied to the treatment beds using an agricultural sprayer and diffused via a pipework system.

Depending upon the physical nature of the soils it was possible to modify the availability of contaminants by the addition of chemicals such as surfactants which break the physical bond between the contaminant and soil particles. The treated soil was periodically tested until it reached the required safe levels for redistribution around the site as backfill material.

River water monitoring since the completion of the decontamination works in August 1995 has produced encouraging results, with most effluent discharge quality standards set by the NRA fully met by January 1996. Further improvement has occurred over time but, as a final safeguard, a reed bed system was built to deal with any residual contamination.

A challenge of a different kind was presented by the spoil heap at the opposite end of the site. An unstable, steep-sided 60 m high dome containing 2 million cubic metres of potentially combustible colliery waste. The barely accessible heap was hemmed in on three sides by a meander of the river and on the fourth by an ancient semi-natural woodland. With off-site disposal ruled out on cost grounds the only practical means of reshaping and consolidating the heap was to redistribute the surplus material around the rest of the site. The Council's ground modelling computer software proved invaluable as an aid to the considerable task of landform design which this entailed.

In addition to the earthworks an afteruse infrastructure including drainage, ducting to facilitate the provision of services and a footpath and bridleway network have been constructed.

As shale removal, regrading, clay capping and soiling proceeded on each section of the spoil heap, grass seed was sown on the area in preparation for a landscape contract. Work on this contract, originally programmed to be completed in March 1996, was brought to a successful conclusion in spite of the worst possible combination of weather conditions throughout the winter and early spring of 1995/96.

The entire scheme will be completed in March 2004 at a total cost of £14.0 million – an average of £280,000 per hectare.

Proposals for afteruse were developed following a comprehensive public consultation process. They take into account the site's Green Belt location, the considerable landscape and nature conservation value of the lower Derwent Valley and the discovery during the reclamation works of important industrial archaeological remains associated with Winlaton Iron Mill – founded in 1697 and the earliest known integrated iron works in the world.

The general theme of the afteruse proposals was one of creating opportunities for active and passive recreation in a wooded environment, balanced with income generation through local authority/private sector partnerships or franchises in order to offset the costs of long term management and maintenance.

The site provides first class facilities for football, cricket, tennis, (all managed by private clubs), and angling. As a major extension to the existing Derwent Walk Country Park it also provides opportunities for walking, cycling, horse riding and picnicking. The natural history value of the site has increased dramatically with woodpeckers, deer and otters regularly seen on site.

As part of it's official opening, in 1998, the "Great North Walk" was based in the park and attracted 6000 people.

An annual cycling event is based in the park and regularly attracts hundreds of people to the site. Extending the use of the site by horse riders is currently being tried out. This will be monitored and reviewed.

The archaeological remains – a massive stone dam and mill race have been consolidated and preserved as an impressive feature within the new landscape. It is hoped that, if funding becomes available, that the development of interpretation facilities will be possible.

4 WATERGATE COLLIERY

The site was used as a colliery between 1924 and 1964, the coal being transported to Dunston Staiths, (adj Staiths South Bank development above), via the Tanfield branch line of the London and NE Railway. After closure of the colliery, the site was left derelict and the disused railway became a public footpath.

In 1992 one of the spoil heaps caught fire. Attempts by British Coal to control the burning had limited effect and the fire and consequent smoke got worse. The Tanfield Railway Footpath was closed due to dust and smoke, and collapsed mineworkings.

Gateshead MBC bought the land from British Coal and started emergency works (funded by the DOE and carried out by Hall Construction Services Ltd between 1993 and 1995) by removing the burning material from the spoil heap, spreading it and compacting it between layers of clay in order to extinguish the fire and permanently exclude oxygen to stop future fires from occurring. The old mine-workings were stabilised or removed and the shafts capped. Some of the nearby agricultural land was used for spreading the burning material, and later returned to agriculture.

Once the burning material had been extinguished the land was remodelled, (funded by English Partnerships), the site capped with clay, subsoil and topsoil, and a new landscape was formed to create a varied terrain, including a lake, bridlepaths, sports pitches and different ecological niches. The culverted Black Burn, which originates in Washingwell Wood was uncovered. The new lake was filled up from the burn via reed beds and via settling ponds fed by a stream running through nearby agricultural land.

The finishing works have been funded by One North East, with tree planting 80% funded by The Forestry Commission.

More than 120,000 trees have been planted around the site to create mixed woodland, hazel coppice with standards, heathland, and an arboretum planted with trees having unusual bark.

The lake has been colonised naturally by vegetation and wildfowl, including coots, moorhens, tufted ducks and a breeding pair of swans. More than 160 varieties of willows have been planted around the lake and adjacent to ponds around the site and Gateshead Council intends to acquire National Status for this collection of Salix.

The park is a Gateway Site to the Great North Forest, and includes the sculptures "Greenheart" by William Pym and "Floating Rocks" by Colin Rose. The site is seen as an educational resource and the landform and interpretation allows "access for all".

Reclamation works also included the Gateshead Central Nursery Showground which hosts the Spring and Summer Flower Shows, new sports pitches and a meadow on the site of a Roman Fort. The Flower Shows typically attract 20,000 visitors for each event.

On 26 April 2002 part of the Pirelli Rally was held in the park and attracted in excess of 6000 spectators.

The site is used by cyclists, horseriders, joggers, walkers, local schools.

The use of the site for craft courses and demonstrations, such as willow basket making, has started. This use of the site is expected to develop as the hazel coppice and willow areas start to produce useful material.

Gateshead Harriers held a race on the site in 2002 which attracted 600 people to the site.

The reclamation and establishment works will come to a total of £5.2 million – an average of £104,000 per hectare. The scheme is currently in its establishment phase and will be enhanced over the years as money becomes available.

5   BROWNS QUARRY/HEWORTH CEMETERY EXTENSION

Ordnance Survey Maps and Parish records show this site to have been occupied by a sandstone quarry used intermittently between the 18th Century and 1970. The site covers an area of 5 ha.

The quarry was infilled with unlicensed domestic waste and builder's rubble.

Initial reclamation took place in 1973, and the site was used as a public open space.

Results of a site investigation carried out in 1983 concluded that the fill material was in a loose state of compaction and had visibly suffered from settlement. Internal combustion was also occurring.

Further site investigations, carried out in 1998, found that gas protection measures would need to be incorporated into any new structural developments. Trial pits were excavated and sampling and analysis showed levels of sulphate, zinc and copper to be above the ICRCL trigger thresholds in some areas of the site.

With the adjacent cemetery nearly full to capacity, the Council decided to develop the site for burials.

It was decided to encapsulate the area of landfill, forming a 4 m cut-off wall around the edge of the 1 m^3 clay cap. This required more than 20,000 m^3 of clay to be imported.

In order to achieve the minimum level of 2.4 m depth of uncontaminated soil needed for burials a further 45,000 m^3 of sub-soil were imported and laid above the clay-cap. This was followed by a 300 mm depth of topsoil.

The importation of material helped to determine the final profile of the site which changed from a bowl to a mound. The peripheral contours were tied into existing levels.

New fencing and a shelter belt of trees were placed around the perimeter of the site, a network of paths was formed and apart from a few shrub-beds, the remainder of the site was seeded.

A small garden has been created adjacent to the cemetery.

The works cost £733,000 – an average of £146,600 per hectare.

There are 2850 adult plots, (some double), and 2000 plots for children.

A reception building was constructed at a cost of £145,000.

6   FELLING RIVERSIDE

The remediation of a contaminated site at Felling Riverside, Gateshead, (not far from the International Stadium), at an estimated total cost of £7.5 million – an average of £300,000 per hectare – will be completed in 2003. The site covers an area of 25 ha.

Site investigations and archival studies showed that the site had been used, by several large manufacturing companies, to produce caustic soda, sulphuric and phosphate-based fertilisers. Unfortunately, when these plants closed in the 1920's, the decommissioning works only included limited removal of hazardous wastes within the operational site boundaries. The adjacent chemical tips remained and became the main source of soil contamination and leachate problems throughout the site.

Contamination is in the form of inorganic compounds such as calcium sulphate, calcium sulphite and calcium sulphide. Arsenic, cyanide and lead "hotspots" also occur indiscriminately throughout the site. Contaminated material is found very close to the surface which allows rainwater to easily percolate through contaminated material to form leachate which runs north towards the River Tyne.

Based on site investigation results, laboratory-scale studies were undertaken to provide site-specific information on remediation techniques to treat the leachate. The studies indicated that chemical and biological oxidation would only partially treat the leachate and that a filtration process would be required. Microfiltration techniques and final screening by reverse osmosis have been used.

For practical reasons the site was split into two areas. Phase 1, north of Tyne Main Road, and Phase 2 south of Tyne Main Road.

6.1 Phase 1 (7 ha)

In 1995, Phase 1 was regraded in order to remove a dangerously steep chemical face to the contaminated site and reprofile the chemical waste heaps. Works were then carried out to stop the infiltration of rainwater and subsequent leachate generation. The area was capped with clay and isolated from the surrounding land with a clay cut-off wall. A drainage network was installed on and around the waste tip in order to intercept surface water. The capping system was designed to minimise any impact on the current landscape and visual features, in particular, the Friars Goose Pumping Station, an important monument to industrial heritage.

6.2 Leachate treatment plant

In 1998, a leachate treatment plant was constructed. The treated leachate has been closely monitored to fine tune the process and ensure that the Environment Agency discharge consent is met.

A microfiltration plant removes most of the toxic metals. This is augmented by the reverse osmosis plant which reduces the concentration of arsenic in the final effluent to less than 0.05 mg/l.

The filtrate, (sludge), generated by the leachate treatment process is deposited in an engineered encapsulation area along with any excavated contaminated waste which cannot be retained in-situ.

6.3 Phase 2 (18 ha)

Site investigation showed contamination present in the majority of trial pits and boreholes excavated. Generally, the contaminated ground consisted of a blue/black dry consolidated layer underlain by a blue/black thixotropic slurry.

Laboratory analysis confirmed that the dry layer is calcium sulphate/sulphite and the slurry calcium carbonate and calcium sulphide. Hydrogen sulphide gas was also found. Elevated levels of arsenic, cyanide, sulphate and sulphide dictated that site remediation was necessary.

Based on archival studies, the natural clay profile of the site, thickness of contaminated material and perched water table, it is known that an elevated wagonway had been constructed around the central bowl area of the site to enable the even deposition of chemical waste. It is likely that the waste deposited in the "bowl" area is predominantly lime mud residue, from the sulphur recovery "Chance" process, with a significant calcium sulphide content. Lime mud was created by forming a slurry from the solid calcium sulphide residue of the Leblanc process. (Prior to the instigation of the "Chance" process in 1890, the removal of the solid residue of the Leblanc process resulted in large heaps of waste such as that found in Phase 1).

Taking in to consideration a hydrogeological study and other investigations, a design for the final major stages of the reclamation was drawn up. The works commenced in September 2000. Clay capping, a clay cut-off wall, and drainage works were constructed in order to eliminate the ingress of water and hence the creation of leachate. The leachate treatment plant will remain active until leachate creation has ceased.

Development platforms have been formed for sports facilities construction.

Landscape finishing works are programmed to be carried out during 2002/03.

The reclamation of this site will allow the creation of new sports facilities on what was once a potential hazard to the residents of Gateshead. These will consist of: 2 grass full size soccer pitches; 8 netball courts; 2 5-a-side pitches; Floodlit throws area (Hammer, Javelin, Discus); Car parking.

7 DUNSTON RIVERSIDE

A residential redevelopment site approx. 5.2 ha, with associated landscape works, riverside promenade and artwork.

The site has an industrial and coal mining past.

The "CWS Hide & Skin Works", a listed reinforced concrete building, was demolished in November 2000. (Government Office North East granted "Listed Building Consent" on 20.06.96 to allow the demolition to go ahead. Consent was conditional on a proposal for development being in place.)

Demolition rubble was subsequently crushed on site for re-use in the reclamation scheme. Reclamation works started in February 2001.

The reclamation consisted of a new piled river wall, a piled concrete slab over a backfilled lagoon, excavation and consolidation of suitable fill, installation of capillary breaks, removal of unsuitable material from site, drainage works, capping a mine shaft and generally making the ground suitable for housing construction.

The riverside housing scheme was designed by Ian Darby Partnership for Miller Homes. It contains 262 dwellings and includes, spread throughout the scheme, 30 units of "Social Housing". It includes a 5 storey block of luxury flats overlooking the River Tyne. Miller Homes started the first phase of house construction, 79 units, in June 2001. The second phase started in April 2002. 27 units had been built by November 2002, with a further 156 units to be built. Miller Homes have 5 years to complete the approved housing scheme, but due to the popularity of the scheme, the development is expected to be completed well before this.

This development has created housing and public open space. Works to the river edge will create a new riverside promenade. This will relate the housing to the river, allow public access and create a further link in the Keelman's Way, (an extensive cycle network). The reclamation works have been part funded by One North East, acting on behalf of English Partnerships.

8 COWEN ROAD

An old tip overlying peat deposits of varying depth, adjacent to a landfill site, which is still producing landfill gases. A "cut-off wall" with a gas control membrane was constructed to isolate the site from the adjacent landfill material. The ground was excavated and consolidated to provide a suitable platform for industrial units. The reclamation works took approximately 18 months.

A triangular portion of the site has been used to create a recycling and civic amenity site. The north east corner has become a salt store. The remainder of this 4.5 ha. site was split into 4 plots for industrial units and allowed the extension of an existing Business Centre. The project cost £1.8 million – an average of £400,000 per hectare.

9 GATESHEAD QUAYS

A total of 7 ha has had to be reclaimed for this development to take place. The development contains:-
The Baltic Centre for Contemporary Art; Gateshead Millennium Bridge. Work is currently progressing on The Sage, Gateshead, Housing developments, a Hotel development, Retail Developments. Further Housing and Hotel developments will be starting on site in the near future.

10 CONCLUSION

The schemes described here have achieved the reclamation of approximately 220 ha of derelict land. Other reclamation schemes carried out during the last 16 years bring the total area reclaimed within the Borough to more than 245 ha. In the process the Borough Council has amply demonstrated its ability to conceive, plan, programme and implement large scale reclamation projects of a highly technical and complex nature, working in partnership with consultants, contractors, developers and central government to achieve lasting and cost effective improvements to the Borough's environment.

Gateshead's reclamation and regeneration programmes illustrate the Council's commitment to residents and its' intention to attract businesses and people to the region. Land reclamation forms an integral part of this strategy.

If the above achievements were not enough, the Council still has its sights firmly set on future challenges. Further dereliction is still being created as traditional industries continue to contract and the implementation of the Contaminated Land Strategy is expected to identify sites in need of reclamation.

Among current initiatives, such as the regeneration of the Town Centre, Gateshead MBC is currently heavily involved in the NewcastleGateshead bid for Capital for Culture 2008.

Gateshead has a track record of doing more than just the basic essentials. The Council has continually made the most of existing opportunities and created them wherever possible.

The optimism and vision of Councillors, Council officers and residents coupled with support from Central Government and Europe has conspired with events to raise the profile of the region and achieve International status in different areas. Brendan Foster's World Record 3000 m run at Gateshead International Stadium in 1974, The National Garden Festival in 1990, The Gateshead Angel, erected in February 1998, (constructed on a reclamation site formerly known as "Team Colliery, Pit Head Baths), Jonathan Edwards" Olympic Gold Medal in Sydney in 2000, are all examples.

Cause and effect is always difficult to prove, but the stadium is currently undergoing refurbishment with a new track, and new sports facilities are to be built on the Felling Riverside site. Gateshead's commitment to its' Public Art Programme has culminated in the Gateshead Angel and the Baltic Centre for Contemporary Arts.

The main beneficiaries of this regeneration are the residents of and visitors to Gateshead. As regeneration continues these numbers are expected to increase. All of these people will be able to live, work, play, (and be buried), on a range of reclaimed sites.

Land Reclamation – Moore, Fox & Elliott (eds)
© *2003 Swets & Zeitlinger, Lisse, ISBN 90 5809 562 2*

Newlands – a strategic approach to derelict, underused and neglected land in the Community Forests of North West England

L.A. Dudley
The Mersey Forest Project Risley Moss, Ordnance Avenue, Birchwood, Warrington, UK

ABSTRACT: Newlands is a strategic approach to socio-economic regeneration of derelict, underused and neglected land in north west England. The project utilises a system for measuring public benefit where proposed community woodlands could be developed. The Newlands project will deliver 507 ha of community woodland on derelict, underused and neglected land, making a significant contribution to the successful delivery of the Community Forests.

1 INTRODUCTION

1.1 England's Community Forests

There are twelve Community Forest projects which were established in England during the early 1990s. The creation of these forest programmes commenced following central government research into the urban fringe environments located in and around towns, covering 450,000 ha and reaching 24 million people. The Countryside Agency and the Forestry Commission

Figure 1. The twelve Community Forests in England and location of Mersey and Red Rose Forests.

are national partners with a local partnership comprising of local government authorities.

The objectives across the twelve Community Forests are consistent in that they focus on the four main objectives of improving the landscape in highly populated and industrial areas, rejuvenating urban areas, providing an environment in which socio-economic renaissance can take place and increasing recreational facilities for all. There is active community participation, increased access, recreational and health benefits, giving new education opportunities, nature conservation benefits and a better place to work and live. (Mersey Forest 2001 and Red Rose 1994).

1.2 The Community Forests and derelict land.

The two Community Forests in the north west of England are situated where the Nineteenth Century Industrial Revolution took place; such pioneering industry was unaware of its polluting legacy, which was largely ignored until the 1970s. Increasing industrial and social decline during recent times, epitomized by the Toxteth riots in Liverpool during July 1981, reflected a need to link social, economic and environmental benefits. During the late 1980s work by the Countryside Commission (later the Countryside Agency) and the Forestry Commission combined socio-economic and environmental issues with a changing attitude towards British forestry. Upland coniferous afforestation was being viewed negatively and so community forests were proposed in urban areas (Smith 1997). The Community Forests have a role to play in drawing together economics, people and enhancing habitats. During the early development of the Community Forests the legacy of derelict land was seen as an important opportunity in achieving all aspects of sustainable development. This was reflected in the individual Community Forest plan documents across England.

Pilot projects for derelict land remediation developed from approval of the Community Forest programmes. Initially it was considered that there was sufficient research to assist in woodland creation on most land types, particularly from the Forestry Commission research establishments (Dobson and Moffat 1995), however the valid science based approach was not ideal for the practitioner and led to a lack of woodland creation on landfill sites (MacKay and Hesketh 1998). Consequently, The Mersey and Red Rose Forests established specialist working groups for landfill and brownfield sites. The landfill group published guidance for a staged pathway leading to community woodland on landfill sites (Nolan 1998). Other professionals had identified the need to improve the image of post-industrial England at a national level and incorporate such proposals into the urban regeneration sector (Handley and Perry 1998; MacGillivray 1998). Reclamation of brownfield land, those parcels of land that are derelict, underused, neglected or contaminated, was approached in a similar way to closed landfill sites, because research work already carried out was only weakly linked to the work of the practitioner. The brownfield group have not published guidelines, however their work has led to a socio-economic approach to tackling the problems of brownfield land, (known as Public Benefit Recording System; PBRS (*infra)*). The approach has subsequently attracted a significant amount of investment funding to commence the Newlands project. In parallel with the Community Forest led landfill and brownfield projects, The Mersey Forest and Red Rose Forest Projects developed mechanisms for the third party acquisition of land by other partners (at that time the Community Forests had no mechanism to own land). These three elements combined to support a bid with the Forestry Commission to UK Governments' Capital Modernisation Fund for woodland creation projects preferably on derelict, under-utilised and neglected (DUN) land.

1.3 Capital Modernisation Fund in the Community Forests

The three-year Capital Modernisation Fund (CMF) was not strictly focused on DUN land. The CMF project did pilot Forest Enterprise's approach to land acquisition, management and public consultation in urban and peri-urban areas of three Community Forests (Thames Chase, Mersey Forest and Red Rose Forest). The project attracted £5.8 million from central Government and achieved match funding of £3.3 million (Oliver 2002). The outputs from the project include 1,016 ha of land acquired for community woodland end use, 27 communities involved and 40,640 m of access created.

Although the CMF project was successful in modernising the Forest Enterprise approach to community forestry and the socio-economic agenda, the CMF strategy did not have a mechanism of targeting areas where community forestry could deliver maximum public benefit. As CMF approached completion a new strategy was called for that delivered community woodland outputs while more keenly addressing socio-economic aspects. This new programme is called Newlands, derived from New Environments via Woodlands it builds upon the success of CMF and has established a close working relationship with the North West Development Agency (NWDA).

1.4 Development agenda and Community Forestry

The NWDA strategy was particularly forward looking in recognising the value of the environment and its place amongst business investment and retention, and communities (NWDA 2000). It was from this strategy that the Mersey Belt, encompassing Liverpool, St Helens, Warrington and Greater Manchester was identified as a regeneration priority for urban renaissance. The strategy identified the importance of the environment and landscape in projecting a positive image and restoring the environmental deficit that the north west has experienced for several generations. The strategy also sought to address the issues of dereliction as a central task and proposed increased resources to deal with a problem that was identified as long term (NWDA 2000). As part of the implementation of the North West Regional strategy, the Forestry Commission and Community Forests Partnerships were encouraged to bid for investment funding for the Newlands project.

1.5 Newlands inception

A number of elements coalesced to form the Newlands project, two strong partnership led Community Forest projects, an important strategic axis of the Mersey Belt, brownfield and landfill research, an acquisition mechanism, the successful CMF project and a new mutually beneficial relationship with the Regional Development Agency.

2 NEWLANDS PROCESS

2.1 Newlands and Public Benefit Recording System (PBRS)

The Newlands project is based around the delivery of community woodlands that deliver the maximum public benefit to the north west region of England. It has two strategic selection criteria; the PBRS, which evaluates a socio-economic gain and the spatial analysis of DUN sites and major transport corridors (road, rail and canal) and the positive visual improvement that would result from the creation of a community woodland.

PBRS was a product of the Community Forests Brownfield Project Working Group and the Derelict Underused and Neglected Land Group. Research commissioned by the groups had indicated that there were insufficient resources to remediate most derelict, underused and neglected (DUN) land. It was therefore necessary to prioritise DUN land remediation. As the research concentrated on the Community Forests the objective lay in the delivery of maximum public benefit through habitat creation. This objective also achieved the priorities of the England Forestry Strategy (Forestry Commission, 1998). Consultants were appointed to devise a methodology that could rigorously assess DUN land in the Mersey Belt for its potential to deliver benefits to the public if community woodland habitat was created. The categories of DUN land were defined as:

- Derelict: "land so damaged by industrial and other development that it is incapable of beneficial use without treatment".
- Underused: "land that is periodically or intermittently managed and there is little evidence of consistent use either by the public or as agricultural land".
- Neglected: "Land that is uncared for, unmanaged and often subject to fly tipping yet may be in active use by the public".

The above definitions are UK Government recognized definitions. (DOE 1995)

The first task for the consultants was to identify the extent of DUN land, which was carried out by surveying aerial photographs and desk top surveys which revealed approximately 16,000 ha of DUN land within 19 local authority areas within or adjacent to the Mersey Belt. Following the identification of sites it was necessary to consult with partners and landowners to filter inappropriate sites that were either wrongly identified as DUN land, already developed, had development plans, were less than 1 ha in area or had other constraints. This filter reduced the original 1,471 DUN sites to 743 covering 8,904 ha. (TEP 2001). The filtered sites were mapped as polygons on a geographic information system and recorded on a linked database. The filtered DUN sites were then assessed against the PBRS criteria, this scoring identified that the majority of sites that could deliver high public benefit if developed as community woodlands were greater than 30 ha in area further implying that the physical impact of DUN land is extensive.

2.2 *PBRS methodology*

There are four independently scored parameters within the PBRS assessment.

- Social benefit
- Public access
- Economic benefits
- Environmental benefits.

2.3 *Social benefit*

The DETR data based Index of Multiple Deprivation had previously been used by the Community Forests for planning and monitoring purposes. The index scores income, health, employment, education, housing, access and child poverty, giving a robust method of identifying priority areas for socio-economic projects at a sub-local government level (ward level). This is one of the main existing measures of social deprivation used in the UK which when used in the DUN survey highlighted the strong relationship between social deprivation and the location of DUN land.

There are four other measures that assess social benefits. Firstly, there is the relationship between population density and public open space. Normally, high population densities tend to lack adequate recreational provision, thus the opportunity for creating new community woodlands would potentially benefit many residents and may be more appreciated if packaged with other socio-economic improvements. The data for PBRS was extrapolated on the basis of housing density around a 500 m buffer of a potential site. (TEP 2001). Secondly, site size was considered where large sites were rated higher because of their visual and physical impact for users and in contributing greater delivery in terms of hectares of land improved. Thirdly, the sites' linkages to other local or national initiatives in deprived areas including Health Action Zones, New Deal for Communities, Education Action and Employment Zones were considered important. Creation of new community woodlands can achieve outputs and outcomes for these initiatives. Finally, there was a need to identify and measure the educative role community woodland establishment could have if schools near to identified DUN land were involved in their development, here a 1 km site perimeter buffer was used to identify such schools.

2.4 *Public access*

The UK Ordnance Survey maps indicate the location of a Public Rights of Way (PROW). DUN sites that incorporated a PROW were scored highly as the possibility of providing access on the whole site is therefore easily achieved. The spatial relationship between DUN sites, primary roads, car parks, cycleways and bridlepaths was also recorded. Assessments were made of *de facto* access as these clearly illustrate a demand to use the site whether it is negative or otherwise. Research into public transport linkages was also applied as within most deprived areas car ownership is not as high as in more affluent areas. DUN sites with open public access could provide new opportunities for people to travel to school, work, shop or recreational facilities. This measure also supports policies that seek to reduce private car use and the associated pollution. The location of other public open space was examined and scored as a measure of demand, for example little public open space provision scored highly as the potential new community woodland would provide public benefit.

2.5 *Economic benefits*

The Community Forests have historically recognised that landscape enhancement can have a number of economic effects, not least of these is the creation of an attractive landscape to enhance potential inward investment and retain existing business. The lack of attractive landscape in the Mersey Belt also drives away successful, skilled employees (NWDA, 2000). Furthermore the skilful design of woodlands and other habitats can screen visually blighted areas or separate land uses. Economic regeneration measures around a DUN site were assessed, including the proximity to planned and existing business parks, retail and industrial areas. House price data was measured and compared to national figures.

2.6 *Environmental benefits*

Some may consider the environmental benefits in converting DUN land into community woodland as self evident, however the PBRS process assessed the land adjacent to the site and scored points based on the proximity of ancient semi-natural woodland and other woodland as separately scored attributes. DUN sites that linked these existing assets scored highly. The same approach was applied to identified areas of ecological interest and water bodies outside the DUN site, if there was existing ecological value on a DUN site then that site would not be scored highly as there would be no value in converting the site to community woodland. Sites inside Air Quality Management Areas (AQMAs) identified by local authorities would score highly as the presence of new woodland would improve air quality in these areas by removing particulates and airbourne contaminants, because of similar air quality issues along transport corridors DUN sites adjacent to major roads scored highly.

2.7 *County and regional plans*

The PBRS process sites were assessed in relation to the local authority unitary development plans, policies and strategies. Along with landowners, Community Forests and the Forestry Commission, the local authority plans are important mechanism for the identification of potential for DUN land and whether it is to be converted to community woodland end use or not.

3 NEWLANDS IMPLEMENTATION

3.1 *Project mechanisms*

The targeted approach described (*supra*) is designed to meet socio-economic and environment agendas while operating at a strategic level, initially in the sub regional area known as the Mersey Belt, and later across the whole of the north west region. The Newlands vision is "to improve the regions working and living environment and to make the north west a more exciting and viable choice for investment opportunities" (Forestry Commission 2002)

NWDA supports the programme based on its vision for a step change in the way DUN land is remediated and the value and strength of sub-regional partnerships. The Forestry Commission will be the managing agent for the regional allocation of NWDA funds, while the Forest Enterprise, Community Forests and landowners would be the main delivery agents, ensuring local partnership implementation. The outputs will be:

– The re-use of brownfield land.
– Improving the image of the region.
– Encouraging of inward investment.
– Creating or improving community, tourism and recreational facilities.
– Increasing woodland cover in one of the least wooded regions in England.
– Improving the quality of life.

The vision outlined above will be realised through the existing Community Forest partnership working towards county delivery plans that have a primary theme of improving gateways and transport corridors, have their own strategic county level objectives and have identified potential sites that have scored significantly in the PBRS research. Newlands is initially to take place over five years with the possibility of running beyond 2008 for a further five years. As with many projects there is a regional advisory board that will steer the projects and a partner working group that will ensure a high standard of project delivery and community engagement. This steering group will also monitor and report to fund managers and grant giving organizations.

The mechanism for a Newlands site is sequential. Sites are identified and assessed against the primary theme (transport corridors), ranked using PBRS, assessed against funding opportunities and achievability, surveyed, possibly acquired by the Forest Enterprise, designed with public consultation and implemented. At any point sites may become inappropriate for the programme, however this does not necessarily mean that the site should then remain derelict, underused or neglected. The Community Forest partnerships would then seek alternative mechanisms to improve such sites in the medium or long term, or when resources become available.

3.2 Project implementation

At implementation, the Newlands programme will require resources and consequently two project officers, one for each Community Forest area, will be employed to project manage the programme. Their role will commence with identifying the sites, managing surveys, acquisition processes and ensuring full and complete implementation, identify and secure matching funds and community consultation with the landowner. It is expected that initially the Community Forests Project Officers would be involved with community consultation and engagement, but it is proposed that this responsibility will pass to a community development officer in the long term.

Once a site has been identified as a Newlands site, the Forestry Commission's Land Regeneration Unit, with an expertise in land ownership and brownfield remediation, will assess the site for both potential acquisition and capacity to sustain woodland and other habitat work. The surveys are combined desk-top and site based that review above and below ground conditions, including physical and chemical attributes. Other assessments are made of ecological, archaeological, landscape planning, recreational and community attributes. The surveys produced are highly detailed to reduce both liabilities but also to fully inform the design and implementation processes. The acquisition of land is not an essential part of the programme, land may be retained by the owner, leased or purchased by the Forest Enterprise.

The nature of these sites implies that there will be liabilities associated with the land. The acquisition process would only acquire land on a freehold basis where liabilities are limited. Leasehold land would enter into a "pie-crust" lease where only the first few vertical centimetres (perhaps only as deep as the ploughing depth) would be leased, therefore the majority of liability would remain with the present landowner. At the time of writing NWDA have awarded the pre-programme expenditure at £486,477 to commence the surveying of sites initially identified by the Community Forest Partnership.

As well as acting as the managing agent, the Forestry Commission's role in the programme will also assure the quality of the project outputs, this being achieved through the existing role of Forestry Commission Woodland Officers. Designs and proposals will be submitted by the project partnership to the Newlands regional project officer who will lead the appraisal and assessment process. The woodland design plans will be further assessed by the Forestry Commission local Woodland Officer based on existing Woodland Grant Scheme and England Forest Strategy criteria. The Community Forest Project Officer and Woodland Officer will work together to ensure full professional consultation with local authorities, statutory and non-statutory bodies to reduce the possibility of conflicting interests.

3.3 Financial management

The funding of projects will be linked to outputs and outcomes and as the work continues income and expenditure will run chronologically for an individual site. The funding will be released by two accountable bodies, the Forestry Commission and NWDA to the site developers, Community Forest Trusts or other delivery agents. The five phases are described separately below.

3.3.1 Site appraisal, investigation and planning

Most of these costs are accrued by the Forestry Commission Land Regeneration Unit including site surveys and negotiations on land ownership; some costs may rest with landowners, and Community Forest Teams in developing initial community liaison and outline designs. Each site is treated as an individual project and will follow an appraisal process, for which the Forestry Commission will hold the responsibility. The process for appraisal will differ based on anticipated project costs. Projects costing less than £250,000 will be championed by the local Community Forest project officer, appraised by the Forestry Commission project officer and decided upon by a Mersey Belt sub-group of the steering group. If the cost is between £250,000 and £1 million, then the NWDA will make the final approval decision and if it is above £1 million then the NWDA will seek increased levels of appraisal at an early stage in the project.

3.3.2 Acquisition

This brief phase covers the actual costs and professional fees for acquiring land where a transfer of ownership has taken place.

3.3.3 Site restoration and ground works

This phase covers all aspects of remediation and construction for the site, including soil moving, importation, ripping, fencing, furniture and other infrastructure, such as ponds, and will also include professional fees where applicable. The Forestry Commission is developing a computer program (ROOTS) that will produce specifications for site restoration to community woodland based on current best practice. This program will be used to produce specifications for each site to be restored under the Newlands program.

3.3.4 Establishment
This item refers directly to the woodland and other habitat work and would include items such as procurement of plants, planting, grass seeding, mowing and weeding.

3.3.5 Long-term management and endowment
This part of the funding is unusual in that it is being paid for directly from the regional development agency rather than the accountable body, Forestry Commission. The fund will pay for ongoing management of the site as part of a committed contract between the developer and NWDA. (Forestry Commission, 2002).

3.4 Endowments
Most DUN land has been in a poor condition for a long period of time. It is therefore recognised amongst the Newlands partnership that to ensure a high level of public benefit long term secure funding is required. Newlands seeks to achieve this by delegating funds to the Community Forest Trusts who would treat the fund as an endowment that will secure long term public benefit by maintaining the site. The developer will draw these funds on completion of management work. At the time of writing, this proposal is unapproved and if this remains the case issues will be raised about the sustainability of public benefit through long-term operations such as the creation of community woodlands. The Community Forest Partnership will continue to seek opportunities for secured long term funding for brownfield and other sites.

3.5 Newlands output, outcomes and impact on derelict land
The combined socio-economic and environmental role of Newlands could create an array of benefits, some simple to measure and others not. A summary of the quantitative outputs is included in Table 1.

In The Mersey Forest Plan the partnership seeks to create 250 ha of community woodland on DUN land over the next five years (there are no specific targets for Red Rose Forest). The Newlands project commits to remediating 507 ha of DUN land in the Mersey Belt over 5 years. The DUN land survey identified 8,904 ha of sites for potential community woodland end use. In five years Newlands will remediate 6% of the brownfield legacy, at this rate of amelioration and assuming no more DUN land is created it will take over 80 years to address the problem and cost in the region of £400 million. The Community Forests Partnership, which includes the Forestry Commission, remains committed to the remediation of brownfield sites and Newlands is an important and significant contribution to ameliorating the blight of DUN land.

4 CONCLUSIONS

4.1 Findings and future development
The project is scheduled for full approval and implementation in April 2003. Newlands is an invaluable project in the Mersey Belt. It has identified DUN land and sought positive mechanisms and partnerships to make a step change in our approach to DUN land. It has proved the value of partnership and has secured a realistic level of funding to tackle the problem. The delivery of the project can only be assessed in the future but the commitment of partners suggests that the project will succeed in improving significant areas of derelict land.

Development areas for the future are described below.

4.2 Outstanding DUN land
The Mersey Forest Project is a 30-year project, while Red Rose will run for 40 years. Within these programmes there are challenging targets in terms of DUN land remediation, as described above. However successful Newlands is there will be sites that will not pass the appraisal process. The Community Forests cannot dismiss these sites and must seek other mechanisms to remediate the most difficult (and therefore

Table 1. Newlands outputs.

Core outputs	Total	Manchester	Merseyside	Cheshire
DUN land reclaimed (hectares)	507	245	137	125
Footpath & cycleway new and restored (km)	20.3	9.8	5.5	5.0
Jobs safeguarded or created	26	–	–	–

(Shiletto, 2002).

most expensive) sites. The Community Forest Partnership must continue to seek methods of reducing implementation costs, maintaining design and implementation standards and seeking additional funding.

4.3 Endowments and long term management

Based on the predicted Newlands expenditure per hectare of £43,453 there is at least a £400 million industrial legacy from our previous generations, underlining the modern thinking behind sustainability. This and future generations are paying for a lack of regard for the environment in these costs. Impressive though the £19.8 million Newlands project is, it does only solve part of the problem. The costs of remediating DUN land require serious consideration as to where future and matching funding may come from and how funds will be managed. It is anticipated that as inward investment occurs opportunities for amelioration of DUN land will take place and the creation of new DUN land will cease to happen.

4.4 PBRS and socio-economic improvement

Community Forests are about society, people and the environments in which they live and work. It could be argued that the habitat work is a by-product of the socio-economic outputs Community Forests strive towards. PBRS gives a good indication of the benefits of a proposed community woodland, which is a valuable starting point. The monitoring of accrued benefits will be a test of Newlands and development of PBRS to assess these changes would be a supporting tool in emphasizing the value of woodlands in an urban or industrial setting.

ACKNOWLEDGEMENTS

The partnership approach

Newlands is a project built around the many skills of both individuals and organisations. This paper would be incomplete without the contributions made by Keith Jones (Forestry Commission), Paul Nolan (The Mersey Forest Project), Francis Hesketh (The Environment Partnership), Martin Reynolds (NWDA), Chris Robinson, Mark Street (Forestry Commission Land Regeneration Unit) Chris Waterfield (Red Rose Forest, now with Forestry Commission), Tony Hothersal (Red Rose Forest) and Tina Shiletto (Forestry Commission).

REFERENCES

Department of the Environment 1995. *Survey of derelict land in England 1993 Volumes 1 & 2*. London: HMSO.

Dobson M.C. & Moffat A.J. 1995. Site capability assessment for woodland creation on landfill, *Research Information Note 263*. Edinburgh: Forestry Commission.

Forestry Commission 1998. *England forestry strategy – a new focus for England's woodlands*. Cambridge: Forestry Commission.

Forestry Commission 2002. *Newlands, new environments via woodlands – regional business plan*. Delamere: Forestry Commission. Unpubl.

Handley J.F. & Perry D. 1998. Woodland expansion on damaged land: reviewing the potential. *Quarterly Journal of Forestry* 92(4): 297–306.

MacGillivray S. 1998. Woodland expansion on damaged land – the Scottish experience. *Quarterly Journal of Forestry* 92(4): 307–317.

MacKay J.M. & Hesketh F.B. 1998. The Mersey Forest and Red Rose Forest landfill woodlands project, *Land reclamation: achieving sustainable benefits*: 65–71. Fox, Moore & McIntosh (eds). Rotterdam: Balkema.

Mersey Forest 2001. *The Mersey Forest Plan August 2001* Warrington: The Mersey Forest Project.

Nolan P.A.J. 1999. *Creating community woodlands on closed landfill sites*. Warrington: The Mersey Forest Project.

NWDA 2000. *England's North West – a strategy towards 2020*. Warrington: NWDA.

Oliver T. 2002. Pers. Comm.

Red Rose 1994. *Forest Plan August 1994*. Salford: Red Rose Forest.

Shiletto T. 2002. Pers. Comm.

Smith C. 1997. Community Forests – the first years. *Quarterly Journal of Forestry* 91(1): 21–26.

TEP 2002. *A report summarising the establishment of a derelict, underused and neglected land database and the application of a public benefit scoring system*. Warrington: The Environment Partnership (commissioned by the Forestry Commission). Unpubl.

Land Reclamation – Moore, Fox & Elliott (eds)
© 2003 Swets & Zeitlinger, Lisse, ISBN 90 5809 562 2

Contaminated land – land affected by contaminants; an approach for Building Regulations

Michael Johnson
Building Regulations Division, Office of the Deputy Prime Minister, London, UK

Appropriate development of previously used land is a key strategy in current planning and land use objectives. If these policies are to work there is a need for co-ordinated regulatory frameworks and consistent guidance. Such an approach improves the confidence and competence of practitioners and gives some assurance to future occupiers.

Different regulations have distinct purposes but where they impinge on each other they should be able to produce a satisfactory outcome overall. This is the reason for the differences in terminology between regulations made under Part IIA of the Environmental Protection Act 1990 (EPA) and the Building Regulations. However, the final result, for both policies, is the satisfactory use of previously used or damaged land.

Part IIA of the EPA is concerned with the identification and remediation of contaminated land. It defines Contaminated land as:

"Any land which appears to the local authority in whose area the land is situated to be in such a condition, by reason of substances in, on or under the land, that

– significant harm is being caused or there is a significant risk of such harm being caused; or
– pollution of controlled waters is being, or is likely to be caused."

The Building Regulations are enable by section 1 of the Building Act 1984. This allows regulations to be made for the purposes of

– securing the health, safety, welfare and convenience of persons in or about buildings and of others who may be affected by buildings or matters connected with buildings.
– furthering the conservation of fuel and power, and
– preventing the waste, undue consumption, misuse or contamination of water. Sub-section (c) is not used for making Water Regulations under the Building Act, they are made under powers in the Water Industry Act 1991.

Contamination is dealt with in Part C of the Building Regulations. The requirement and guidance is based on ensuring health and safety.

Building Regulations and the construction process are the latter stages of development activity and convert plans into a reality through remedial and building activities. This is part of the reason of using the term "land affected by contaminants" rather than "contaminated land." Land affected by contaminants is a more positive statement. It indicates that contamination is an aspect of the land not its intrinsic nature and that, following remediation, contamination is something from the past, not a continuing burden.

The current requirement in Part C, "Site preparation and resistance to moisture of the Building Regulations" dates from 1992. The requirement is headed "Dangerous and offensive substances" and can be linked to section 29, "Site containing offensive material of the Building Act" 1984 (Building Act 1984). This mentions faecal matter or offensive anima or vegetable matter. It was carried forward from the Public Health Act 1936 and is more related to building over privies and stables than dealing with industrial contamination. There has been a comprehensive review of Part C. Contamination has been the major topic; this is reflected in the new title of the Approved Document, "Site preparation and resistance to contaminants and moisture."

The new guidance for the Building Regulations goes a long way in complementing the aims of Part IIA of the EPA (1990) in that it uses the same risk assessment process and remediation. However, there may be some situations where the levels of remediation may not quite match, as the Building Regulations only require work to be done for the sake of health and safety of people in about buildings rather than the environment generally. This will cover human needs and the necessity to protect the building and its services from the possible effects of contaminants. However, the guidance in the Approved Document reminds developers and designers of the need to keep planners, environmental health officers and other regulators informed of their actions.

Although Building Regulations and environmental protection legislation have different purposes they aim to bring a common outcome, the safe re-use of potentially damaged land. In reaching such an outcome they share similar preparatory processes and treatment techniques.

These processes for dealing with land affected by contamination follow the same logic as analysing any problem:

Is there a problem?	Hazard identification
Will the problem have any adverse effects?	Hazard assessment
How do those problems manifest themselves? i.e. What is the likelihood of the hazard/problem occurring and achieving its potential?	Risk estimation
What is the significance of the hazard and the estimated risk?	Risk evaluation

These stages are set out in greater detail in the Environment Agency/NHBC document (EA/NHBC Pub. No 66) on housing on land affected by contamination.

– The general approach of risk assessment is dependent on the possibility that the hazard can reach and affect a vulnerable subject. When dealing with contaminants the process for determining whether harm may occur is founded on the concept of "source-pathway-receptor" relationship or pollutant linkage. This is illustrated by the conceptual model in Figure 1. When land affected by contaminants is developed, receptors (i.e. buildings, building materials and building services, as well as people) are brought on to the site and so it is necessary to break the pollution linkages (DETR 02/2000). This can be achieved by:
– treating the contaminant (e.g. use of physical, chemical or biological processes to eliminate or reduce the contaminant's toxicity or harmful properties),
– blocking or removing the pathway (e.g. isolating the contaminant beneath protective layers or installing barriers to prevent migration),

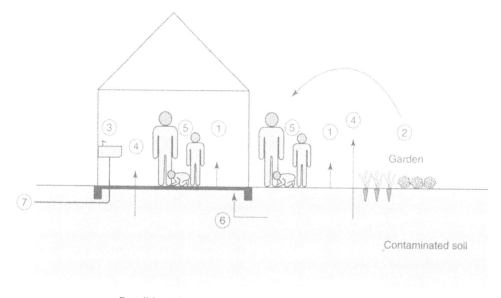

Possible pathways

Ingestion: of contaminants in soil/dust ①
of contaminants in food ②
of contaminants in water ③
Inhalation: of contaminants in soil particles/dust/vapours ④

Direct contact: with contaminants in soil/dust or water ⑤

Attack on building structures ⑥

Attack on services ⑦

Figure 1. Example of a conceptual model for a site showing source-pathway-receptor.

- protecting or removing the receptor (e.g. changing the form or layout of the development, using appropriately designed building materials), or
- removing the contaminants (e.g. excavating contaminated material).

The earliest stages of site assessment are the desk study and walk over survey. The desk study involves finding out as much as possible about the history of the site from documentary sources including maps. Old series geological maps are particularly useful as they were usually printed on a 1910 series one inch to a mile base. It is often possible to pick up details such as "works" or "mine"; these can be valuable sources of earlier uses of the site. The walk over survey involves visiting the site and seeing the lie of the land. There are some useful checklists in the NHBC guide on Land Quality (NHBC 2000). Even street and place names can provide evidence of previous use; "Lead Mill Lane" should start alarm bells ringing.

The preliminary site assessment will usually enable potential hazards on the site to be identified. This can be linked to the outline development proposals and the general conceptual model to produce a site-specific model showing the potential linkages. The next stage is to determine if these findings are a reasonable representation of the hazards in order to decide what further investigation is required. These processes are common to assessment prior to development or to determine how to deal with pollutants that could affect the environment. Check lists can be useful for site evaluation and hazard assessment (CIRIA 2001).

Previous uses of the site may give indications of substances that may be found on the site. These were listed in early documents such as the "Industry profiles" (DOE 1996). There is now more comprehensive guidance in the CLR 8 (DEFRA/EA 2002). Some aspects of physical appearance of the site such as the colour of the soil or surface irregularities. In order to obtain the extent and severity of the contamination a geo-environmental site investigation is almost certainly required. Standards are now available to facilitate a structured approach (AGGS, BSI 2001). It is usually necessary to inform regulatory authorities prior to any intrusive investigation.

The findings of the risk assessment process will determine the remedial actions necessary to break pathways so that contaminants cannot reach identified receptors and cause harm. The degree of remediation may be influenced by the proposed or actual uses within the site. In the case of retail and industrial sites there may be large parking areas. Providing there is not any significant ground water movement it may be possible to have a lower level of treatment if there is a high degree of confidence that the contaminants will be permanently covered by the paving.

For Building Regulations the earlier presumption that remediation is only necessary for the ground beneath the footprint of the building has been modified (ODPM 2002). It is more likely that, for building control as well as environment protection reasons, the building and all its associated land will need some form of treatment. This should deal with the need to protect persons in and around buildings completely from contaminants on a development site. There is also a proposal regarding the provision of protective and remedial measures when buildings are converted to any use that may include sleeping accommodation. This does not exempt other activities from a duty to protect existing buildings when a change of use occurs it simply excludes action under the Building Regulations. Any new occupation as a work place will tend to be covered by general health and safety legislation or environmental controls.

Building Regulations and the contaminated land regime under Part IIA of the Environmental Protection Act 1991 (CLR 2000) follow similar process for determining the risk to receptors from contaminants. The difference lies in detail and the nature and level of remediation. Building Regulations can only be concerned with the building, its occupiers and services. The scope now includes the land associated with the buildings as there can be no long term certainty of what occupiers may do to land close to their buildings. The CLR reports 7, 8, 9 and 10 are now a key elements in assessing land quality and the remedial action for both development and environmental protection.

Building development can also use the same guidance as that relating to environmental protection for determining remediation processes. Some of the key guidance was in the CIRIA special report series in the mid-1990's. In the interest of sustainability more prominence is being given to treatment processes. Land treatments reduce the burden on landfill sites and the disruption associated with haulage of contaminated material. Better processes and monitoring systems have significantly improved confidence in treatment systems. Increased demands for site improvement have drive demands for appropriate remedial processes. CIRIA has produced a guide summarising the processes available (CIRIA 2001).

The choice of remediation process for Building Regulations needs to include robustness as well as effectiveness. Gas protective systems that use monitoring systems, sometimes linked to powered ventilation systems can be used on commercial developments where the landlord has an on-site building management team. However, systems that require on going maintenance and management are unlikely to be suitable for owner occupied housing sites due to the risks of interference or disrepair. Similarly the risk of damage by occupiers should be carefully considered if barrier systems are a possible means of containment for contaminants.

The range of assessment processes and treatment systems have been described at length in many guidance documents. However, the most important piece of advice in such publications is that chance reading of several well-respected technical papers does not make the reader an instant expert. Most developments will need advice from a geo-environmental specialist. The role of such specialists will not cease at tender stage. Most remediation work will need on site surveillance and often post-completion monitoring. Complex schemes will usually need a completion report, often supported by laboratory reports.

From identification to new use, the development process, including building control can use the same practices as control of pollution and limitation of harm as Part IIA. Many of the procedures are extensions of good practice for the construction process but with some detail added. Most work needs a site investigation and any divergence from the norm requires a risk assessment and evaluation of the problem. Whether it be the use of similar processes such as excavation and use of barriers or the added value from a more profitable land use development is one of the most important ways of paying for and achieving site clean up. Well structured guidance for building regulations or other land use processes facilitates the process of getting damaged land back into a use that serve the community.

REFERENCES

BS 10175:2001, Investigation of potentially contaminated land. Code of practice. BSI 2001, London.

Building Act 1984, 1984 – Chapter 55, UK Government, 1984 HMSO, London.

Building Regulations 2000, Proposals for amending Part C – Site preparation and resistance to contamination and moisture, A consultation paper, Office of the Deputy Prime Minister (ODPM), 2002, London.

Chapter 4.1, Land quality, NHBC standards, 2000, Amersham

CIRIA report C549, Remedial processes for contaminated land, Principle and practice, CIRIA 2001, London.

CIRIA report C552, Contaminated land risk assessment, A guide to good practice, CIRIA, 2001, London.

Contaminated Land Research Report, CLR 8 Priority contaminants for the assessment of land DEFRA/Environment Agency, 2002, TSO, London.

Contaminated Land (England) Regulations 2000, 2000 SI No 227, & Statutory Guidance, DETR, 2000, TSO, London

Department of the Environment, Industry Profiles, 1996.

DETR Circular 02/2000: Environmental Protection Act 1990, 2000, TSO London.

Guide for combined geo-environmental and geotechnical investigations, Association of Geotechnical and Geo-environmental Specialists (AGGS).

Guidance for the safe development of housing on contaminated land Environment Agency/NHBC R & D Publication 66, 2000, TSO, London.

Land Reclamation – Moore, Fox & Elliott (eds)
© 2003 Swets & Zeitlinger, Lisse, ISBN 90 5809 562 2

Contaminated land reports (CLR 7–10, CLEA): applications and implications

S. Ruzicka
Waterman Environmental, London, UK

ABSTRACT: In the UK a new framework for the assessment of the risk to human health from land contamination was published in March 2002 by the Department for Environment, Food and Rural Affairs and the Environment Agency. The main purpose of the framework is to provide technical material that can be used to support statutory regimes addressing land contamination, particularly Part IIA of the Environmental Protection Act, and assist development control under the Town and Country Planning Acts. This paper summarises the key issues arising from the implementation of the framework and presents a case study of its application for small residential properties. The case study emphasises the need for a comprehensive design of the investigation, higher level of chemical testing frequency and more detailed data analysis when compared to previous best practice common in the UK industry.

1 INTRODUCTION

In March 2002 the Department for Environment, Food and Rural Affairs (DEFRA) and the Scottish Executive, together with the Environment Agency (EA) and the Scottish Environment Protection Agency (SEPA), published a series of Contaminated Land Reports (CLR 7–10), which provide a framework for the assessment of risk to human health from land contamination (DEFRA 2002a, b, c, d). A further report, CLR 11, setting out model procedures, is yet to be published.

The main purpose of the CLR framework is to provide technical material that can be used to support UK statutory regimes addressing land contamination, particularly Part IIA of the Environmental Protection Act, and assist development control under the Town and Country Planning Acts. Currently, the framework does not extend to the risks to controlled waters.

The framework utilises a Contaminated Land Exposure Assessment (CLEA) model and proposes new generic screening levels for surface soil contamination. Where applicable, these generic screening levels or Soil Guideline Values (SGVs) take precedence over other existing guidelines on human health risk including the ICRCL values which have been widely used in the UK industry over the past 15 years. Currently, only seven SGVs have been published. These apply to only three standard scenarios of land-use and site conditions. A rolling program should see as many as 40 SGVs published by 2004 with the possibility of further standard land-uses being addressed. The standard land-uses include residential, allotments and commercial land-uses. Currently, the published SGVs can not be applied to alternative land-uses (e.g. schools, parks, playing areas) where bespoke modelling is necessary to address the risks to human health.

This paper uses a case study to demonstrate the application of the guidelines in the real environment.

2 CASE STUDY

The case study comprises nineteen residential properties which form a part of a housing estate in southern England. The estate was built in the late 1960s on land which was formerly occupied by a large industrial facility. The extent of the remedial works carried out during the development of the estate was not known, nor was the origin of the materials used for the construction of the gardens in the residential properties, which form the most sensitive land-use on the estate. Consequently, a study was commissioned to carry out an assessment of potential human health risks arising from the use of the properties.

The example given below will concentrate on substances currently covered by the CLR guidelines although, naturally, other potential contaminants were tested and assessed. Throughout the case study reference is made to various principles and terminology used in the CLR guidelines. As it is beyond the scope of this paper to provide detail explanation the reader is encouraged to make reference to the CLR documents which are freely available from DEFRA website (www.defra.gov.uk/environment).

2.1 Methods

In accordance with the CLR guidelines the assessment had to ensure that all relevant contaminants had been identified, the appropriate conceptual model was validated and that the concentrations and prevalence of contaminants had been estimated to the extent that they allow a robust comparison with soil guideline values (SGVs). This was achieved by selecting the appropriate sampling design, fieldwork methodology and chemical analysis of the samples.

The risk assessment methodology is detailed in Section 2.4.

2.1.1 Sampling design

The objective of the field investigation was to estimate the distribution of the concentrations of potential contaminants in the surface soils in the subject gardens. In order to achieve an accurate estimate, a hierarchical stratified random sampling design was adopted (Department of the Environment 1994, Environment Agency 2000). Each garden was stratified into three cells, with three replicate samples taken from random locations in each of the cells (Fig. 1), giving a total of nine replicates per subject garden.

Six generic land-uses were encountered in the gardens. The conceptual model of potential exposure for each land-use was formulated. Its implications for the sampling design are summarised in Table 1.

The land-use in each garden was recorded during the sampling in order to allow for potential assessment of its effect on the sample data.

The location of the samples was determined using random coordinates. Areas excluded from investigation, e.g. hardstanding, were not sampled (see Table 1). Where a selected random location was located in a hardstanding area, the process was repeated until a suitable location was obtained.

The depth of the sampling was chosen as 0–20 cm, considering the land-use of the gardens was established (Environment Agency 2000). This depth was considered to provide a sufficient breadth of information, while minimising the physical disruption to the subject gardens. Apart from the primary exposure routes such as ingestion of soil and inhalation of soil-borne dust, this sampling depth also allowed for the estimation of risks associated with growing and consumption of homegrown vegetables, where this activity was encountered. It should be noted that this sampling depth does not allow for the assessment of

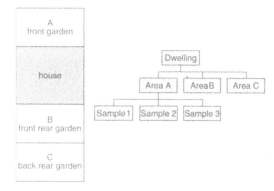

Figure 1. Schematic representation of the hierarchical stratified random sampling design of the field investigation.

Table 1. Conceptual model of land-uses of the gardens and the implications for the investigation.

Land-use	Implications
Hardstanding (patios, paths)	Excluded from the investigation, significant exposure not envisaged
Grass	Sampled, land-use recorded
Flower beds	Sampled, land-use recorded
Vegetable patches	Sampled, land-use recorded
Water features (ponds etc.)	Excluded from the investigation (water features are likely to be lined)
Structures (garden sheds etc.)	Sampled only if set on bare soil, land-use recorded

risks arising from future uncontrolled activities in the gardens such as digging, extensive landscaping or consumption of fruits from deep rooting plants.

2.1.2 Fieldwork

The fieldwork comprised a brief walkover of the gardens, to record the existing land-uses, and to collect the soil samples. The sampling was carried out manually with a spade and trowel. Approximately 1 litre (1.5 kg) of soil was taken from each sampling location. The samples were stored in airtight plastic soil containers specifically provided for the purpose by the laboratory. The samples were consequently handed over under a chain of custody procedure to a courier and were transported (refrigerated) to the receiving laboratory.

2.1.3 Chemical analysis

All chemical analyses were undertaken by a laboratory specialising in the analysis of contaminated land. The laboratory holds all the necessary QA/QC accreditations (UKAS, ISO 17025).

The limits of detection were chosen in order to minimise the possibility of overlooking significant contamination whilst maintaining the cost effectiveness of the testing.

All laboratory analyses were to be carried out using current standard industry methodology. The precision and accuracy of the testing was further verified by examining appropriate QC data (Shewhart charts) provided by the laboratory.

2.1.4 Statistics

Prior to the calculation of basic statistics the following assumptions were tested:

- normality of errors – some datasets were transformed [$y = \ln x$ when all $x > 1$ for that variable; $y = \ln(x + 1)$; when $\exists\, x \leq 1$ within that variable]; mean $\bar{x}$ and standard deviation s were computed from the transformed mean $\bar{x}_T$ and transformed standard deviation S_T using the following equations:

$$\bar{x} = e^{\bar{x}_T} \text{ or } \bar{x} = e^{\bar{x}_T} - 1, \quad s_1 = e^{\bar{x}_T + S_T} - e^{\bar{x}_T}$$

and

$$s_2 = e^{\bar{x}_T} - e^{\bar{x}_T - S_T}$$

where s_1 corresponds to the standard deviation of upper tail and s_2 the standard deviation of the lower tail of the distribution;
- homogeneity of variance – Cochran's C-test was used;
- homoscedacity (equal variance of errors or zero correlation between errors) – Scheffe's test.

All data were found to follow either normal or lognormal distribution. The 95th percentiles for upper tails of the distribution were calculated following the above transformation where necessary.

2.2 Works

During the sampling exercise the properties were surveyed for existing land-uses and a model for each garden was developed. An example is shown in Figure 2.

2.3 Results

In total 168 samples were collected from the gardens of 19 residential properties. Nine samples were taken from each garden as per proposed methodology. The exception was a single garden where half of the rear garden was formed by a concrete patio, and, consequently, only six samples were taken.

The objective of data analysis was to determine the distribution of each of the contaminants within the

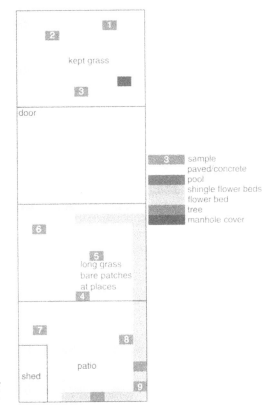

Figure 2. Example of a land-use model for residential property.

surface soils in the subject gardens. The analysis comprised the calculation of basic statistics for the concentration of each contaminant in each garden (e.g. arithmetic mean, standard deviation, 95th percentile). This included testing the distributions of the sample data and application of appropriate data transformation where necessary (Table 2). In addition, outliers of the estimated distributions were determined using normal probability plots.

As shown in Table 2, the majority of the analytes were lognormally distributed in the soils of the subject gardens. The calculated standard deviations suggest a low to medium variability in the concentrations which, along with the small number of outliers detected, points to generally low heterogeneity of the soil material.

The highest number of outliers, two, was encountered in the distribution of mercury. This is generally a very low proportion of outliers, which suggests that these resulted from random irregularities rather than that they form statistically distinct populations, which would indicate potential hotspots of contamination.

Table 2. Estimated distribution and basic statistics for each analyte in the sampled soils; only normal and lognormal distributions were considered. All concentrations are in mg kg^{-1} unless stated otherwise.

Analyte	distribution	average	st.dev[a]	95th per.[b]	outliers
pH (units)	normal	7.29	0.48	8.34	
Arsenic	log	20.05	6.28	21.76	
Cadmium	log	0.73	0.28	1.76	2.8
Chromiumc	normal	11.44	6.52	11.99	38
Lead	log	183.0	132.2	197.2	7680
Mercury	log	1.22	1.04	2.33	12, 12
Selenium	normal	0.85	0.44	1.86	
Nickel	log	18.32	5.90	19.99	99
Benzo[a]pyrene	log	0.60	0.58	1.67	5.5

[a] standard deviation of the mean; for lognormal distribution the standard deviation of the upper tail.
[b] upper confidence level for the 95th percentile of the mean.
[c] total chromium, no Cr^{VI} was detected.

2.4 Risk estimation

The CLR guidelines advocate a tiered approach to the risk estimation. Firstly, a conceptual model of a site is compared with the generic conceptual model used by the Contaminated Land Exposure Assessment Model (CLEA) for the derivation of Soil Guideline Values for the given land-use. In addition, the guidelines introduce the concept of "averaging area", the area or volume of soil that represents the "source" within the source-pathway-receptor framework. Where the generic conceptual model is considered appropriate, the mean concentration of the potentially harmful substances within the averaging area can be compared with the CLEA-derived SGVs.

A second tier assessment is carried out where either significant breaches of the assumptions of the generic conceptual model are observed, or the measured concentration of a particular substance within the averaging area is likely to exceed the generic and more conservative SGV. In such cases, site-specific SGVs are determined using CLEA's probabilistic model.

2.4.1 Conceptual models

The conceptual models used in the assessment of human health risks from contaminants in soils are based on three elements: land-use, fate and transport of contaminants and contaminant toxicology. As already noted, the conceptual models of the land-use were prepared for each subject garden (Fig. 1). The averaging area, i.e. the potential source of contamination was set as the top 20 cm of the soft-landscaping of each garden (see section 2.1).

The models of fate and transport properties and toxicological information for the seven contaminants and toxicological information on benzo[a]pyrene are given in the CLR guidelines. Basic toxicological principles are outlined in the section below.

The generic conceptual model can be applied appropriately to a specific site when either any deviation from the conceptual model is likely to be insignificant or any such deviation would result in a more conservative assessment of the risk. The main assumptions of the generic conceptual model and potential site-specific deviations for the subject properties are listed in Table 3.

As shown in Table 3, there are minimal deviations from the main assumptions of the generic conceptual model used by CLEA to derive SGVs for residential land-use. This, along with the generally conservative exposure frequencies for both oral and dermal contact and for inhalation used by the model, makes the generic SGVs suitable for screening of soils in the subject gardens.

2.4.2 Basic toxicological principles

Potentially harmful substances can be grouped into two categories, according to their toxic action: those for which there is considered to be a threshold below which they fail to induce any discernible adverse health effect, and those for which a threshold cannot be assumed (DEFRA 2002c). The first group of chemicals generally comprises non-carcinogens and non-genotoxic carcinogens, i.e. carcinogens that operate through chronic cell damage. The second group comprises mutagenic and genotoxic carcinogens which have the potential to affect DNA and initiate genetic mutations or cancer.

For chemicals for which a threshold for the onset of adverse health effects can be assumed, a tolerable daily intake (TDI) is determined. It is derived from no observed adverse effect level (NOAEL) or the lowest

Table 3. Main assumptions of the generic conceptual model for residential land-use used by CLEA and site specific deviations.

No.	Assumption of generic conceptual model	Site specific deviation
1	Residential single property within a square shaped area, two storeys high.	No significant deviation.
2	The household has ready access to private garden.	No significant deviation.
3	The property receives drinking and bathing water from mains supply.	No significant deviation.
4	The household grows vegetables for its own consumption (Brussels sprouts, cabbage, carrot, leafy salads, onion, leak, shallots and potato).	This activity was encountered only in one household out of the 19 properties visited, however, may have occurred in other households in the past. This deviation would be significant for most contaminants and would result in more assessment conservative of risk.
5	There is no significant exposure via ingestion of produce other than the common garden vegetables (see above) grown or reared in the garden, e.g. fruits or livestock.	No such activity was observed. Significant deviation is unlikely.
6	Critical receptor for the potential contamination is assumed to be a young female child with the duration of exposure covering the first six years of life.	Make up of the households and activity patterns were not investigated, however, at the time of sampling, the majority of the households appeared to be occupied only by adults. Any deviation from this assumption would result in a more conservative assessment of risk.
7	The exposure frequencies and soil ingestion rates do not account for geophagia, a pathological condition that refers to persistent and purposeful consumption of soil, primarily by children. Condition often known as "pica child".	There is no information to suggest that a child with this condition is resident at any of the subject properties. CLR guidelines recommend that protective measures are best implemented for each individual suffering geophagia. Consequently, this condition was not considered during this study.

dose at which an adverse effect is observed (LOAEL). An exceedence of the TDI is undesirable although it does not inevitably lead to increased health risk. The implications of exceedence need to be assessed on a case-by-case basis.

For chemicals for which a threshold of adverse health effects cannot be presumed, an index dose is determined. It represents the level of exposure at which the associated risk is considered minimal. The index dose is generally based on a nationally or internationally agreed exposure standard from a single source. An exceedence of index dose indicates an increased risk to health. The significance of this risk may require expert judgement. CLR9 advises that: "Despite the wide-ranging values reported by different authoritative organisations, there is some consensus for selecting a figure of 10^{-4} as the upper bound of "acceptable" additional lifetime risk from exposure to environmental contamination from any one source (such as, for example, a contaminated site). This corresponds to an annual excess risk of cancer of about 10^{-6} (one in a million per year). However,

making a decision as to what is an acceptable level of risk to individual members of the public from exposure to ambient levels of an environmental pollutant, is a value judgement. Although knowledge of some background scientific and technical information is necessary, the essential decision is one that involves socio-political judgements."

2.4.3 *Tier 1 assessment*
With the exception of cadmium and benzo[a]pyrene, all SGVs used for the primary screening are those derived by CLEA using the generic conceptual model and published in the relevant CLR10SGV documents. The SGV for cadmium is pH dependent and was therefore derived by entering a site-specific pH value into the CLEA model. An SGV for benzo[a]pyrene has not been published at the time of writing this report and was therefore derived by entering the published index doses of $ID_{oral} = 0.02 \, \mu g \, kg^{-1}$ bw day^{-1} and $ID_{inh} = 0.00007 \, \mu g \, kg^{-1}$ bw day^{-1} into the CLEA model. The results of the screening are shown in Table 4.

Table 4. Screening of potential contamination in the subject gardens using generic SGVs. All values are upper confidence level for the 95th percentile of the mean for each garden in mg kg^{-1}; values that exceed SGV are in italics.

Garden No.	Arsenic	Cadmium	Chromium (Total)	Lead	Mercury	Selenium	Nickel	Benzo[a]pyrene
SGV	20	2.8[a]	130	450	8	35	50	1[b]
1	22.2	1.7	14.3	198.9	3.8	0.6	23.8	2.5
2	21.8	1.3	11.5	155.8	3.2	0.6	19.6	1.6
3	26.6	1.5	13.8	263.5	3.6	0.6	23.5	1.5
4	25.4	2.0	22.0	277.9	3.5	0.9	22.2	2.5
5	23.2	2.1	11.2	280.0	2.5	1.5	19.2	2.0
6	20.9	1.3	7.0	172.7	1.7	1.6	15.3	1.6
7	20.6	1.4	6.0	142.0	2.7	1.8	15.3	2.0
8	26.4	1.3	15.7	171.4	1.5	0.8	15.6	1.2
9	21.5	1.5	13.9	127.3	1.6	0.9	16.6	1.7
10	20.2	1.4	13.3	129.1	1.7	0.7	17.0	1.3
11	16.3	1.5	9.8	131.2	1.8	0.6	18.3	1.6
12	21.5	2.2	22.2	291.2	2.4	0.5	22.9	1.8
13	22.4	1.4	20.7	308.9	2.2	1.2	21.3	2.1
14	23.3	1.6	21.5	206.8	2.1	0.7	24.3	1.6
15	18.9	1.4	8.6	149.0	1.9	0.5	16.9	1.6
16	17.2	1.5	13.1	334.8	1.7	1.0	18.4	1.9
17	17.9	1.6	8.9	125.6	1.9	1.4	21.1	1.7
18	18.4	1.3	12.6	120.7	2.0	1.6	18.7	1.5
19	16.4	1.3	10.5	150.3	1.5	1.1	17.7	2.4

[a] derived for site specific pH using CLEA
[b] derived for recently published toxicological information using CLEA.

Table 5. Tier 2 assessment of arsenic (As) and benzo[a]pyrene (B[a]p).

Assumption	Effect on SGV		Effect on risk	
	As	B[a]p	As	B[a]p
Consumption of home-grown vegetables. This was observed only in one garden (CL20). If this was taken into account:	23	1.2	risk in 6 gardens	risk in all gardens
Critical receptor (female child aged between 0–6 years). The majority of the households were occupied by adults. However, acceptance of this assumption would need to be corroborated by further study. If adult (16–59) used as the critical receptor:	230	7.4	no risk	no risk

CLR guidelines establish a tolerable daily intake or an index dose for each appropriate contaminant which is then used as an input variable into the CLEA model.

It is apparent from Table 4 that the only contaminants that exceed the generic SGV in the subject gardens are arsenic and benzo[a]pyrene.

2.4.4 Tier 2 assessment
Both arsenic and benzo[a]pyrene are considered non-threshold contaminants. Consequently, the risk models used by CLEA for arsenic and benzo[a]pyrene are based on index doses rather than tolerable daily intakes, the exceedance of which is not acceptable.

CLR guidelines postulate that for potential hazardous substances that are not screened out during the

Tier 1 assessment, site specific SGV can be derived. This is achieved by refining the generic assumptions using the default CLEA conceptual model to reflect site-specific information. There are a number of assumptions that would appear too conservative for the conditions observed in the majority of the subject gardens. These, along with their effect on the risk estimation, are summarised in Table 5.

It is apparent from Table 5 that even when the exposure route via ingestion of home-grown vegetables is discounted, arsenic and benzo[a]pyrene still potentially pose a risk to the critical receptor of a female child.

This critical receptor for the contamination encountered may not be appropriate for all subject gardens. Even in cases where children of this age group reside,

128

their exposure frequency may not be as high as that used by the generic CLEA model. For soil and indoor dust the ingestion exposure frequency is 365 day yr^{-1} for the critical receptor. Exposure frequency for dermal contact indoor is 365 day yr^{-1} and outdoor is 160 day yr^{-1} for the critical receptor. Any deviation from this assumption would result in a less conservative assessment of risk. However, it is stressed that before any of these assumptions are refined, a study of the exposure activities of the household residents would be required.

Another key assumption used by the generic CLEA model is that the Soil Guideline Values are based on contaminant intakes and it is therefore assumed that 100% of the benzo[a]pyrene and arsenic in soil is taken up by the systemic circulation.

There is some evidence, particularly for arsenic, that where it is either strongly bound to the surface of soil particles or present in an insoluble form, then its bioavailability to the human body may be less than 100%. Work carried out on another site showed bioavailability of arsenic to range between 17 and 25% in five soils tested. However, further site specific testing would be required before this assumption can be refined. This has not yet been carried out.

2.5 Risk evaluation

The results of the risk estimation suggest that, potentially, the concentrations of benzo[a]pyrene and arsenic, only in a proportion of the properties, may pose unacceptable risk to the occupants of the subject properties. The highest risk estimated using the CLR guidelines appears to be no more than 2.5-fold of what is considered to be acceptable by the UK statutory authorities. The risk was calculated on the assumptions of a female child being exposed to the soils during the first six years of life. In their current conditions, the gardens appear to be safe for adult residents.

Both contaminants appear to be relatively evenly distributed in the soils investigated. With few exceptions, there were no statistically significant differences between the concentrations detected in individual gardens, despite the fact that several gardens had significant amounts of clean soil imported during the past 30 years. This suggests that both arsenic and benzo[a]pyrene are not contained within hotspots of contamination but, rather, seem to appear in the subject gardens in relatively elevated background concentrations.

3 CONCLUSIONS

The above case study illustrated that adherence to the new CLR guidelines requires a much more thorough assessment than what has been the industry practice to date. Much more emphasis is given to planning and design of the investigation and the chemical testing regime is also more extensive.

Based on the above observations the main implication of the new frameworks appears to be the requirement for a much more comprehensive site assessment and risk analysis for human health risk. Although there is still a danger of overdesign of remedial solutions through misuse of the new SGVs, the new framework should enable risk-based site specific solutions for brownfield site with less engineering and, ultimately, lower remediation costs. At the same time, the transparency of the new regime should simplify regulatory compliance and improve public confidence in brownfield developments.

REFERENCES

Department for Environment, Food and Rural Affairs and the Environment Agency (2002a) Assessment of Risks to Human Health from Land Contamination: An Overview of the Development of Soil Guideline Values and Related Research, Report CLR 7, Environment Agency, Bristol.

Department for Environment, Food and Rural Affairs and the Environment Agency (2002b) Priority Contaminants Report, Report CLR 8, Environment Agency, Bristol.

Department for Environment, Food and Rural Affairs and the Environment Agency (2002c) Contaminants in Soil: Collation of Toxicological Date and Intake Values for Humans, Report CLR 9, TOX1–10, Environment Agency, Bristol.

Department for Environment, Food and Rural Affairs and the Environment Agency (2002d) The Contaminated Land Exposure Assessment Model (CLEA): Technical Basis and Algorithms, Report CLR 10, SGV1–10, Environment Agency, Bristol.

Department of the Environment (1994). Sampling strategies in contaminated land. London, Department of the Environment.

Environment Agency (2000). Secondary model procedure for the development of appropriate soil sampling strategies for land contamination. Research and Development, Technical Report P5-066/TR. Environment Agency, Bristol.

Land Reclamation – Moore, Fox & Elliott (eds)
© 2003 Swets & Zeitlinger, Lisse, ISBN 90 5809 562 2

Planning policy and land contamination

D. Brook
Office of the Deputy Prime Minister, London, UK

ABSTRACT: Government gives high priority to the sustainable re-use of previously developed land including that affected by contamination. While contamination posing unacceptable risks to existing use of land is dealt with through the contaminated land regime, the majority of remediation activities will involve redevelopment. The planning system controls such redevelopment and needs to take full account of the potential for contamination due to past industrial activities as well as the natural presence of potentially harmful substances. Following consultation on draft guidance, a formal policy statement is expected to be accompanied by technical advice during 2003 to enable the planning system to encourage redevelopment and beneficial use.

1 INTRODUCTION

Successive governments have given high priority to the restoration of derelict and despoiled land and the regeneration of run-down urban areas. The use of derelict land grant and a variety of other forms of financial assistance have helped to restore and regenerate previously developed land. This continues and planning policy guidance gives priority for housing on previously developed land and re-use of existing buildings and for retail development in town centres rather than out-of-town.

In his statement on *Sustainable communities, housing and planning* on 18 July 2002, the Deputy Prime Minister said – "we need to make better use of land, by improving design, increasing densities and using brown-field sites to the full. In 1998 I committed the government to a target that 60% of new homes should be on brown-field land. We have met that target 8 years early but we need to keep up the pressure. To help with this I will establish a register of brown-field land held by government and public bodies."

The national land-use database, using returns from local authorities, has identified about 66,000 Ha of previously developed land in England that is suitable for development. This equates to an area half the size of London. Unfortunately much of this land is affected by contamination that may affect both its current and future use.

2 PLANNING CONTROL

Under the Town and Country Planning Act 1990, the planning system controls development and the use of land in the wider public interest. New planning legislation is currently being enacted but the essential elements of policy remain unchanged.

Land contamination has long been recognised as a consideration that may be material to planning decisions because of its potential effect on public health and safety, on the natural and built environment and on economic activities.

The Inter-Departmental Committee on the Redevelopment of Contaminated Land (DOE, 1977) was established in 1977 to provide general guidance on assessment and redevelopment of contaminated sites. Advice was also given in planning circulars dealing with structure and local plans, planning conditions and derelict land (DOE, 1984, 1985 a, b). This was consolidated and updated in Department of the Environment (DOE, 1987) Circular 21/87 *Development of contaminated land* and later in Planning policy guidance note (PPG) 23 *Planning and pollution control* (DOE, 1994).

Since the latter was published, there was an abortive attempt to introduce registers of potentially contaminative uses and a new contaminated land regime has been introduced under Part IIA of the Environmental Protection Act (EPA) 1990. This requires local authorities to identify contaminated land and to take such action as is necessary to remove unacceptable risks to human health and the environment.

To take account of these changes, the then Department for Transport, Local Government and the Regions issued in February 2002 a consultation document on draft planning technical advice on *Development on land affected by contamination*. The key objectives for the planning system were identified as being to:

– encourage the redevelopment and beneficial use of previously developed land, both to bring about the social, economic and environmental regeneration of that land and areas around it and to reduce unnecessary development pressure on green-field sites; and

– ensure that any unacceptable risks to human health, buildings and the environment from contamination are identified and properly dealt with as new development and land-uses proceed.

2.1 Relationship between planning control and the contaminated land regime

The substantial legacy of past contaminated land is being addressed through Part IIA of EPA 1990. The approach is based on the principles of risk assessment, including the concept of a contaminant, receptor and pathway being combined as a pollutant linkage that is causing or is likely to cause significant harm or pollution of controlled waters. In broad terms it applies where land is causing unacceptable risks assessed on the basis of the current use and circumstances of the land and requires remediation to a standard that is suitable for that use. It does not apply to possible future uses of land. Nor does it apply to potentially harmful elements that are present for natural reasons.

The land-use planning system complements this regime in a number of ways.

Remediation is controlled by the planning system alongside relevant pollution control legislation. Planning permission will normally be required for the engineering works necessary to remove unacceptable risks from contaminated land. This requirement will still remain even if a single remediation permit is introduced as recommended by the Urban task force (Remediation Permit Working Group, 2002).

A key feature of Part IIA is that it is designed to encourage voluntary remediation. Where appropriate remediation activity is being or will be done without serving a remediation notice, then it cannot serve such a notice. Since remediation is costly, it is likely that a large proportion of that undertaken will be funded from the profits of new development.

Development or new use of land could create pollutant linkages on land that is not contaminated under Part IIA, in that the existing use is not threatened. In such circumstances, as with voluntary remediation funded by redevelopment, it is necessary to consider the risks on the basis of the proposed new use and circumstances of the land.

Finally, the planning system operates on a wider basis since it is immaterial whether the presence of potentially harmful substances arises from human activities or they are present naturally. In addition, contamination may be material in planning terms even where the potential harm is not significant under the terms of Part IIA. For example, the perception of contamination may directly impact on the amenity and economic value of neighbouring land.

3 FORWARD PLANNING

Government policy is translated into forward plans for development and the use of land through regional planning guidance, which will be replaced by regional strategies under the new planning legislation. The need for regeneration of urban areas features strongly in such guidance.

Statutory development plans then set out land-use policies of individual local authorities. PPG 12 *Development plans* (DETR, 1999) emphasises the need to "take environmental considerations comprehensively and consistently into account." In drawing up development plans, local authorities need to take into account any potential implications of land contamination.

Indeed, there is a positive role for development plans in that steering development onto previously developed land, some of which may be affected by contamination, will protect green-field sites from development. It can assist in progressively improving the condition of land as a whole, provided the contamination is identified and dealt with appropriately.

Local authorities, therefore, need to take the likely presence of contamination into account when considering proposals for development in their area. Information acquired during the inspections required under Part IIA of EPA 1990 clearly feed into this process, together with any other information on previous uses. They should also take account of the natural presence in some areas of potentially harmful substances. Where land contamination is likely to be a significant issue, the development plan should include the considerations that will be taken into account when determining planning applications.

4 DEVELOPMENT CONTROL

Where development is proposed on land that may be affected by contamination, any unacceptable risk to or arising from the development should be identified and dealt with effectively through the development process. Those proposing development should carry out any necessary investigations to identify and assess the risks from contamination. They should provide sufficient information to satisfy the local planning authority that any contamination can be successfully remediated with the minimum adverse environmental effect to ensure the safe development and secure occupancy of the site.

Particular attention is needed where the proposed use would be more vulnerable to past contamination or where the current circumstances or past use of the land suggest that contamination may be present. Residential development, for example, is clearly more vulnerable to contamination than commercial and industrial development. Similarly, certain past industrial uses, sites of pollution incidents and recontoured or filled ground are more likely to be affected by contamination. It should be recognised that even some green-field sites may be affected by contamination and that there may be potentially harmful substances present naturally in the land.

4.1 Information from the applicant

Applicants for planning permission need to determine whether their site is affected by contamination and provide sufficient evidence to satisfy the local planning authority that any unacceptable risks will be dealt with. This requires a staged investigation and risk assessment. Since the expertise of both developers and local planning authorities on contamination issues is very variable, it is essential that such investigations are carried out by an appropriate competent person on whose conclusions and recommendations reliance can be placed.

Depending on the particular circumstances the level of information required to determine whether the development is appropriate and should be granted planning permission will vary. However, the minimum requirement is likely to be a comprehensive desk study and site reconnaissance to identify:

- The history of the site, including the nature of any industrial process or other activities that are likely to have contaminated the site;
- The geological and hydrogeological setting; and
- Present and future potential receptors (i.e. proposed future users of the site).

A desk-study and site reconnaissance will assist in determining the likely nature and extent of any possible contamination. In particular, it will identify the need for and scope of further investigations, the problems that might require remediation and whether that remediation could be secured satisfactorily by means of planning conditions.

Further detailed site investigations and risk assessment will be required where there is evidence of potential contamination and in order to prepare an appropriate remediation scheme. In many cases, however, it would be reasonable to grant permission subject to conditions that such investigations and any necessary remediation scheme are carried out before the development take place.

4.2 Determining applications

Since the English planning system is plan-led, applications must be determined in accordance with the provisions of the development plan in so far as they are material to the application in question unless other material considerations dictate otherwise. This emphasises the need for consideration of contamination issues at that earlier stage where practicable.

When considering the application it is important that there is close liaison between those responsible for the planning system and those responsible for Part IIA activities. Indeed, the remediation proposed as part of a development may need to satisfy the requirements of a remediation scheme under Part IIA. A number of bodies have responsibility for the protection of key receptors, such as the Environment Agency for controlled waters, English Nature and English Heritage, for designated sites of natural and cultural heritage. Local planning authorities will need to consult these along with other appropriate organisations such as water companies and local community groups.

Where it is satisfied that the proposed development is appropriate having regard to the information available about contamination, the standard of any remediation necessary, and other material considerations, the local planning authority should grant permission subject to any conditions that may be required for further investigations and remediation.

An appropriate development will not present or incur unacceptable risks, will not create new pollutant linkages, will ensure that unacceptable concentrations of contaminants are not left in place or pollutant pathways left open and will not impede any necessary remediation for contamination that does not impact on land use. For planning purposes, what constitutes an unacceptable risk differs from that under the contaminated land regime of the EPA 1990. The range of receptors and effects is wider, the risk assessment is primarily concerned with the specific development proposed as well as current use and other circumstances and the scope of control should not be limited to contamination likely to cause "significant harm."

Planning conditions can be used to cover a range of issues, including the collection and reporting of further information through site investigation and risk assessment, the submission of a remediation scheme and the carrying out of remediation before development proceeds. They must be imposed for planning purposes and be fairly and reasonably related to the proposed development and be capable of enforcement. Where conditions are not appropriate, planning agreements can provide an effective legal mechanism to ensure that works or other steps to deal with contamination are undertaken satisfactorily.

When land affected by contamination is treated to enable development to proceed, it is important that a record should be maintained of what was done. This can be achieved by a condition requiring the submission of a completion report, which is agreed by all parties. This should detail the conclusions and actions at each stage of the process. In particular it should include the investigations and assessment of risks, the remediation proposed and actually undertaken, the verification of its effectiveness and any future monitoring or in-use management provisions.

Where a local planning authority is not satisfied that the development proposed is not appropriate, it should refuse permission. In particular, where circumstances suggest there is a potential risk and insufficient information has been provided to exclude the reasonable possibility of such risk, where an unacceptable risk cannot be dealt with adequately or where the steps needed to deliver an appropriate development are not already in place, permission should be refused.

5 CONCLUDING REMARKS

The government wishes to encourage the full and effective use of land in accordance with the principals of sustainable development that address social, economic and environmental issues. Priority is given to the re-use of previously developed land. However, such land is frequently subject to a number of constraints including contamination resulting from previous industrial activity.

The impact of contamination on existing use has been addressed through Part IIA of the Environmental Protection Act 1990. It is recognised, however, that this does not address future uses and that the most cost-effective of dealing with contamination is to fund it through redevelopment. It is the role of the planning system to consider contamination issues alongside other planning considerations to determine whether proposed development is appropriate and that it will not lead to unacceptable risks.

Planning policy has evolved over time, the latest development being the issue in 2002 of draft guidance on land affected by contamination. The essentials remain that when development is proposed on land

that may be affected by contamination, the risks involved should be identified and dealt with as part of the development and a full record of what was done should be maintained.

Following the 2002 consultation exercise and in line with the government's planning reform agenda, policy guidance on land affected by contamination will be issued in a revised planning policy statement on planning and pollution control. The technical issues are not within the normal range of expertise of land-use planners and this statement will be accompanied by technical advice that takes account of responses to consultation.

REFERENCES

DETR (Department for the Environment, Transport & the Regions), 1999. *Planning policy guidance note 12. Development plans*, London, The Stationery Office.

DOE, 1977. *Department of the Environment Circular 49/77. The inter-departmental committee on redevelopment of contaminated land.*

DOE, 1984. *Department of the Environment Circular 22/84. Memorandum on structure and local plans.* London, HMSO.

DOE, 1985a. *Department of the Environment Circular 28/85. Derelict land grant.* London, HMSO.

DOE, 1985b. *Department of the Environment Circular 30/85. The use of conditions on planning permissions.* London, HMSO.

Remediation Permit Working Group, 2002. *The remediation permit: towards a single regeneration licence.* London, Urban Task Force Working Group.

DOE, 1987. *Department of the Environment Circular 21/87. Development of contaminated land.* London, HMSO.

DOE, 1994. *Planning policy guidance note 23. Planning and pollution control.* London, HMSO.

Land Reclamation – Moore, Fox & Elliott (eds)
© 2003 Swets & Zeitlinger, Lisse, ISBN 90 5809 562 2

Environmental insurance – from contamination to confidence

D. Brierley
Bridge Risk Management Ltd, Manchester, UK

ABSTRACT: The objective of this paper is to encourage developers to investigate environmental insurance packages to assist them in becoming champions of reclamation and regeneration. Insurance protection addresses the contradiction between the objectively desirable Brownfield Agenda and an increasingly tough legal and regulatory regime in this country and indeed throughout the European Union. The growing environmental insurance market is currently worth around £50 million per annum in the UK and is largely dedicated to addressing the costs and liabilities arising from the development of contaminated sites. Most covers are placed to facilitate property transactions, although there is a growing market to protect businesses from ongoing operational risks. We will examine insurances which cap the cost of clean-up operations and others which protect against future environmental liabilities, presenting these products against a background of new guidelines and technology which actually help Insurers to provide effective insurance solutions at a realistic price.

1 THE CENTRAL PARADOX

The Government's Sustainable Development Commission stated that "sustainable development should be the organising principle of all democratic societies, underpinning all other goals, policies and processes". For sustainable, read brownfield.

The Government has stipulated that 60% of residential development should take place on brownfield sites. The fact that a target has been set (and tax incentives offered) is to be applauded, but the figure may be inadequate. From every point of view, from quality of life through biodiversity to regeneration of blighted areas, it is probably time for a total moratorium on greenfield development.

Be that as it may, the government's target is extremely challenging in view of the legal and regulatory framework that has evolved, along with a number of other factors. The much discussed Part IIa of the Environmental Protection Act contains a statutory definition of contaminated land. Sites falling within this definition are to be cleaned up by "the polluter" or, if they cannot be traced, the present owner or occupier, known as the "Class B person". The fact that *former* industrial sites are available for development would suggest that if the polluters were still in business, they would still be operating from the site. In other words, the "Class B" person is highly likely to be liable for remediation of land falling within the definition. This alone is bound to make potential purchasers wary and strike fear into the hearts of their funders, who would inherit contamination liabilities as Mortgagees in Possession in the event of foreclosure. Despite the potential impact of this legislation, the Environment Agency reported in September 2002 that only 40 sites had been designated as contaminated under Part IIa. In the same report, it was estimated that 20,000 hectares of land need action. It is, however, the existence of the legislation rather than its impact to date (which may change) that could deter potential purchasers of brownfield sites. There is evidence that local authorities sell surplus contaminated sites "with information" at a generous discount and that owners of similar sites offer them to the authorities free of charge. This uncommercial behaviour would normally take place only in an atmosphere of fear.

There are nevertheless other fear factors than Part IIa, which at this point in time are more likely to give developers pause for thought as they look to purchase former industrial land:

- Litigiousness; people and businesses are far more likely these days to bring an action due to bodily injury or property damage arising from pollution either on site or migrating from adjacent land; and
- Corporate Reputation; the good publicity generated by the development of a former industrial site may be cancelled out if a pollution incident occurs in the future.

Unlike bottles and newspaper, the recycling of land is not without pitfalls. There is therefore an obvious need for a mechanism to bridge the gap between the pros and cons of re-using former industrial sites, thus ensuring that a low cost plot is truly a bargain and that a PR opportunity does not backfire. The mechanism is insurance.

2 ENVIRONMENTAL INSURANCE – THEORY AND PRACTICE

Confining the discussion as we are to the historic pollution that potential purchasers of brownfield sites may encounter, there are three main products which provide comfort at all stages of development, from remediation design through to new occupiers moving onto the site and beyond. Before discussing these, four essential points need to be made:

1. Public Liability Policies specifically exclude cover in respect of gradual pollution and proof of sudden and unforeseen cause may be difficult if not impossible;
2. Consultants' and Contractors' Warranties can only be successfully invoked if negligence is proved in court;
3. A remediation strategy may be signed off by the Environment Agency. This does not remove liability for future pollution incidents; and
4. The value of Environmental Warranties and Indemnities provided by vendors or purchasers is directly proportional to their current and future financial standing.

2.1 *Clean-up cost cap*

It seems logical that remediation contractors who guarantee their final estimates would factor in an element for unforeseen cost overruns and that developers carrying out their own clean-up operations would err on the side of caution in compiling their budgets. It is possible to insure against substantial changes to the original estimate, thus ensuring that the project is not abandoned and that capital is not tied up against the possibility of an increased contract price.

2.2 *Contractor's pollution liability*

A pollution incident may occur as a result of contractors' operations either during remediation or development works. These two stages are potentially hazardous as contaminants will inevitably be disturbed and may travel. It may take some time for an incident of this nature to come to light. CPL policies are designed to cover these risks.

2.3 *Environmental impairment liability*

These policies cover the cost of on and offsite remediation costs, arising from regulatory intervention or

from a third party claim for bodily injury or property damage. They also address accompanying Legal Expenses which may be substantial and, very importantly, changes in the law during the policy period. The amount of recent legislation in respect of contaminated land makes this one of the most important features of this insurance. It is possible to extend the policy to include Business Interruption and Additional Working Costs, which may become a factor if occupiers need to be moved to other premises following a pollution incident. In this eventuality, there may also be a risk of lost revenue in the form of rental income. In situations where developers are looking to sell the site following construction works, cover may be assigned to one or more subsequent purchasers.

There follows a selection of case studies of situations where we as a company have been able to provide insurance solutions which enabled acquisitions and disposals of brownfield sites to go ahead, by providing comfort to purchasers, their lenders and their legal advisors.

2.3.1 *PFI*

The government was disposing of a large site for redevelopment as a "new town" with areas for residential, retail and soft end uses. A number of bidders put proposals forward for this, one of whom approached us for a risk management package which was to be incorporated into their bid. We obtained terms for Clean-up Cost Cap, which meant that an enhanced bid was put forward, as there was no need to make provision for cost overruns. The Environmental Impairment Liability policy was designed to address residual issues on site following remediation works and to be assignable to the various developers who would carry out the successful bidder's design plans.

2.3.2 *Capital release*

A housing estate consisting of several hundred units was constructed some years ago on the site of a former gas works. The contaminated ground was capped to a height of several metres and monitoring of gas and groundwater has continued ever since. The owners had been making increasing financial provision against the eventuality of a future pollution claim but wanted to release this fund for other purposes. We provided a long term insurance package for twice the financial provision at a premium equivalent to less than 3% of it.

2.3.3 *Staged site disposal*

A large industrial concern decided to dispose of a large site in stages. Site investigations indicated that the first parcel, which was designated for residential use, would need to be extensively remediated. Insurance was provided for offsite migration during remediation works with full Environmental Impairment

Insurance in place to address subsequent residual issues. The policy was written in the names of the purchaser and the vendor in order to address the future liabilities of both parties and effectively enabled the disposal to go ahead.

3 STEPS IN THE RIGHT DIRECTION – CLEA AND SiLC

Whilst some recent developments may have acted as a deterrent to anyone wishing to invest in brownfield land, two initiatives could well assist in improving confidence among developers, planners, funders and Insurers:

3.1 *Contaminated land exposure assessment (CLEA)*

Environmental consultants have long had to contend with the inappropriateness of intervention "trigger" values for contaminants which were demonstrably inappropriate to the UK in the 21st century. The CLEA model, which came into effect last year, stipulates Soil Guideline Values (SGVs) for a number of major contaminants, with others in preparation. As it only relates to soil contamination risks to human health, the model is limited, but it creates an excellent starting point for the future. Insurers and planners are unlikely to consider site investigation reports that have been compiled without reference to CLEA. This suggests a degree of confidence in the model, even at this early stage, which may lead to wider availability of insurance cover and more competitive premiums.

3.2 *Specialist in land condition (SiLC)*

A large proportion of the daily work of environmental insurance brokers and underwriters involves wading (sometimes literally) through pages of environmental data. This process is essential to the assessment of the site specific risks requiring insurance.

The Government's Urban Task Force produced a report in 1999 recommending a standardised document which could be used by all parties to a property transaction. This emerged in 2000 in the form of the Land Condition Record (LCR) which needs to be signed off by a rigorously accredited Specialist in Land condition (SiLC). This new documentation should certainly help to simplify the brownfield development process and assist greatly in the provision of appropriate insurance.

4 EMERGING TECHNOLOGIES

At the time of writing, 70% of remediation schemes still involve the expensive transfer of contaminated ground to a landfill, affectionately known as "dig and dump", or in the case of groundwater, "pump and treat". This is totally counter-productive as it involves the positive benefit of brownfield development being cancelled out by transferring the problem elsewhere.

Fortunately, a host of new sustainable technologies are here to help. Whilst these may take longer than the traditional methods and involve more extensive site investigations, the financial savings are considerable and the technologies themselves often lend themselves to insurance.

A good example is Monitored Natural Attenuation (MNA). This is a passive remediation technique for groundwater, whereby degenerating pollutant plumes are monitored until they are neutralised. Let us assume that this process would take place over a period of ten years, during which it may be possible for development to take place, depending on the depth of the plume. Insurance would be effected for the predicted attenuation period and premiums would take into account the reducing risk over the period.

This is not the place to go into detail on all of the available techniques, but suffice it to say that the future of remediation lies in sustainable technologies in which the Insurers and regulators are showing increasing confidence.

5 MOVING FORWARD

There are few brownfield development scenarios where insurance should not be seriously considered. Consultants and legal advisors may be your first source of advice, but subsequently, it is essential to obtain the services of a specialist environmental insurance broker. If insurance is being seriously considered, it is dangerous to use a broker without extensive experience in these specialist covers and imprudent to obtain cover directly from an insurer. The former may fail to understand the implications of the site investigation reports and the latter has no incentive to offer the most appropriate cover at the most competitive price.

The development of former industrial sites is essential to our future. It breathes life into areas killed by our heritage and ultimately protects our future by preserving green space. These two objectives can be achieved with confidence by insurance.

Land Reclamation – Moore, Fox & Elliott (eds)
© 2003 Swets & Zeitlinger, Lisse, ISBN 90 5809 562 2

Understanding the motivation behind the end use of mineral sites

R.L. Curzon
University of Central England, Birmingham, UK

ABSTRACT: The research examines the factors (environmental, economic and social) that influence the end use priorities of mineral sites, in order to identify the primary motivation for different after use regimes. The complexity of the decision making process led to the consideration of the potential influence of both primary and secondary stakeholders. It is set in the context that, historically, restoration to previous use (usually agriculture) was customary, but that more recently a wider range of afteruse options have been realised.

The research found that the mineral industry seem to be taking the lead in advancing restoration. The primary driving forces behind end use choice were found to be environmental and social although were increasingly underpinned by the influence of secondary stakeholders.

1   INTRODUCTION

1.1   *The research context*

England and Wales are rich in minerals (including coal, sand, gravel, iron, salt and limestone (Wallwork, 1974)) but suffer from pressure on the land through competing claims on space for housing, industry, agriculture, recreation, nature conservation and mineral extraction (Department of the Environment, 1989). This has led to a need to ensure land taken for mineral extraction is reclaimed as soon as feasibly possible to an acceptable and beneficial end use. (End use is defined as the ultimate use after mineral working has ceased and can include agriculture, forestry, amenity (including nature conservation), industry or other development (Department of the Environment, 1996)).

Before 1981, restoration was nearly always to agriculture, which had typically been the previous use before extraction (Ward, 1997). However, since the late 1980's a variety of end uses have been realized (Box, 1996). This paper intends to develop an understanding of the motivation behind end use choice, through a critical assessment of the attitudes of the local authority, mineral operators and other stakeholders.

1.2   *Methodology and data collection*

The research had particular reference to soft (nature conservation, forestry and recreation) and mixed end use and was based in England. It incorporated two methods of data capture namely postal questionnaires and case studies (utilizing stakeholder interviews).

- Postal Questionnaire: The postal questionnaire surveyed Local Authorities and Mineral Companies throughout England and served to identify general trends and highlight issues that would be examined further during the case studies. The response rate was quite high given the nature of questionnaire distribution with 24 out of 50 mineral companies responding and 36 out of 65 local authorities. The questionnaire collected information on the type of end uses, the reasons for choosing single or mixed use sites and factors influencing end use. It also provided contact details, which were utilized during the case study process.
- Case Studies: Selection criteria were developed to select six case study sites in England. The criteria considered location, mineral type, original use of land, site size, end use type and the number of stakeholders involved. The chosen sites included one in Tyne and Wear, two in Nottinghamshire, two in Staffordshire and one on the Wiltshire/ Gloucestershire border. Detailed information was collected through the case study process, which incorporated site visits and interviews with key stakeholders.
- Stakeholder Interviews: The term stakeholder has been in existence since the 1960's and is defined by Freeman (1984, cited in Andriof et al, 2002) as "any group or individual who can affect or is affected by the achievement of the organization's objectives" (p. 44) and may include employees, customers, communities and even competitors (Goodpaster et al, 2002). For the purpose of this research

stakeholders were taken to be any parties who have an interest in the decision making process and the end use of a site. Primary stakeholders were considered as having a direct influence on the decision making process and included mineral companies, landowners and local authorities (district or county level). Environmental organizations and any other groups, for example charities, non governmental organizations and local action groups were classed as secondary stakeholders as, although they are outside the planning network, they were considered to be potentially able to influence decisions on end use.

Thirty-two interviews were completed with representatives from both primary and secondary stakeholder categories. This incorporated 14 local authorities, 9 mineral companies, 5 environmental organizations and 4 who were categorized as "other" due to the mixture of groups represented. The interviews were semi-structured and sought to gather information on reasons for end use option and attitudes of different stakeholder groups. All but 4 of the interviews were completed in person and recorded using a dictation machine taking between one and three hours. The remaining four people were interviewed by phone due to the preference of the respondent or because of the need for excessive travel to meet in person. Throughout the paper the findings from both the questionnaire and interviews will be discussed, however to protect the identity of respondents, all quotations will be anonymous and simply labeled according to stakeholder group. The author wishes to acknowledge the help of all stakeholders interviewed, with particular thanks to the mineral company representatives.

Due to the complexity of the decision making process, it is important to understand the motivation behind end use choice perhaps through consideration of the various factors acting upon the primary stakeholders and therefore influencing the decision making process. Key categories identified for consideration include policy and planning regulations, site parameters and environmental, economic and social factors, the attitudes and potential influence of secondary stakeholders will be considered throughout. However, the following section will discuss the popularity of various end use options, as the most common end use category may provide some understanding of the motivation behind the decision making.

2 COMMON END USE OPTIONS

The 1994 Survey of Land for Mineral Workings in England found that between 1988 and 1994, 58.5% of land was reclaimed to agriculture, 26.3% to amenity, 5.2% to forestry and 10% to other development (including built development) (Simpson 1996). The popularity of agriculture as an end use was supported

by the questionnaire findings with over 92% of local authorities and 87% of mineral company respondents being able to give examples of mineral sites which had been restored to agriculture in the last 20 years. The second most common type of end use identified through the survey was nature conservation with over 75% of Mineral Companies and 92% of Local Authorities having examples.

The stakeholder interviews also identified agriculture and nature conservation as popular end uses, added to this was provision for informal leisure, water recreation and tourism, which were all, mentioned a number of times. The general feeling amongst the mineral company representatives was "you name it, we'll try it" (Mineral Company 3) as most of the operators boasted examples of a variety of uses including a rifle range, an arboretum, a graveyard, a fish farm, holiday homes, an energy plantation and an education center.

Amongst those interviewed there was an awareness of a change in restoration fashions over time (Spreull, 1992). There was general agreement that the current trend involves a movement away from agricultural restoration towards "a phase where most sites are put back to ecological use" (Mineral Company 4). The explanation most frequently offered was that nature conservation had become a high priority and had been "pushed up the agenda both externally and internally" (Mineral Company 2).

When selecting an end use for a site it does not need to be limited to a single end use, as a mixture of two or more activities on a site is becoming increasingly popular. Examples in the literature of successful combinations include multi purpose woodland (Stapleton, 1999) and a combination of nature conservation with education or landfill (Box et al, 1996). Other common combinations identified during the research included nature conservation and agriculture and nature conservation with leisure or recreation such as water sports. All 32 respondents believed mixed use to be a popular option perhaps summarized by the comment from one mineral company "most sites do have a mixed use, there's always quite a lot of nature conservation built into afteruse" (Mineral Company 5).

The following sections will examine a number of influential factors in order to discover the reasons why certain end uses, like nature conservation, are popular and why there is a tendency towards mixed use.

3 THE IMPACTS OF PLANNING AND POLICY REGULATIONS

3.1 Agricultural policies

This section will consider the potential effect of national and local policies on the decision making process, beginning with agricultural policy. The shift in agricultural policies in the UK over the past decade which although continuing to protect the best and most versatile land has offered the chance for rural diversification (Ward, 1997) and therefore opened up opportunities for land use change through mineral extraction. During the interviews, none of the respondents mentioned either an agricultural or rural economy policy by name however amongst the nine mineral company representatives there was an understanding that the best and most versatile land would be returned to agriculture on most occasions. As many mineral operations occur in rural areas on agricultural land this may go some way to explain the popularity of this option.

3.2 Forestry policy

The Forestry Policy for Great Britain 1991, and the White Paper "Rural England – A nation committed to a living countryside", 1995, both seek to expand woodland in England with the hope of doubling it by 2045 (Department of the Environment, 1996). One means of achieving this target is through the restoration of mineral sites to forestry. Initiatives such as the Community Woodland, Great North Forest and National Forest (Department of the Environment, 1996) were referred to numerous times during the interviews and seemed to have encouraged tree planting as all of the mineral companies questioned could cite numerous examples of woodland planting on their sites. However, although many respondents seemed keen on tree planting as a small part of a large site they were not favorable of forestry end use.

Secondary stakeholders can influence end use by helping to create policies that actively encourage or support the option they wish to promote for example, one of the non-governmental organizations questioned believed the best way to encourage forestry was to "get policies relating to the forests written into the local plans" as such "positive, sympathetic policies would encourage restoration and end uses of the type we want" (Other 1).

3.3 Sustainability and sustainable development

Since 1987, the concept of sustainability or sustainable development has become an integral part of modern life and there is likely to be a multitude of policies, all with some sustainability content, acting upon stakeholders. Interview respondents were asked whether they felt the concept had been influential on end use choice. One respondent replied that sustainability is "written through so many policies … it's a concern now for all organizations, particularly the Local Authority" (Other 1). Many of the 32 respondents used the term freely throughout the interview to refer to a multitude of issues including "a sustainable approach to the delivery of aggregates" (Local

Authority 1), "economic sustainability" (Mineral Company 6) and "sustainable land use" (Mineral Company 7). This has resulted in a belief that "everything we do has to be sustainable" (Mineral Company 4) with one company representative expressing that the company was "entirely driven by it" (Mineral Company 5). The research suggests that sustainability was a major concern for the 23 primary stakeholders when considering end use, perhaps best summarized by one mineral company representative who stated sustainable development is "always mentioned in our introductions to planning permissions" (Mineral Company 2).

3.4 Biodiversity Action Plan

The Biodiversity Action Plan (BAP) was produced by the UK government in 1994 in a response to Rio 1992 (Box & Hill, 1999). From reference to the national BAP, local authorities have produced their own local BAPs that have appropriate priorities and targets in terms of habitats and species for the local area (Humphries et al 2000). Mineral extraction can potentially contribute to local and national BAP targets through nature conservation end use (McDougall, 1997).

All but one of the interviewees were aware of the BAP, a further two respondents believed it had not been very influential as "it hasn't made us do anything more than what we were already doing" (Mineral Company 7). However the remaining 29 stakeholders all supported the notion that it had a positive influence on end use. The BAP was described as being "helpful" (Mineral Company 1) in identifying possibilities that could be achieved through mineral extraction with many using the local BAP to identify appropriate species and habitats for the area. The BAP seems to have fired the imagination of mineral operators with many keen to introduce target habitats or species to their sites. The most commonly mentioned contribution to BAP habitat was reedbeds and associated wildfowl, there was a belief held by some stakeholders that the introduction of the BAP had had a direct influence on the development of reedbed areas.

It may therefore be possible when considering the research findings, to make a link between the increasing inclusion of nature conservation in end use and the desire of stakeholders, particularly mineral companies, to achieve local and national BAP targets. This raises a further question of why primary stakeholders are keen to make a contribution to biodiversity; this will be examined further in later sections.

4 THE LOCAL AUTHORITY AND PLANNING CONSENT

Local authorities provide guidance on appropriate land uses within the Minerals Local Plan (Department of the Environment, 1996) this document should therefore offer strong guidance towards the local authority's attitude towards what is an appropriate end use. This was supported in the research findings as one local authority admitted, "you can gain planning permission relatively easily provided you're within the confines of the development plan" (Local Authority 1).

The literature suggests that local authorities could actively encourage some end uses through their support for rural diversification (RPS Clouston and Wye College, 1996), desire to meet BAP targets (Humphries et al, 2001), or need to increase amenities in the local area (Boyes, 1983). The research identified nature conservation, recreation and tourism as end uses that respondents believed the local authority encouraged or were likely to grant permission for. There was a feeling that "the planning authority have their own agenda" and "are pushing in certain directions" (Mineral Company 2) with a desire for economic investment being the most often cited reason for encouraging certain end uses.

One end use, which was identified as undesirable during the research, was open water as "more and more planners are saying no more holes in the ground – we don't want lakes" (Mineral Company 7). This was confirmed by a number of the 14 local authority representatives who reasoned that there had been so much restoration to open water in the past that villages were becoming islands and such an end use was no longer welcome in some localities.

Many of the mineral companies questioned felt under pressure to "put something in that's acceptable so they give us consent" (Mineral Company 3). This was reiterated by the majority of interviewees who believed that the ultimate decision on end use lay with the local authority. The local authority could therefore influence end use through refusing permission to an applicant for a certain type of end use. Therefore one motivating factor for end use selection is likely to be a desire to please the local authority due to the need to gain permission. The attitude of local authorities and desirability of certain end uses may therefore influence end use choice through either active encouragement or refusal of certain afteruses.

5   SITE PARAMETERS

5.1   Physical constraints

The literature suggests that end use choice can be limited by a number of physical factors including: mineral type, site location (McDougall, 1997), availability of fill, site size, land form, hydrology (Humphries et al, 2001), quality and quantity of soil material, gradient and exposure (Box & Hill, 1999). A number of

physical constraints highlighted during the case study visits will be discussed below.

The availability of material to fill the void left by mineral extraction may affect the options for end use. Fill may be in the form of controlled waste, inert material (Department of the Environment, 1996) or as was often used on the case study sites, pulverized fuel ash (PFA). A third of interview respondents made particular reference to the importance of fill material with regard to its availability and whether it could be transported to the site by road or pipeline. When applying for planning consent the applicant may need to include forecasts from a supplier of how much fill will be available for restoration and how it will be transported to the site. The local authority could refuse permission for the fill to be transported to the site by road, as was the situation on one of the case study sites, if a pipeline is not a feasible option, another end use may have to be accepted.

Site location may be an influential factor as respondents suggested that a site would be considered for different end uses depending on whether it was urban or rural. Possibilities for urban areas included recreation, amenity and built development. It was thought permission for built development was more likely in the proximity of other buildings and there was also a belief that the local authority would "push more heavily for public access" (Mineral Company 2) in an urban area and therefore lead to recreational provision being incorporated into the end use. The importance of blending any new land use into the landscape was highlighted as a key issue in rural areas as it was thought desirable to "create a landform that would fit harmoniously with the surrounding landscape" (Mineral Company 2). Uses that were thought to be acceptable for rural areas included agriculture, leisure and tourism.

5.2   Landownership

A further constraint on end use choice may be due to the introduction of a landowner; this is particularly relevant as Bishop (1991), believes there is a shift away from freehold acquisition of surface rights towards leasehold. The landlord is considered a primary stakeholder and will be viewed as a site parameter as the influence on end use will be determined by personal opinion and therefore site specific. The majority of landowners are farmers and most would like an end use which is profit making and cheap and easy to maintain, often settling for good quality agricultural land (Glen, 1994 and Boyes, 1983) however during the research examples were found of landowner farmers wanting a fishery lake and water based recreation. The landowners views must be considered at an early stage of the decision making process as there is "no point going to the trouble of creating a land use that the landlord is not going to maintain … it's demor-alizing" (Mineral Company 4). Although in some cases the landowner may be determined to have the land returned to its previous use, most mineral companies questioned had found there was room for negotiation.

To summarize the influence of site parameters, despite the existence of challenging circumstances such as fill availability, site location or a determined land-lord, there are such a plethora of possibilities for end use it may still be possible to have a choice. The follow-ing section will consider the level of influence of envi-ronmental factors on end use including the desire to achieve environmental improvements and benefits and the potential for involving environmental organizations.

6 ENVIRONMENTAL FACTORS

In the past mineral extraction often had a detrimental effect on the environment for many years until nature healed the scars, however in the present day mineral extraction, through restoration is believed to be capa-ble of enhancing the environment (Bradshaw et al, 1987). The potential for mineral extraction to improve the environment or offer environmental benefits, per-haps through habitat creation, could be considered as having an effect on the choice of end use, the follow-ing will consider whether this is feasible.

The survey of Local Authorities revealed a third of the 36 respondents believed environmental improve-ment was the reason for end use choice. All nine mineral companies interviewed believed their sites had environmental benefits and believed it was important to minimize the effect of extraction by offering envi-ronmental improvements on completion. Many of the environmental organizations and Local Authorities questioned agreed that the mineral industry were "more broadly aware now of their responsibilities to protect habitats" (Other 1). Environmental improvement was considered a high priority by over 90% of interviewees. These findings suggest that environmental improve-ment seems to be an accepted goal and environmental benefit highly desirable by all stakeholders. It could be argued that environmental improvement could be achieved through a number of end uses however when considering environmental benefit, this seems to sit well with nature conservation. Given the trend towards including nature conservation on a site, even as a minor element, one of the motivations behind it may be envi-ronmental benefit and improvement.

6.1 *The influence of environmental organizations*

As the discussion is currently focusing on environ-mental factors it seems appropriate to include the potential influence of environmental organizations.

Environmental organizations can have a varying level of involvement with end use for example, English Nature "want to help industry make decisions" so sites will include the maximum number of wildlife habitats and geological exposures (English Nature, 2001).

The research uncovered various levels of influence of environmental organizations over end use. This ranged from the environmental organizations play-ing a "large part in the decision making process" (Mineral Company 2) where as soon as a site is provi-sionally identified as a possibility for mineral extrac-tion, discussions begin with various environmental groups. One of the reasons given for involving such organizations early was "we need them on board from the start or we end up in a battle" (Mineral Company 8). It tended to be the larger mineral companies who approached environmental organizations at an early stage. At the other end of the scale the environmental organization's involvement was limited to advice about site management or project design once the decision on end use type had been made.

The five environmental organizations questioned perceived themselves as having little influence over end use choice although they acknowledged they were often consulted regarding project design and site manage-ment. One environmental organization representative believed to achieve their goals of environmental improve-ment and increased habitat it was preferable to "lobby the district and county councils about the economic good of having a good environment" (Environmental Organization 1). This he felt was the best means to influence end use. Some of those questioned believed environmental organizations "would like to see all mineral workings going back to conservation areas" (Mineral Company 2) which was reinforced by one environmental organization representative who stated he would prefer a single use site for nature conserva-tion "with big fences all round and limited access" (Environmental Organization 2). This suggests that secondary stakeholders such as environmental organi-zations hope to achieve different goals through end use than the primary stakeholders.

There is potential for environmental factors to influence end use either through the applicants desire to provide environmental benefits, perhaps to impress the local authority or through an environmental organ-ization being included in the decision making process. Both of these rely heavily on the applicant wanting to do something for the environment and being willing to involve another group in the talks about end use.

7 ECONOMIC FACTORS

7.1 *Restoration costs*

During the stakeholder interviews, respondents were asked whether the cost of restoration influenced the choice of end use, this resulted in a range of responses. At one end of the scale were comments including "economics come into the equation but they are not the single factor" (Mineral Company 4) and the term economic viability was mentioned a number of times. One company representative explained that their company writes the cost of restoration off across the whole of the mineral reserves by working out the full life restoration costs then basing it on a price per ton, usually 25–40 p per ton, effectively passing the cost onto the customer. He argued that this meant the cost of restoration does not influence the end use choice whereas another respondent when asked whether cost had an influence on end use said, "always got to say that it does" (Mineral Company 7). This suggests that the cost of end use is a consideration when choosing an option.

Another factor that was highlighted during the research was the influence of company size on ability or willingness to meet restoration costs. Bondesen (2002) believes that since the 1970s large mineral companies have become increasingly willing to spend money on restoration in order to recover their blighted image. However, one local authority representative believed the mineral company "resented the fact it cost them money" (Local Authority 3). Of the nine companies questioned, seven were large companies, one was a small county based company and the last, although a large company had suffered a decline in product value over recent years and may therefore be more concerned about the amount spent on restoration. One of the large companies had taken over a small company's site and found that "restoration may not have been that high on the agenda" (Mineral Company 9) and promptly applied to change the proposed end use from agriculture to nature conservation and informal leisure and found much time and money needed to be spent getting the site to the large company's usual standard.

Cost of restoration varies between different end uses and may therefore encourage stakeholders to opt for a particular type. For example, the Land Capability Consultants (1989, cited in Moffat & McNeill, 1994) found the cheapest option for reclamation was nature conservation at £4100 per hectare followed by woodland at £10,000 and agriculture at £16,700. The major categories of end use will be looked at in turn in terms of initial cost, return value and maintenance and management costs.

7.2 *Nature conservation*

Nature conservation as an end use is one of the most cost effective options with low capital costs and low maintenance costs (Box, 1996). However, nature conservation will not generate income and often requires long term funding to pay for maintenance and management (Land Use Consultants & Wardell Armstrong, 1996). A number of respondents believed that nature conservation was selected as an end use because it was a low cost option. However, this opinion was clearly resented by the mineral companies "there is an opinion that nature conservation is an easy option, a cheap option, it's not. Certainly not if it's done properly and managed into the future the way it should be. It's quite expensive because there's no revenue" (Mineral Company 7).

The mineral companies were keen to stress that they receive no monetary value from restoration to nature conservation and have the added financial responsibility of maintenance and management costs throughout the aftercare period. The 1981 Town and Country Planning Act requires an aftercare period of 5 years, however an increasing number of Local Authorities are seeking 10 year aftercare periods, especially for those end uses with a nature conservation element (Glen, 1994). This was found to be true during the research with four of the case study sites having 10 year aftercare periods and some of the mineral company representatives knew of sites with 15 year aftercare periods. The research found that nature conservation sites are often given to environmental organizations such as county wildlife trusts or in some cases the RSPB following the aftercare phase. Not only are the environmental organizations given the land they are also often given a lump sum to help pay for the maintenance and management of the site. This is a further cost that must be added to the economic equation.

One way that mineral companies have developed of overcoming the financial burden of nature conservation is to combine it with another use. One mineral company cited an example where nature conservation had been combined with agriculture; the agriculture generated revenue and helped support the nature conservation aspects. Another method of generating money to help towards the costs of upkeep of a site is to sell fishing permits to local anglers.

7.3 *Agriculture*

Restoration to agriculture continues to be a popular end use (McDougall, 1997), which could be due to its ability to provide an income (Stapleton, 1999). Maintenance costs for agriculture are also low at approximately £200 per hectare per year compared to recreation at £950 (Land Capability Consultants cited in Moffat & McNeill, 1994). None of the case study sites had been restored entirely to agriculture, however 5 of the 6 had an agricultural element, either arable or grazing land. In four out of the five cases the land was rented to tenant farmers, in the final case the mineral company director kept the agricultural land and described himself as a "hobby farmer" (Mineral

Company 6). It may be assumed that it is economically advantageous for the landlord to have some agricultural land to generate an income perhaps, as previously mentioned to support other aspects of the site.

7.4 Woodland and commercial forestry

According to the literature, reclamation to woodland is often cheaper in terms of both initial costs and maintenance than a return to agriculture (Bishop, 1991 and Forestry Commission, 1983). The reason for this lower cost, as suggested by Moffat & McNeill (1994) is that site leveling, importation of topsoil and drainage are usually not required.

During the interviews, respondents were asked whether they would consider commercial forestry as an afteruse, the response was universal amongst the mineral companies that it was not economically viable at present as the UK market could not compete with overseas prices. The only method of making it financially beneficial was through specializing for example; a walnut grove had been successfully developed on one former mineral site.

Woodland not only has low initial costs, it also has low maintenance costs, the cheapest option found by the Land Capability Consultants at £160 per hectare per year (Moffat & McNeill, 1994). Although forestry is seen as a "cost effective and sustainable afteruse" (p.34, Dutton and Bradshaw 1982 in Rollinson et al, 2000) it is still not a popular option as only 5% of mineral sites were restored to forestry between 1988–1994 (Simpson, 1996). Explanations in the literature include: woodland may be seen as a barrier to future development (Rollinson et al, 2000), responsibilities for management (Stapleton, 1999) and the time frame involved (Spreull, 1992) as aftercare is likely to be 10 years or more. However during the interviews although woodland planting was seen as a desirable part of a site, commercial forestry was discarded as lacking in return value.

To summarize the economic section, nature conservation and woodland although cheaper options should not be seen as easy options (McDougall, 1997). Long term arrangements for maintenance and management costs are important considerations if a site is to be successful (Humphries et al, 2001). Perhaps it could be viewed that stakeholders will consider any afteruse that is economically viable and it should be remembered, "the bottom line is winning the mineral and making a profit ... restoration comes as part of the package" (Mineral Company 9). Economic influences do appear to be an important factor in the decision making process and perhaps could be viewed as "the balancing factor" (p. 24 Lucas, 1983) when searching for motivation for end use.

8 SOCIAL FACTORS

8.1 Public demand

The social factors section will cover public demand for end uses and public relations, including public attitude towards mineral extraction and the importance of image.

Restored mineral sites can be an opportunity to meet the needs of local people. Public demand for high quality countryside and water areas for a variety of recreational uses (Lucas, 1983) is popular at the moment. However, the greatest demand is often simply for access onto restored sites (Moffat & McNeill, 1994). Hertfordshire County Council, in 1991, asked the local populace what end use they would like to see on former mineral sites, the most popular choices were nature reserves, recreation areas/parks and outdoor sports facilities (Spreull 1992).

Interview respondents were asked whether there was a public demand for certain end uses. Popular suggestions included mixed use sites, recreation, conservation or simply a desire for access. One mineral company representative explained the demand for leisure orientated sites as "people have got much more leisure time now ... people look to us to provide leisure time for them ... not just motorized water sports but bird watching" (Mineral Company 5). These suggestions of end use demanded by the public reflect demands from the general population however; perhaps potentially even more influential is the needs of the local community. The following will consider whether the indigenous population can influence end use option perhaps by becoming involved in the decision making process.

8.2 Influence of the local community on end use

There appears to be a number of levels on which the local community can be involved in end use decision making. Firstly, as practiced by three of the large companies questioned, consultations with the public before submitting an application in the hope that any concerns about the mineral operations or the end use can be discussed and solved before planning permission is sought. Consultation may take the form of an open meeting where ideas about end use are discussed. One local authority, of the 14 questioned, had tried to involve the public in end use selection by using a "Planning for Real" exercise in a public meeting however it was unsuccessful as "people are not aware of the possibilities ... they just want it to be "nice" or "better than it is" (Local Authority 3). The remaining mineral companies and local authorities were not keen to have open meetings with the public, reasons included: "wouldn't want to involve the public too much ... nothing worse than a public meeting"

(Mineral Company 3) and "it's like going into the lion's den" (Mineral Company 9).

Secondly, consultation prior to submitting an application may be in the form of an exhibition where the mineral company present their ideas for end use but the public still have the opportunity to offer suggestions and get the proposals amended. However in this instance the mineral company have already decided upon their chosen end use option and the public influence is likely to be limited to project design details such as the addition of paths, a bridleway or a bird hide etc.

In other cases, the application for planning consent, including end use option, would be submitted to the local authority, the local villagers would be informed by letter and given 21 days to comment. If any complaints were perceived as valid by the local authority then the applicant may be told to amend their plans. In some instances this results in action groups being set up to lobby for change. All stakeholders interviewed seemed more inclined to discuss issues with the local community than to deal with action groups. Such groups were described as having "their own agenda" (Local Authority 1) and "generally opposed to everything" (Environmental Organization 1).

The public in some cases may be able to influence end use if they are consulted before a planning application has been submitted, however, it is more likely that they will be encouraged to be involved in project design rather than end use choice. It should be noted that the local authority are "servants to members of the public" (Local Authority 4) and it is their responsibility to consider the effect of both mineral operations and end use on the local community, so in that, the public do have a representative in the decision making process. There was a feeling amongst many respondents from all groups that if too many people were involved in the decision making process from the start then nothing would be achieved, perhaps because different stakeholders have different interests.

8.3 Public attitude to the mineral industry

The public's "lingering fear" (p. 68, Bradshaw et al, 1987) towards the mineral industry has been caused by past poor quality or non-existent restoration until the late 1970s (Glen, 1994). Mineral companies may hope to improve their image through developing relations with the public and making them aware of examples of good restoration, such as the provision of recreational facilities to meet local need (Spreull, 1992). There are a number of reasons why the mineral sector would wish to improve its image, most notably to reduce objections to planning applications (Spreull, 1992). It should be considered that the desire to please the public and improve their image might result in a certain end use category. For example, Bishop (1991) suggests tree planting may improve a company's public image whilst Bate et al, (1998) believes a mineral company's involvement in BAP targets would be recognized by the public.

All respondents were asked what they thought the public's attitude was towards the mineral industry, responses varied from "they still hate them" (Local Authority 2) to regarding the public as "still prejudiced"(Mineral Company 4) to more positive feelings of "attitudes have changed dramatically over the last decade" (Mineral Company 9) and talk of "enlightenment" (Local Authority 5) where the public has seen what can be achieved through restoration. The general impression from responses was although public attitude towards the mineral industry may have softened in some areas, generally there is room for improvement. This raises the issue of whether a mineral company to improve its image would opt for a certain end use to please the public by providing a facility, meeting a demand or to impress them with quality restoration so as to ease them through the next planning application.

There is more pressure on large companies to improve their image and maintain a good standard of restoration, as they need to protect a national or international reputation. A representative from a large mineral company said "big organizations have a challenge to put their money where their mouth is – we're only as good as our last site" (Mineral Company 8). The large well known mineral companies may feel under increased pressure to produce impressive, good quality end uses to maintain or improve the company's reputation.

9 THE DECISION MAKING PROCESS

The responsibility for deciding on the intended after use of a mineral working rests jointly with the mineral company, the landowner and the local authority, whilst also taking into account national policy and planning regulations (Department of the Environment, 1989). Hence the reason these groups were classified as primary stakeholders. This was confirmed through the research, as there was a general consensus between all respondents that the mineral company and the local authority compiled a list of possible end use options for a site. However the decision making process regarding the end use of mineral sites is not as straight forward as it may seem as there are various other influences acting upon the primary stakeholders. Many of the Mineral Companies questioned stated that "lots of influences" (Mineral Company 1) from "various bodies" (Mineral Company 2) could influence the final choice. These other influences include secondary stakeholders, policy and planning regulations, site parameters and environmental, economic and social factors.

All sites have a set of individual circumstances, which will influence the end use choice, however, through the research it has been possible to identify a few factors that seem to be consistently capable of influencing end use choice.

Both policy and planning were found to be capable of encouraging certain end uses through various initiatives and the ease of gaining planning consent. The Biodiversity Action Plan was found to be particularly popular with primary and secondary stakeholders and may be one motivating factor for the inclusion of nature conservation on sites. The attitude of the local authority as expressed through the local plans was found capable of influencing end use choice, as applicants were only willing to submit options likely to gain consent. The physical constraints of some sites although causing more of a challenge in restoration could be overcome to allow at least a small number of end use options and were therefore not deemed a primary motivational factor. Economic factors were considered to be part of the decision making equation as sites need to be economically viable. The influence of the local community and environmental organizations over end use was found to be limited although they are often actively encouraged to be involved with project design. However, both these groups may be able to exert some influence through lobbying for changes in national or local policy.

10   CONCLUSIONS

Through the identification of the pure attitude of primary stakeholders and the determination of the potential effect exerted by secondary stakeholders, key areas of motivation can be surmised which have caused the trend in nature conservation and mixed end use.

The trend towards mixed use and nature conservation end use may be symptomatic of those involved in the decision making process. Where secondary stakeholders have been encouraged to become involved, end use may be more likely to incorporate public access, recreational pursuits and a wildlife element. The local authority and the mineral company have the most influence over end use, the motivating factors behind the decision made are likely to include a reference to policy, restoration cost in both the short and long term and the potential to improve both reputation in the local area and national image. There is potential influence from secondary stakeholders where the primary stakeholders have an understanding of what they demand from a site. Individual mineral companies and local authorities

determine the level of involvement of these secondary stakeholders.

All stakeholders were found to be environmentally aware and due to this strived to create good quality sites. The mineral industry seems to be taking the lead in advancing restoration as they have the most to lose and gain through the perception of a site. The primary driving forces behind end use choice were found to be environmental and social, although they were increasingly underpinned by the influence of secondary stakeholders.

The findings suggest that for the further advancement of end use and an increased ability to please a variety of groups of stakeholders, secondary stakeholders need to be involved in the decision making process from an early stage. This requires the primary stakeholders to develop acceptable means of including secondary stakeholders in the process that will lead to democratic negotiations and not cause them to be viewed as a nuisance because it makes the decision process longer.

REFERENCES

Alker, S., Joy, V., Roberts, P. & Smith, N. 2000 The Definition of Brownfield *Journal of Environmental Planning and Management* 43 (1) 49–69.

Andriof, J., Waddock, S., Husted, B. & Sutherland Rahman, S. 2002 *Unfolding Stakeholder Thinking: Theory, Responsibility and Engagement* Sheffield: Greenleaf Publishing Ltd.

Bate, R., Bate, J., Bradley, C., Peel, H. & Wilkinson, J. 1998 *The Potential contribution of the Mineral Extraction Industries to the UK Biodiversity Action Plan* English Nature Research Report No. 279.

Bishop, K. 1991 Community Forests: Implementing the concepts *The Planner* 77 (18) 6–10.

Bondesen, E. 2000 Development of Restoration of Raw Material Pits in Denmark *Journal of Environmental Planning and Management* 45 (1) 141–148.

Box, J., Mills, J. & Coppin, N. 1996 Natural Legacies: Mineral Workings and Nature Conservation *Mineral Planning* 68 24–27.

Box, J. 1996 Post-industrial Landscapes: Challenge and Opportunity *Landscapes – Perception, Recognition and Management: Reconciling the Impossible?* Conference Proceedings, Sheffield, 2–4 April, 117–124.

Box, J. & Hill, A. 1999 Mineral Extraction and Heathland Restoration *Mineral Planning* 80 (September) 5–8.

Boyes, K. 1983 Restoration of opencast coal sites *Quarry Management and Products* 10 (January) 25–28.

Bradshaw, A., Parry, G. & Johnson, M. 1987 Land Restoration – What is the future? *Transactions of the Institution of Mining and Metallurgy* A96 68–72.

Department of the Environment 1989 *Minerals Planning Guidance 7: The Reclamation of Mineral Workings* London: HMSO.

Department of the Environment 1996 *Minerals Planning Guidance 7: The Reclamation of Mineral Workings* London: HMSO.

English Nature 2001 Position Statement Aggregate Extraction and Nature Conservation URL: www.english-nature.org.uk 20/7/01.

Forestry Commission Research and Development Paper 132 1983 *Reclamation of Mineral Workings to Forestry* Edinburgh: Forestry Commission.

Glen, M. 1994 The Changing Role of Opencast Coal Restoration in Landscape Regeneration *Mineral Planning* 58 29–31.

Goodpaster, K., Maines, T. & Rovang, M. 2002 Chapter 2 in Andriof, J., Waddock, S., Husted, B. & Sutherland Rahman, S. *Unfolding Stakeholder Thinking: Theory, Responsibility and Engagement* Sheffield: Greenleaf Publishing Ltd.

Humphries, R., Foster, R. & Horton, P. 2000 Curse or Opportunity *Mining, Quarrying and Recycling* December 29 (10) 26–29.

Land Use Consultants & Wardell Armstrong 1996 *Reclamation of Damaged Land for Nature Conservation* London: HMSO.

Lucas, O. 1983 Design of Landform and Planting in Forestry Commission Research and Development Paper 132 *Reclamation of Mineral Workings to Forestry* Edinburgh: Forestry Commission.

McDougall, D. 1997 Agriculture or Alternative Use? *Mineral Planning* 73 (December) 27–30.

Moffat, A. & McNeill, J. 1994 *Reclaiming Disturbed Land for Forestry* London: HMSO.

Rollinson, T., Price, G. & Heslegrave, B. 2000 Forestry for Sustainable Regeneration *Mining, Quarrying and Recycling* November 29 (9) 33–36.

RPS Clouston & Wye College 1996 *The Reclamation of Mineral Workings: Executive Summaries of two new Research Reports from the Department of the Environment* London: HMSO.

Simpson, T. 1996 Survey of Land for Mineral Workings in England *Mineral Planning* 69 (December) 35–37.

Spreull, J. 1992 Restoration and the Community *Quarry Management* (June) 37–40.

Stapleton, C. 1999 Restoration to Forestry in England *Mineral Planning* 80 (September) 9–11.

Wallwork, K.L. 1974 *Derelict Land* Devon: David and Charles Ltd.

Ward, A. 1997 The Restoration and Aftercare of Mineral Sites: Retrospect and Prospects *Mineral Planning* 71 June 25–29.

Addressing Difficult Sites

Land Reclamation – Moore, Fox & Elliott (eds)
© *2003 Swets & Zeitlinger, Lisse, ISBN 90 5809 562 2*

Stabilisation of Galligu

D. Johnson
Lafarge Envirocem Solutions, Nottingham, UK

ABSTRACT: Native to North West England, especially in Runcorn and Widnes, *Galligu* is a soil of poor engineering quality containing heavy metals and other hazardous species. A novel solidification/stabilization method, known as Accelerated Carbonation Technology (ACT) converted the unstable soil into fill conforming with the UK Specification for highways, thereby avoiding the need to transport the material to landfill and to import clean fill. The treatment method used cement under ambient conditions of temperature and pressure to physically and chemically stabilize contaminants and simultaneously improve the engineering qualities of the soil. Unlike the usual hydration reactions that take place when cement is combined with water, ACT involves the mixing of cement with gaseous carbon dioxide to form relatively insoluble salts.

1 INTRODUCTION

The demonstration trial was one of a series of trials commissioned by Halton Borough Council in 2001 to treat potentially contaminated soil near Widnes. The intention of the Council was to redevelop the land in order to attract small industrial companies and possibly for domestic use. Two stabilisation/solidification (S/S) techniques would be compared – the first simply mixing soil with binder and the second using a recently patented treatment method known as Accelerated Carbonation Technology (ACT). Treatability studies were required to identify the optimum binders and additives for treating the soil using each technique.

1.1 *Galligu*

Galligu is a waste by product of the manufacture of sodium carbonate from the Leblanc process (Oxford University Press 1999) dating back to the nineteenth century; the salt was subsequently used in the production of dyes, soap and glass. The Leblanc process used coke, calcium carbonate, sulfuric acid and sodium chloride to produce the pure sodium carbonate; however, waste by products were numerous, the chief being calcium sulfide. Geographically, Lancashire supplied the coke whilst Cheshire supplied sodium chloride, the two meeting in the Runcorn/Widnes area (now the responsibility of Halton Borough Council), to become the centre of the early world Chemical Industry.

The waste by products were usually in the form of sulphurous muds, which were dumped into deep pits and covered with ash. The chemical composition and physical properties of this material have changed over the years to produce a wide variety of materials, ranging from tars to glass-like solids. In addition, impurities from soap residues have produced layers with different colours. The waste, therefore, is a mixture of *Galligu*, burnt pyrites (iron sulfide), metal processing residues, furnace ash clinker, general works rubbish and fatty residues from soap manufacture. Owing to the historical nature of the site, other chemical wastes, such as asbestos and mustard gas (phosgene) were thought to be present.

"Pure" *Galligu* is black in appearance before oxidation, changing to yellow-green and finally to white. Calcium sulfide oxidizes on weathering to form calcium sulfate and calcium carbonate plus sulfate/ carbonate complex salts; the degree of oxidation depends on the degree of weathering. The waste may be present as a hard fused mass or a soft thixotropic material with a consistency of toothpaste. From an engineering perspective, the land is unstable in places and moves under loading, as the pits were often up to 15 metres in depth.

Chemical analysis of the soil and groundwater showed that there were localized areas where contamination was very high, and areas where little contamination existed. Limited groundwater data indicated high levels of heavy metals, especially arsenic, lead, chromium, copper, nickel and zinc, although again, these were localised. Chemical problems associated with the area have included extremes of pH, ranging from acidic through to alkaline, hydrogen sulfide odours, soil capable of self combustion, ground heave caused by formation of sulfates and extremes of physical properties, ranging from soft to hard concrete.

1.2 The technology

The use of hydraulic cements (i.e. those, which chemically react with water to form a hard mass) for soil treatment is well established (Glasser 1997), especially in the USA (British Cement Association 2001). S/S is a combination of two processes, each defined as follows (Conner 1991):

- Stabilisation, which renders soil constituents less leachable; and
- Solidification, which improves the structural integrity of the soil.

"Traditional" S/S technology utilizes the hydration of Portland cement to chemically stabilise and physically encapsulate soil components. However, the complex nature of many waste materials is known to inhibit the chemical reactions that are responsible for effective cement-based S/S (Hills & Pollard 1997).

Exposure of cement matrices to carbon dioxide at early age, i.e. when cement is in a "green" state or, prior to the normal hydration reactions proceeding to any degree, has been found to overcome these deleterious interactions (Sweeney et al 1998). Atmospheric carbonation is however, a slow, natural phenomenon of cement and concrete that are exposed to air. During ACT, this process is accelerated by exposure of waste/cement mixtures to gaseous carbon dioxide at atmospheric pressure.

Processes based on accelerated carbonation date back to the mid-nineteenth century (Maries 1992) where it was used primarily as a means of hardening cement. Several researchers have studied the role of carbonation in stabilising metallic wastes (Bonen & Sarker 1994) and have shown that the process significantly alters the initial chemistry of a cement matrix (Smith & Walton 1991). Calcium carbonate is precipitated within the pore structure of the matrix by reaction of calcium hydroxide with carbon dioxide, which significantly affects the pH of the pore fluid.

The improved leaching characteristics observed with a carbonated cement system in comparison to a hydrated cement system is partly attributed to the effect on the pH of the pore fluid, which drops from about 12.5 to 8.3 in a fully carbonated system. At this pH, most metal hydroxides are close to their minimum solubility (Conner 1991). Furthermore, isomorphic substitution can result in the formation of metal carbonates that are less soluble than the corresponding metal hydroxides. The precipitation of calcium-metal-carbonate double salts and sorption of components

into the solid phases within the cement matrix further improves the performance of the carbonated system.

Although the nature of reactions involving anhydrous cement with carbon dioxide in the presence of water are still under investigation, it is clear that there are significant differences in comparison with hydrated cement. The following mechanism has been postulated (Maries 1998) for the process:

1. Solvation of the gaseous carbon dioxide to form aqueous carbon dioxide.
2. Hydration of aqueous carbon dioxide to form carbonic acid.
3. Ionisation of carbonic acid to form various species, which lower the pH.
4. Oxidation and reduction of metals to form metal cations.
5. Dissolution of the cement hydration phases.
6. Nucleation of calcium carbonate, calcium-metal silicate hydrates and calcium-metal carbonates.
7. Precipitation of solid phases and conversion of calcium silicate hydrate to metal silicates and calcium carbonate.

The above reactions take place in a short time period, in as little as a few minutes following a brief induction period. The consumption of CO_2 is significant; as much as 30% by weight of binder of CO_2 has been observed in the laboratory and during a previous field demonstration.

2 TREATABILITY STUDY

The number of hydraulic binders is large and it was necessary to identify the optimum binder(s) for both S/S and ACT, which would provide the desired effects. Historical data had indicated that the degree of contamination and physical properties were likely to be variable and therefore material for this study would need to be representative of the material, which would be available for the site trial. Since a site characterisation study was outside the scope of the trial, several samples were taken from the site and combined to produce a composite sample (approximately 100 kg).

Stones and bricks larger than 6 mm were removed in order to avoid damaging mixing equipment. The pilot plant, based at Lafarge Envirocem Solutions, allowed ACT treatment of 25 kg batches of soil, thereby providing processing conditions closer to site compared to those in the laboratory. Chemical contamination was found to be relatively minor (Table 1), with many contaminants being below their respective ICRCL limits (Interdepartmental Committee on the Redevelopment of Contaminated Land 1987) with respect to use for parks and playing fields; only arsenic and copper exceeded the threshold limit.

Table 1. Contaminants present in the soil compared with ICRCL intervention values for parks and playing fields.

Species	Soil (mg/kg)	Threshold (mg/kg)
Arsenic	130	40
Cadmium	<0.5	15
Chromium	12	1000
Lead	300	2000
Mercury	1	20
Selenium	1.6	6
Copper	320	130
Nickel	20	70
Zinc	220	300
Boron	0.1	3

Table 2. Effect of S/S binders and additives on *Galligu*.

Material	Binder	Level (%)	Additive	Dosage (%)
Galligu	None	0	None	0
TSS1	None	0	PFA	30
TSS2	E650SR	10	E-HL	10
TSS3	E650SR	5	PFA	30
TSS4	E650SR	20	E-HL	10
TSS5	SRPC	20	E-HL	10
TSS6	EHB	10	None	0
TSS7	EHB	10	PFA	20
Soil	None	0	None	0

Table 3. Effect of binders on grading.

Material	Consistency	Particles larger than			
		10 mm (%)	5 mm (%)	2 mm (%)	1 mm (%)
TSS1	Damp powder	10.4	4.1	9.8	17.1
TSS2	Hard lumps	73.3	13.4	6.7	2.8
TSS3	Hard lumps	52.1	21.2	17.5	5.1
TSS4	Hard lumps	70.6	14.2	8.3	2.9
TSS5	Hard lumps	20.4	21.8	43.5	8.1
TSS6	Hard lumps	14.6	14.1	13.3	12.5
TSS7	Granular	5.7	6.5	16.4	13.0
Soil	Granular	0	40.0	21.7	11.1

Therefore, all decisions relating to the optimum binder/additive were based on the physical effects of the treatment methods, as this was considered to be the more relevant. However, pH was monitored before and after treatment as this is known to influence the mobility of metals (Conner 1991). The optimum treatment in each case would be based on the ability to transform *Galligu* into material similar to soil, using the local material as a typical example since soil varies considerably around the country.

The chief problem with *Galligu* is the relatively high water content for such a gritty material (25%) and a means of reducing excess water was required. Physical methods such as centrifugation or filter presses were known to be expensive and time consuming. Chemical methods based on the use of sorbents and special binders were examined. Two water-absorbing additives commonly used in the UK Civil Engineering industry are pulverized fuel ash (PFA) and EnvirOceM HL (E-HL). Special binders based on calcium sulfoaluminate (CSA) chemically combine with water to form stable hydrates (Brown 1993).

2.1 Stabilisation/Solidification (S/S)

Galligu was mixed with additive and binder in the pilot plant mixer for 5 minutes prior to a 24-hour curing period under ambient conditions. The additive was mixed for 2 minutes prior to addition of the binder as detailed in Table 2. Samples before and after treatment were subjected to a number of physical and chemical tests to identify the optimum treatment method for producing granular material, which could be re-used on the site.

PFA produced a heavy wet powder (TSS1), with poor durability whilst EnvirOceM HL produced a semi-dry granular material. The use of a binder was found to be necessary to produce a durable material, which would not break down on exposure to weathering conditions.

Binders based on Portland cement (E650SR and SRPC) produced hard lumps (Table 3), even with the use of additives, which would require further processing, such as shredding, on site to convert into a usable material. The use of the CSA-based binder (EHB) without PFA produced a hard monolith, which was difficult to reduce in size. The CSA-based binder was not tested with EnvirOceM HL, as there were known to be problems with respect to chemical compatibility.

The pH of the treated material was dependent upon the binder, with those based on Portland cement generally giving values in excess of 12 and CSA-based binder giving values below 10 (Table 4), the exception being TSS3.

The optimum binder in terms of physical and chemical properties was EnvirOceM HB (EHB) in combination with PFA (TSS7), which produced a granular material with density and particle size distribution similar to that of the local soil. This binder/additive combination was, therefore, chosen for the S/S demonstration trial.

2.2 Accelerated Carbonation Technology (ACT)

Galligu was mixed with additive and binder in the pilot plant mixer for 5 minutes prior to 20 minute

Table 4. Effect of binders on pH.

Material	pH
TSS1	8.6
TSS2	12.1
TSS3	8.8
TSS4	12.0
TSS5	12.1
TSS6	8.8
TSS7	9.3

Table 5. Effect of ACT binders and additives on *Galligu*.

Material	Binder	Level (%)	Additive	Dosage (%)	Density (kg/m^3)
Galligu	None	0	None	0	1769
TACT1	None	0	PFA	30	1918
TACT2	E650SR	10	E-HL	10	1346
TACT3	E650SR	5	PFA	30	1412
TACT4	E650SR	20	E-HL	10	1239
TACT5	SRPC	20	E-HL	10	1103
Soil	None	0	None	0	1393

Table 6. Effect of ACT on pH.

Material	pH initial	pH after ACT
TACT1	8.6	8.6
TACT2	12.1	9.9
TACT3	8.8	8.6
TACT4	12.0	11.1
TACT5	12.1	12.0

Table 7. Effect of ACT on grading.

Material	Consistency	Particles larger than			
		10 mm (%)	5 mm (%)	2 mm (%)	1 mm (%)
TACT1	Damp powder	4.6	4.2	21.4	6.6
TACT2	Granular	1.1	19.9	26.6	15.0
TACT3	Granular	4.5	4.4	29.6	20.3
TACT4	Granular	26.1	12.3	30.0	16.1
TACT5	Hard lumps	59.5	16.5	17.1	4.3
Soil	Granular	0	40.0	21.7	11.1

Table 8. Binders/additives chosen for demonstration trials.

Code	Materials
S/S	EnvirOceM HB + PFA
ACT1	EnvirOceM 650 SR + EnvirOceM HL
ACT2	EnvirOceM 650 SR + PFA

mixing in a carbon dioxide atmosphere. The additive was mixed for 2 minutes prior to addition of the binder. The high-speed enclosed mixer had been modified to allow the injection of the gas and removal of exhaust gases through a carbon filter. Samples before and after treatment were subjected to a number of physical and chemical tests to identify the optimum treatment method for producing granular material, which could be re-used on the site (Table 5).

Carbonation was accompanied in each case by a rapid increase in temperature from about 15°C to over 50°C in some instances (e.g. TACT2). ACT treatment produced densities similar to local soil, especially TACT2 and TACT3; the latter however, showed reduced degree of carbonation (as characterised by the lower rise in temperature, from 23°C to 34°C).

As discussed above, carbonation reduced the pH of the treated material to below 10 (Table 6), which may be beneficial in some instances.

The particle size distributions of TACT2 and TACT3 (Table 7) were similar to that of local soil and as a consequence, these treatments were chosen for the ACT demonstration trial.

3 RESULTS

3.1 Demonstration trials

In accordance with local environmental regulations and stringent site safety practice, a completely contained area was designed for conducting the trials and for the construction of a series of cells/units in which materials were placed for longer-term monitoring following treatment. The test units were metal skips, capable of storing the material, which could be moved from the premises when the owner needed to use the site. The laboratory-based treatability study provided the mix designs for both the S/S and ACT trials (Table 8). Material for treatment was taken from several areas across the depot site and mixed to form a single stockpile.

Treatments were carried out on a batch-mixing basis with each batch comprising of approximately 4 m^3 of material. A computerised Pugmill blended soil with the required dosage of binder and additive before transferring into the rotating carbonation chamber for exposure to carbon dioxide for approximately 20 minutes. The gas was added at a controlled rate to the slowly rotating carbonation drum, first purging the air then maintaining a CO_2 saturated atmosphere at normal atmospheric pressure for the duration of the carbonation reaction. The drum rotated at approximately 10 revolutions per minute to agitate and keep material continuously exposed to gas. S/S treated batches were transferred from the Pugmill mixer directly into the cell.

The carbonation reaction was observed to be very vigorous and characterised by a rapid rise in temperature of 50–60°C. In comparison, the S/S batch showed moderate temperature increase of 10°C, which may have been due to mass mixing effects. The exhaust gases were monitored at the chamber door, in the exhaust itself and in the chamber to assess quality and type of emission and to ensure that all emissions were removed from the exhaust stream to maintain ambient air quality. Following carbonation, the treated materials were, in turn, placed within the test cells.

3.2 Geotechnical test results

Untreated *Galligu* in its various stages of weathering is an industrial waste product; however, in order to compare the treated and untreated materials from an engineering perspective, all samples were tested in accordance with the appropriate standard for soils, BS 1377 (British Standards Institution 1990).

3.2.1 Description of soils

The material investigated at the site varied considerably in physical consistency both laterally and in depth. The greatest influence on the strength and consistency were the degree of weathering and its proximity to groundwater. The former caused the *Galligu* to become more granular and soil-like, whilst the latter had a softening effect. The weathered and unweathered materials behaved in a thixotropic manner when wet, forming paste-like materials.

Both the ACT and S/S treatment processes produced soil-like material, which demonstrated typical granular characteristics when subjected to the standard 2.5 kg compaction test under BS 1377 (Table 9). The maximum dry densities in the range 1.03 to 1.28 Mg/m³ were low compared with figures for "normal" soils, which are typically in the range 1.56 to 2.07 Mg/m³ (Transport and Road Research Laboratory 1974). Optimum moisture contents were high (30–36%) relative to "normal" soil (typically 9–28%). This behaviour is probably indicative of the formative constituents of the untreated *Galligu*.

3.2.2 Compaction tests

The results of the 10% Fines value (Table 10) showed that ACT1 would be suitable as a capping layer in road construction, whilst ACT2 and the S/S treated material might be used as a capping layer fill (Department of Transport 1991a).

Exceptionally high Californian Bearing Ratio (CBR) values were obtained for samples treated by ACT and S/S, with values in excess of 15 recorded (Table 11), thereby allowing utilisation as a capping layer for road construction (Department of Transport 1991b).

Table 9. Soil characteristics of treated *Galligu*.

Treatment	Moisture content (%)	Specific gravity (Mg/m³)	Maximum dry density (Mg/m³)
S/S	34	2.25	1.25
ACT1	30	2.23	1.03
ACT2	36	2.22	1.28

Table 10. 10% Fines value of treated *Galligu*.

Treatment	Value (kN)	Class	Use
S/S	40	F1	Capping layer
ACT1	50	B	Starter layer
ACT2	25	F1	Capping layer

Table 11. Californian Bearing Ratio of treated *Galligu*.

Treatment	CBR	Use
S/S	35	Capping layer
ACT1	33	Capping layer
ACT2	78	Capping layer

A frost heave value of 8.7 mm was obtained for ACT1, which indicated that the material could be defined as being non-frost susceptible (Department of Transport 1991c).

3.3 Chemical test results

Three samples were taken from each test cell at 0.3 m depth and subjected to soils analysis and leachate testing (Lewin et al. 1994). ICRCL Guidelines and UK Drinking Water Limits (The Surface Waters Regulations 1996) were employed to assess the effectiveness of the remediation treatments.

3.3.1 Chemical composition

Chemical analysis of the treated materials showed that the heavy metal content was either reduced or remained unchanged (Table 12).

Contamination with respect to heavy metals was relatively low, with only arsenic above the ICRCL threshold limit; mercury, copper and zinc were below but close to the limit. ACT treatment (ACT1) reduced metal contents, with arsenic being reduced to below the threshold limit for use in parks and playing fields. S/S had little effect on heavy metal content, although zinc was reduced. The arsenic figure was sufficiently high to preclude use in parks and playing fields.

Table 12. Effects of S/S and ACT on chemical composition.

Species	None (mg/kg)	S/S (mg/kg)	ACT1 (mg/kg)	ACT2 (mg/kg)
Arsenic	84.3	80.0	36.7	57.0
Cadmium	1.0	0.9	0.6	0.5
Chromium	13.3	20.3	12.0	14.3
Lead	216.7	180.0	111.3	106.7
Mercury	1.5	1.3	0.4	0.7
Selenium	1.4	1.5	0.5	0.7
Copper	120.0	113.3	67.0	78.0
Nickel	18.0	22.3	14.7	17.7
Zinc	273.3	230.0	216.7	190.0
Boron	1.7	1.4	< 0.1	1.1

Table 13. Effect of treatment on leaching of metals.

Species	None μg/l	S/S μg/l	ACT1 μg/l	ACT2 μg/l	DW1 μg/l
Arsenic	10.3	12.3	<10	<10	50
Cadmium	<0.5	<0.5	<0.5	<0.5	5
Chromium	<10	113.3	16	49	50
Lead	<10	<10	<10	<10	50
Mercury	<0.2	<0.2	<0.2	<0.2	1
Selenium	<2	<2	<2	<2	10
Copper	<10	<10	<10	<10	50
Nickel	<10	<10	<10	<10	*NL
Zinc	26	18.7	11	21	3000
Boron	<50	<50	<50	<50	*NL

*NL = No limit

3.3.2 Leachate analysis

Leachate analysis showed little release of heavy metals, values in general being below their respective detection limits, with the exception of arsenic and zinc (Table 13). ACT reduced mobility of these metals, but increased chromium, which may be related to the presence of the metal in the binder or additive, since the chromium level in the *Galligu* was not of any great significance (Table 12). Chromium mobility was increased further by S/S to a figure above the limit for drinking water (DW1).

4 DISCUSSION

The purpose of the demonstration trial was to assess the effects of two solidification/stabilisation techniques on *Galligu*. The ultimate aim of the study was to assess the potential for re-using treated material on the site; consequently, the engineering capability, chemical composition and durability of the treated material were considered in detail. It is important to note that both techniques are quite different in their respective mechanisms and should not be directly compared as

Table 14. Effect of weathering on pH of treated material.

Treatment	Initial	3 months
S/S	8.7	8.4
ACT1	>12.0	11.3
ACT2	9.9	8.2

different binders were employed. In addition, both techniques produce different materials; thus, whereas S/S produces a material, which is normally, compacted to produce a solid mass, ACT generally produces granular material suitable as engineering fill.

Galligu was known to be a particularly difficult material in terms of its load bearing capability, varying from a hard rigid material to a soft paste. Both the ACT and S/S processes produced very significant physical changes, transforming in most cases a lumpy slurry-like material into one suitable for use as a starter layer fill material in accordance with Department of Transport Specifications. In addition, chemical data showed little tendency for heavy metals to leach into groundwater, with the exception of chromium, which was thought to be present in the binder or additive.

S/S initially produced a granular material, which hardened when compacted within one hour of placement in the test cell. Such a monolith would be expected to resist flow of water and may be suitable as cement-bound material for construction of road sub bases (Department of Transport 1991c).

ACT1 appeared to be the optimum treatment method for producing granular fill based on the physical and chemical properties of the treated material. The initial high pH of the treated material decreased during a 3-month weathering period (Table 14).

The carbonation process was very rapid, involving contact times of 20 minutes or less; since time is an important factor in any remediation scheme, this would not only impact upon the length of the project but also on the cost. The reaction was highly exothermic, which could also be beneficial in removing water and fumes. This exotherm could potentially be employed to provide an alternative heat source to the electric vaporisers, which are used to convert the liquid carbon dioxide to the gaseous form. The carbonation chamber, being self contained, reduced release of fumes and dust to the atmosphere.

From the financial perspective, the ACT process used similar equipment to that employed for the stabilisation of contaminated soil, with the additional use of a carbonation chamber and a carbon dioxide delivery system. Balanced against these potential extra costs would be the savings obtained by avoiding the use of imported fill and landfilling the *Galligu*. The ACT process lends itself well to a wide range of project sizes, as the basic treatment plant can be mobilised and set-up rapidly on a relatively small footprint and excavated materials can be pretreated (i.e. crushed, sorted, shredded as necessary), loaded, treated and re-deposited as a continuous operation.

The ACT plant is licensable as an ex-situ treatment process under the Environment Agency MPL (Mobile Plant License) system and as such all potential environment impacts of the process were dealt with as part of the site-specific working plan. ACT did not require the use of any intrinsically hazardous materials and most potential impacts were dealt with by using conventional noise and dust abatement measures. Any adverse gas emissions were identified by the laboratory benchmarking programme and necessary mitigation measures implemented prior to operations commencing on site.

The Environment Agency view was that the trial indicated that ACT was a potentially suitable treatment for the physical stabilisation of alkaline *Galligu* waste. However, more detailed work would be needed to demonstrate that the material could meet a design specification. There was uncertainty regarding the wider use of ACT for the immobilisation of metals on this site, owing to the relatively narrow range of compositions tested during the trial. In general data should be interpreted with caution because of the small sample size. Any treatment proposals would need to examine the potential for contact with acidic groundwater and the implications for both strength of the material and mobility of the metals.

5 CONCLUSIONS

1. Both ACT and S/S converted poor quality material into granular aggregate, which could be reused on the site as engineering fill conforming to regulatory specifications. In addition, S/S could potentially be used to construct solid sub-bases for roads.
2. The two ACT treatments used different additives, which affected the physical and chemical properties of the end product. Thus, whilst ACT1 produced a hard aggregate suitable as granular fill, ACT2 produced a weaker material suitable as a capping layer fill. Similarly, ACT1 leached less metals in the groundwater compared with ACT2, but the higher initial pH exhibited by the former may preclude usage under certain circumstances.
3. Portland cement mixed with *Galligu* produced hard monoliths, which would require further processing to produce a usable material. Alternative binders based on calcium sulfoaluminate produced weaker material, which allowed conversion of the soil into a fill suitable as a capping layer.
4. Although *Galligu* was a difficult and large volume material the trial did demonstrate the viability of a processed-based solution. It would, however, be prudent to commission more trials, following a comprehensive risk study, to reduce some of the uncertainty associated with the range of characteristics exhibited by this and associated co-disposed wastes on a particular site.

REFERENCES

Bonen D. & Sarker S. 1994. The present state of the art of immobilisation of hazardous heavy metals in cement-based materials. *Advances in cement and concrete.* (eds) M.W. Gnitzeck & S.D. Harker. American Society of Engineers.

British Cement Association. 2001. Cement-based stabilisation and solidification for the remediation of contaminated land. *The British Cement Association.*

British Standards Institution. 1990. Methods of test for soils for civil engineering purposes. Classification tests. *British Standards Institution, London:* BS 1377:Part 2.

Brown A D R. 1993. Applications of calcium sulfoaluminate cements in the 21st century. *Proceedings of the International Conference, Dundee.* E & F N Spon. Concrete 2000: 1773–1785.

Conner, J. 1991. Solidification and Stabilisation of Wastes using Portland Cement. *Portland Cement Association.*

Department of Transport. 1991a. Specification for Highway Works. Series 600 Earthworks. *HMSO London.*

Department of Transport. 1991b. Design manual for roads and bridges. Series 2000. *HMSO London.*

Department of Transport. 1991c. Manual of Contract Document for Highway Works. Volume 1. Specification for Highway Works. Series 1000. Road pavements – Concrete and Cement Bound Materials. *HMSO London.*

Glasser, F.P. 1997. Fundamental aspects of cement solidification and stabilisation *Journal of Hazardous Materials* Vol 52: 151–170.

Hills C.D. & Pollard S.J.T. 1997. The influence of interference effects on the mechanical, microstructural and fixation characteristics of cement-solidified hazardous waste forms. *Journal of Hazardous Material.* Vol. 52: 171–191.

Interdepartmental Committee on the Redevelopment of Contaminated Land. 1987. Guidance on the assessment of redevelopment of contaminated land. *Department of the Environment, London.* 2nd Edition. ICRCL Paper 59/83.

Lewin K et al. 1994. Leaching tests for the assessment of contaminated land: Interim NRA guidance. *Bristol, UK:* R & D Note 301.

Maries A. 1992. The Activation of Portland Cement by Carbon Dioxide, *Conference in Cement and Concrete Science,* Institute of Materials, University of Oxford, UK, 21–22 Sept.

Maries A. 1998. Utilisation of carbon dioxide in concrete construction. *CANMET/ACI International symposium on sustainable development of the cement and concrete industry:* 433–443.

Oxford University Press. 1999. A Dictionary of Science. Market House Books Ltd.

Smith R.W. & Walton J.C. 1991. The effects of calcite solid solution formation on the transient release of radionuclides from concrete barriers. *Scientific Basis for Nuclear waste management*. Vol. XIV (212): 403–409.

Sweeney R.E.H. et al. 1998. Investigation into the carbonation of stabilised/solidifed synthetic waste. *Environmental Technology*. Volume 19 : 893–902.

The Surface Waters (Abstraction for Drinking Water) (Classification) Regulations. 1996. *HMSO, London*.

Transport and Road Research Laboratory. 1974. Soil mechanics for road engineers. *HMSO London*: 166.

Land Reclamation – Moore, Fox & Elliott (eds)
© 2003 Swets & Zeitlinger, Lisse, ISBN 90 5809 562 2

The migration of volatile organic compounds in soil vapour to indoor air and the assessment of the associated impacts on human receptors

G. Digges La Touche & L. Heasman
M J Carter Associates, Atherstone, Warwickshire, UK

ABSTRACT: The migration of volatile organic compounds (VOCs) from contaminated soil and groundwater to indoor air often represents the main pathway for exposure to VOCs in contaminated soil and groundwater. Traditionally risk assessment models are based on simple two phase equilibrium partitioning between either the soil vapour and soil or between soil vapour and groundwater. No account is taken of the complexity of partitioning between the three phases and vapour migration from each phase is assessed separately. The methodology presented in this paper makes use of soil vapour survey data to overcome the limitations of the two phase equilibrium partitioning approach. Examples of applications of the methodology at a variety of industrial and residential sites in England and France are presented to illustrate the practical application.

1   INTRODUCTION

Predicting the migration of volatile organic compounds (VOCs) from the ground to indoor air is important for estimating potential risks to human health from soil and groundwater contaminated with VOCs. VOCs include compounds such as benzene and trichloroethene and are common contaminants on industrial sites with a history of manufacture, use or storage of fuels and or solvents. With the United Kingdom Governments ambition of encouraging residential housing development on previously used or contaminated land, otherwise known as brownfield sites, the development of effective methodologies for estimating risks to human health is of increasing importance.

In this paper we present a methodology for assessing the potential risks from the migration of soil vapour to indoor air using data from soil vapour surveys rather than the results of the analysis of soil and or groundwater samples.

2   BACKGROUND

Standard site investigation methodologies for contaminated land sites as embodied in standards such as BS 10175:2001 (BSI 2001) recommend that soil vapour surveys for VOCs should be used only as part of a screening process to identify areas of VOC contaminated soil and groundwater and that the soil vapour surveys should be followed up by obtaining samples of soil and groundwater in the areas of VOC contamination identified during the soil vapour survey in order to evaluate the risk. The collection of soil and groundwater samples in areas of VOC contamination is important in the assessment of the risk to groundwater and to human receptors through ingestion, dermal contact and dust inhalation. However, the most commonly used risk assessment models rely on simple two phase equilibrium partitioning between either the soil and the soil vapour or groundwater and the soil vapour. In these models no account is taken of the complexity of partitioning between the three phases and vapour migration from soil and from groundwater is assessed separately. In this paper we present a methodology which in our opinion and experience overcomes some of the major limitations of the simple two phase equilibrium partitioning approach.

The migration of VOCs from contaminated soil and from groundwater into indoor air often represent the main pathways for exposure to VOCs. In an industrial setting exposure by dermal contact, accidental ingestion or inhalation of site derived dust are with the exception of construction or remediation workers typically minimal as many sites are covered by concrete hardstanding or tarmac. In a residential setting exposure to contaminants by dermal contact accidental ingestion or inhalation of contaminated dust can be minimised by the placement of clean soil over the surface to a depth of 1m or more. The migration of vapour from deeper contaminated soil or from ground-water may form the principal exposure pathway. In circumstances where an off site source of VOC contamination results in the presence of VOC contaminated groundwater beneath properties as at Sawston in Cambridgeshire (Cambridge County Council 2000) the ingress of vapours to the properties can represent a significant exposure pathway.

3 DATA COLLECTION METHODOLOGY

The methodology proposed uses a combination of data from the analysis of soil and groundwater samples and data from soil vapour surveys. VOC concentrations in soil and groundwater are determined by gas chromatography and mass spectrometry (GC/MS) and the total concentration of VOCs in soil vapour are determined by the analysis of soil vapour. Samples of soil vapour are obtained either by purging soil vapour from monitoring wells installed in boreholes where the well is screened in the vadose or unsaturated zone or from narrow diameter probe holes drilled using either a hand driven spike bar or a hand held rotary drill to approximately one metre below ground level. Once the hole is drilled a narrow diameter pipe is inserted into the hole and vapour is pumped into the analytical device. Typically either a photo-ionization detector (PID) or a flame ionization detector (FID) is used for the analysis of VOC concentrations in the field as analysis effectively is instantaneous. Portable gas chromatographs (GC) are available and may be used as can absorption tubes which are subsequently analysed in the laboratory. The latter options have the benefit of allowing speciation of the potential compounds of concern. It is imperative that samples are collected at low flow rates to avoid depletion and dilution. When samples are collected on absorption tubes the volume of gas sampled must be recorded in order to facilitate conversion to a concentration from the mass absorbed on the tube.

4 VAPOUR RISK ASSESSMENT METHODOLOGY

In most risk assessment methodologies such as the Environment Agency (EA) and the Department for Environment, Food & Rural Affairs (DEFRA) Contaminated Land Exposure Assessment (CLEA) model (Environment Agency 2002) or the American Society for Testing and Materials (ASTM) risk based corrective action (RBCA) model (ASTM 1995) indoor air concentrations are predicted for soil and groundwater separately. The CLEA model does not consider groundwater as a potential contaminant source for the migration of soil vapour to indoor air.

For each phase whether soil or groundwater equilibrium vapour concentrations are calculated. The migration of soil vapour to indoor air is then calculated to estimate the indoor air concentration. The calculated indoor air concentration is used in a calculation of the potential exposure of occupants of the building and the health risks associated with the exposure are assessed. Similar risk assessment methodologies have been published by the Scottish & Northern Ireland Forum for Environmental Research (2000) and the United States Environmental Protection Agency (1996).

The main problem with the models or methodologies described is that the soil to indoor air and groundwater to indoor air pathways are considered separately. While this approach is in many cases pragmatic it is fundamentally flawed as the concentration of VOCs in soil vapour is dependant on the concentration of VOCs in both the soil and the groundwater.

We propose that for the assessment of risk from the migration of VOCs from soil or groundwater to indoor air the measured soil vapour concentration is used as the input to the risk assessment model rather than calculating theoretical soil vapour concentrations based on the concentrations of contaminants in the soil and groundwater. It is suggested that where only total VOC concentrations are available from soil vapour survey data the primary risk driving compound is identified on the basis of the results of the analysis of VOCs in soil and groundwater and that as a conservative assumption it is assumed that the total VOC concentration in the soil vapour is the risk driving compound. For example if 100 mg/kg benzene and 50 mg/kg toluene were identified in a soil sample and a total VOC concentration of 100 ppm was recorded in the soil vapour survey for the purpose of the risk assessment a soil vapour concentration of 100 ppm benzene would be used. It is acknowledged that this approach is extremely conservative and it is recommended that where it is possible to speciate VOCs in soil vapour that this should be undertaken.

Indoor vapour concentrations are calculated using the algorithms for migration through a concrete slab or brick walls given in Johnson & Ettinger (1991). Other algorithms such as those used in the CLEA model (Environment Agency 2002) may be used but should be used with caution as they have not been calibrated against field data unlike those presented by Johnson & Ettinger (1991).

5 CASE STUDIES

5.1 Bitumen coating plant, North West England

The site was previously used for rubber vulcanising and coating and impregnation of asbestos products with bitumen. Gasoline and diesel were previously stored on the site in underground storage tanks. The site is currently vacant and comprises a series of low rise warehouses constructed on a 0.2 m thick concrete slab. Samples of soil and groundwater were obtained from boreholes drilled at the site in 1993 and analysed for a range of volatile and semi-volatile organic compounds. A soil vapour survey was undertaken in 2002 prior to divestiture of the site. On the basis of the soil and groundwater data benzene was identified as the risk driving compound and was used as a surrogate compound for the VOCs as it is highly mobile with low health threshold values. Total VOC concentrations in soil vapour at the site ranged from 0.8 ppm to 39.4 ppm. Based on a VOC concentration of 39.4 ppm in the soil vapour under the main building at the site an indoor air concentration of 0.002 ppm was calculated using the Johnson & Ettinger concrete slab model (Environmental Protection Agency 2000). The target air quality criteria for benzene as an eight hour time weighted average is 5 ppm (Health & Safety Executive, 1999). The calculated indoor vapour concentration is three orders of magnitude less than the target criteria. As the calculated indoor vapour concentration was significantly below the target air quality criteria it is considered that there is no significant risk to human health from VOCs for commercial occupancy of the site.

5.2 *Paint factory, Northern France*

The site was used formerly for the manufacture and storage of specialist paints. Solvents including toluene were stored in drums and underground storage tanks. The site ceased operation in the early 1990s and has been subject to a series of site investigations to obtain samples of soil, groundwater and soil vapour and to install groundwater monitoring wells. The site currently is used for storage and it is proposed that in the future residential apartment blocks may be built on the site.

Soil gas samples were obtained from twelve locations across the site from probe holes drilled to a depth of approximately one metre below ground level using a narrow diameter hand held power drill. Soil vapour was drawn through a series of Tenax TA cartridges and sorbed VOCs were analysed by GC/MS. The maximum concentration of toluene recorded was 13.3 $\mu g/m^3$ and the average was 2.2 $\mu g/m^3$.

Measured maximum soil vapour concentrations were used as input to the Johnson & Ettinger concrete slab model (US Environmental Protection Agency 2000). An indoor air concentration of 0.00025 $\mu g/m^3$ was predicted compared with a risk based screening level of 0.392 $\mu g/m^3$ for residents (ASTM 1995) and

0.493 $\mu g/m^3$ for commercial occupation (ASTM 1995). The risk assessment showed that there were no risks as a result of indoor exposure of residents to vapour.

6 CONCLUSIONS

In the opinion and experience of the authors the use of measured soil vapour data rather than the use of calculated soil vapour concentrations derived from partitioning coefficients enables a more robust assessment of potential health risks to human receptors. The use of soil vapour surveys to identify areas of VOC contamination is recommended at sites with a history of solvent use in order that the collection of soil and groundwater data can be targeted.

The use of soil and groundwater data alone to predict indoor vapour concentrations may lead to an overestimation of risk. While such an approach may be conservative the use of soil vapour survey data enables a more refined assessment of risk.

REFERENCES

ASTM 1995. Standard Guide for Risk-Based Corrective Action Applied at Petroleum Release Sites. E 1739–95.

BSI 2001. Investigation of potentially contaminated sites – Code of practice. BS 10175:2001.

Cambridge County Council 2000. Final Report for Cambridge County Council prepared by Dames & Moore. Groundwater and indoor air quality monitoring at Sawston, Cambridgeshire on behalf of South Cambridgeshire District Council. Reference 2739R/43347-002-401/JDW/MQ. 7 February 2000.

Environment Agency 2002. The Contaminated Land Exposure Assessment Model (CLEA): Technical basis and algorithms. R & D Publication CLR 10.

Health & Safety Executive 1999. EH40/1999 Occupational Exposure Limits, 1999.

Johnson, P.C. & Ettinger, R.A. 1991. Heuristic Model for Predicting the Intrusion Rate of Contaminant Vapors into Buildings. Environmental Science and Technology 25: 1445–1452.

Scottish & Northern Ireland Forum for Environmental Research 2000. Framework for Deriving Numeric Targets to Minimise the Adverse Human Health Effects of Long-term Exposure to Contaminants in Soil. Final Report SR 99(02) F. Prepared by Land Quality Management, SChEME, The University of Nottingham. January 2000.

US Environmental Protection Agency 1996. Soil Screening Guidance: Technical Background Document. EPA/540/R95/128. May 1996.

US Environmental Protection Agency 2000. User's guide for the Johnson and Ettinger (1991) model for subsurface vapor intrusion into buildings (revised). December 2000.

Land Reclamation – Moore, Fox & Elliott (eds)
© 2003 Swets & Zeitlinger, Lisse, ISBN 90 5809 562 2

Reclamation of Brymbo Steelworks, Wrexham

P.S. Roberts
Director of Development Services, Wrexham County Borough Council

Abstract: This paper considers the challenges of the reclamation and regeneration of the former Brymbo Steelworks site from that perspective. It focuses largely on the financial, commercial, marketing and partnership issues for land reclamation, given that the technical issues at Brymbo although challenging are not unique, although it has to be said that the technical solution is quite innovative.

The project has also been demanding because of its long gestation and the frustration felt by the local communities who had depended so much for Brymbo for employment and wealth generation in the area. Whilst the circumstances of its demise were largely overshadowed by the headline closures of Ravenscraig and Shotton it nevertheless created a huge challenge for the Local Authority, the Welsh Assembly Government and the Welsh Development Agency and of course its private sector owner BDL Ltd.

Brymbo is a unique and amazing location, a 95 ha site situated at a 800 feet above sea level immediately to the west of Wrexham, surrounded by urban village communities but with far reaching views eastwards out over the Cheshire plain. Its life began 200 years ago when John Wilkinson established iron making on the site using its local resources and it evolved over the years into a modern, engineering steels plant until its closure in 1990. At its peak Brymbo had employed over 3000 but in its latter days employment had diminished to around 1200.

Following its closure and the sale of the plant and machinery, the site fell into dilapidation until it was purchased in 1994 by the present owner who acquired it directly from United Engineering Steels. A scheme for regeneration and after-use was eventually granted planning permission by Wrexham County Borough Council in 1997. The 95 ha site was to be significantly re-profiled to create new development plateaux for housing (300 houses in the original permission) and 12 ha of employment land. This did not include what is known as the heritage area which contains many of the original buildings which are scheduled monuments.

The reclamation scheme itself is relatively complex given the fact that the site was created by gradual tipping of blast furnace steel, electric arc slags and refractory material over the lifetime of the site, some of which has buried former mineworkings and which have resulted in high, steep and unstable slopes on the eastern perimeter of the site.

The site overlies coal measures and the proposal is to initially create a void of 1.6 m cubic metres which will allow the removal of some 156,000 tonnes of coal and 1m tonnes of clean overburden which will be used in the reclamation. Other than contaminated material and slags which may fail a leaching test all material remains on site and is re-worked. Of huge importance however is the inclusion in the after-uses of a major spine road which would not only open up the site for development but would provide a crucial and strategic linkage between the village of Brymbo and the modern road network, the Steelworks itself having been linked with a new road built by the then Clwyd County Council in the early 1980s.

The economics of the reclamation are marginal i.e. the balance between its cost and the residual land values is significant and the owners had been involved in protracted discussions with the Welsh Development Agency for several years with a view to obtaining land reclamation grant which was offered at the usual 80% rate. Modifications had been made to the scheme including reducing the thickness of the overall cover of clean material but increased levels of grant support were not possible and it was not considered viable, even at the 80% level.

The solution, therefore, was for the public sector i.e. Wrexham County Borough Council to acquire the more difficult and costly parts of the site but with the underwriting of 100% land reclamation grant from the Welsh Development Agency. Wrexham Council readily agreed to this but strictly on a no cost to the Council basis. The intention then is for the Council to acquire the easterly and northern sections of the site including the steep slopes and the heritage area but for the scheme to progress as an integral scheme over some 2½ years, and in a tri-partite partnerships: WCBC, WDA and the site owner, BDL.

Although the land ownership split is somewhat arbitrary it will be complicated by the need for the Local Authority to undertake a competitive contract procurement or partnering route whilst BDL will be undertaking its own contracting work. Simply put, BDL will create the void, dispose of the coal and stockpile sufficient clean overburden to complete the works after the Council puts its unwanted slag materials into the void.

The heritage area itself is an interesting project and much consideration and study has been undertaken as to its future. What is clear is that the listed buildings will need to be preserved. The uses are being explored. Inevitably the site contains much interesting ecology; Great Crested Newts and Lesser Horseshoe Bats are present.

In 2001 the West Wrexham area was selected as an EU Urban II project, the only one in Wales, which would address the issues of social exclusion and regeneration in its urban villages. Brymbo is at the heart of this area and hopefully the European funding (£6 m grant, £6 m match funding) can provide much needed learning and training opportunities possibly in re-used buildings on the Steelworks site in the fullness of time.

Although at this time a start has been made on site with the demolition of all of the super-structures of the former Steelworks buildings being undertaken by the Local Authority there have been protracted discussions between the parties arising from the due diligence work being undertaken by the Local Authority. It is further complicated by the fact that the Local Authority will not take ownership of its part of the site until sufficient void has been created and sufficient clean overburden generated to enable the Council work to start and this is likely to run some 14–18 months behind the BDL programme. Therefore much time has been spent with the lawyers, not just on the preparation of the land transfer arrangements but particularly with protecting the position of the Council and the local environment by investigating the various forms of environmental insurance and contractor's pollution insurance which will be necessary, but principally establishing whether such insurance is affordable within the scheme.

However, towards the end of 2002 further complications arose due to the Consultant's continuing investigations regarding the reclamation methodology. The scheme which had received planning permission provided for the re-use of slag materials as engineering fill. Severe doubts have been raised regarding the expansive nature of many of the slags, the difficulty of viably separating them given the huge quantities involved and the knock-on effect this has in terms of the viability of the land to be sold for housing and employment uses, let alone the construction of the spine road. It is to be emphasised that this is a commercial rather than an environmental issue.

A solution has been devised which will involve placing all of the dubious slag and refractory materials in the void and using clean overburden, but in thinner layers to cap off the remainder of the site. Issues remain then about the long term viability of the void itself and whether careful placement and compaction of the slag material in the void can give some short to medium term certainty of development. Obviously consideration will need to be given to the construction corridor for the road through the void area and the points where the road crosses the walls of the void.

The proposal also redistributes the housing land onto the more viable areas in the southern half of the site leaving the employment land on the viable areas to the north and on the void itself. This has obviously created concerns in the minds of local people and politicians that the original objective of jobs and prosperity could be put in jeopardy. Furthermore, it is intended to allow an increase in the housing density on the same housing footprint from 300 to 450 which is also in accord with current government policy of concentrating development in brownfield sites. The increased density will also increase the residual land values and contribute to the high infrastructure costs. Inevitably though, the public sector will have to contribute to the cost of the spine road, given its importance in local regeneration terms. Both of these changes i.e. the new reclamation methodology and the changed after-uses, particularly the housing change, require variations to the existing planning conditions.

In conclusion then, the proposal has produced mixed blessings, but overall it has achieved a measure of certainty, where none existed. The public/private partnership approach is productive in that it adds value both from a financial and expertise standpoint, it draws in both political and community support and it will, I'm sure, give added confidence to the prospective purchasers of the development land.

Land Reclamation – Moore, Fox & Elliott (eds)
© 2003 Swets & Zeitlinger, Lisse, ISBN 90 5809 562 2

The Mersey Forest Brownfield Research Project

P.D. Putwain & H.A. Rawlinson
School of Biological Sciences, The University of Liverpool, Jones Building, Liverpool, England

C.J. French & N.M. Dickinson,
School of Biological and Earth Sciences, John Moores University, Byrom Street, Liverpool, England

P. Nolan
The Mersey Forest, Risley Moss, Ordnance Avenue, Birchwood, Warrington, England

ABSTRACT: The Brownfield Research Project was set up to assess the feasibility of utilising closed landfill sites and contaminated brownfield land for the establishment of Community Woodlands which will significantly contribute to the development of the Mersey and Red Rose Forest areas. The field trials encompassed a wide range of tree species and cultivars planted in 50 experimental plots. Percentage mortality of trees, vertical growth and stem diameter were recorded over a period of three years during the first phase of the project. Patterns of tree survival and growth varied widely over the 11 landfill sites. Survival and growth of tree species were often negatively correlated, species with poor survival rates grew well and *vice versa*. All but three landfill sites were suitable for Community Forestry without additional remedial treatment. On contaminated brownfield sites, establishment and growth of short-rotation coppice cultivars of *Salix* and *Populus* was excellent. Relatively high biomass yields were achieved after three years and except at a single plot, there was no evidence that elevated concentrations of metals in the soil depressed tree growth. The prospects for the development of Community Forestry on derelict brownfield sites in the Mersey and Red Rose Forest areas are generally good.

1 INTRODUCTION

1.1 *Community Forests*

The Mersey Forest Brownfield Research Project commenced in October 1997 and combines two sequential but linked areas of research. The initiative for the two projects was stimulated by the national strategy to utilise lowland forestry for the revitalization of urban and urban fringe areas. This is currently a high priority for the regeneration of land. In 1989, the Countryside Commission and the Forestry Commission launched an initiative to create twelve community forests in England, across an area of 4,500 km^2. This was probably one of the most significant environmental projects to be launched during the 20th century, potentially bringing the benefits of amenity woodland to more than 20 million people, encouraging urban renaissance and regional investment. The aim of the Community Forest Initiative is to create substantial opportunities for recreation, employment and education, with additional environmental and aesthetic benefits.

Additionally, in England the Forestry Strategy is to increase woodland coverage by 50% in 50 years (Her Majesty's Government 1995).

In the Northwest, the planting of the Mersey Forest and the Red Rose Forest (two of the twelve Community Forests) are likely to result in the most radical landscape changes since the Industrial Revolution. Their combined area is over 700 square miles and contains a population of more than 3.2 million people (Mersey Forest 2000). Both regions had a very low proportion of tree cover, about 4% at the start of the project, compared to 10% nationally (Red Rose Forest 1998). However, in the Mersey Forest area the proportion of tree cover has increased to 6% today as a direct result of the Community Forest Initiative.

1.2 *Opportunities for planting Brownfield land*

Reclaiming derelict and contaminated land (including landfill sites) to forestry has been questioned in the past due to negative governmental guidance, predominantly Waste Management Paper 26, Dept. of Environment

(1986), and insufficient research involving the relevant issues. Nevertheless, the opportunities presented by a large stock of derelict land for woodland establishment have been considered to be an opportunity (Handley & Perry 1998). A national figure of 40,000 ha of derelict land has remained constant for 25 years with the rate of new dereliction being equivalent to the rate of land reclamation (Handley & Perry, 1998). Up to 5,300 ha of derelict land in the UK consists of landfill, of which closed landfills account for some 3,000 ha. Approximately 1,060 ha of derelict and neglected land within the Mersey and Red Rose Forest Regions have been identified as closed landfill sites (Bending & Moffat 1997). Clearly, the reforestation of closed landfills potentially could provide an important contribution to Community Forests.

Assessment of the ecotoxic potential of soil at Brownfield sites in the UK is recognized to be inadequate and lacking in relation to planting trees. Until very recently guidance was based mainly on trigger and threshold concentration levels such as those contained in the ICRCL (1987) guidelines (Clifton *et al.* 1999). These guidelines have been superseded by the Contaminated Land Exposure Assessment (CLEA) model made available by DEFRA and the Environment Agency in 2002.

Thresholds for phytotoxic metal contamination are often quoted for Zn, Cu, Ni and B. However, Dickinson *et al.* (2000) suggested that the published values assumed that the metal content of soils was mainly bioavailable when in practice it is not. It is difficult to find convincing evidence that contamination of urban soils has a detrimental effect on trees. Hence, the initial project which examined the potential for growth of trees on closed landfill sites was extended to encompass metal contaminated brownfield sites in Merseyside. It will be important to evaluate the mobility of metals in soils and the metal uptake by trees, develop metal budgets in relation to strategies of phytostabilisation and phytoextraction and apply resulting models to risk assessment. It will also be important to measure the growth and production of trees on brownfield sites in Merseyside in order to develop a strategy for planting such sites.

2 THE LANDFILL RESEARCH PROJECT

2.1 The staged pathway methodology

The initial aim of the project was to produce a standard methodology that could be used to assess the capability for the establishment of Community Forest on large areas of closed landfill in the Mersey and Red Rose Forest region. It was then intended to extend the methodology to encompass other types of derelict, damaged and unused land including metal contaminated sites. The methodology was published by the Mersey and Red Rose Forests (Mersey Forest 2000) in a technical document and was described as the staged pathway approach. This is a route from inception (a proposal that a site may be suitable for Community Woodland use) through to producing master plan designs for the site. The pragmatic methodology incorporates an assessment of the significance of utilities and installations, planning, landscape, archaeological, ecological, community and safety issues as well as a technical assessment of soil characteristics, site drainage, topography, landfill gas emission and leachate from the site. The staged pathway approach is an improved and extended methodology that supercedes the earlier Forestry Commission guidance in the form of Research Information Note 263 (Dobson & Moffat 1995). The project was mainly funded by English Partnerships (North West Development Agency) and the Forestry Commission.

2.2 The closed landfill sites

Large-scale woodland research plots were established across the Mersey Forest and Red Rose Forest regions. The research involved the investigation of tree establishment on closed landfill sites that have had minimal remediation treatment. This was to assess what can be achieved at a site with minimal cost and to determine where investment should be targeted to produce at least naturalistic woodland and preferably productive woodland (Mersey Forest 2000). Experimental sites were chosen on the basis that they were representative of a wide range of waste types, history, ownership and potential constraints to tree survival and growth. This resulted in the establishment in 1997 of 22 ha of experimental trial woodlands at 11 sites, five located in Merseyside, two in Cheshire, two in Lancashire and two in Manchester.

The landfills were closed between the 1960s and 1980s, inadequately capped, and restoration was below the standard of modern landfill design in terms of soil quality, soil depth and pollution control. Soil cover is characteristically shallow, compacted and poor in structure, low fertility, low in organic matter content, prone to waterlogging or drought, and exposed to landfill gases. Thirty-nine experimental plots, generally each 0.19 ha, were planted with 21 native and non-native trees and shrub species.

Various factors likely to affect tree performance at closed landfill sites were investigated including:

- Soil physical condition; compaction (penetration resistance) and stoniness
- Landfill gas concentration in soil
- Soil moisture availability.

Other parameters such as soil chemistry, heavy metal contamination and mineral nutrient availability were assessed as a component of the staged pathway methodology but were not investigated in detail.

2.3 Experimental methods-landfill sites

Each of the selected sites was subject to a preliminary assessment including a soil quality survey based on trial pits and including measurements of soil oxygen, carbon dioxide and methane concentration at various depths using a spike bar method.

Thirty-nine experimental plots were planted in January and February 1998, across the eleven sites, their location being carefully selected to encompass a range of environmental conditions across each site (as well as between different sites). For example, some plots had known areas of waterlogging and compaction, shallow and relatively deep soils, visual "hotspots" of assumed high methane concentration, steeply angled and undulating topography and variable aspect. In most cases landscape design objectives also influenced the location decisions. At each site, three or four replicate experimental plots were planted with trees, each plot covering approximately 0.22 ha (49.5 m $\times$ 43.5 m). A full size plot (882 trees in total) consisted of four columns, each with 32 rows 1.5 m apart. Twenty-one species of trees were planted in six rows of each plot. Each row consisted of seven individuals of the same species spaced 1.5 m apart.

All the replicate rows were allocated using computer generated random numbers with a separate set for each of the 39 experimental plots. At least 1.2 ha of non-experimental buffer woodland was planted on each site to minimise edge effects, improve the local landscape and to achieve the target area of planted woodland. At six of the sites, the area available for experimental planting was constrained by site factors. In which case the plot size was reduced and the number of replicate rows for each species was limited to four or five.

The tree species selected for planting included a mixture of short and long-lived species selected for their probable suitability for landfill sites (Dobson & Moffat 1993. Mitchell 1981). The species mixture was intended to contribute to structural diversity, provide a range of habitats for fauna and to include some productive species. There were 14 species native to the British Isles (*Alnus glutinosa, Betula pendula, B. pubescens, Crataegus monogyna, Fraxinus excelsior, Malus sylvestris, Pinus sylvestris, Prunus padus, P. spinosa, Quercus petraea, Salix caprea, S. fragilis, S.viminalis, Sorbus aucuparia*) and seven non-native species (*Acer psuedoplatanus, Alnus cordata, A.incana, A.viridis, Larix europaea, Pinus nigra, Quercus rubra*). Cell grown trees measuring 40–60 cm in height were planted in the experimental plots, although certain species *(P. nigra, Salix caprea, S. viminalis* and *A. viridis* were smaller, 20–40 cm stock).

Trees were kept as weed free as possible throughout each growing season by means of two applications of glyphosate. Fertiliser was not applied in order to obtain an evaluation of the constraints imposed by environmental conditions at each site, without intervention.

2.4 Monitoring tree growth and survival

Tree growth and survival were monitored with an annual census at the end of each growing season in September and October in 1998, 1999 and 2000. In the first growing season, tree survival was recorded for all trees on three occasions in May, July and October 1998, and in October 1999 and 2000. A tree was defined as dead when devoid of green leaves and green tissue was undetected on the leading shoot 5 cm above ground level.

Over 40% of the experimental trees were examined in greater detail. Four replicate rows of each species in each plot were randomly selected in Year 1.

Four of the central five trees from the seven in each selected row were assessed as follows:

1. Total tree height.
2. Tree diameter (20 mm above ground).
3. Leading shoot growth increment (cm).

2.5 Monitoring landfill gas and soil moisture

Four sites were selected where high concentrations of CO_2 and CH_4 and low concentrations of oxygen were recorded in the preliminary site investigations. The concentration of CH_4, CO_2 and O_2 was recorded over a series of census dates, within the soil zone where the majority of tree roots occur. A 4 $\times$ 4 grid of probe locations was installed with probes at three depths (25 cm, 50 cm and 75 cm).

Analysis of soil moisture was undertaken within the plots also selected for monitoring landfill gas. Two additional sites were investigated because distinct areas of waterlogging were observed within the experimental plots.

3 THE CONTAMINATED BROWNFIELD RESEARCH PROJECT SITE

3.1 Experimental methods – Brownfield sites

Field trials were established on six brownfield sites in Merseyside. Preliminary surveys had established that these sites were contaminated with a range of metals including As, Cu, Ni, Pb and Zn (Table 1). Intensive sampling of soils at each site enables contaminated hotspots to be identified and this information was used to aid the specific location of the experimental plots and provide detailed baseline data for experimental work. The tree trials were situated across areas of increased contamination incorporating one or more hotspots. This approach is illustrated by Figure 1, which

Table 1. Minimum and maximum total metal concentration (mg kg^{-1}) at three experimental Brownfield plots in Merseyside.

Metal (n)	Merton Bank St Helens (plot 3) (72)	Cromdale Grove St Helens (plot 1) (71)	Sugar Brook Liverpool (plot 1) (14)
As Min	4.9	nd	<dl
Max	2888.7	nd	<dl
% > ICRCL	100%	–	0%
Cu Min	29.1	10.4	38.6
Max	484.5	883.5	75.7
% > ICRCL	43%	27%	0%
Ni Min	6.9	10.1	19.1
Max	57.5	109.4	38.4
% > ICRCL	0%	38%	0%
Pb Min	70.4	62.2	108.8
Max	1770.4	560.1	484.8
% > ICRCL	29%	3%	0%
Zn Min	2.8	56.2	127.5
Max	1303.3	826.6	372.8
% > ICRCL	4%	38%	7%

% > ICRCL = % of plot samples (n) above ICRCL (1987) guidelines for phytotoxicity.
nd = no data available, <dl = concentration below the limit of detection.

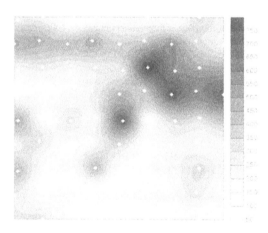

Figure 1. A two-dimensional contour representation of total zinc concentration at an experimental plot, Cromdale Grove, St. Helens.

shows the total soil zinc concentration (mg kg^{-1} air dry soil) at an experimental plot situated at Cromdale Grove, St. Helens.

The experimental design at each site was based on a 30 m × 30 m plot. Ten species were selected for planting; these included 5 *Salix* and 2 *Populus* cultivars, known to be fast growing and listed as potential species for biomass coppice production (Tabbush & Parfitt 1999). The three additional species were *Alnus incana, Betula pendula* and *Larix eurolepis*. There were 100 blocks of tree species/cultivars where each block contained 12 individuals, in two rows of six trees with a spacing of 1.5 m between the rows and 0.5 m within the rows, thus each species was replicated in 10 blocks randomly allocated within the plot. The plot design was modified at some sites because a smaller plot size (21 m × 21 m) was used in order to cope with various site constraints. There was a total of 11 field trial plots on six brownfield sites containing a total of 10,380 individual trees.

The sites were prepared for planting by ripping in locations where there were perceived drainage problems. Plots were cultivated to a depth of 300 mm via soil turning using a JCB and rotavated to provide the final fine tilth for planting. All plots were sprayed with glyphosate before cultivation took place.

Planting took place in April 2000, *Salix* and *Populus* were 25 cm unrooted cuttings and all other species were 20 cm commercially available cell grown stock. After one growing season all *Salix* and *Populus* cultivars were cut back to 10–15 cm above ground level following standard coppicing practice to promote the production of multiple stems. Annual assessments were made at the end of each growing season for tree height and mortality.

3.2 Metal uptake and mobility – sampling

Maps showing contamination data (Figure 1) were used as a basis for targeted investigation for metal uptake and mobility by selecting "hotspots" of contamination for each metal and allocating individual trees within these areas for detailed analysis. Twenty to 25 trees, encompassing all species used, were selected by this method for each plot. These trees were subjected to foliar sampling at the end of seasons 2 and 3 and stem and soil sampling at the conclusion of season 3.

Foliar and stem samples were digested by microwave digestion in HNO_3 and H_2O_2. Soil samples were extracted using $CaCl_2$ and EDTA as a 2-stage process (Maiz & Esnaola 1997). All samples were analysed for a range of heavy metals using ICP-AES.

Once complete these data will provide details of uptake rates and mobilization to develop metal budgets in relation to strategies of phytostabilisation and phytoextraction of metals within a field production system.

4 RESULTS

4.1 Growth and survival – landfill sites

Over three growth seasons the mean cumulative percentage mortality, (pooled over all tree species), varied

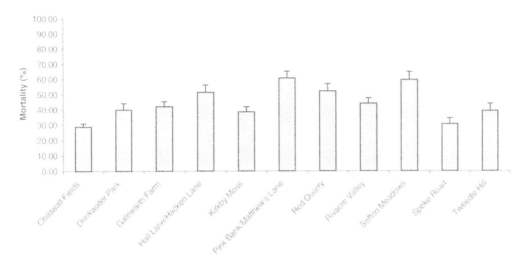

Figure 2. The mean cumulative percentage mortality (pooled over all tree species) comparing sites.

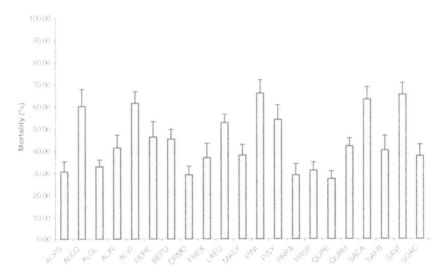

Figure 3. The mean cumulative percentage mortality (pooled over all sites) comparing tree species.

between 29% (Childwell Fields, Liverpool) and 60%, (Pink Bank/Matthew's Lane, Manchester) (Figure 2). Mean mortality was greater than 40% at six sites. However the mean response masks considerable variation between plots within sites. For example, at Speke Road (Widnes) plots 2, 1 and 4 ranked first, third and fourth respectively for the lowest mortality, but plot 3 ranked 24th (significantly different P < 0.01). At the Hall Lane/Hacken Lane (Bolton) site where tree mortality was relatively high overall, the plot rankings were plot 3 (19th), plot 1 (25th), plot 2 (26th) and plot 4 (34th).

At some sites there was an obvious cause of differences between plots in mean mortality pooled over

species. At Gatewarth Farm (Warrington) plot 1 had the highest visual presence of waterlogged soils and survival was lowest in this plot. Waterlogging was also a causal factor in tree death on part of the Red Quarry site (St. Helens). At Rivacre Valley (Ellesmere Port) tree survival was much higher on the plots located on the sloping side of the site. The plot on the plateau area was waterlogged in parts and mortality was higher.

The mean cumulative percentage mortality (pooled over all experimental sites) for each tree species is shown in Figure 3 (See Table 2 for the key to species codes). Ranked from least mortality to increasing mortality the top ten species (with the exception of *Acer pseudoplatanus*) were all native to the British Isles.

Table 2. Key to species shown in Figures 3 and 5.

Tree species	Common name	Code
Acer pseudoplatanus	Sycamore	ACPS
Alnus cordata	Italian Alder	ALCO
Alnus glutinosa	Common Alder	ALGL
Alnus incana	Grey Alder	ALIN
Alnus viridis	Green Alder	ALVI
Betula pendula	Silver Birch	BEPE
Betula pubescens	Downy Birch	BEPU
Crataegus monogyna	Hawthorn	CRMO
Fraxinus excelsior	Ash	FREX
Larix europaea	European Larch	LAEU
Malus sylvestris	Crab Apple	MASY
Pinus nigra	Corsican Pine	PINI
Quercus rubra	Red Oak	QURU
Salix caprea	Goat Willow	SACA
Salix fragilis	Crack Willow	SAFR
Salix viminalis	Common Osier	SAVI
Sorbus acuparia	Rowan	SOAC

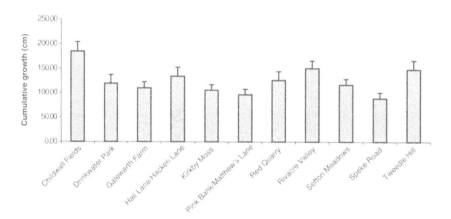

Figure 4. The mean cumulative extension growth (pooled over all tree species) comparing sites.

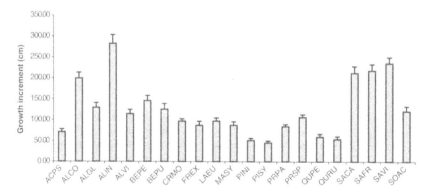

Figure 5. The mean cumulative extension growth (pooled over all sites) comparing tree species.

Quercus petraea had the least mortality and in rank order *Prunus padus, Crateagus monogyna, Prunus spinosa, Alnus glutinosa* and *Fraxinus excelsior* also exhibited lower than average mortality. At the other end of the scale, in rank order, *Salix viminalis, Pinus nigra* and *Salix caprea* showed the highest mortality, greater than 60% losses.

Variation in the growth of species between and within sites was substantial (Figure 4). At certain sites almost all species grew well, above the mean growth pooled over all sites. For example at Childwall Fields, Rivacre Valley and Tweedle Hill (Manchester) twenty or more species grew above average. Other sites, such as Speke Road, Pink Bank/Matthew's Lane and Kirkby Moss (Knowsley) exhibited below average growth for the majority of species. A comparison of species cumulative extension growth (Figure 5) shows that although several species experienced high mortality, nevertheless the survivors grew very well (e.g. *Salix caprea, S. viminalis, Alnus incarna, A. cordata*).

Within each species, potential suitability for planting was defined as a combination of less than 60% mortality and at least 80% of mean growth of the same species pooled over all sites. Ranking of the species on this basis showed that the majority of non-pioneer late – successional broadleaved species appeared to show potential for planting whereas less than half of the pioneer species met the same criteria. However, at Speke Road just four species were suitable for planting and these were all pioneer species, at Kirkby Moss 13 species in total were suitable whilst at Childwall Fields all 21 species were suitable.

It is concluded that all but three sites were suitable for Community Forestry, at least for naturalistic woodland, without further remedial treatment.

4.2 *Biomass production – contaminated sites*

Coppice species of *Salix* and *Populus* (and to a limited extent *Alnus*) have been grown extensively for production of biomass crops that have a variety of uses such as fuel in biomass power stations. (exSite 2001) Achieving high yields is important in the context of this project for dual reasons: that increased yield will maximize potential metal uptake and that good-quality yields may be sold on commercial markets, thus introducing an element of economic regeneration to brownfield sites.

At the end of season 3 an assessment of yield was made on all plots. Each species block was felled using chainsaws and all stems weighed to 0.1 kg in the field using a parallel weigh beam apparatus. Sub-samples were taken for moisture correction and data were collated for each species block to produce a total yield for each species in Oven Dry Tonnes per hectare (ODT ha^{-1}).

Table 3. % Mortality and yield in ODT ha^{-1} for 2 year old trees on Cromdale Grove.

Species	% Mortality	ODT ha^{-1}
S. cap x cin x vim "Calodendron"	8.3	15.5
S. cap x vim "Coles"	6.8	25.7
S. burjatica "Germany"	14.2	22.2
S. viminalis "Orm"	4.2	27.7
S. vim x schwerinni "Tora"	7.5	23.1
P. deltoides x nigra "Ghoy"	14.8	13.4
P. trichocarpa "Trichobel"	20.8	17.0
A. incana	20.0	13.5

Table 3 indicates relatively low mortalities coupled with high yields for all species. Where data exist for comparison, species such as *S. burjatica* "Germany" at 22.2 ODT ha^{-1} compare favorably with a range of 12.5–20.1 ODT ha^{-1} for a similar age/species combination in other trials (Tubby & Armstrong 2002). Such trials are conducted on mainly ex-agricultural land with none of the contamination problems present at this site.

5 DISCUSSION

The influence of site was the major factor in determining the overall tree performance in terms of growth and mortality on the closed landfill sites. At certain sites the majority of species exhibited good growth and relatively low mortality. At other sites the reverse was the case with the majority of species performing poorly whilst at many sites the response was mixed with some species above average and some below for both growth and mortality. It is difficult to isolate specific factors determining growth and mortality although differences in performance between plots within sites could be attributed to waterlogging. At one site very poor growth was clearly the outcome of nutrient poor, shallow soil cover. There was evidence that "hotspots" of high CO_2, high CH_4 and low O_2 concentrations at depths of 0.25 m and 0.5 m, reduced the growth of trees in local patches. However, these areas were relatively small and occurred at just a few sites. Taking a broad assessment, the effect of landfill gases on the growth and mortality of trees was negligible over all the experimental sites.

In the case of the contaminated brownfield sites performance of the SRC willow cultivars was excellent on all sites with generally low mortality and high biomass yield per hectare two years after initial coppicing. Similarly although total mortality was somewhat greater at some sites the *Populus* cultivar "Trichobel" and *Alnus incarna* produced good growth and relatively high biomass yields. In relative terms *Alnus* grew well at all sites and even better than the willow cultivars in some plots. In contrast *Larix eurolepis* performed poorly at all sites with both low survivorship and poor growth. Clearly correct choice of species is very important when planning community woodland on derelict, possibly contaminated brownfield sites. Preliminary trials are an essential pre-requisite to ensure that the risk of failure is minimized.

Although potentially toxic metals were elevated at many of the landfill sites and at the majority of the contaminated brownfield sites, there was no measurable effect on the establishment of trees (survival) or growth over a period of three years with the exception of one site (Merton Bank, plot 1) where a high concentration of arsenic may have been the cause of reduced growth. Soil pH was relatively high (>7.0) at most sites and this will reduce the bioavailability of many metals. It appears that the ICRCL threshold concentrations are very conservative with respect to possible phytotoxic effects of metals in derelict land soils.

REFERENCES

Bending, N.A.D. & Moffat, A.J. 1997. *Tree establishment on Landfill Sites – Research and updated guidance.* Edinburgh: DETR, Forest Research, Forestry Commission.

Clifton, A., Boyd, M. & Rhodes, S. 1999. Assessing the risks. *Land Contamination and Reclamation.* 7: 27–32.

Department of the Environment (DoE) 1986. Landfilling Wastes. *Waste Management Paper 26.* London: HMSO.

Dickinson, N.M., MacKay, J.M., Goodman, A, & Putwain, P.D. 2000. Planting trees on contaminated soils: Issues and guidelines. *Land Contamination and Reclamation* 8: 87–101.

Dobson, M.C. & Moffat, A.J. 1993. *The Potential for Woodland Establishment on Landfill Sites.* London: HMSO.

Dobson, M.C. & Moffat, A.J. 1995. *Site capability assessment for woodland creation on landfill sites.* Research Information Note 263, Edinburgh: Forestry Commission.

exSite Research (2001). *Marginal land Restoration Scoping Study: Information review and Feasibility Study.* Nottingham, UK, Land Quality Press.

Handley, J.F. & Perry, D. 1998. Woodland expansion on damaged land. Reviewing the potential. *Quarterly Journal of Forestry 92,* 297–306.

Her Majesty's Government 1995. *Rural England.* CM 3016. London: HMSO.

ICRCL 1987. Inter-Departmental Committee on the Redevelopment of Contaminated Land: Guidance Notes (59/83). *In Guidance on the Assessment and Redevelopment of Contaminated Land* (2nd edition). London: HMSO.

Maiz, I. & Esnaola M.V. 1997. "Evaluation of heavy metal availability in contaminated soils by a short sequential extraction procedure." *The Science of the Total Environment* 206: 107–115.

Mersey Forest 2000. *Creating Community Woodlands on Closed Landfill Sites.* Warrington: Mersey Forest.

Mitchell, A.F. 1981. *The Native and Exotic Trees in Britain.* Arboricultural Research Note 29/81/SILS, Edinburgh: Forestry Commission.

Red Rose Forest 1998. *A Plan for Life.* Manchester: Red Rose.

Tabbush, P. P. & Parfitt, R. 1999. *Poplar and willow varieties for short rotation coppice.* FC Information Note 17. Edinburgh: Forestry Commission.

Tubby, I. & Armstrong, A. 2002. *Establishment and management of short rotation coppice.* FC Practice Note 7. Edinburgh: Forestry Commission.

Land Reclamation – Moore, Fox & Elliott (eds)
© *2003 Swets & Zeitlinger, Lisse, ISBN 90 5809 562 2*

Japanese Knotweed and land reclamation in Stoke-on-Trent

Sally Evans
Senior Environment Officer, Stoke-on-Trent City Council

Nicola Farrin
Senior Environment Officer, Stoke-on-Trent City Council

1 THE ISSUE

Japanese Knotweed is a highly invasive non-native species, which colonises former lava flows in its native Japan. In the United Kingdom it spreads by means of stem or rhizome (a root like underground stem) fragments, which may reach a depth of three metres and extend up to seven metres away from the parent plant (Cornwall County Council 2001). Once established, rhizomes can spread underground at a rate of six metres in one season.

Development on land colonised by Knotweed can result in costly structural damage. Rhizomes can grow through asphalt and have been recorded penetrating foundations, floors, walls and land drainage works etc (Child and Wade 2000). Prevention is better than cure - a programme for appropriate treatment and disposal of material from and within infested sites will prove cost effective in the long term.

Japanese Knotweed in public open spaces reduces visibility, and dead canes (which take up to three years to decompose) trap litter and become unsightly. The growth of native flora is also suppressed by Knotweed.

Legislation puts the onus on the landowner to control Japanese Knotweed. It is an offence to plant or otherwise cause the species to grow in the wild under the Wildlife and Countryside Act 1981. Japanese Knotweed is also classed as controlled waste under the Environmental Protection Act 1990 and as such must be disposed of safely at a licensed landfill site according to the Environmental Protection Act (Duty of Care) Regulations 1991 (Child and Wade 2000).

2 THE CHALLENGE

The first official recording of Japanese Knotweed in Stoke-on-Trent dates back to 1955, although Knotweed is thought to have invaded the City during the beginning of the last century. Japanese Knotweed is likely to become more widespread and a programme for dealing with this well-established invasive plant requires concerted city-wide commitment linked closely to similar regional programmes.

Japanese Knotweed control is high on the reclamation and regeneration agenda for the City of Stoke-on-Trent. The majority of the Priority Investment Areas (PIA's) in the City include extensive brownfield development sites which have been colonised by Japanese Knotweed. The quality of their physical and structural regeneration therefore relies on the eradication of Japanese Knotweed. The identification, treatment method and cost of Knotweed control should be preliminary considerations in the land reclamation and regeneration process, in line with other principal issues such as the treatment of contaminated land.

The key to Japanese Knotweed control on development sites is early identification and developer/contractor awareness and co-operation. Experts agree that successful control of the plant can be achieved by the early and continued application of an appropriate herbicide prior (ideally three years before) to any land disturbance.

Advice on appropriate site specific methods of control should be sought from the Environment Agency and the Local Authority. Appropriate permissions are also required in environmentally sensitive areas or sites near water.

3 JAPANESE KNOTWEED IN STOKE-ON-TRENT

In November 2000, the Environment Agency commissioned a study of the status of Japanese Knotweed in Stoke-on-Trent (Ecology-first 2000). The study area included the River Trent within Stoke-on-Trent and six of its major tributaries.

Watercourses (approximately ten metres either side) were surveyed on foot and all patches of Knotweed colonisation were mapped. The survey area was extended in suitable habitats or where larger infestations occurred.

Japanese Knotweed was found to be widespread and common throughout the study area, especially within urban areas'. The majority of patches recorded were between 5 and 20 m². Very few of the Knotweed patches identified in the survey were attributed to river dispersal. However, river dispersal of Japanese Knotweed can be a problem where the earth banks on which the plant is established are eroded and fragments of rhizome (root) are carried downstream and deposited in new locations.

The majority of patches recorded in Stoke-on-Trent were located along the City's disused railways, the canal system and brownfield sites with a history of industrial development and waste disposal. Many infestations in these areas are well established and new colonisation is generally associated with earth moving operations and development of colonised sites (Ecology First 2000).

4 COUNTING THE COST

In causing the spread of Japanese Knotweed, an offence is committed under the Wildlife and Countryside Act 1981 which could result in criminal prosecution. Violation of the Environmental Protection Act 1990 can lead to enforcement action being taken by the Environment Agency with a potentially unlimited fine. Case law also demonstrates that third parties can be held liable for costs incurred from the spread of Knotweed across adjacent properties (RPS Group 2002).

Unfortunately, Japanese Knotweed control is often tackled at short notice, particularly on urban brownfield development sites where late identification reduces the options for control i.e. there is not enough time to conduct and complete an appropriate in situ spraying programme. The on-site burial or removal and disposal methods are therefore the next best legal options, however the latter can have significant cost implications. A valuable reference document for the control of Japanese Knotweed, by the use of both chemical herbicides and non-herbicidal methods, has been produced by the Welsh Development Agency (1998).

Examples of Knotweed control both in Stoke-on-Trent and nationally suggest that chemical control is the preferred and most cost effectiveness option. On-site burial is an option where smaller stands of Knotweed are found within sites immediately prior to development. However, where sites infected with

Figure 1.   Japanese Knotweed in Hartshill Park Stoke-on-Trent.

Knotweed require an immediate solution prior to disturbance, the removal and disposal method is the only option, and it comes at a high price. For example, a 1 m² stand of knotweed contaminates c. 675 m³ of material, which would result in landfill charges alone of approximately £20,000 (RPS Group 2002). On a whole site basis, ten hectares of removal and disposal could cost in the region of £1 million to excavate and dispose.

Hartshill Park in Stoke-on-Trent was heavily infested with Knotweed, with established stands of up to 50 m² (Figure 1). An annual spraying programme, with three treatments a year, has now been in operation for two years. Although the extent of infestation has lessened considerably, the high level of contamination will take more than the average three years to eradicate, even with the use of the strongest chemicals. This requirement for on-going control therefore imposes a significant burden on the City's greenspace maintenance budget.

5 CONCLUSION

There are a number of lessons to be learnt when encountering Japanese Knotweed contamination on land reclamation sites. The key to successful, cost effective treatment is early identification of Japanese Knotweed, as this will substantially reduce the cost of treatment. Appropriate supervision and monitoring are also essential to ensure that treatment activities and follow up surveys are undertaken, and that the illegal spread of Japanese Knotweed does not result from the inadvertent movement of contaminated material both on and offsite.

6 REFERENCES

Child L and Wade M (2000) The Japanese Knotweed Manual.

Cornwall County Council and the Environment Agency (May 2001) Japanese Knotweed, Guidance for Householders and Landowners.

Ecology-first (November 2000) Stoke-on-Trent Invasive Weeds Survey.

Palmer C (2001) Knotweed: manual assists control programmes. In: Forestry and British Timber.

RPS Group (2002) RPS advises on Japanese Knotweed – a developer's nightmare. RPS Group plc news, 15 July 2002. www.rpsplc.co.uk

Welsh Development Agency (1998) The eradication of Japanese Knotweed. WDA: Cardiff.

BSSS/IPSS – Soils and Land Reclamation

Land Reclamation – Moore, Fox & Elliott (eds)
© 2003 Swets & Zeitlinger, Lisse, ISBN 90 5809 562 2

Phytoextraction of Ni and Zn from moderately contaminated soils

J. Scullion

Soil Science Unit, Institute of Biological Sciences, University of Wales, Penglais Aberystwyth, UK

ABSTRACT: Physically degraded soils from two disused landfarms had elevated Ni, Zn and oil hydrocarbon concentrations. Pot trials evaluated amendments (three rates of EDTA or sulphur) for increasing metal uptake by plants (*Brassica juncea*, *Lolium perenne* and *Alyssum argenteum*). Plant growth, foliar metal concentrations, extractable soil and 'leachate' metal concentrations were measured.

Plants did not grow well (<30% of yield in compost) in landfarm soil. Without amendments, foliar Ni and Zn concentrations, metal availability and contamination of drainage water were low. EDTA and sulphur increased metal availability and uptake by plants. However, highest input rates reduced yields and metal offtake. Highest metal offtakes were equivalent to reductions less than $3 \, mg \, kg^{-1}$ soil Ni and Zn. *Alyssum argenteum* contained Ni concentrations five times those of *L. perenne*, but growth was poor and concentrations low in unamended soils. EDTA and the highest rate of sulphur caused persistent increases in Ni and Zn leaching.

1 INTRODUCTION

Soils moderately polluted by metals often pose a risk to the environment or to human health that does not justify expensive technology-based remediation approaches. For such soils, metal removal by plants has been proposed as a low cost, "green remediation" (Baker et al 1994) solution to the problem. Two broad approaches have been adopted using either hyper-accumulator plant species (Baker et al 1994, Robinson et al 1997, McGrath et al 1997) or soil amendments to increase metal availability to tolerant agricultural crop species (Blaylock et al 1997, Epstein et al 1999, Romkens et al 2002).

In order to be effective, phytoextraction techniques must achieve several objectives. Plants must have high productivity combined with elevated metal concentrations in harvestable foliage. Productivity can be increased by appropriate nutrient inputs but this does not necessarily ensure a corresponding increase in metal offtake. Plant roots must be capable of accessing reserves of often sparingly soluble metals or the solubility of these metals must be enhanced by soil amendments. High concentrations of metals must not reduce plant growth.

Research into the potential for phytoextraction of metals has been carried out on soils with high naturally occurring concentrations of metals using plant species adapted to these conditions (Robinson et al 1997).

Other studies (Blaylock et al 1997) have artificially contaminated soils with metal salts, albeit of low solubility. Another common substrate for phytoextraction studies has been fertile agricultural soils receiving metal contaminated sewage sludges (McGrath et al 1997). Only a few investigations have been included soils suffering industrial contamination (Blaylock et al 1997, McGrath et al 1997, Romkens et al 2002). There is even less information on the potential of this technique for soils affected by a range of contaminants, with poor physical condition and limited nutrient fertility.

Soils on two disused landfarms at the TOTAL-FINAELF oil refinery in Milford Haven, used previously for the aerobic degradation of oily sludges, had elevated Ni and Zn concentrations by comparison with guideline values. Metal contaminants originated principally from precipitated effluent lagoon sludge that was spread on these landfarms. Ni is used in catalysts for part of the refining process and was unknowingly overdosed for an extended period allowing a build up in the effluent lagoon sludge. A series of pot trials was undertaken to evaluate the potential of phytoextraction techniques for the remediation of these soils and risks of metal leaching arising from treatments aiming to promote phytoextraction.

Soils collected from the landfarm sites were mixed, amended as appropriate to increase metal availability, then planted with a succession of "crops" to evaluate potential metal offtake.

2 METHODS

An initial survey of soils and vegetation was carried out on the landfarm sites. Soil pH was in the range 6.1–7.2. The soil had a clay loam texture and contained between 0.5 and 2.3% oil hydrocarbons based on a diethyl ether soxhlet extraction. Of the total metal concentrations measured, only Ni (160–200 mg kg^{-1}) and Zn (240–400 mg kg^{-1}) showed values markedly higher than those of surrounding, untreated soils. If the site were to be redeveloped, Ni concentrations in all samples would exceed UK ICRCL guideline values (ICRCL 1987); only 30% of samples had Zn concentrations exceeding these guidelines and many did so by a small amount. Thus, reducing soil Ni concentration would be a main priority of any remediation process. Readily available metal concentrations (CaCl$_2$ extractable Ni 1.8 mg kg^{-1} and Zn 2.1 mg kg^{-1}) represented a very small fraction of these total metal pools. EDTA extractable Ni and Zn averaged 8.9 and 16.2 mg kg^{-1}, respectively. Comparisons of foliar metal concentrations showed low values (<15 mg kg^{-1} Ni and <30 mg kg^{-1} Zn) with no significant differences between plants growing on landfarm and on undisturbed soils. Indeed metal concentrations in plants growing on the undisturbed soils were slightly higher.

2.1 Experimental details

Commercially available varieties of Indian mustard (*Brassica juncea* ori 060), *Alyssum argenteum* and perennial ryegrass (*Lolium perenne* cv Talbot BA 10915) were grown in the same landfarm soils under standard glasshouse conditions in a series of pot trials. All treatments consisted of 5 replicates. Previous studies (Blaylock et al 1997) had indicated that *B. juncea* might be a suitable crop for accumulation of metals such as Ni and Zn. *A. argenteum* has been found to hyperaccumulate Ni (Baker & Brooks 1989). *L. perenne* was included in the trials as it was found growing on site and provided a local agricultural reference crop against which to compare the other two species. The initial phase of the experiment (0–6 months after amendments applied) included *B. juncea* and *L. perenne* only. Following several harvests and plantings of these species the soil was left fallow. The experiment was then continued (9–12 months after amendments applied) using the same soils but growing *L. perenne* and *A. argenteum* only.

Soils were collected from the surface 20 cm at four randomly located points on each landfarm. These soils were partially air-dried, mixed in a cement mixer and passed through a 4 mm sieve. Prior to any amendment, the mixed soil had a pH (water) of 6.7 with very low concentrations of extractable P (6 mg kg^{-1}), but extractable K (82 mg kg^{-1}) concentrations that were adequate for plant growth.

As initial investigations suggested that metals present in these soils were unlikely to be plant available, various amendments were added to soils prior to planting to increase this availability. In addition to an untreated landfarm soil control, one of three levels (1, 2 and 3 g kg^{-1} soil applied as a potassium salt in solution) of EDTA (ethylenediaminetetra acetic acid) was mixed into soils prior to their being placed in pots. Previous studies (Blaylock et al 1997) had shown that this complexing agent increased uptake of metals, including Ni and Zn, by *B. juncea*. Application rates used here were comparable to those of the above study.

Soil pH is an important factor affecting solubility of metals in soils and, for Ni and Zn, the existing soil pH was likely to limit this solubility (Adams & Sanders 1984). Sulphur was available on the refinery site and therefore offered a practicable option for pH reduction. Laboratory elemental sulphur powder was mixed into soils at three rates (1.2, 2.4 and 3.6 g kg^{-1} soil), based on results from an initial study of soil pH responses to varying S additions.

A set of replicates was also grown in John Innes No 2 potting compost to provide an indication of growth potential under ambient conditions.

All treatments, including the compost, received regular nutrient inputs at rates considered optimum, based on results from preliminary trials. Nutrients were applied in solution to individual pots (50 cm^3 containing 2.5 mg of N, P$_2$O$_5$ and K$_2$O) initially and then at roughly two-week intervals. Over each cropping phase of the experiment, inputs of N were equivalent to 20–25 mg pot^{-1}.

Plants were grown in 500 cm^3 pots partially filled with soil or compost, each sitting in a tray to collect drainage water and prevent rapid loss of soil amendments. Water or nutrient solutions were added to avoid moisture conditions causing any limitations to plant growth. Plants were grown under standard environmental conditions (20–25°C) and a photoperiod of 16 hours.

2.2 Measurements

Plant growth and foliar metal concentrations/uptake were measured after 2 months and at several intervals thereafter. Extractable soil metals (CaCl$_2$) and "leachate" metal concentrations were also measured periodically throughout the experimental period. Leachates were produced by removing pots from their trays, washing the base of these pots, then flushing moist soils with 100 cm^3 of distilled water; drainage water was collected for subsequent analysis.

At harvest, plants were cut to the soil surface and washed to remove any soil adhering to harvested material. For "total" metals, soil and plant materials were digested for 20 hours in hot, concentrated nitric acid digests (MAFF 1986). For extractable metals

(McGrath & Cegara 1992), soils were shaken (Griffen reciprocal shaker) for 16 hours in 0.1 M CaCl$_2$. Extract metal concentrations were measured (MAFF, 1986) by atomic absorption spectroscopy (Pye-Unicam SP9). Soil texture, pH and nutrient contents were obtained using standard (MAFF 1986) procedures.

Results from each phase of the experiment and for each species were analysed (Statgraphics 1993) separately using Oneway analysis of variance.

3 RESULTS

Soil, leachate and plant samples were collected on a number of occasions during the course of the pot trials. Only representative data are presented here.

3.2 Initial phase of the experiment (Brassica juncea and Lolium perenne)

Plants were harvested after 2 months. Soil and leachate samples were also taken at this stage. The soil in B. juncea pots was then "cultivated" and reseeded; L. perenne was allowed to re-grow. B. juncea failed to establish in the most acid (high S) soils. This stage of the experiment was then continued for a further 4 months at which point all plant materials were removed from the soils.

All plants grown in landfarm soils showed weak growth irrespective of treatments. Composts yielded on average 12.4 and 11.7 g dry matter pot^{-1} of B. juncea over the first and second growing periods respectively. Corresponding values for L. perenne were 9.7 and 10.4 g dry matter pot^{-1}. Thus even for control soils where metal uptake was low, neither plant species exhibited growth close to its potential when planted in landfarm soils.

Highly significant treatment responses were obtained for most parameters. EDTA increased foliar Ni and Zn concentrations over the first growth period for both plant species (Table 1). These increases were not always related to the rate of application. EDTA tended to decrease plant growth. As a result, there was no consistent increase in the total uptake of metals. Sulphur inputs caused a progressive decrease in plant growth, a trend more pronounced for B. juncea. However, this growth decrease was balanced by marked increases in metal, particularly Ni, concentrations. As a result, total foliar uptake of Ni was higher with added sulphur although this trend was not sustained at the highest input rate. Metal concentrations tended to be higher in L. perenne compared with B. juncea with EDTA but lower for sulphur.

Over the second growth period (Table 2), EDTA effects on plant metals were less pronounced for B. juncea. However, uptake and concentrations of both metals were higher for L. perenne. All sulphur

Table 1. Effect of soil amendments after 2 months on plant growth (g pot^{-1}), foliar uptake (mg pot^{-1}) and metal concentration (mg kg^{-1}) – phase 1.

Treatment	Dry matter	Ni uptake	Zn uptake	Ni conc.	Zn conc.
B. juncea					
Control	3.38	0.04	0.19	13.5	56.1
EDTA 1	1.72	0.05	0.11	24.3	64.0
EDTA 2	0.98	0.04	0.10	53.1	101.8
EDTA 3	2.09	0.04	0.13	21.9	62.3
Sulphur1	2.22	0.05	0.34	23.8	153.3
Sulphur2	1.83	0.14	0.50	69.2	271.9
Sulphur3	0.71	0.08	0.18	103.9	254.7
L.S.D. 5%	1.05	0.09	0.31	37.9	117.2
L. perenne					
Control	2.97	0.03	0.20	10.1	61.1
EDTA 1	2.78	0.13	0.41	46.8	200.6
EDTA 2	1.57	0.06	0.29	38.2	186.1
EDTA 3	2.61	0.13	0.62	49.7	236.5
Sulphur 1	2.14	0.06	0.22	28.0	84.2
Sulphur 2	2.39	0.14	0.27	58.6	85.1
Sulphur 3	1.93	0.20	0.27	103.6	142.8
L.S.D. 5%	0.87	0.16	0.38	42.3	86.7

Table 2. Effect of soil amendments on plant growth between 3–6 months, foliar uptake (mg pot^{-1}) and metal concentration (mg kg^{-1}) – phase 1.

Treatment	Dry matter	Ni uptake	Zn uptake	Ni conc.	Zn conc.
B. juncea					
Control	3.18	0.03	0.09	8.9	29.4
EDTA 1	1.98	0.11	0.06	55.6	27.2
EDTA 2	2.37	0.03	0.07	12.4	27.6
EDTA 3	2.19	0.03	0.06	13.4	29.3
Sulphur1	1.65	0.04	0.09	26.5	63.5
Sulphur2	0.42	0.03	0.04	73.4	101.0
Sulphur3	–	–	–	–	–
L.S.D. 5%	1.02	0.16	0.11	42.7	55.6
L. perenne					
Control	3.41	0.04	0.26	15.9	98.2
EDTA 1	2.37	0.56	0.29	251.1	335.2
EDTA 2	2.61	0.75	0.37	270.0	386.8
EDTA 3	3.34	1.11	0.38	216.4	353.4
Sulphur 1	3.39	0.38	0.22	95.9	95.5
Sulphur 2	3.58	0.33	0.27	88.0	92.1
Sulphur 3	1.51	0.22	0.27	205.3	125.8
L.S.D. 5%	0.98	1.60	0.54	207.0	185.7

treatments decreased growth of B. juncea (failure at high sulphur input) but affected L. perenne only at the highest input rate. Sulphur increased foliar Ni concentrations in both species but effects for Zn were significant only in B. juncea.

The marked adverse effect of sulphur on growth of B. juncea meant that metal uptake was unaffected. Foliar metal concentrations in L. perenne were generally

181

higher in the second compared with the first period of growth and often exceed those in *B. juncea*.

Plant species had no effect on soil pH or metal extractability so soil data were averaged and means given for amendments only. Both EDTA and sulphur treatments increased the pool of readily extractable Ni and Zn (Table 3). For sulphur, increasing metal extractability was associated with a progressive decrease in soil pH. In contrast, EDTA had no effect on pH. EDTA tended to have a more pronounced effect on Zn extractability whereas Ni showed a more marked response to sulphur.

Leachate metal concentrations and pH are given in Table 4. Treatment responses followed the same pattern to that for soil analyses and were broadly similar 2 and 6 months after the application of treatments.

Again there was no effect of plant species and species differences were ignored in presenting data.

EDTA did not affect leachate pH but did increase metal concentrations, particularly for Zn, where higher inputs progressively increased metal concentrations at 2 but not 6 months. Increasing inputs of sulphur led to a progressive decrease in leachate pH and increases in metal concentrations. Sulphur inputs had less effect on metal concentrations than EDTA, with only the highest input rate consistently causing a significant increase. There was little evidence of any reduction in leaching between 2 and 6 months.

3.2 Second phase of the experiment (L. perenne and A. argenteum)

Following completion of the previous experiment, all plant materials were removed from the soil. Soil from replicates of each amendment treatment was bulked, thoroughly mixed and stored air dry for several months. Soils were then planted to either *L. perenne* or *A. argenteum* and the experiment continued for a further three months.

A. argenteum grew very poorly on both compost and landfarm soils compared with *L. perenne* (Table 5); it failed to establish in high sulphur soils. Whereas inputs of both EDTA and sulphur decreased growth of *A. argenteum*, only high inputs of sulphur had this effect on *L. perenne*. Concentrations of Ni, and to a lesser extent Zn, were higher in *A. argenteum* compared with *L. perenne*, at every soil amendment level. But the poor growth of *A. argenteum* meant that uptake levels were low compared with *L. perenne*. Most EDTA and sulphur inputs increased foliar metal concentrations for

Table 3. Effect of treatments on soil pH and CaCl$_2$ extractable soil metal concentrations (mg kg^{-1}) 2 months after amendments applied.

Treatment	pH	Ni	Zn
Control	6.4	2.0	2.3
EDTA 1	6.6	13.2	17.8
EDTA 2	6.5	21.6	26.4
EDTA 3	6.5	27.4	39.3
Sulphur 1	5.7	7.7	5.2
Sulphur 2	5.1	44.8	32.5
Sulphur 3	4.8	66.3	45.2
L.S.D. 5%	0.53	17.6	11.2

Table 4. Effect of soil amendments on pH and metal concentrations (mg l^{-1}) in leachate 2 and 6 months after soil amendment.

Treatment	pH	Ni	Zn
2 months			
Control	6.7	0.02	0.02
EDTA 1	6.8	1.1	2.4
EDTA 2	6.8	1.8	4.5
EDTA 3	6.9	1.5	6.8
Sulphur 1	6.6	0.1	0.1
Sulphur 2	5.4	0.5	0.4
Sulphur 3	4.2	2.0	1.8
L.S.D. 5%	0.33	0.77	0.49
6 months			
Control	6.5	0.1	0.1
EDTA 1	6.6	1.8	5.1
EDTA 2	6.5	1.6	5.0
EDTA 3	6.6	1.5	4.8
Sulphur 1	6.4	0.1	0.2
Sulphur 2	5.7	0.7	0.8
Sulphur 3	5.1	3.9	0.9
L.S.D. 5%	0.26	1.55	0.54

Table 5. Effect of soil amendments on plant growth, foliar uptake (mg pot^{-1}) and metal concentration (mg kg^{-1}) – phase 2.

Treatment	Dry matter	Ni uptake	Zn uptake	Ni conc.	Zn conc.
A. argenteum					
Control	0.41	0.005	0.053	10.9	141.3
EDTA 1	0.31	0.017	0.032	55.8	178.6
EDTA 2	0.10	0.013	0.029	89.7	230.0
EDTA 3	0.24	0.009	0.037	35.5	139.4
Sulphur 1	0.37	0.014	0.057	43.9	173.9
Sulphur 2	0.09	0.010	0.024	104.8	283.5
Sulphur 3	–	–	–	–	–
L.S.D. 5%	0.192	0.012	0.030	33.75	118.91
L. perenne					
Control	5.89	0.02	0.30	2.7	50.4
EDTA 1	4.78	0.05	0.52	12.2	110.3
EDTA 2	6.84	0.08	0.63	12.2	95.3
EDTA 3	6.80	0.07	0.65	11.5	96.8
Sulphur 1	7.30	0.06	0.75	8.6	103.0
Sulphur 2	6.00	0.12	0.93	19.6	105.2
Sulphur 3	3.07	0.12	0.40	39.6	131.9
L.S.D. 5%	1.79	0.053	0.265	8.82	41.96

both species. However, the EDTA effect was not related to input rate. Sulphur had such a response pattern but only the highest rate consistently had a significant effect on metal concentrations.

4 DISCUSSION

Under existing conditions and usage, the former land-farm soils pose little risk of metal loss in drainage water or of transfer of metals into vegetation. Maintenance of soil pH close to neutrality would continue to control the risk of metal toxicity or mobility. Therefore, any requirement for remediation of metal levels can be justified only in terms of planning conditions, contaminated land requirements or of potential future land uses. This is a situation common to many lightly contaminated industrial sites.

One of the major constraints on phytoextraction of these soils was their difficult growing condition. Soils tended to slake on wetting and to be waterlogged as a consequence. Preliminary work indicated that plant available N was negligible, a problem exacerbated by the residual oil content which would be expected to generate plant-microbe competition for this nutrient. Growth of all plants, even on control soils was 30% or less of the potential indicated by productivity in compost. Although there was evidence of preferential uptake of Ni by *A. argenteum*, even for this species Ni concentrations were low in unamended soils.

Some effects of a single input of EDTA on metal extractability, leaching or plant uptake persisted for up to one year. As found here, EDTA may reduce plant growth (Blaylock et al 1997), possibly by reducing transpiration rates (Epstein et al 1999). Since EDTA did not affect soil or water pH, it is assumed that most of the metal taken up by plants and in leachate were present as EDTA complexes (Epstein et al 1999). The relatively high pH of landfarm soils may have limited the effectiveness of EDTA, as found by Blaylock et al (1997) for Pb. Although there was some variability with time and between plant species, EDTA tended to affect Zn behaviour to a greater extent than Ni. Blaylock et al (1997) also found that EDTA had a more pronounced effect on Zn, compared with Ni, uptake by *B. juncea*. Enhanced risk of metal leaching with applications of chelating agents to soil has been identified (Romkens et al 2002) as an unwanted side effect. Data given here were consistent with those of the above study for EDGA, particularly for Zn, in that EDTA increased leaching to a greater extent than it enhanced plant uptake. It is possible that this leaching reduced the pool of plant available metals.

Sulphur inputs rendered soils more acidic and increased the availability of metals. The highest rate of sulphur input caused a larger than predicted fall in pH such that conditions were unfavourable for growth of any of the species tested. However, the lowest input rate often did not reduce growth markedly and caused only a small increase in leachate metal concentrations. Solubility of metals such as Ni and Zn increase markedly below certain pH thresholds (Adams & Sanders 1984). In the present study, concentrations of Ni and Zn increased linearly below leachate pH's of 5.8 and 5.6, respectively. Management of pH around these thresholds may offer an opportunity for maximizing plant uptake of metals without markedly increasing metal leaching.

Although inputs of EDTA and of sulphur generally increased foliar metal concentrations, associated reductions in plant growth often resulted in only modest increases in metal offtake. It is not clear whether growth reductions resulting from soil amendments were attributable only to phytotoxic effects of the metals measured. Zn concentrations were not generally excessive compared with published data (Marschner 1995) and there were no obvious Zn toxicity symptoms; in some cases Ni concentrations were at the upper end of the range for moderately tolerant species.

Individual pots contained approximately 400 g of dry soil. On this basis, highest rates of cumulative metal offtake in aerial parts of plants were equivalent to reductions of around $3\,mg\,kg^{-1}$ soil in Ni and Zn concentrations. However, harvesting in this pot trial was to ground level and any field operation would not achieve this level of plant removal. The above reductions in soil metal concentrations therefore represent a maximum that is unlikely to be achieved in practice.

Only *L. perenne* was grown throughout the experimental period. There was no clear evidence of metal concentrations in this species declining over successive harvests, although values were generally highest in later harvests during phase 1. It is not clear therefore whether rates of metal offtake obtained could be sustained over further harvests.

Results from this study illustrate some of the problems associated with phytoextraction of metals from soils where slightly elevated metal concentrations are present in a form not available for plant uptake and where growing conditions are generally poor. There may be scope for enhancements in metal removal by better selection of crop species/variety, improved nutrient management and more careful manipulation of metal availability. However, it remains uncertain whether some of the very high extraction rates quoted in the literature (Baker et al 1994, Blaylock et al 1997) from previous investigations can be achieved in soils such as those studied here.

ACKNOWLEDGEMENTS

This work was supported financially by TOTALFINAELF; B. Brown of TOTALFINAELF provided background information on the site and advice on developing the project. The author also wishes to acknowledge the practical inputs of H Gustaffson, T Bisol and R Jones.

REFERENCES

Adams, T.McM. & Sanders, J.R. 1984. The effect of pH on the release to solution of zinc, Cu and nickel from metal-loaded sewage sludges. *Environmental Pollution Series B* 8: 85–99.

Baker, A.J.M. & Brooks, R.R. 1989. Terrestrial higher plants which hyperaccumulate metallic elements – a review of their distribution, ecology and phytochemistry. *Biorecovery* 1: 81–126.

Baker, A.J.M., McGrath, S.P., Sidoli, C.M.D. & Reeves, R.D. 1994. The possibility of in situ heavy metal decontamination of polluted soils using crops of metal-accumulating plants. *Resources Conservation and Recycling* 11: 41–49.

Blaylock, M.J., Salt, D.E., Dushenkov, S., Zakharova, O., Gussman, C., Kapulnik, Y., Ensley, B.D. & Raskin I. 1997. Enhanced accumulation of Pb in Indian Mustard by soil applied chelating agents. *Environmental Science Technology* 31: 860–865.

Epstein, A.L., Gussman, C., Blaylock, M.J., Yermiyahu, U., Huang, J.W., Kapulnik, Y. & Orser, C.S. 1999. EDTA and Pb-EDTA accumulation in *Brassica juncea* grown in Pb-amended soil. *Plant and Soil* 208: 87–94.

ICRCL 1987. Guidelines on the assessment and redevelopment of contaminated land. ICRCL 59/83 Second Edition. London: Department of Environment.

MAFF 1986. *The Analysis of Agricultural Materials*. Ministry of Agriculture, Fisheries and Food Technical Bulletin 27. London: Her Majesties Stationary Office.

Marschner, H. 1995. *Mineral Nutrition of Higher Plants 2nd Edition*. London: Academic Press.

McGrath, S. P. & Cegarra, J. 1992. Chemical extractability of heavy metals during and after long-term applications of sewage sludge to soil. *Journal of Soil Science* 43: 313–321.

McGrath, S.P., Shen, Z.C. & Zhao, F.J. 1997. Heavy metal uptake and chemical changes in the rhizosphere of *Thlaspi caerulescens* and *Thlaspi ochroleucum* grown in contaminated soils. *Plant and Soil* 188: 153–159.

Robinson, B.H., Chiarucci, A., Brooks, R.R., Petit, D., Kirkman, J.H., Gregg, P.E.H. & Dominicis, V. de. 1997. The nickel hyperaccumulator plant *Alyssum bertolonii* as a potential agent for phytoremediation and phytomining of nickel. *Journal of Geochemical Exploration* 59: 75–86.

Romkens, P., Bouwman, L., Japenga, J. & Draaisma C. 2002. Potentials and drawbacks of chelate enhanced phytoremediation of soils. *Environmental Pollution* 116: 109–121.

Statgraphics 1993. Statgraphics Statistical Graphics System, Version 7.0. Manugistics Inc. Rockville, USA.

Land Reclamation – Moore, Fox & Elliott (eds)
© 2003 Swets & Zeitlinger, Lisse, ISBN 90 5809 562 2

Erosion risk assessment on disturbed and reclaimed land

R.J. Rickson
National Soil Resources Institute, Cranfield University at Silsoe, Silsoe, Bedfordshire, UK

ABSTRACT: Land disturbance, ground engineering and subsequent reclamation can increase soil erosion risk due to over-steepened slopes, exposure of highly erodible soils and lack of vegetation cover. Soil erosion on-site leads to management difficulties such vegetation establishment, because of the thin, infertile soils and wash off of seeds, young plants and fertilisers. When eroded soil is transported off-site the impact on the environment can be even more devastating. Eroded sediments (and any contaminants adsorbed onto them) are increasingly regarded as major pollutants, which can seriously damage aquatic ecosystems. The methodology of erosion risk assessment is well established, but rarely applied to disturbed or reclaimed sites. Erosion risk can be assessed with techniques such as factorial scoring, empirical and physical-process modeling, and GIS. These methods can improve the process of land reclamation by identifying areas of high erosion risk, and evaluate proposed land use and land management strategies to reduce erosion risk.

1 INTRODUCTION

Natural landscapes can be regarded to be in a state of "dynamic equilibrium", where the quality or capability of the land resource is reflected by the natural land cover. However, this balance can be adversely affected during and after land disturbance by human activity. Such disturbance can be found in a variety of sites – mine spoils, quarries, open cast mines, pipeline corridors, landfill and waste disposal sites, sand and gravel extraction pits, recreational areas and construction sites (for highways, railways and urban development). One indicator of such an imbalance is an increase in soil erosion risk within the disturbed landscape. This paper will review the evidence for increased soil erosion risk on such disturbed sites, the consequences of allowing erosion rates to accelerate and the methods available to assess the erosion risks associated with disturbed landscapes. The objective of the paper is to identify the factors which make such sites so susceptible to erosion, so that management strategies can be developed to mitigate against the adverse environmental effects of land disturbance. Knowing the rates and consequences of erosion from such sites helps to justify both the need for, and expense of, land reclamation techniques.

2 THE PROBLEM

In the literature, there is a wealth of experimental and field data quantifying erosion and sediment rates from disturbed land. Much of the literature relates to agricultural land use, which constitutes land disturbance from the natural state. However, this paper will concentrate on land disturbed by non-agricultural activities, especially those related to former industrial land use such as mining, and infrastructural development (such as highway construction). Most of the data refer to hillslope erosion by water, but wind erosion and mass movements can also occur on these sites (eg Amponsah-Dacosta & Blight 2002, http://www.nuff.ox.ac.uk/politics/aberfan/home2. htm)

Much of the data on erosion rates from disturbed sites originates from the USA (Toy & Hadley 1987). The US Environmental Protection Agency (EPA) quote data which compare a mined and unmined catchment. The latter had 69 times more sediment per unit area of catchment than the former. The same paper reports sediment volume from an unrestored spoil bank to be 968 times that of the surrounding, undisturbed area. The haul roads associated with the mining activity produced an incredible 2,065 times more sediment than the surrounding slopes (EPA 1976a). Other selected data from the literature are shown in Table 1.

Curtis (1981) makes the point that although surface coal mining is the most efficient method of mining, it is also the most destructive of other natural resources. Toy & Hadley (1987) review the different processes employed in surface mining, and the degree of disturbance caused by each method.

Knowledge of the factors influencing such high erosion rates is important, as this is the first step in

Table 1. Selected erosion rates from disturbed lands.

Site description	Erosion/sediment rates	Source
Coal mine spoil, Colorado, USA	102 t/ha/yr	Renner, 2002
Gold mine tailings, South Africa	500 t/ha/yr	Bright, 1991
Surface mined basin, Wyoming, USA	2.38 t/ha/yr (11 × greater than undisturbed land)	Ringen et al., 1979
Highway construction	480 t/ha/yr	Diseker & Richardson, 1962
Highway construction, Virginia, USA	338 t/ha/yr	Vice et al., 1969
Urbanisation	226 t/ha/yr	Yorke & Herb, 1976
Off road recreation	5.52 t/ha/yr	Snyder et al, 1976
Surface mine spoils, Wyoming	1.1–9.5 t/ha over 45 min, with rainfall of 38.1 mm	Lusby & Toy, 1976
Construction	17,000 t/km^2/yr	US EPA, 1973
Actively mined watershed	46,400 p.p.m. sediment concentrations in runoff	Curtis, 1971
Abandoned coal refuse sites	61.5 kg/m^2	Mandel et al., 1982
Surface coal mine dumps	2.1–5.9 mm/yr ground loss	Haigh, 1979
Abandoned surface mines	850 t/km^2/yr	US EPA, 1973
Active surface mines	17,000 t/km^2/yr	US EPA, 1973

devising management strategies to minimize erosion risk on such disturbed lands. The following section analyses why artificially engineered slopes produce such high erosion rates.

2.1 Slope characteristics

Hillslope engineering, land disturbance and consequent reclamation will affect the land slope character of an area. For mine spoil placement, slopes are often over-steepened in order to minimize land take. An increase in slope steepness can be related to an increase in erosion rate (Zingg 1940, Musgrave 1947, Wischmeier & Smith 1978, Carson & Kirkby 1972). These relationships can be expressed in various forms:

$$A = s^n \qquad (1)$$

Where A = soil loss, s = slope gradient and the exponent, n, ranges from 0.7 to 2.0 according to a review by Hadley et al. (1985). Smith & Wischmeier (1957) developed the following equation relating slope gradient to erosion rate:

$$A = 0.43 + 0.30S + 0.043S^2 \qquad (2)$$

where A = the soil loss and S is the slope gradient.

Overland flow on steeper slopes has greater velocity, which increases the erosivity of the flow, both for detachment of soil particles (varies with v^3) and for the transport of already detachment particles (varies with v^5). This is important to know when devising reclamation programmes, because any small increase or decrease in overland flow velocity has a dramatic effect on the erosive power of the runoff.

Slopes are often artificially lengthened too, during ground engineering works. Such changes directly

impact on erosion processes operating on the slopes (Carson & Kirkby 1972, Wischmeier & Smith 1978). These relationships can be expressed thus:

$$A = 1^m \qquad (3)$$

Where A = soil loss, 1 = slope length and exponent m varies from 0 to 2.0, according to Kirkby (1969).

Landscape re-grading during disturbance or as part of the reclamation process will also affect slope curvature in profile and in plan, creating three-dimensional changes in slope character. As with slope gradient and length, these changes affect erosion risk on slopes. Meyer & Kramer (1968) found that sediment production and transport were affected by the shape of the slope profile. Slopes with uniform, concave, convex or convexo-concave profiles produced different erosion rates, all other factors being equal. Convex slopes are relatively most susceptible to erosion because the steepest section of the slope is at the toe, where runoff volume and velocity are greatest. Concave slopes are associated with relatively lower rates of erosion, because the steepest section is at the top of the slope, where runoff is limited in both volume and velocity. The lower slope segment on concave slopes is flatter, so any sediment entrained in the runoff may be deposited in this lower velocity environment (Meyer & Romkens 1976). Hadley et al., (1985) argue that profile form can be as much or even more important than slope gradient in affecting erosion rates.

In addition to these direct effects between slope character and erosion, changes in slope also affect the hydrology of the area. Landscape engineering commonly changes natural catchment boundaries, so creating new zones of flow divergence or convergence.

Table 2. The erodibility or K factor of disturbed v. undisturbed soils.

| | K factor values | | |
Site description	Disturbed land	Undisturbed land	Source
Land disposal facility	0.43	–	US EPA, 1976a
Mine spoil, sandy clay loam	0.1	–	Gilley et al., 1977
Reclaimed mine spoil	0.37	0.48	Mitchell et al., 1983
Solar Sources Mine	0.110	0.059	Stein et al., 1983
Mine spoil	0.38	–	McIntosh et al., 1989
Mine spoil, Italy	0.58	–	http://www.ldd.go.th/Wcss2002/ papers/0798.pdf

Flow discharges generated may be increased or decreased, depending on the nature of the landscape change. This may result in an increased flooding risk (if time of concentrations are reduced and peak flows are increased, as would occur on oversteepened slopes), or drought conditions (where time of concentration of flow was lengthened and peak discharges reduced). Both scenarios have implications for the success of any subsequent reclamation scheme, especially during the critical re-vegetation stages.

Knowledge of slope effects on erosion and hydrology is important in land reclamation schemes, especially during the planning stages. It is important to consider how proposed final slope morphology of a reclaimed site can affect erosion processes, at the same time as designing the aesthetics of the proposed reclaimed landscape.

2.2 Exposure of highly erodible material

Soils from disturbed sites are notoriously susceptible to erosion. The "erodibility" of any slope forming material relates to its susceptibility to erosive forces – rainfall, runoff, wind and wave action. Erodibility usually refers to top- or sub-soil, and is highly correlated to the stability of soil aggregates against dispersion by these erosive forces. Factors affecting aggregate stability include the soil texture (especially clay content), structure, organic matter content, and chemical content of the soil.

Land disturbance can impact on these soil properties. Soil structure is adversely affected when compaction occurs through machinery traffic, mechanical handling, or stockpiling of soil. Organic matter is lost through mineralisation when the relatively organic rich topsoil is exposed. Removal of the topsoil also exposes the less organic rich, and thus poorly aggregated subsoil, which has higher erodibility.

Erodibility has been expressed in many ways (Bryan 1968). For example, the K factor of the Universal Soil Loss Equation (Wischmeier & Mannering 1969) is commonly quoted in the literature (Table 2). Data on K values illustrate the increased risk of erosion on disturbed sites. It should be noted that K factors are dynamic over time, whereby erodibility tends to decrease over time, following cessation of disturbance. This is because without further disturbance, soil structure can re-develop over time, although development of a fully structured soil profile may take many years. This results in improved porosity, better infiltration characteristics, and increases in aggregate stability. Bulk density may also increase as soil consolidation takes place as individual particles and small aggregates "pack" more efficiently under gravity and raindrop impact. This increases soil cohesion, so that more energy from the erosive forces is required to detach soil particles.

From a management point of view, very little can be done to improve the inherent soil properties which affect erodibility, notably aggregate stability. However, there are examples where soil amendments are blended with the disturbed soil in an attempt to reduce K values. The amendments are in the form of sand-sized particles of alkali materials which help to change textural class and control acidity at the same time. Compost amendments have also been used to reduce soil erodibility, but the benefits of this technique tend to be long term. Organic and inorganic soil conditioners have also been used to reduce soil erodibility (Wallace & Wallace 1986, De Boodt 1972).

Toy & Hadley (1987) conclude that soil handling techniques during land disturbance (such as topsoil removal, compaction and re-working) and length of time of stockpiling affect K values. Both factors are related to management practice, and can be manipulated to ensure soil erodibility on disturbed sites is minimised as much as possible.

2.3 Lack of cover

Disturbed lands are often devoid of surface cover, either because of recent disturbance or because the slope forming material is not conducive to plant growth. The latter is commonly a consequence of shallow soil depth for root development, limited water holding capacity, soil compaction or toxicity. One demonstration of the importance of surface cover in reducing erosion rates can be seen when comparing soil loss from a bare soil to that with a uniform sward of grass.

If the erosion observed from the bare soil plot is assigned a dimensionless value of 1.0 (= the C factor in the Universal Soil Loss Equation), the erosion rate from a well established, vegetated plot can be as low as 0.004, all other factors being equal (Wischmeier & Smith 1978, Morgan 1986). This is because the vegetation canopy, stems and roots all protect the surface soil from rainfall and runoff through rainfall interception and storage, stem roughness imparted to overland flow, improved infiltration and greater cohesion in the sub-surface soil/root matrix (Rickson & Morgan 1988). Conversely, the lack of vegetation cover and/or surface protection can significantly increase erosion rates from disturbed ground.

Another illustration of the importance of vegetation cover on erosion is given in Thornes (1990). He proposes that the infiltration capacity of a soil increases exponentially with increasing percentage vegetation cover, as a function of an increase in soil organic matter content and a decrease in soil bulk density. This relationship was first expressed by Holtan (1961):

$$K_{sveg} = K_s \, 1/1(1 - PBASE) \qquad (4)$$

where K_{sveg} = saturated hydraulic conductivity of the soil with vegetation (mm min^{-1}); PBASE = the total area of the base of plant stems expressed as a proportion of the total area of the plane.

On disturbed sites, the effect of vegetation removal can thus have significant effects on slope hydrology and erosion risk. From a land management and reclamation standpoint, the cover factor, C, is the easiest to manage and improve on disturbed sites. The problem, however, is the critical time period between completion of land engineering work and the establishment of 65–75% cover, which is the critical threshold at which vegetation reduces erosion rates significantly. The duration of this period will depend on the environmental conditions (moisture, temperature, species selected), but immediate surface protection at this time can be afforded by erosion control geotextiles, hydroseeding or mulching (Kramer & Meyer 1969, Rickson 1995).

3 CONSEQUENCES OF EROSION ON DISTURBED LANDS

The consequences of soil erosion can be classified as being "on-site" and "off-site". At the sites where erosion has taken place, eroded channel features such as rills and gullies may appear, which concentrate future overland flow events, thus accelerating subsequent erosion rates. Removal of topsoil inhibits vegetation growth (an essential component of site reclamation) due to limited depth of the rooting medium, limited available water storage capacity and paucity of plant nutrients. Indeed, soil erosion is a selective process, preferentially removing N (in runoff) and P (adsorbed onto the sediment), as evidenced by higher concentrations of these nutrients in eroded sediments and slope runoff when compared with in-situ (uneroded) soil. This selectivity is quantified as "enrichment ratios".

Off-site impacts of erosion are regarded as even more damaging to the environment. Sediment acts as a pollutant, increasing turbidity in watercourses, which affects light penetration, bio-available oxygen levels and water temperature, so having detrimental effects on aquatic flora and fauna (Cooper 1993). In lower velocity environments lower down in the catchment, the eroded material is deposited and sedimentation can reduce the capacity of water bodies such as lakes, reservoirs and canals. If the sediments have high concentrations of P (see above), then the slow release of this nutrient can lead to eutrophication of water bodies (Sharpley & Menzel 1987).

The on- and off-site problems associated with uncontrolled erosion highlight the need for land managers to be able to assess the severity of the erosion risk, so that erosion control strategies are soundly designed to tackle the processes operating on site. Knowing the consequences of erosion, both environmental and economic, also justifies the costs of an effective erosion control plan for the site. At a time when the current paradigm is one of "the polluter pays", this is a lesson to be learnt.

4 ASSESSMENT OF THE PROBLEM

To assess the severity of the problem, data from the a) pre-disturbance, b) active disturbance and c) post disturbance phases of land engineering have to be compared (Toy 1984). Comparing a) and b) will assess the impact of land disturbance on the environment. Comparing a) and c) will ensure that any reclamation techniques are effective in bringing erosion rates down to initial, "baseline", levels.

Toy & Hadley (1987) consider the following factors when assessing the problem of erosion risk on disturbed lands: a) areal extent of disturbance; b) intensity of disturbance; c) duration of disturbance d) the nature of disturbance; e) spatial aspects of disturbance and f) cumulative impacts.

4.1 *Factorial scoring*

This semi-quantified approach to erosion assessment considers all the factors considered to affect erosion risk at the site. This allows for some flexibility

188

because it allows for the fact that not all disturbed sites will have the same factors affecting erosion risk. It could be argued that it is difficult to ascertain which factors are significant until the risk assessment has been carried out. In any case, experience of erosion studies show that it is highly likely that rainfall aggressiveness (erosivity), soil susceptibility to erosion (erodibility), slope length, slope steepness, degree of vegetation cover and existing management practices are usually the most significant in affecting erosion risk.

In factorial scoring the site is divided into land planning units (usually based on simple interpretation of the local topography, using hydrological sub-catchments as the basic land planning unit). All factors affecting erosion at the site are then assigned a numerical value for each land planning unit   commonly ranging from 0 (= no risk) to 5 (= severe risk) for example. Thus, absolute quantification of each factor is not required, which is the advantage (but also the disadvantage!) of this approach. All factors are then summed (or multiplied) to calculate a factorial score for that land unit. The range of scores can then be classified arbitrarily into classes, which reflect relative erosion risk on the site.

The resulting factorial scores can be compared between the different land planning units, identifying areas with high erosion risk and those with low erosion risk. Management strategies can be assigned to the different classes of erosion risk, targeting areas where erosion risk is greatest. Morgan (1986) is cautious about this approach as it a) may be sensitive to the different scores assigned to each factor affecting erosion, b) treats all the factors as independent, and yet in reality many of the factors are highly interrelated and c) there is no justification for summing the factor scores, as opposed to multiplying them and d) each factor is given equal weight which may overestimate the importance of some factors, and underestimate the significance of others.

4.2 Geographical Information Systems

Geographical Information Systems may be used to develop the methodology of factorial scoring. Here, the factors identified as affecting erosion risk are assigned their erosion risk class, which can be displayed spatially in a series of thematic maps. The resultant summation (or multiplication) of the factors affecting erosion risk can also be displayed within the GIS (Briggs & Giordano 1992, De Jong & Riezebos 1992). This approach is particularly suited to assessing changes in erosion risk over time, using a number of thematic maps produced at different dates.

The same approach can be taken further to include quantified estimates of the factors affecting erosion, namely, erosivity, erodibility, slope gradient and length, vegetation cover and existing management factors

(Jäger 1994). Existing choropleth maps of these physical attributes are digitised, and placed in the GIS. Quantification of these factors is also needed in any modeling approaches to erosion risk assessment.

4.3 Modeling

Running erosion prediction models is another way to differentiate areas of high or low erosion risk on disturbed lands. A number of erosion prediction models have been used for this purpose: the Universal Soil Loss Equation – USLE (Wischmeier & Smith 1978), the Revised USLE (Renard et al. 1997), the Morgan, Morgan and Finney model (Morgan et al. 1984) and the European Soil Erosion Model – EUROSEM (Morgan et al. 1998).

In the USLE, rainfall erosivity, soil erodibility, slope gradient and length, cover characteristics and existing land management practices are quantified as a series of factors such that:

$$A = R.K.LS.C.P \qquad (5)$$

where A = annual soil loss (t/ha/yr); R = rainfall erosivity factor; K = soil erodibility index; LS = slope factor; C = cover factor; and P = land management factor.

If sufficient data are available (and confidence in the applicability and accuracy of the model exists – see Wischmeier, 1978), then each factor can be quantified, and the calculation of annual soil loss is possible, as shown in equation 5. The outcome is an annual soil loss estimation, which will vary spatially, so highlighting areas of different predicted erosion rates.

In a development of the USLE, the RUSLE considers slopes to be a series of segments, which allows for slopes longer than experienced in an agricultural context (as shown in USLE). This consideration makes the RUSLE more applicable to mined, construction and reclaimed lands where artificially engineered slopes are created.

The USLE is also useful at assessing the effectiveness of different management scenarios in reducing erosion rates. This is achieved by modifying the salient input parameters. For example, revegetating the site during reclamation can reduce the C factor by two orders of magnitude (see above), so reducing the annual soil loss amount (A) considerably.

The major criticisms of the USLE in Europe are: a) it has not been accurately validated beyond the geographical region where the equation was developed (east of the Rocky Mountains in the US); b) it is empirically based, with no explanation of the interrelationships between the input factors nor simulation of the physical processes of soil erosion. In an attempt to respond to these concerns, Morgan et al. (1984) developed the Morgan, Morgan and Finney model of soil erosion prediction. This is still partially an empirical model, but it does consider the different phases of the erosion process, namely splash detachment by rainfall and transport of detached material by overland flow or runoff.

The input data required to run the Morgan, Morgan and Finney model are discussed elsewhere (Morgan et al. 1984), but there is provision to reflect the site characteristics associated with disturbed lands such as changes in bulk density (due to compaction, consolidation and machinery trafficking), limited rooting depth (due to topsoil removal), increased soil susceptibility to detachment (due to soil structure disturbance and breakdown of inherent aggregate stability), increase in slope gradient (due to slope regrading), and removal of vegetation (and re-growth during reclamation).

As with the USLE, model output from the Morgan, Morgan and Finney model demonstrates spatial variability in predicted erosion rates, highlighting areas of severe risk. These areas need soil protection measures to be designed and implemented as a matter of priority for sensible land management.

The major limitations of the Morgan, Morgan and Finney model are that it still relies on some empirical relationships which have not been validated universally and very little validation has been carried out on the model for non-agricultural land. The model also has no consideration of slope length, which on overlengthened, artificial slopes can be a major factor affecting erosion rates in terms of the velocity and volume of runoff generated (see above).

Attempts to develop a physically based erosion model are ongoing, including the development of the European Soil Erosion Model, EUROSEM (Morgan et al. 1998). The advantage of a truly physically based model is that erosion processes are universal – accurately modeling the physical process of splash detachment should be applicable anywhere in the world.

The EUROSEM model simulates sediment transport, erosion and deposition over the land surface by rill and interrill processes in single storms for both individual field and small catchments. Although the operating functions of this physically based model are more complicated than those in the USLE and MMF models, input parameters can be selected to reflect conditions of disturbed lands. Hence the parameters of slope length and gradient, sediment detachment rate, vegetation cover, hydraulic conductivity of the soil, infiltration capacity, rockiness, soil surface condition, can be assigned values which reflect site conditions. Where field measurements of these input parameters are difficult, guide values are given.

These input parameters will also change as a result of reclamation techniques (e.g. revegetation of the site). Hence the model can not only be used for risk assessment: it can also be used to predict the effectiveness of different proposed erosion control and reclamation strategies.

Today, erosion prediction models and GIS can be fully integrated, due to the size, speed and general availability of computers. Interfacing programs allow input of data from the GIS into the erosion model, which is then run to produce erosion prediction estimates, which are then displayed by the GIS (De Roo et al., 1989).

5 CONCLUSION AND RECOMMENDATIONS

Understanding the processes operating and risks of erosion on engineered slopes, disturbed sites and reclaimed land will aid effective and sustainable land management practices. The methodology used in assessing risk of damage by soil erosion can also be used to assess the likelihood of success of different reclamation and management scenarios in the future.

REFERENCES

Amponsah-Dacosta, F. & Blight, G.E. 2002. The effects of wind on the surfaces of mine tailings dams. In *Adventures in erosion education. Proceedings of the International Erosion Control Association Annual Conference*, 33: 3–16. ISSN 1092 2806.

Briggs, D.A. & Giordano, A. 1992. *CORINE soil erosion risk and important land resources in the southern regions of the European Community.* CEC Publication EUR 13233 EN.

Bright, G.E. 1991. Erosion and anti-erosion measures for abandoned gold tailings dams. In *Proceedings, National meeting of the American Society for Surface Mining and Reclamation, 1991.* Durango, Colorado, USA.

Bryan, R.B. 1968. The development, use and efficiency of indices of soil erodibility. *Geoderma*, 2: 5–26.

Carson, M.A. & Kirkby, M.J. 1972. *Hillslope form and process.* Cambridge University Press, 475 pp.

Cooper, C.M. 1993. Biological effects of agriculturally derived surface water pollutants on aquatic systems – a review. *J. Environ. Qual.* 22: 402–408. July–September 1993.

Curtis, W.R. 1971. Surface mining, erosion and sedimentation. *Trans. Am Soc. Agr. Engrs.* 14: 434–436.

Curtis, W.R. 1981. Reclamation research needs in relation to Public Law 95–87. In *Proceedings, Convention of the Society of American Foresters, 1981, Society of American Foresters*, 70–76.

De Boodt, M. 1972. Improvement of soil structure by chemical means. In Hillel, D. (ed.) *Optimising the soil physical environment toward greater crop yields.* 43–54. Academic Press.

De Jong, S.M. & Riezebos, H Th, 1992. *Assessment of erosion risk using multi-temporal remote sensing data and an empirical erosion model.* Department of Physical Geography, University of Utrecht.

De Roo, A.P.J., Hazelhoff, L. & Burrough, P.A. 1989. Soil erosion modeling using "ANSWERS" and GIS. *Earth Surface Processes and Landforms*, 14(6 and 7): September–November 1989.

Diseker, E.G. & Richardson, E.C. 1962. Erosion rates and control methods on highway cuts. *Trans. Am. Soc. Agr. Engrs.* 5: 153–155.

Gilley, J.E., Gee, G.W., Bauer, A., Willis, W.O. & Young, R.A. 1977. Runoff and erosion characteristics of surface mined sites in western North Dakota. *Trans. Am. Soc. Agr. Engrs.* 20: 697–700; 704.

Hadley, R.F., Lal, R., Onstad, C.A., Walling, D.E. , & Yair, A. 1985. Recent development in erosion and sediment yield studies. *Technical Documents In Hydrology*. UNESCO. Paris. 127 pp.

Haigh, M.J. 1979. Ground retreat and slope evolution on regraded surface mine dumps, Waunafon, Gwent. *Earth Surface Processes and Landforms*, 6: 183–189.

Holtan, H.N. 1961. A concept for infiltration estimates in watershed engineering. *USDA Agricultural Research Service, Publication* ARS – 41–51.

http://web.umv.edu/~somm/Chapters/chapter13.html

http://www.nuff.ox.ac.uk/politics/aberfan /home2.html

http://www.ldd.go.th/Wcss2002/papers/0798.pdf

Jäger, S. 1994. Modeling regional soil erosion susceptibility using the Universal Soil Loss equation and GIS. In Rickson, R.J. (ed.), *Conserving soil resources: European perspectives:* 161–177. CAB International. Wallingford, Oxon.

Kirkby, M.J. 1969. Erosion by water on hillslopes. In Chorley, R.J. (ed.), *Water, earth and man:* 239–238. Methuen.

Kramer, L.A. & Meyer, L.D. 1969. Small amount of surface mulch reduces surface erosion and runoff velocity: *Transactions American Society of Agricultural Engineers.* 12: 638–641, 645.

Lusby, G.C. & Toy, T.J. 1976. An evaluation of surface mine spoils area restoration in Wyoming using rainfall simulation. *Earth Surface Processes and Landforms*, 1: 375–386.

Mandel, R.D., Sorensen, C.J. & Jackson, D. 1982. A study of erosion-sedimentation processes on abandoned coal mine refuse piles in south eastern Kansas. In Graves, D.H. (ed.), *Proceedings of a Symposium on surface mining, hydrology, sedimentology, and reclamation, 1982.* 663–669. Lexington, Kentucky, Office of Engineering Services, University of Kentucky.

McIntosh, J.E., Barnhisel, R.I. & Powell. J.L. 1989. Erodibility and sediment yield of reconstructed mine soil and spoil materials. *Green Lands* 19(3): 24–27.

Meyer, L.D. & Kramer, L.A. 1968. Relation between landscape shaper and soil erosion. *Transactions Am. Soc. Agri. Engs.* 11: 1–14.

Meyer, L.D. & Romkens, M.J.M. 1976. Erosion and sediment control on reshaped land. In *Proceedings, Third Interagency sediment conference, PB-245-100.* 2–65; 2–76. Water Resources Council, Washington, D.C.

Mitchell, J.K., Moldenhauer, W.C. & Gustavson, D.D. 1983. Erodibility of selected reclaimed surface mined soils. *Trans. Am. Soc. Agr. Engrs.* 26(5): 1413-1417; 1421.

Morgan, R.P.C. 1986. *Soil erosion and conservation*. Longmans UK Ltd.

Morgan, R.P.C., Morgan, D.D.V. & Finney, H.J. 1984. A predictive model for the assessment of soil erosion risk. *Journal Ag. Eng. Res.* 30: 245–253.

Morgan, R.P.C., Quinton, J.N., Smith, R.E., Govers, G., Poesen, J.W.A., Auerswald, K., Chisci, G., Torri, D. & Styczen, M.E. 1998. The European soil erosion model (EUROSEM): A dynamic approach for predicting sediment transport from fields and small catchments. *Earth Surface Proc. and Landforms.* 23: 527–544.

Musgrave, G.W. 1947. The quantitative evaluation of factors in water erosion: a first approximation. *Journal of Soil and Water Conservation.* 2: 133–138.

Renard, K.G., Foster, G.R., Weesies, G.A., McCool, D.K. & Yoder, D.C. 1997. Predicting soil erosion by water: A guide to conservation planning with the revised Universal Soil Loss Equation (RUSLE). *USDA Agricultural Research Service, Agricultural handbook* No. 703, pp 384.

Renner, S.G. 2002. Coal basin mine reclamation case study. In *Adventures in erosion education. Proceedings of the International Erosion Control Association Annual Conference,* 33: 473–483. ISSN 1092 2806.

Rickson, R.J. 1995. Simulated vegetation and geotextiles. In Morgan, R.P.C. & Rickson, R.J. (eds.), *Slope stabilisation and erosion control: a bioengineering approach.* E & FN Spon, London.

Rickson, R.J. & Morgan, R.P.C. 1988. Approaches to modeling the effects of vegetation on soil erosion by water. In Morgan, R.P.C. & Rickson, R.J. (eds.), *Erosion assessment and modeling.* 237–254. C.E.C., DG VI, EUR 10860 –EN.

Ringen, B.H., Shown, L.M., Hadley, R.F. & Hinckley, T.K. 1979. Effect of sediment yield and water quality of a non-rehabilitated surface mine in north central Wyoming. *U.S. Geological Survey Water Resources Investigations*, 79–47: 23.

Sharpley, A.N. & Menzel, R.G. 1987. The impact of soil and fertiliser phosphorus on the environment. *Advances in Agronomy*, 41: 297–324.

Smith, D.D. & Wischmeier, W.H. 1957. Factors affecting sheet and rill erosion. *Trans. Am. Geophysical Union.* 38: 889–896.

Snyder, C.T., Frickel, D.G., Hadley, F. & Miller, R.F. 1976. Effects of off-road vehicle use on the hydrology and landscape of arid environments in central and southern California. *U.S. Geological Survey Water Resources Investigations*, 76–99, pp 45.

Stein, O.R., Roth, C.B., Moldenhauer, W.C. & Hahn, D.T. 1983. Erodibility of selected Indiana reclaimed strip mined soils. In Graves, D.H. (ed.), *Proceedings of a Symposium on surface mining, hydrology, sedimentology, and reclamation, 1982.* 101–106. Lexington, Kentucky, Office of Engineering Services, University of Kentucky.

Thornes, J.B. 1990. The interaction of vegetational and erosional dynamics in land degradation: spatial outcomes. In Thornes, J.B. (ed.), *Vegetation and erosion.* 41–53. Wiley, Chichester.

Toy, T.J. 1984. Geomorphology of surface mined lands in the western United States. In Costa, J.E. & Fleischer, P.J. (eds.), *Developments and applications of geomorphology*, 133–170. Berlin, Springer Verlag.

Toy, T.J. & Hadley, R.F.1987. *Geomorphology and reclamation of disturbed lands*. Academic Press, pp 480.

US Environmental Protection Agency, 1973. *Methods for identifying and evaluating the nature and extent of non-point sources of pollutants*. EPA-4030/9-73-014, pp 261.

US Environmental Protection Agency, 1976a. *Erosion and sediment control: Surface mining in the Eastern US. Volume 1: "Planning".* EPA Technology Transfer Seminar Publication, EPA 625-13-76-006, pp102.

US Environmental Protection Agency, 1976b. *Erosion and sediment control: Surface mining in the Eastern US. Volume 2. "Design"* EPA Technology Transfer Seminar Publication, EPA, pp136.

Vice, R.B., Guy, H.B. & Ferguson, G.E. 1969. Sediment movement in an area of suburban highway construction, Scott Run basin, Fairfax county, Virginia, 1961-64. *U.S. Geological Survey Water Supply paper*, 1591-E, pp 41.

Wallace, G.A. & Wallace, A. 1986. Control of soil erosion by polymeric soil conditioners. *Soil Science*, 141: 363–367.

Wischmeier, W.H. 1978. Use and misuse of the Universal Soil Loss Equation. *Journal of Soil and Water Conservation*, 31: 5–9.

Wischmeier, W.H. & Mannering, J.V. 1969. Relation of soil properties to its erodibility. *Soil Sci. Soc. Am.*, 23: 131–137.

Wischmeier, W.H. & Smith, D.D. 1978. Predicting rainfall erosion losses. *USDA Agricultural Research Service Handbook No. 537.*

Yorke, T.H. & Herb, W.J. 1976. Urban area sediment yield affects of construction site conditions and sediment control methods. In *Proceedings, Third Interagency sediment conference, PB-245-100.* 2-52; 2–64. Water Resources Council, Washington, D.C.

Zingg, A.W. 1940. Degree and length of land slope as it affects soil loss in runoff. *Agricultural Engineering.* 21(2): 59–64.

Land Reclamation – Moore, Fox & Elliott (eds)
© *2003 Swets & Zeitlinger, Lisse, ISBN 90 5809 562 2*

Paper mill sludge as a soil amendment: the performance of field beans on a site restored with Gault Clay

G. Sellers & H.F. Cook
Imperial College London, Wye, Ashford, Kent, UK

ABSTRACT: Results are presented from a field trial investigating the use of Gault Clay and potentially soil forming materials in the restoration of a Landfill site at Small Dole, West Sussex, UK to arable agriculture. Tertiary paper mill sludge was investigated as a substrate amendment, with field beans as the crop. After one year the paper mill sludge had improved the organic content of both substrates however, the sludge seemed to sequester N from the substrates reducing yield substantially on the Gault Clay. Furthermore, it contained no other mineral nutrients, which may have exacerbated the P deficiency in the Gault Clay. Also, Gault Clay proved difficult to cultivate in wet autumn conditions. Crop performance was superior on the soil forming material probably because it did not stay waterlogged for so long and the beans became inoculated with *Rhizobium* bacteria so lack of N wasn't such a factor compared to the Gault Clay.

1 INTRODUCTION

A typical UK household may produce up to one ton of waste per year (Maltbaek 1999) and in the UK landfills remain the most economic form of disposal (El-Fadel 1999). It is a requirement to restore the landfill sites to a beneficial afteruse and up to 60% of UK landfill sites have been, or are, restored to agriculture (both arable and grassland). This has been seen as the most appropriate and least controversial afteruse, especially in rural areas (Reeve et al. 1998) although, amenity and wildlife restorations are becoming increasingly common (McRae 1998).

On many sites the original soils were not retained and the restoration has had to be based on either imported soils or on locally available soil forming materials; but these are not always available in sufficient quantities, at reasonable cost, or of the right quality. A possible alternative is to use amendments to improve the quality of the soil forming material.

Gault Clay, like many other "potential soil forming materials" (Bending et al. 1999), contains little organic matter, nitrogen or phosphorus. These important factors limit the growth of vegetation on substrates such as Gault Clay (Chu & Bradshaw 1990). Gault Clay displays little tendency to develop soil structure, becoming easily compacted and difficult to work, particularly in terms of obtaining a good seedbed and uniform sowing depth.

To date, the standard reclamation procedure at reclaimed sites, including landfill sites, has involved adding mineral fertiliser to the substrate used for restoration and sowing a grass crop (King & Evans 1989). This produces an almost immediate vegetation cover, reduces soil erosion and protects exposed areas (Larney et al. 1995). However, restoration to grassland generates little income for the company and the application of mineral fertiliser alone does not immediately improve the organic matter status of the restored land, which is often very low.

The study reported here was conducted at Horton household waste landfill site at Small Dole, West Sussex, England (Grid Ref TQ 212 122), which is being restored for potential arable agricultural production as a condition of it's planning consent. Little or no indigenous topsoil was available for restoration and to import topsoils in the quantities needed for restoration was an economically unviable option. Consequently, the restoration material used was Gault Clay from the bottom of an adjacent clay pit to act as a subsoil forming layer (300 mm depth) overlain with a 700 mm layer of potentially soil forming material (SFM) consisting mainly of screened construction waste material, to act as a topsoil forming layer.

As the quality, availability and nutrient content of the SFM was variable the emphasis of the research was to improve the substrate characteristics of Gault Clay to see if it would be possible to sow directly into

Gault Clay rather than having to import material for a topsoil layer.

A second research aim was to improve the revegetation potential of the site by investigating the use of de-inked tertiary paper mill sludge as a possible organic soil improver for both the Gault Clay and the SFM. Paper mill sludges present a disposal problem for the industry and are usually landfilled or incinerated. Increasingly however, they are being used as an organic soil improver on reclaimed or agricultural land (Baggs et al. 2002) but despite this their effects on soil properties and plant growth have not been extensively researched (Zibilske et al. 2000). Three types of paper mill sludges are defined by Demeyer & Verloo (1999) and the type used in this investigation was tertiary sludge, defined as "The residue of mechanical transformation of used or recycled paper after the removal of ink" which is usually dewatered leaving it with a dry weight of around 40% and a high organic matter fraction (Demeyer & Verloo 1999). It also has virtually no N which means that it has a very high C:N ratio and it has been shown to produce net immobilisation of N at least in the short term (Aitken 1998). The tertiary sludge used in this investigation had an N and P content that were low enough to be detrimental to plant growth but a very high C content. This can be compared to secondary sludge for example where the N and P content can be higher due to microbial degradation processes used in their production (Demeyer & Verloo 1999).

Two trials were initiated: one on 1 m depth of Gault Clay and one where a cover of 700 mm of SFM was added to 300 mm depth of Clay. Four treatments were compared which were paper mill sludge, mineral fertiliser at MAFF recommended rates (MAFF 2000) and mineral fertiliser together with paper mill sludge. The control was no amendment. The crop chosen was field beans (*Vicia faba*) to investigate whether the field beans could begin to start adding N to these two N poor substrates.

2 MATERIALS AND METHODS

Ground preparation on the soil forming material (SFM) and Gault Clay (GC) began on 29th August 2001 by applying herbicide with a knapsack sprayer to the trial sites. This was followed on the 12th September by cultivation using tines to remove the larger weeds and to break up the surface to improve discing.

On the 2nd November each trial site was harrowed and laid out. Both trials were fully randomised with 4 replicates. Each plot measured 4 m × 8 m (32 m^2).

The treatments and codes for the Gault Clay trial were:

– Gault Clay substrate with no amendment (GC)
– GC amended with 100 t ha^{-1} paper mill sludge wet weight, about 40 t ha^{-1} dry weight (GCP)

– GC amended with 85 kg ha^{-1} P + K fertiliser (GCF)
– GC + above paper mill sludge application + above fertiliser application rate (GCPF).

The treatments for the trial on Soil Forming Material (SFM) were:

– Soil Forming Material with no amendment (SFM)
– SFM amended with 100 t ha^{-1} paper mill sludge wet weight, about 40 t ha^{-1} dry weight (SFMP)
– SFM amended with 85 kg ha^{-1} P + K fertiliser (SFMF)
– SFM + above paper mill sludge application + above fertiliser application rate (SFMPF).

The paper mill waste was applied as a layer by hand across the designated plots. The autumn crop sown was winter bean (var. Target) at 250 kg ha^{-1} and was broadcast onto the plots by hand. Following this, all the plots were cultivated using tines, which incorporated the paper mill waste and covered the seed to a depth of about 70 mm. The mineral fertiliser was then broadcast onto the surface of the designated cultivated plots.

Due to problems with the crop through winter, notably removal of many seedlings by rooks, but also water logging, it was decided on the 9th March 2002 to re-sow the plots on both substrates with spring beans (var. Scirocco). A tracked digger with a ½ m wide bucket dug three 10 cm deep trenches on each plot and the beans were broad cast into the three trenches at a rate of 250 kg ha^{-1}. The trenches were then covered, raked over and firmed down manually. P + K fertiliser was broadcast onto the surface at a rate of 85 kg ha^{-1}. Bird twine was then placed around and in the plots. The beans were sown this way because the land was too wet to cultivate with a tractor.

Harvesting commenced on the 26th July 2002 by taking 6 individual 0.5 m^2 quadrats from each plot, two from each row. These samples were then dried at 60°C for 48 hours. Each sample was weighed to give an above ground biomass. The seed was then weighed separately to give a yield and a thousand seed weight (TSW).

Chemical analyses used ADAS standard methods (ADAS 1986) and regular measurements of substrate temperature and percentage substrate water content were taken at regular intervals.

3 RESULTS

3.1 *Substrate analysis prior to sowing*

Substrate analysis prior to sowing (Table 1) indicated that neither substrate or paper mill sludge had a pH that would restrict the growth of beans or would lock up available nutrients such as phosphorus making them unavailable for crop growth. None of the materials

Table 1.   Substrate analysis prior to sowing.*

	GC	Paper mill sludge	SFM	GC + Paper mill sludge @ 100 t ha^{-1} wet weight	SFM + Paper mill sludge @ 100 t ha^{-1} wet weight
pH	8.41	7.71	7.44	8.31	8.12
ECE	2.78	2.35	1.41	3.02	0.92
(mS cm^{-1})	(0 Salt free)	(0 Salt free)	(0 Salt free)	(0 Salt free)	(0 Salt free)
Total N	395.4	1733.3	471	782.8	781.8
(mg l^{-1})	(Very low)	(Low)	(Very low)	(Very low)	(Very low)
Available P	7.12	8.08	53.2	9.82	33.01
(mg l^{-1})	(0)	(0)	(4)	(1)	(3)
Available K	420	30	149	526.25	235.63
(mg l^{-1})	(3)	(0)	(2−)	(4)	(2+)
Available Mg	349.76	56.9	98.8	317.91	78.98
(mg l^{-1})	(5)	(2)	(2)	(5)	(2)
Available Ca	10958	10500	6938	11937	14093
(mg l^{-1})					
% CaCo$_3$	6.39	47.59	4.3	31.04	13.9
	(calcareous)	(Very calcareous)	(Slightly calcareous)	(Very calcareous)	(Very calcareous)
% Organic	0.86	21.69	0.99	5.75	3.22
carbon	(Very low)	(Very high)	(Very low)	(Medium)	(Low)
% Sand	6.61		48.37		
% Silt	54.17		32.4		
% Clay	39.22		19.24		
Soil type	Silty clay		Sandy loam		

* Indicies (where applicable) in brackets beneath result.
Index for K, P, and Mg taken from MAFF Fertiliser recommendations 2000.
ECE, index taken from Booker Tropical Soil Manual (Landon 1991) adapted from FAO-UNESCO 1973.
Organic carbon and N indices taken from Booker Tropical Soil Manual (Landon 1991) adapted from Metson 1961.

have EC levels that would cause salinity problems. Both substrates and the paper mill sludge were low or very low in total N so cannot be expected to act as a nitrogen source. Available P was low in the Gault Clay and the paper mill sludge but readily available in SFM with an index of 3 and would therefore not need any additional P. Available K is low in the paper mill sludge but an index between 3 and 4 for the substrates suggests that K is readily available for plant growth. The Ca availability and % CaCO$_3$ suggest that both substrates are calcareous, which is consistent with the pH values. The organic carbon content of both substrates was Very Low and may affect nutrient retention but it is high in the paper mill sludge. The addition of paper mill sludge to both substrates improved the organic carbon content, the Gault Clay improving to a medium rating from very low and the SFM improving to a low rating from very low. None of the metals analysed for (Table 2) were present at levels that breached or approached the Maximum Permissible Limit (MAFF 1998) allowed in arable soils.

3.2   Substrate analysis after one year's cropping

The Gault Clay contained little P after one year of cropping (Table 3) and any P from the mineral fertiliser

Table 2.   Heavy metal analysis of the substrates.

	GC	P	SFM	GCP	SFMP
Zn (mg kg^{-1})	34.6	268	35	33.6	27.6
Cu (mg kg^{-1})	9.6	9.6	9.2	9.6	5.4
Pb (mg kg^{-1})	6.6	5.8	21	7.8	16
Cd (mg kg^{-1})	0	0	0	0	0
As (mg kg^{-1})	22.4	4	12.1	16.1	11.2
Hg (mg kg^{-1})	0.07	0.29	0.15	0.07	0.21
Se (mg kg^{-1})	0.34	0.4	0.32	0.16	0.32
Cr (mg kg^{-1})	55.1	5.7	50.9	61.5	87.2

See above for treatment key.

had been utilised. The paper mill sludge had no effect on substrate P content. The SFM after one year still contains reasonable levels of available P (Index 3). The results for K show that both substrates contain reasonable levels of available K. As expected the addition of paper mill sludge, which is low in K as well as P has not affected the K levels in either substrate. After one year of cropping the % OC levels are not high in either of the substrates but is slightly higher in the SFM. The addition of paper mill sludge has moved the Gault Clay from an index of very low to low. There was little N in the Gault Clay after one year's cropping. The

Table 3. Substrate soil analysis after one year of cropping (September 2002).*

	pH	EC (mS cm^{-1})	Total N (mg l^{-1})	Available K (mg l^{-1})	Available P (mg l^{-1})	% Organic carbon
GC	8.23	1.7 (0 Salt free)	400 (Very low)	450 (4)	0 (0)	1.7 (Very low)
SFM	7.75	0.18 (0 Salt free)	1100 (Low)	180 (2−)	28 (3)	3.58 (Low)
GCF	8.21	2.4 (0 Salt free)	400 (Very low)	450 (4)	4 (0)	1.83 (Very low)
SFMF	7.93	0.53 (0 Salt free)	1500 (Low)	230 (2+)	38 (3)	2.7 (Low)
GCP	7.98	3.3 (0 Salt free)	450 (Very low)	460 (4)	0 (0)	1.73 (Very low)
SFMP	7.93	0.37 (0 Salt free)	1250 (Low)	180 (2−)	28 (3)	3.24 (Low)
GCPF	8.11	2.2 (0 Salt free)	450 (Very low)	450 (4)	6 (0)	2.05 (Low)
SFMPF	7.97	0.18 (0 Salt free)	1100 (Low)	180 (2−)	34 (3)	2.9 (Low)

* Indicies in brackets beneath result.
Index for K, P, and Mg taken from MAFF Fertiliser recommendations 2000.
ECE, index taken from Booker Tropical Soil Manual (Landon 1991) adapted from FAO-UNESCO 1973.
Organic carbon and N indices taken from Booker Tropical Soil Manual (Landon 1991) adapted from Metson 1961.
(See above for treatment key.)

Table 4. Crop mineral analysis.

	Total N (mg l^{-1})	Total P (mg l^{-1})	Total K (mg l^{-1})	Total Ca (mg l^{-1})	Zn (mg kg^{-1})	As (mg kg^{-1})	Hg (mg kg^{-1})	Se (mg kg^{-1})	Cr (mg kg^{-1})
GC	56.86	2	34.05	11.37	167.25	3.57	0.04	0.21	12.8
SFM	73.33	4.13	38.38	10.81	117.83	0.88	0.03	0.12	11.2
GCP	43.88	2.35	35.99	12.86	96.75	7.82	0.05	0.17	14.3
SFMP	77.08	4	59.38	14.52	126.75	1.21	0.04	0.09	11.6

(See above for treatment key.)

Table 5. Substrate temperatures (°C) and percentage soil water content at 50 mm depth.*

Treatment	GC	GCP	GCF	GCPF	SFM	SFMP	SFMF	SFMPF
27.5.02	18.19	18.55	18.53	17.83	16.79	16.90	17.56	16.84
	(33.63)	(31.18)	(34.36)	(31.79)	(20.72)	(27.74)	(24.03)	(32.54)
26.6.02	24.89	23.02	24.26	23.57	23.76	24.05	24.56	23.4
	(15.58)	(19.39)	(16.14)	(18.18)	(6.37)	(5.73)	(4.79)	(4.88)
16.7.02	24.93	24.53	23.48	23.94	22.36	22.44	22.47	22.88
	(21.94)	(25.01)	(23.25)	(23.85)	(13.28)	(13.28)	(10.78)	(14.67)

* Figure in brackets shows % soil water content. (See above for treatment key.)

addition of paper mill sludge caused a slight drop in N compared to GC but all of the GC treatments have an index of very low.

For SFM, the N content is higher than Gault Clay with and Index of Low for all treatments. The paper mill sludge seems not to have affected N content in the SFM substrate.

3.3 Crop mineral analysis

Table 4 indicates that beans grown on SFM contained significantly more N ($p < 0.001$), P ($p < 0.001$) K ($p < 0.001$) and Ca ($p = 0.005$) than beans grown on

Gault Clay. However, the addition of paper mill sludge produced no significant changes in plant nutrient content on either substrate. No detectable levels of copper, lead or cadmium were found in any of the samples and zinc, arsenic, mercury, selenium and chromium levels in all cases did not surpass Maximum Permissible Limits for those elements.

3.4 Substrate temperatures and percentage soil water content at 50 mm depth

There were no significant differences in soil temperature (Table 5) at 50 mm depth between the substrates for any of the dates recorded. Gault Clay had a significantly higher soil water content (p < 0.001) than the soil forming material for all dates tested.

3.5 Emergence (plantsm22)

There was no significant difference in field bean emergence (Table 6) between substrates or treatments by the 27th May. Furthermore, emergence was above what is considered the minimum agriculturally acceptable emergence rate of 8 plants m^{-2} (Jellings & Fuller 1995). However earlier on the 9th and 16th May there was significantly higher emergence (p < 0.001) on SFM compared to Gault Clay.

3.6 Total monthly rainfall from November 2001–August 2002

Rainfall figures (Figure 1) indicate that in the winter months of November–January rainfall was lower than the average for the previous six years but from February onwards rainfall was higher than average, considerably so for the months of February, May, July and August.

3.7 Above ground weight and yield

Figure 2 shows that there is a large significant increase in plant growth and yield (p < 0.001) when beans are grown on SFM compared to Gault Clay. There was no significant difference between treatments for beans grown on SFM (p = 0.48) but there was a significant difference between treatments grown on Gault Clay. The GCF treatment produce significantly greater above ground weight (p = 0.002) and yield (p = 0.007) compared to the other treatments.

3.8 Thousand seed weight (TSW)

The Thousand Seed Weight (TSW) (Figure 3) for beans was significantly greater on the SFM than on the Gault Clay (p < 0.001). There was a significant difference between the treatments on the Gault Clay (p = 0.04) but not on the SFM. On the Gault Clay, TSW for the GCF treatment was significantly greater than for the GC and GCP treatments and TSW on GCPF was significantly greater than GCP.

Table 6.   Emergence (plants m^{-2}).

Date	GC	GCF	GCP	GCPF	SFM	SFMF	SFMP	SFMPF
9.5.02	27.44	25.22	9.89	23.78	63.35	66.89	68.56	73.11
16.5.02	46.22	61.44	24	45.11	70.77	75.67	80	74.45
27.5.02	74	83.89	60.11	68.22	78.44	75.45	81.78	81.89

(See above for treatment key.)

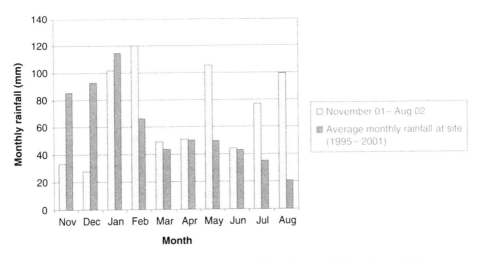

Figure 1.   Total monthly rainfall at the site between November 2001 and August 2002 together with the average monthly rainfall for the previous 6 years.

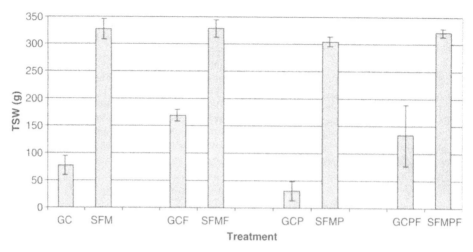

Figure 2. Above ground weight and yield for beans. (See above for treatment key.)

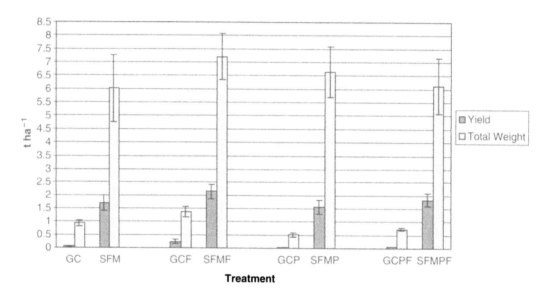

Figure 3. Thousand seed weight. (See above for treatment key.)

4 DISCUSSION

The addition of paper mill sludge did lead to an increase in organic matter in both substrates (Table 1) but had no effect on mineral nutrient content of the substrates, particularly N and P where the low levels in Gault Clay affected growth. Unlike sewage sludge (for example), it cannot be described as a soil nutrient improver as well as a soil organic matter improver. Furthermore, due to the low phosphorus levels in the Gault Clay it became necessary during the growing season to add more than the recommended rate at a mid season application when the spring beans were sown to improve growth. After one year of cropping (Table 3) the paper mill sludge had maintained an organic carbon improvement in the Gault Clay. However, this had dropped from an index of medium to low. On the SFM, organic carbon improved over the season. This probably reflects the better growth of the beans on this substrate compared to the growth on Gault Clay, leading to greater below ground biomass production, rather than the addition of the paper mill sludge because organic carbon improved for all SFM treatments not just those with paper mill sludge. Nutrient levels had not changed compared to the beginning of the season (Table 1). The N results after the first season showed that the N content in both substrates remained at a level that would be detrimental to plant growth without additional N fertiliser. On the Gault Clay the beans grew poorly and did not become nodulated so added nothing to the N content of the Gault Clay. On the plots where paper mill sludge was added the N content of the Gault Clay dropped by almost a half. This may be a reflection of the lack of N input by the beans but also may indicate N sequestration for paper mill sludge decomposition. However on SFM there was an improvement in N with content doubling compared to the beginning of the season (although this is still an index of low). The nodulated beans maybe releasing N into the SFM. This suggests that on the SFM, N cycling has commenced and gradual improvement in substrate N might be expected over time, a finding that is not clear for the Gault Clay. Furthermore, the addition of paper mill sludge does not seem to have had a detrimental effect on N content so possibly the beans are releasing sufficient N for both microbial breakdown of the sludge and for bean (and weed) growth.

As expected the paper mill sludge after one application did not improve P levels in the Gault Clay and the higher P levels in the SFM was maintained after one year. This may be an important consideration since P seems to be a factor reducing crop growth on Gault Clay. The paper mill sludge did not present any heavy metal problems in either the substrates or the crops. (Table 4).

The paper mill sludge had no adverse effects on bean emergence by the 27th May. However the results do suggest that the beans emerged faster on the SFM which probably helped improve the growth and yield on SFM. This may become more important where ground conditions, such as waterlogging, and low substrate temperatures in the Gault Clay restrict emergence further.

Agriculturally acceptable emergence rates did not, however, translate into acceptable yields. Average field bean yields in the UK are about $2.5–3\,t\,ha^{-1}$ (Loss & Siddique 1997). The yield was minimal on the Gault Clay but was close to the average on all SFM treatments. The TSW was also significantly greater on the SFM (Figure 3) and the improved growth was reflected in the raised % OC content of the SFM due to the extra root biomass in the substrate. The reason for the good performance of the field beans on the SFM was due to the fact that the SFM contained *rhizobium* innoculum and so the beans became nodulated, which could be visibly seen on plants examined during the growing season. The beans never became inoculated in the Gault Clay so were unable to fix nitrogen. When the beans were given nitrogen in large container glasshouse trials (results not reported) the beans responded with good growth. Why, a year after placement, there was no *rhizobium* inoculate in the Gault Clay is unclear. However, the rainfall (Figure 1) from February to August was much greater than average and the clay remained very wet until late in the season (Table 5) leading to possibly anaerobic conditions in the clay which does not favour the growth of nitrogen fixing bacteria.

Suppression of yield by the paper mill sludge was probably due to sequestration of the little available N in either substrate for further microbial sludge decomposition, a problem documented for both paper mill sludges and other organic waste products (Aitkin 1998; Garcia et al. 1991).

5 CONCLUSION

Unlike other substrate forming materials such as London Clay (Sellers et al. 2002) it seems that a top-layer of Soil Forming Material is needed to get crops to grow mainly due to the poorly draining nature of Gault Clay, and the lack of organic carbon, N and P.

Paper mill sludge in the first year after application seems to suppress bean growth compared to not applying it due probably to the sequestration of N for decomposition. Furthermore, the paper mill waste is low in all other major nutrient minerals such as P and K. This may exacerbate any lack of major nutrients in the substrate such as lack of P in Gault Clay and will need to be alleviated by applying mineral fertiliser at above recommended rates.

Paper mill sludge does provide organic matter, which is important for the long term prospects for soil creation. Furthermore, there seems to be improvement in organic matter even after a season of cropping although this is reduced.

These results suggest that further investigation would be useful in several areas. Quantifying the N loss to sequestration so that the application of mineral N fertiliser can be accurately targeted is essential. This would reduce the danger of increased nitrate leaching due to over application of N fertiliser or reduce financial expenditure on fertiliser that is not needed. The lack of *rhizobium* innoculum, even after a year, in the Gault Clay could also be investigated as this is essential in the soil forming process and does not seem to be happening on the Gault Clay. Further studies of the longer term effects of paper mill application to land reclamation substrates would show, for example, how long yield suppression or the desirable improvements in organic matter last. Is it possible to give yearly applications of paper mill sludge or does the yield suppression become cumulative making regular applications uneconomic eventually? Finally, the Gault Clay clearly has a P deficiency problem which research into alternative waste product amendments may be able to solve.

ACKNOWLEDGEMENTS

We acknowledge the financial support and cooperation given by Viridor Waste Management Ltd and to M-Real New Thames Ltd for providing the paper mill sludge.

REFERENCES

ADAS. 1986. Analysis of Agricultural Materials. 3rd Edition. Reference Book 427. London: HMSO.

Aitken, M.N. 1998. Effect of applying paper mill sludge to arable land on soil fertility and crop yields. Soil Use and Management 14: 215–222.

Baggs, E.M., Rees, R.M., Castle, K., Scott, A., Smith, K.A. & Vinten, A.J.A. 2002. Nitrous oxide release from soils receiving N-rich crop residues and paper mill sludges in eastern Scotland. Agriculture, Ecosystems and Environment 90: 109–123.

Bending, N., McRae, S.G. & Moffat, A.J. 1999. The use of soil forming materials, their use in land reclamation. London: The Stationary Office, Department of the Environment, Transport and the Regions.

Chu, L.M. & Bradshaw, A.D. 1990. The effects of pulverised refuse fines (PRF) as a soil material on plant growth. Resources, Conservation and Recycling 4: 257–269.

Demeyer, A. & Verloo, M. 1999. Evaluation of paper sludge as organic fertilizer for the growth of Rye grass on a Belgian clay soil. Agrichimica 43 (5–6): 244–250.

El-Fadel, M. 1999. Leachate recirculation effects on settlement and biodegradation rates in MSW landfills. Environmental Technology 20: 121–133.

Garcia, C., Hernandez, T. & Costa, F. 1991. Agronomic value of urban waste and the growth of ryegrass (Lolium perenne) in a calciorthid soil amended with this waste. Journal of the Science of Food and Agriculture 56: 457–467.

Jellings, A.J. & Fuller, M.P. 1995. Arable Cropping. In R.J. Soffe (ed), Primrose McConnell's The Agricultural Notebook. 150–193, Blackwell Science London.

King, J.A. & Evans, E.J. 1989. The growth of spring barley related to soil tilth produced on restored opencast and unmined land. Soil and Tillage Research 14: 115–130.

Landon, J.R. (ed). 1991. Booker Tropical Soil Manual. Harlow England: Longman Scientific and Technical.

Larney, F.J., Janzen, H.H. & Olson, B.M. 1995. Efficacy of inorganic fertilisers in restoring wheat yields on artificially eroded soils. Canadian Journal of Soil Science 75: 369–377.

Loss, S.P. & Siddique, K.H.M. 1997. Adaption of Faba Bean (Vicia faba) to dryland mediterranean type environments I: Seed yield and yield components. Field Crops Research 52: 17–28.

MAFF. 1998. Code of Good Agricultural Practice For The Protection of Soils: The Soils Code. London: MAFF Publications.

MAFF. 2000. Fertiliser Recommendations for Agricultural and Horticultural Crops, 7th Edition. London: The Stationary Office.

Maltbaek, C.S. 1999. MRFs and municipal waste – bold success or heroic failure. Proceedings of the Institution of Civil Engineers and Municipal Engineers 133: 19–24.

McRae, S.G. 1998. Land reclamation after open-pit mineral extraction in Britain. In A.H. Wong M.H, A.M. Wong & A. Baker (eds.), Remediation and Management of Degraded Lands: 47–62. Boca Raton: Lewis Publishers.

Sellers, G., McRae, & Cook, H.F. 2002. Ryegrass, fescue and clover growth on London Clay amended with waste materials. Land Contamination & Reclamation 10 (2): 79–89.

Zibilske, L.M., Clapham, W.M. & Rourke, R.V. 2000. Multiple applications of paper mill sludge in an agricultural system: Soil effects. Journal of Environmental Quality 29: 1975–1981.

Land Reclamation – Moore, Fox & Elliott (eds)
© 2003 Swets & Zeitlinger, Lisse, ISBN 90 5809 562 2

Microbiological tools for monitoring and managing restoration progress

Jim Harris

Institute of Water and Environment, Cranfield University, Silsoe, Bedfordshire

ABSTRACT: Terrestrial civil engineering programmes for mineral extraction and landscape re-instatement inevitably lead to significant degradation of the soil ecosystem. Principally, there are large shifts in the functional and phenotypic characteristics of the soil microbial community, which are consistent and predictable for the type of degradative activity applied. Similarly, as the system recovers (or not as the case may be) during the course of reclamation and restoration programmes there are significant shifts in the community structure towards the type of profile found in the desired target system.

With modern methods of rapid and reliable assay of the soil microbial community, timely determinations of the success and progress of the restoration/management regime are emerging as a real possibility. Furthermore, prescriptions for re-establishing the structural and functional links between the biology, chemistry and physics of the soil system can be made. This will facilitate the production of decision support tools encompassing the full range of variables associated with soil handling and re-establishment of the soil-plant system. In the long term this will enable significant reductions in failure rates of restorations, and bring about the restoration of systems in the truest sense.

1   INTRODUCTION

1.1   *Background*

The handling and re-instatement of soils and soil forming materials is usually one of the principal activities of the land reclamation process. There is little consideration given to the soil as a living material, rather it is treated as an engineering material with some unfortunately recalcitrant characteristics. What is meant by restoration and reclamation is central to the assessment of the success or otherwise of such programmes, and still the subject of much debate (van Diggelen *et al*, 2001). What is clear, however, is that objective measurement of restoration success will become critical for all land managers when the new European Union Directive on environmental liability becomes enacted (Commission of the European Communities, 2002).

For reclamation and restoration schemes to be successful it must be recognised that soil changes minute by minute, and has both *in situ* and inherent characteristics which must be monitored and managed.

1.2   *Avoiding failure by adaptive management*

If we are to avoid failure in reclamation and restoration programmes, we must have early indications of emerging problems. In Figure 1 we can see a hypothetical ecosystem attribute changing with time as a result of natural regeneration or active restoration intervention, with a target range being set.

We need to be able to monitor such attributes consistently and effectively if we are to avoid failures as illustrated. In order to achieve this we need ecosystem attributes which address the following criteria

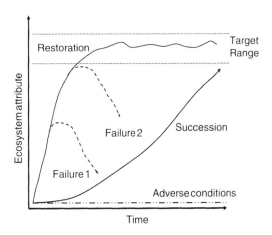

Figure 1.   Monitoring is essential to avoid failure.

(Andreasen *et al*, 2001):

- Are they relevant to the ecosystem(s) under study and to the objectives of the assessment programmes?
- Are they sensitive to anthropogenic changes?
- Can they provide a response that can be differentiated from natural variation?
- Are they environmentally benign?
- Are they cost effective to measure?

Most current methodologies meet criteria 4 and 5; analysis of the soil microbial community meets all five. Once key attributes are identified and readily measured, we can use them to plan, monitor and adapt management regimes.

I suggest that we can use measurements of the soil microbial community, particularly in circumstances where the degradation is initially physical, as found in the majority of civil engineering programmes. What follows is an illustration of how such measures have been used, rather than a comprehensive review.

1.3   *The soil microbial community*

The microbial community is at the heart of organic matter transformation and stability in the soil. In pasture systems the total biomass of the soil biological community exceeds the standing crop of animals above it.

The soil microbial community has three principal functions:

- Nutrient cycling
- Organic matter storage and transformation
- Genesis and maintenance of soil structural stability.

The biochemical decomposition of plant and animal residues is largely carried out by the soil microbial community, and some functions are exclusively microbial, particularly nitrogen transformations (Insam, 2001; Panikov, 1999).

2   MEASURING THE SOIL MICROBIAL COMMUNITY

2.1   *Introduction*

Measurement of the soil microbial community has been advocated by several authors, over the last 10–15 years (e.g. White *et al*, 1998; Bentham *et al*, 1992; Harris and Birch, 1992). For a fuller review of the methodologies available see Harris and Steer (2003) NATO.

Methods of measurement

- Size
- Composition
- Activity
- Physical arrangement.

2.2   *Size*

This is the "standing crop" of the microbial biomass and is usually expressed at biomass-C and may be determined in a number of ways, directly and indirectly. Fumigation of soils with chloroform followed by extraction and determination of soluble C, in comparison with a non-fumigated control is commonly used, as is determination of the soils adenosine triphosphate content (ATP). This latter method is useful in a restoration context as it is sensitive to small changes in biomass, and represents viable biomass very accurately as it is metabolised almost instantaneously on cell death.

2.3   *Composition*

The composition of communities has been commonly used in the sense of species numbers and their abundance. This is not easy to determine, as the community is incredibly diverse, with estimates of 10,000 species of soil bacteria alone (Torsvik *et al*, 1996), with in excess of 5000 in a single gram. Traditional methods of culturing "plate-counts" seriously underestimate this diversity, with as little as 0.1% of soil microbial species being culturable. Modern approaches rely on biochemical methods such as the extraction of phospho-lipid fatty acid profiles (Peacock *et al*, 2001; Steer and Harris, 2000), which provide a community "finger-print" which can be used to ascertain the effects of management, pollution, ecosystem health, and plant growth, or molecular and approaches which indicate the genotypic diversity of the micro-organisms in a sample (Torsvik and Ovreas, 2002).

2.4   *Activity*

Some measurements are carried out *in situ* and are particularly useful when considering changes over short time spans. However, most measurements on restored sites are those carried out on samples returned to the laboratory. Common measurements include enzymatic assays (such as dehydrogenase – DHA) and respirometry. Recent attempts have been made to determine the ability of the soil microbial community to process a range of carbon substrates. One common approach has been the use of the BIOLOG MicroPlate™ bacterial identification system; a microtitre plate system with 95 carbon substrates – these are inoculated with a soil suspension and substrate utilisation is detected by a colour change. One drawback of this approach is that the culturable fraction of the microbial community is favoured, which may only represent less than 1% of the soil microbial community (Preston-Mafham *et al*, 2002). Degens and Harris (1997) have developed a catabolic community profiling technique in which the substrate is taken to the community, and the response measured over the first 4 hours, before any cell division is likely to have occurred and therefore is more likely to represent the response of the *in situ* community.

2.5 *Physical arrangement*

Determining the physical arrangement of the microbial community, particularly with respect to organic matter resources and soil physical architecture, can be very rewarding in terms of elucidating the recovery of biotic-abiotic linkages and structural stability. It is, however, difficult and time consuming but continues to be very useful in a research context.

3 MICROBIAL COMMUNITY MEASURES IN USE

3.1 *Changes during disturbance*

There are immediate and significant changes in the soil microbial community as a result of the movement of topsoil during civil engineering activities, with the deleterious effects exacerbated by storage for long periods in stockpiles. (Harris *et al*, 1989; 1993). There are large and rapid increases of bacterial numbers because of the large amounts of organic materials, such as fungal hyphae, plant roots and soil animals, killed by the soil lifting and store construction processes. This is followed by a period where the numbers of bacteria decline as these reserves are exhausted, until numbers found in Figure 2. Recovery of total and active bacterial species after mining (redrawn from Yin *et al*, 2000) reference areas are reached. Numbers of fungi decline immediately, as does total microbial biomass. The fungal biomass is providing organic and inorganic substrates for the bacterial explosion.

Invertebrates, particularly earthworms, are severely affected by handling procedures. They are liable to be crushed and unable to find physical refuges from the large compaction and shearing forces being imparted to the soil by the heavy earth-moving equipment, leading to large scale declines in their populations (Scullion, 1994).

Over longer time scales the microbial biomass in stored soils to values less than 5% of the undisturbed values, and, depending on texture, less than 10% of the stored soil recovers to pre-disturbance values (Harris and Birch 1990a). However, this recovery is further destroyed when the soil is re-instated at the end of coaling operations.

3.2 *Recovery after disturbance*

Ruzek *et al*, (2001) demonstrated clear relationships between time since restoration and increases in soil microbial biomass, in reclaimed sites in the Czech

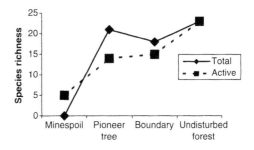

Figure 2. Recovery of total and active bacterial species after mining (redrawn from Yin *et al*, 2000).

republic and Germany. Ruzek and co-workers indicated that this was related to both organic matter content as a starting point in new reclamations, and the textural characteristics of the soils reclaimed, and developed an algorithm which could be applied to different soil types, with modifications appropriate to regions with differing soil resources.

Total biomass is not the only characteristic which changes with time. Yin *et al*, (2000), showed that the total number of bacterial species on a site recovering from mining activity increased fairly quickly around pioneer vegetation, but the proportion of these species which were active (as determined by a radio-label incorporation method) took much longer to increase, and at no point matched the undisturbed reference forest site (Figure 2). This offers a potentially very sensitive approach for determining the efficacy of treatment strategies for restoring function to systems.

3.3 *Relationship between soil microbial community and other parameters*

The soil microbial community has links to several characteristics essential to soil function. One of the most important is soil structural stability, and the resistance of aggregates to water is a commonly employed method for providing an index of such stability.

Figure 3 shows the relationship between water-stable aggregates at 5 opencast site reclamations (Edgerton *et al*, 1995). The relationship appears linear over the first eleven years; longer than this and the relationship becomes a log-linear one. The effects of compaction and water logging are also clearly detectable. This indicates the importance of restoring not only the obvious features of a site, but also the below ground parts. Without the development of aggregation reclaimed soils will always remain, shedding and not infiltrating, and will be droughty in the summer and water-logged in winter. The useful spin-off is that measurement of the microbial community can be used as a surrogate for characteristics such as aggregation, and may be rapidly assessed.

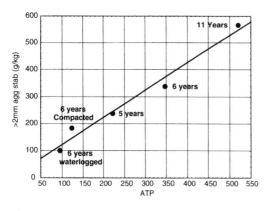

Figure 3. Relationship between water stable aggregates (>2 mm g/kg) and microbial biomass (ATP ng/g) at 5 restored sites (Redrawn from Edgerton *et al*, 1995).

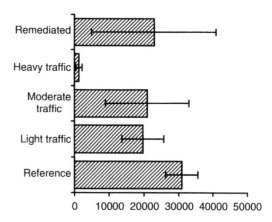

Figure 5. The effect of military traffic and remediation on microbial biomass as determined by PLFA (pmol/g) – Redrawn from Peacock *et al*, 2001.

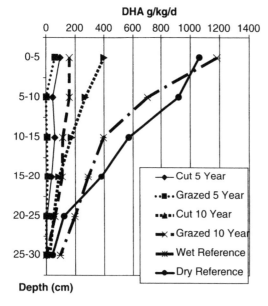

Figure 4. Microbial activity profile development as affected by management regime (redrawn from Harris & Birch 1990b).

3.4 *Effects of management*

The effects of management regime are illustrated in Figure 4 (Harris and Birch, 1990b). Two reference sites were available, "wet" and "dry"and four re-instated fields adjacent to these. The five year re-instated areas had little microbial activity, as indicated by DHA, down the soil profile, but there were higher activities in the 10 year re-instated areas. The results also clearly indicate that cutting and leaving the aftermath was superior to grazing in encouraging microbial activity.

More sophisticated indicators of soil microbial community diversity are available to use routinely, and principal amongst them is the use of biomarkers derived from cell membrane components. The lipids are the major group used for this purpose, and have found widespread application (Zelles, 1999). Mummey and co-workers (2002) investigated the application of one group of lipid biomarkers, the fatty acid methyl esters (FAMEs). They extracted lipids and converted them to FAMEs, using standard procedures, from samples taken from a surface mine reclamations of different ages, and an undisturbed reference site. They were clearly able to distinguish between the different stages of reclamation, and identified a trend towards the undisturbed reference condition. They suggest that the ratio of fungal to bacterial biomarkers is a useful indicator of reclamation progress. This needs further examination with respect to management intervention and the role of soil organic matter.

Quite simple measurements of soil microbial lipid characterstics can yield useful information. Figure 5 shows the effects of varying intensities of military traffic/training on the soil microbial biomass, as determined by PLFAs (Peacock *et al*, 2001). All levels of traffic caused some decrease in biomass, but heaviest traffic (i.e. tank training) caused a decrease by an order of magnitude. In an area which had been remediated (i.e. 10 years after replanting with trees) the microbial biomass had returned to the levels found in light and moderately trafficked areas but note that the variability here was very high indicating the patchy nature of the recovery.

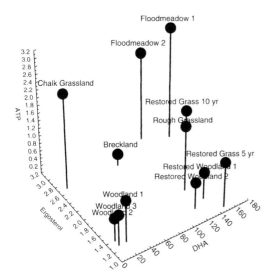

Figure 6. A three dimensional ordination of the size, composition and activity of the soil microbial activity at a number of restored and reference sites (Redrawn from Bentham et al, 1992).

By using a three dimensional ordination, Bentham et al, (1992) demonstrated the effects of time and management type on three characteristics of the soil microbial community (Figure 6). Here sites reclaimed after opencast coal mining are shown as "restored". Points of particular note are:

– The 10 year restored grassland is now clustering with a "rough" (improved) grassland,
– The restored woodlands are following a trajectory from the 5 year grassland restoration in the direction of woodland reference sites.

A more comprehensive database would facilitate interpretation of data of this type, allowing us to clearly identify the effects of management interventions.

4 CONCLUSION: A TOOLKIT FOR LAND MANAGERS

4.1 Technologies ready to use

The methodologies for determining a useful set of soil microbial community characteristics are ready to use. By these measurements long term benefits will be gained in informing the management programmes of severely degraded soil systems. The microbial community is uniquely sensitive to changes in the biology, chemistry and physical characteristics of soil, and provides an early indicator of change.

4.2 Future directions

We now need to consolidate the advances made in soil microbial community measurement and the application of the data. It is important that we establish a common set of community characteristics to be measured, against which to compare reclamation and restoration, coupled to well-characterised reference sites. This could be achieved, in part, by the routine adoption of microbial community measures for adaptive management and compliance.

REFERENCES

Andreasen, J.K., O'Neill, R.V., Noss, R. & Slosser, N.C. (2001) Considerations for the development of a terrestrial index of ecosystem integrity. *Ecological Indicators*, **1**, 21–35.

Bentham, H., Harris, J.A., Birch, P. & Short, K.C. (1992) Habitat classification and soil restoration assessment using analysis of soil microbiological and physicochemical characteristics. *Journal of Applied Ecology*, **29**, 711–718.

Commission of the European Communities (2002) *Proposal for a directive of the European Parliament and of the Council on environmental liability with regard to the prevention and remedying of environmental damage.* COM(2002) 17 final 2002/0021(COD), EU, Brussels.

Degens. B. and Harris J.A (1997) Development of a physiological approach to measuring the catabolic diversity of soil microbial communities, *Soil Biology and Biochemistry*, **29** (9/10), 1309–1320.

Edgerton, D.L., Harris, J.A., Birch, P. & Bullock, P. (1995) Linear relationship between aggregate stability and microbial biomass in three restored soils. *Soil Biology and Biochemistry*, **27**, 1499–1501.

Harris, J.A. and Birch, P. (1990a) The effects of heavy civil engineering and stockpiling on the soil microbial community, *Micro-organisms in Civil Engineering, Proceedings of the Conference*, FEMS Symposium No. 59, P. Howsam (Ed), Pub. E. & F.N. Spon, London, pp 274–286.

Harris, J.A. and Birch, P. (1990b) Application of the principles of microbial ecology to the assessment of surface mine reclamation, *Proceedings of the 1990 Mining and Reclamation Conference*, American Society of Surface Mining and Reclamation, West Virginia, J.Skousen, J.Sencindiver and D.Samuel (Eds), Part I, pp 111–120.

Harris, J.A. & Birch, P. (1992) Land reclamation and restoration, In *Microbial control of pollution*, J.C. Fry, G.M. Gadd, R.A. Herbert, C.W. Jones & I. Watson-Craik. Society for General Microbiology, Symposium 48, Cambridge University Press, pp 269–291.

Harris, J.A., Birch, P. & Short, K.C. (1989) Changes in the microbial community and physico-chemical characteristics of topsoils stockpiled during opencast mining, *Soil Use and Management*, **5**, 161–168.

Harris, J.A., Birch, P. & Short, K.C. (1993) Changes in the microbial community during the construction and subsequent storage of soil stockpiles: A strategist theory interpretation. *Restoration Ecology*, **1**, 88–100.

Harris, J.A., and Steer, J. (2003) Modern methods for estimating soil microbial biomass and diversity: An integrated approach. In *The Utilization of Bioremediation to Reduce Soil Contamination: Problems and Solutions.* Sasek, Glaser and Baveye (eds) Kluwer, for NATO Scientific Affairs Division, Dordrecht (In Press).

Insam, H. (2001) Developments in soil microbiology since the mid 1960s. *Geoderma*, **100**, 389–402.

Mummey, D.L., Stahl, P.D. and Buyer, J.S. (2002) Microbial biomarkers as an indicator of ecosystem recovery following surface mine reclamation. *Applied Soil Ecology* **21**, 251–259.

Panikov, N.S. (1999) Understanding and prediction of soil microbial community dynamics under global change. *Applied Soil Ecology*, **11**, 161–176.

Peacock, A.D., McNaughton, S.J., Cantu, J.M., Dale, V.H. & White, D.C. (2001) Soil microbial biomass and community composition along an anthropogenic disturbance gradient within a long-leaf pine habitat. *Ecological Indicators*, **1**, 113–121.

Preston-Mafham, J., Boddy, L., Randerson, P.F. (2002) Analysis of microbial community functional diversity using sole-carbon-source utilisation profiles – a critique. *FEMS Microbiology Ecology*, 42, 1–14.

Ruzek, L., Vorisek, K., & Sixta, J. 2001. Microbial biomass-C in reclaimed soil of the Rhineland (Germany) and the north Bohemian lignite mining areas (Czech republic): Measured and predicted values. *Restoration Ecology*, **9**, 370 – 377.

Scullion, J. (1994) *Restoring Farmland after Coal: The Bryngwyn Project.* British Coal Opencast, Mansfield.

Torsvik, V., and Ovreas, L. (2002) Microbial diversity and function in soil: from genes to ecosystems. *Current Opinion in Microbiology*, **5**, 240–245.

Torsvik, V., Sørheim, R. and Goksøyr, J. (1996) Total bacterial diversity in soil and sediment communities - a review. *Journal of Industrial Microbiology*, **17**, 170–178.

van Diggelen, R., Grootjans, A.P., and Harris, J.A. (2001) Ecological Restoration: State of the Art or State of the Science? *Restoration Ecology*, **9**, 115–118.

White, D.C., Flemming, C.A., Leung, K.T. & Macnaughton, S.J. (1998) *In situ* microbial ecology for quantitative appraisal, monitoring, and risk assessment of pollution remediation in soils, the subsurface, the rhizosphere and in biofilms. *Journal of Microbiological Methods*, **32**, 93–105.

Yin, B., Crowley, D., Sparovek, G., De Melo, W.J. & Borneman, J. (2000) Bacterial functional redundancy along a soil reclamation gradient. *Applied and Environmental Microbiology*, **66**, 4361–4365.

Zelles, L. (1999) Fatty acid patterns of phospho-lipids and lipopolysaccharides in the characterization of microbial communities in soil: a review. *Biology and Fertility of Soils*, **29**, 111–129.

Land Reclamation – Moore, Fox & Elliott (eds)
© 2003 Swets & Zeitlinger, Lisse, ISBN 90 5809 562 2

Making soil from waste material

S.D. Maslen
Maslen Environmental Ltd, Little Germany, Bradford, West Yorkshire, UK

B.J. Chambers, S.W. Hadden & S.M. Royle
ADAS, Worcester, Whittington Rd, Worcester

ABSTRACT: Following laboratory based trials; experiments were laid down in the field at the Winterton landfill site to investigate the practicalities of soil making using waste materials. The subsoil mixes comprised: 100% clay, 50% clay:50% gypsum and 50% clay:50% foundry sand. The organic amendments incorporated into the surface to form topsoil treatments were: composted green waste, digested sewage sludge cake and peat/compost. The sown grasses grew vigorously on all the experimental treatments, with the organic amendment additions increasing grass dry matter yields compared with the unamended topsoil treatments. The amendments increased levels of soil organic matter, total nitrogen, nitrate-nitrogen (not peat/compost), extractable phosphorus and potassium (composted green waste only), plant available water supply and topsoil porosity. Also, the amendments increased the size of the soil microbial community and its activity, and readily mineralisable organic nitrogen levels. Gypsum increased soil extractable phosphorus levels and only resulted in a small decrease in the pH of the clay/gypsum mixture despite the low pH of the gypsum itself (*c.* 3.0). Fluoride levels in the soil were increased by the gypsum addition, but plant uptake of fluoride was low.

1 INTRODUCTION

Many derelict sites are seriously short of soils for restoration. This applies especially to older industrial areas, such as North Lincolnshire, where ironstone was extracted for the steel industry. It is often possible to identify materials from the local waste streams which have the potential for soil generation, either on their own, or in combination with locally available geological materials (DETR 1999). The wastes are thereby recycled, saving on landfill capacity and removing the need for expensive importation of soils.

The potential for soil generation was recognised at two landfill sites (Winterton and Immingham), operated by Wastewise Ltd., in North Lincolnshire. Previous pot trial studies undertaken by Steve Maslen & Associates and Bradford University (Steve Maslen & Associates 1996a, 1996b) had investigated soil making materials available at the sites in North Lincolnshire and identified suitable mixtures of waste gypsum, foundry sand and composted green waste, in combination with local clay or marine alluvium.

The field experiments reported here extend the previous work; by investigating the practicalities of creating soil making materials on an operating landfill site. The project was originally designed to run for at least three years (1997–2000) and has continued until 2002.

2 MATERIALS AND METHODS

Experiments were laid down in specially prepared treatment cells during June–July 1998 and sown with an agricultural grass seeds mixture in September 1998. The cells had a basal drainage layer of stones and a vertical access pipe in one corner to facilitate the removal of leachate.

The experimental design was a split plot randomised block, with three replicates. The main plots were 20 m × 10 m and the sub-plots 5 m × 10 m. In order to allow for edge effects, the experimental area used for monitoring and harvesting within each sub-plot was 4 m × 8 m. Subsoil mixes formed the main plots, with organic amendments incorporated into the surface of the subplots to form the topsoil mixes.

The subsoil mixes comprised: (a) 100% clay, (b) 50% clay:50% gypsum and (c) 50% clay:50% foundry sand. Organic amendments forming the topsoil treatments were: (a) nil, (b) composted green waste,

(c) digested sewage sludge cake and (d) peat/compost.

The incorporation rate of the composted green waste and peat/compost additions was approximately 33% by volume to a depth of *c*. 150 mm. The sewage sludge cake application was 400 t/ha fresh weight (70 t/ha dry weight).

The waste gypsum was a by-product of phosphoric acid production and was excavated from a settlement lagoon. The peat/compost comprised peat and peat-based compost from bags, which had failed to meet weight specifications and were normally tipped at Winterton. The composted green waste was produced on site at Winterton. Foundry sand was imported from North Hykeham (Lincoln) and the digested sludge cake was provided by Anglian Water from their Newton Marsh works.

The clay at Winterton, a geological material, was readily available. It had a high silt and clay content (47% and 45%, respectively) and was classed as a silty clay.

Representative samples of the mixed materials were collected from each subplot (0–150 mm depth) in August 1998 and analysed for pH, extractable and total nutrient contents, organic matter, conductivity, fluoride and heavy metals, using standard methods (MAFF 1986). ADAS Indices were ascribed to the extractable nutrient levels (MAFF 1994); these indicate the relative amounts of nutrients in the soil that are available to the plant and range from 0 (deficient) to 9 (very large).

Nitrogen fertiliser was applied to all the treatments in spring/summer each year at recommended rates (MAFF 1994). Grass harvests were taken annually during May to July 1999–2001. Dry matter yields were measured and herbage samples analysed for nutrient (N, P, K, Mg, Ca, S), fluoride and heavy metal concentrations. Samples of leachate were collected for chemical analysis from the mainplots.

3 ANALYSES

The sub-plots were analysed for a range of soil physical, biological and chemical properties, including:

- Soil moisture retention and available water capacity
- Soil strength and density
- Aggregate stability (dispersion ratio method)
- Infiltration rate
- Biomass carbon and nitrogen
- Soil respiration
- Readily mineralisable organic nitrogen
- Organic matter
- Total and nitrate nitrogen
- Extractable P, K, Mg
- pH and conductivity
- Total and water soluble fluoride

4 RESULTS

4.1 *Initial*

The main chemical characteristics of the subsoil materials are summarised in Table 1. The addition of gypsum to the clay resulted in a mean reduction in pH of *c*. 0.4 units compared with the 100% clay treatment (pH 7.8 down to 7.4), which was a relatively small decrease given that the gypsum had a pH of 3.0 prior to mixing. The gypsum also increased extractable phosphorus (ADAS Index 3 to 6) and total fluoride concentrations. The fluoride concentrations were increased to levels substantially above the normal range of 20–500 mg/kg for uncontaminated soils (Fuge & Andrews 1988; Weinstein 1997). Conductivity, measured in saturated calcium sulphate solution (MAFF 1986), was moderately high on all treatments (Index 3), largely as a result of elevated sulphate levels in the clay.

The main effects on the organic amendments (Table 2) were to increase extractable nutrient, total nitrogen and organic matter levels, and decrease fluoride concentrations (by dilution). The sludge cake gave the greatest increase in nitrate-nitrogen levels (up to ADAS Index 4), which was consistent with the vigorous early grass growth observed on this treatment. The composted green waste addition increased potassium concentrations (up to ADAS Index 6) and had the high conductivity (Index 6).

4.2 *Overall*

The sown grasses grew vigorously on all the experimental treatments between 1999 and 2001. No symptoms of plant toxicity were observed.

Table 1. Analysis of the subsoil materials at Winterton (August 1998).

	100% Clay	50% Clay: 50% foundry sand	50% Clay: 50% gypsum	SE[1]	p[2]
pH	7.80	8.13	7.37	0.028	0.001
Conductivity ($\mu s\,cm^{-1}$: Index)	2620 (3)	2640 (3)	2650 (3)	55	0.08 (NS)
Extractable phosphorus ($mg\,l^{-1}$: Index)	27 (3)	10(1) (1)	116(6) (6)	4.39	<0.001
Total fluoride ($mg\,kg^{-1}$)	208	132	2217	54	<0.001

[1] Standard error; [2] Probability of treatments being different.

208

Table 2. Analysis of topsoil materials at Winterton (August 1998).

	Nil	Composted green wastep	Peat/Co compost	Sewage sludge	SE	p
pH	7.77	7.80	7.53	7.64	0.03	<0.001
Organic matter (%)	3.47	9.54	12.56	4.71	0.47	<0.001
Total N (%)	0.05	0.45	0.20	0.18	0.02	<0.001
Nitrate-N (mg l^{-1}:Index)	13(0)	81(2)	13(0)	161(4)	12	<0.001
Extractable P (mg l^{-1}:Index)	51(4)	145(7)	101(6)	97(5)	5	<0.001
Extractable K (mg l^{-1}:Index)	128(2)	1003(6)	159(2)	133(2)	34	<0.001
Conductivity (($\mu s\,cm^{-1}$:Index)	2630(3)	3270(6)	2790(4)	2670(3)	79	<0.001
Total fluoride (mg kg^{-1})	852	585	696	680	31	<0.001

Visual assessments of grass growth in spring indicated benefits from all the organic amendments, with no visual differences between the subsoil treatments. The best visual growth was on the sludge cake treatment followed by the composted green waste addition.

At harvest in 1999 to 2001, all the organic additions increased grass dry matter yields compared with unamended topsoil treatments. There were no yield differences between the unamended topsoil treatments.

Conductivity levels, which were initially moderate to high in all the test materials (ADAS Indices 2–7), had reduced to ADAS Indices of 1–4 by 2001. Also, soil pH's had dropped by 2001, approximately 0.2–0.3 units from 1998, but remained within satisfactory ranges on all the treatments (pH 7.2–7.7).

The gypsum additions increased soil concentrations of total and water soluble fluoride, but levels in the plants remained low (2–4 mg/kg) and well within normal ranges (2–20 mg/kg). Fluoride levels in the leachates were increased by the gypsum. There were no treatments to maximum levels of 15.3 mg/l, but concentrations remained well below tolerance limits for coarse fish (75–91 mg/l). All the leachate samples had elevated conductivities, largely resulting from high sulphate concentrations in the clay. Concentrations of most other elements analysed in the leachates were low and generally within acceptable limits for Drinking Water (Water Supply (Water Quality) Regulations 1989).

Soil profiles had developed on all the treatments. Dark coloured topsoil horizons were present on all the organic amendment treatments; these were particularly well defined on the composted green waste and peat/compost treatments. The organic amendments also improved the development of topsoil structures. The gypsum improved structural formation throughout the soil profile and the foundry sand fine structures in the topsoil. Roots were present in most profiles to 60–70 cm depth.

The foundry sand generally had little effect on soil and plant nutrient levels compared with the unamended clay topsoil treatment, apart from some small increases in soil pH and small decreases in soil nutrient concentrations through dilution.

The gypsum slightly lowered soil pH (0.2–0.4 units) and increased extractable phosphorus and total and water soluble fluoride concentrations. Soil organic matter, total nitrogen, extractable potassium and magnesium concentrations were lower than the unamended clay topsoil treatment, largely as a result of dilution. Plant phosphorus concentrations were increased by the gypsum addition.

The composted green waste addition increased soil organic matter, total nitrogen, extractable phosphorus and potassium concentrations. There were also increases in plant phosphorus, potassium and sulphur concentrations.

The waste peat/compost addition gave the largest increase in topsoil organic matter content and also increased total nitrogen and extractable phosphorus concentrations. Plant nutrient concentrations were the same as the unamended clay topsoil treatment.

The sludge cake addition increased soil organic matter, total nitrogen and extractable phosphorus concentrations. Plant nitrogen, phosphorus, magnesium and sulphur concentrations were also increased.

The gypsum and foundry sand increased soil surface strength (measured with a shear vane), which would improve the trafficability of the soil to vehicles and grazing stock.

The sludge cake improved topsoil aggregate stability (measured in 1999), however, none of the organic amendments improved aggregate stability in 2001.

The composted green waste and waste peat/compost reduced the topsoil bulk density, increasing the porosity of the soil for air and water movement and root development. The gypsum addition also reduced bulk density compared with the unamended clay topsoil treatment.

The organic amendments increased plant available water capacity – AWC (measured in 1999), with the increase in AWC positively related to the organic matter addition rates (up to 4% by volume increase). In 2001, the composted green waste and waste peat/compost increased (2–6% by volume) the AWC of the topsoils.

All the organic amendments increased soil biomass carbon and nitrogen (i.e. the size of the soil microbial community), soil respiration rates (i.e. the activity of the soil microbial community) and readily mineralisable organic nitrogen levels. The composted green waste addition gave the largest increases in biomass carbon and nitrogen (97 and 31 mg/kg, respectively), compared with the clay control in 2001, which contrasted with the 1999 results when the sludge cake gave the greatest increases (169 and 44 mg/kg, respectively). Soil respiration rates were increased by the organic amendment additions in 1999, but not 2001. The composted green waste treatment increased readily mineralisable organic nitrogen levels by 35 mg/kg in 2001 compared with 26 mg/kg in 1999, and the sludge cake treatment 22 mg/kg in 2001 compared with 42 mg/kg in 1999.

5 CONCLUSIONS

The study has shown that it is practical and economically viable to use the waste material available at the Winterton landfill site to create soil forming materials. All the potential soil making materials successfully supported vigorous grass growth, with no detrimental or toxic effects observed on plant growth from any of the waste materials used.

The organic amendments added valuable amounts of organic matter and plant available nutrients, with increases measured in soil organic matter, total nitrogen, nitrate-nitrogen, extractable phosphorus and potassium levels. Increases were also measured in plant available water supply and topsoil porosity, and aggregate stability on the sludge cake treatments.

All the organic amendments increased soil biomass carbon and nitrogen (i.e. the size of the soil microbial community), soil respiration rates (i.e. the activity of the microbial community) and readily mineralisable organic nitrogen levels.

Soil profiles had developed on all the treatments. Well defined, dark coloured topsoil horizons were present on all the organic amendment treatments, particularly the composted green waste. Roots were present in most soil profiles to 60–70 cm depth.

REFERENCES

DETR 1999. *Soil-forming Materials, Their Use in Land Reclamation*. London: the Stationery Office.
Steve Maslen & Associates 1996a. *Soil Making Pot Trials – Winterton and Immingham Landfill Sites*. Bradford University report, November 1996.
Steve Maslen & Associates 1996b. *Testing of Soil Making Materials – Lysimeter Trials*. Bradford University report, December 1996.
MAFF 1986. *The Analysis of Agricultural Materials*. MAFF, Reference book 427. London: HMSO.
MAFF 1994. *Fertilizer Recommendations for Agricultural and Horticultural Crops*. MAFF, Reference book 209, sixth edition. London: HMSO.
Fuge, R. & Andrews, J. 1988. Fluoride in the UK environment. *Environmental Geochemistry and Health* 10: 96–104.
Weinstein, L.H. 1997. Fluoride and plant life. *Journal of Occupational Medicine* 19: 49–78.
Water Supply (Water Quality) Regulations 1989. *Statutory Instrument 1147*.

Land Reclamation – Moore, Fox & Elliott (eds)
© 2003 Swets & Zeitlinger, Lisse, ISBN 90 5809 562 2

The effect of composted green waste on tree establishment on landfill

K.J. Foot & M. Hislop
Forest Research, Alice Holt Lodge, Farnham, Surrey, UK

S. McNeilly
Maslen Environmental Ltd, Albion House, 64 Vicar Lane, Little Germany, Bradford, West Yorkshire, UK

ABSTRACT: This study examined the use of composted green waste for tree establishment on three landfill sites in Humberside. Composted green waste was applied at five rates from 0 to 500 t/ha using two techniques, the first involving incorporation of the material to a depth of 0.6 m and the second applying the material as a surface incorporation to 0.1 m into soil which had been cultivated to 0.6 m deep. The survival, growth rates and foliar nutrient condition of newly established alder and sycamore were measured on an annual basis.

During the first three years, rates of tree survival and growth were poor and showed no significant relationships with green waste application. Tree survival rates reflected the harsh site conditions and exposure to intense weed competition, especially on plots treated with surface-incorporated green waste. By Year 4, tree performance showed a significant relationship with the green waste application rate, which was influenced by site conditions and incorporation depth. Improvements in tree growth performance with time appeared to be primarily related to foliar nitrogen levels, with the nitrogen-fixing alder demonstrating more rapid growth. Sycamore showed poorer growth but significantly higher foliar phosphorus and potassium levels and a significant improvement in tree height under high green waste application rates. Recommendations for the use of green waste for tree planting on restored sites are presented.

1 INTRODUCTION

Restored landfill sites soils are usually characterised by poor structure, low organic matter and poor fertility (Dobson and Moffat, 1993). They are very exacting sites on which to successfully establish trees. The Landfill Tax, introduced in October 1996, has encouraged landfill operators to find positive uses for plant residues arising from civic amenity sites and domestic gardens. The addition of composted green waste from such sites may have benefits for the structure and fertility of restored soils on landfill sites and may improve the survival and growth rates of newly planted trees. There is currently a great deal of interest in the use of green waste material (Rainbow, 1998), however very few scientific papers have been published on its potential value for tree establishment on newly restored sites.

The objective of this research was to establish the potential of composted green waste as an environmentally sustainable and cost effective method of soil improvement in the establishment of woodland on reclaimed landfill sites.

2 METHOD

Three Humberside landfill sites of contrasting soil type were selected for the study. Details of site characteristics are presented in Table 1. All sites consisted of capped landfills over which between 0.7 to 1.2 metres

Table 1. Landfill site characteristics.

Site	Winterton	Immingham	Carnaby
Location grid ref.	Scunthorpe SE915200	Grimsby SE207147	Bridlington SE648152
Restored planted	1995 1998	1997 1998	1996–7 1998
Rainfall (mm p.a.)	600	650	700
Exposure	Moderately exposed	Very exposed	Moderately exposed
Soil	Sandy loam	Heavy clay	Medium clay

depth of soil materials had been placed. Immingham soils were very heavy restored clays, Carnaby soils were restored clayey brown earths and Winterton soils were restored sandy brown earths.

Total cultivation by excavator was conducted to relieve compaction in the rooting zone. All plots were worked to a depth of 0.6 m in relatively dry conditions in late October and early November 1997 using the Bending and Moffat (1997) method.

Two methods of incorporating green waste were examined. The aim was to identify the relative importance of concentrating the green waste around the new planted tree roots for nutritional benefit compared to distributing the organic matter throughout the profile to improve the soil structure, moisture retention and longer-term tree performance. Deep incorporation of green waste to 0.6 metres was conducted on selected plots during the total cultivation. Shallow incorporation was conducted by applying the green waste material to the cultivated surface and then rotovating the upper 0.1 metres of the soil.

A range of five green waste application rates between zero and the target forest fertilisation rate for nitrogen were selected. The target forest fertilisation or application rate for nitrogen is 150 kg/ha for a three year period (Taylor, 1991), equivalent to a target rate for the composted green waste of 500 t/ha. Green waste was applied to plots at rates of 0, 50, 100, 250 and the target 500 t/ha, to represent fractions of the recommended nitrogen supply. Tables 2 and 3 give the green waste composition and equivalent nutrient value for the various application rates.

Two tree species, sycamore (*Acer pseudoplatanus*) and the nitrogen-fixing Italian alder (*Alnus cordata*), were selected as being reasonably hardy and tolerant of the harsh conditions associated with landfill sites (Hodge 1995, Moffat & McNeill 1994). Trees were 40–60 cm bare-rooted transplants, located at a spacing of 1.5 m × 1.5 m. Plots were fenced for tree protection against rabbits and hares. Regular applications of Roundup and Kerb and additional mechanical weeding were conducted to minimise weed competition.

The experiment was designed and installed in March 1998 as a randomised split-split plot design with four replicate blocks. The method of green waste application was the main plot treatment with the rate of application the sub-plot treatment. Tree species was the sub-sub plot treatment for each rate and method of green waste application. This blocked design was repeated at each of the three sites.

Annual assessments of tree height (metres) by direct measurement, percentage survival and foliar concentration (as % of dry weight) of nitrogen, phosphorus and potassium (Taylor, 1991) were conducted from 1998 to 2001. Ground vegetation cover was visually assessed in 1999 and 2000 and scored from 1 to 10 as an approximation of percentage cover.

Table 2. Composted green waste composition.

Parameter	Content
Nitrogen (N)	1% (N 125 mg/l)
Phosphorous Pentoxide (P_2O_5)	1% (P 100 mg/l)
Potassium (K_2O)	1.2% (K 1200 mg/l)
Calcium (Ca)	800 mg/l
Magnesium (Mg)	50 mg/l
Manganese (Mn)	5 mg/l
pH	approx. 7.5
C:N ratio	17:1
Electrical conductivity (EC)	0.8 dS/m
Organic matter	30%
Humic acid	25%
CEC	20 meq/100 g
Air filled porosity	14%
Bulk density	420 g/l

Table 3. Calculated equivalent nutrient values for chosen application rates of composted green waste.

Nutrient	Nutrient content (% of dry matter)	Application rate (t/ha)	Equivalent nutrient value (kg/ha)
N	0.030	0	0
		50	15
		100	30
		250	75
		500	150
P	0.024	0	0
		50	12
		100	24
		250	60
		500	120
K	0.300	0	0
		50	150
		100	300
		250	750
		500	1500

3 RESULTS AND DISCUSSION

3.1 Tree survival

Many tree losses were sustained in years 1 and 2 after planting, therefore dead trees were replaced annually by beat-ups from 1998 to 2000. Tree survival in 2001 including beatups was in excess of 70% at all sites. Table 4 indicates the percentage of trees surviving in 2001 from the original March 1998 planting, however. Reasonable survival rates were maintained in 1998 and hence 2001 only for alder at Winterton and sycamore at Immingham and Carnaby. Poor survival and growth appeared to reflect the harsh conditions – drought in the dry months, water-logging in the wet months and exposed windy conditions, particularly at

Table 4. Survival of trees planted in 1998 and present in 2001.

| Site | Tree Survival (%) | |
	Alder	Sycamore
Winterton	74	57
Immingham	11	59
Carnaby	16	83

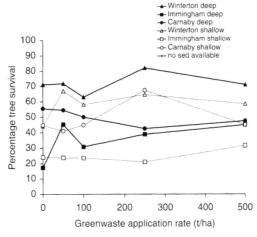

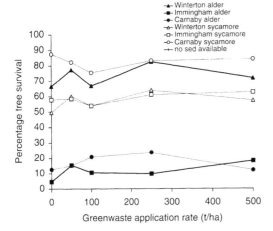

Figure 1. Effect of green waste application rate on survival of alder and sycamore in 2001.

Figure 2. Effect of green waste incorporation depth on tree survival in 2001.

Immingham. Physical stem-breakage from air-borne debris was a common source of damage at Immingham and Carnaby where plots were adjacent to active landfill tipping sites.

There were no significant relationships between the green waste treatments and the survival rate in early years (1998 to 2000). However, by 2001 the alder and sycamore survival rates appeared to have stabilised since the last set of beat-ups were planted in March 2000. Immingham demonstrated the highest survival rates in 2001, followed by Winterton and then Carnaby. The similarity of 2001 survival data to 2000 data suggest that the site conditions in the first few years were most critical to survival and that as the trees become established they are better able to withstand these stresses.

REML analysis on the percentage of trees which survived from 1998 through to 2001 indicated that there was a significant ($p < 0.05$) relationship between percentage survival and site, incorporation depth, species, and with the interaction between site and species.

Figure 1 shows that the rate of green waste application has no effect on initial survival; however, strong differences between site and species are evident, which

may be attributed to climatic and ground conditions and species habit. The results suggest that the initial site conditions are more important in determining survival rates than is the green waste application rate.

Deep incorporation of composted green waste was related to significantly ($p < 0.05$) higher survival rates than shallow incorporation across all species and sites (Figure 2); the data standard error of difference (sed) was too small to show on the graph.

This may reflect a significantly ($p < 0.05$) greater ground vegetation cover, which was observed (using visual scores) on plots with surface-applied green waste, particularly at high rates of application. Ground plants may directly compete with the trees for available resources and threaten their early survival, suggesting that deeper incorporation or burial of green waste is necessary to minimise weed growth. However, this trend was not sufficiently strong to influence survival rate under different green waste application levels.

Secondly, deep incorporation of even low rates of green waste amendment can reduce settlement after cultivation and may have maintained open fissures within the soils to encourage root penetration. Further research into soil structural condition and rooting in the profiles might identify if the latter is a significant factor. Carnaby and Immingham were also affected by the migration of landfill gas into the rooting zone which may distort the results at these sites.

3.2 Tree growth performance

Mean annual tree height results for alder and sycamore for the sites under deep and shallow green waste incorporation methods are presented in Table 5. The

Table 5. Mean annual tree height (m) relative to incorporation method.

Species site	Year	Alder		Sycamore	
		Deep	Shallow	Deep	Shallow
Winterton	1997	0.45	0.45	0.40	0.42
	1998	0.52	0.50	0.39	0.39
	1999	0.84	0.72	0.36	0.37
	2000	1.66	1.31	0.46	0.43
	2001	2.44	1.80	0.64	0.55
Immingham	1997	0.52	0.52	0.36	0.35
	1998	0.35	0.32	0.24	0.20
	1999	0.66	0.61	0.21	0.24
	2000	1.29	1.26	0.24	0.28
	2001	2.25	2.17	0.32	0.37
Carnaby	1997	0.50	0.51	0.33	0.38
	1998	0.27	0.31	0.18	0.21
	1999	0.60	0.61	0.29	0.28
	2000	1.14	1.17	0.56	0.55
	2001	1.73	1.85	0.82	0.81

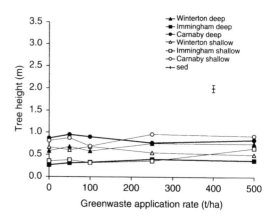

Figure 3. Effect of green waste application on sycamore height in 2001.

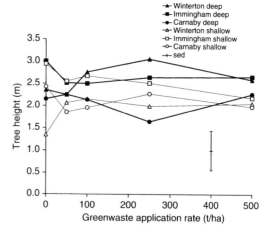

Figure 4. Effect of green waste application on alder height in 2001.

data include measurements from both the original tree plantings and subsequent beat-ups.

Healthy trees would normally be expected to increase in height from year to year. Some decreases in tree height can be observed in the data, which may reflect either the replacement of a tree by a shorter beat-up or dieback of a main stem in one year which is replaced by a new but smaller shoot from the same tree in the following year.

Following a very variable and unsteady start, all treatments planted to alder appeared to be maintaining a steady incremental growth rate in 2001. Sycamore growth was initially very slow but showed a growth spurt in 2001 at Carnaby and Winterton. The maximum annual growth increment for alder in 2001 was 0.96 m at Immingham, compared to 0.27 m for sycamore at Carnaby.

Alder significantly ($p < 0.05$) outperformed sycamore in its mean height throughout the experiment, as Table 5 indicates. The maximum alder height measured in 2001 was over 3 metres at Winterton under the 250 t/ha green waste with deep incorporation treatment, compared with 0.9 metres for sycamore under the 250 t/ha green waste with shallow incorporation at Carnaby.

As for the survival data, no significant relationships between tree height and green waste application rate were identified in the early years, except for significantly ($p < 0.05$) greater alder heights under the deep incorporation treatments at Winterton and Immingham in 2000. In 2001, however, sycamore height was significantly ($p < 0.05$) influenced by the

site and the interaction between site, incorporation depth and green waste application rate.

A general increase in height of sycamore in 2001 occurred with increasing green waste application under most treatments at each of the three sites (Figure 3). However, under alder, only the incorporation depth and the interaction between site and incorporation depth were significant ($p < 0.05$) in 2001 (Figure 4) and no direct relationship between alder height and green waste application rate was observed. Figure 5 illustrates the contrast between alder and sycamore mean height in 2001 and the significant effect of incorporation depth.

The results suggest that physical factors controlling the root growth rate play an important role in successful tree establishment on landfill sites and may be more important determinants of the tree growth response

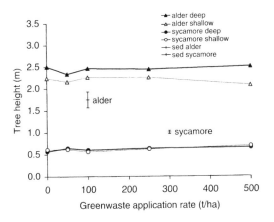

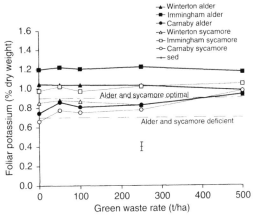

Figure 5. Effect of green waste incorporation depth on alder and sycamore height in 2001.

Figure 6. Mean foliar potassium results in 2001.

than nutrient availability. Deep incorporation of applied green waste may maintain an open-structured soil condition which enhances root penetration and may allow roots to more readily exploit the available nutrient reserve. Greater growth responses of alder to deep incorporation of green waste (Figure 5) may also reflect an improved ability of alder to fix nitrogen where rooting is more advanced.

Differences in the growth rate and habit of the two species mean that it is not possible to directly compare the species heights or growth rates. Strong differences are observed in growth habit of each species between sites, which may reflect site factors such as soil type and initial condition, drought and waterlogging. However, the tree growth results show that alder is significantly out-performing sycamore at all sites, which is likely to be positively related to the increased availability of N to alder.

3.3 Foliar nutrition

Potassium, phosphorus and nitrogen are important constituents of green waste (Table 2) which are released during decomposition and therefore tree foliar nutrition levels were anticipated to be a key factor in explaining differences in tree growth response between the green waste treatments.

3.3.1 Potassium

Foliar potassium (K) levels showed no clear relationship with green waste treatments in early years. It is common for young or unestablished trees to show unusual fluctuations in foliar nutrition which may explain this. In 2001, however, analysis of variance indicated that green waste application rate had a significant ($p < 0.05$) effect on foliar K at all sites, with higher foliar K in sycamore at higher green waste

application rates (Figure 6). Alder exhibited a slight decline in foliar K at higher green waste applications at Immingham and Winterton, although a small increase occurred at Carnaby. Figure 6 also indicates that alder and sycamore at all sites had higher than optimal foliar K in the 500 t/ha green waste treatment in 2001. Only sycamore at Carnaby was deficient in foliar K in 2001 and this was under the zero application rate. Significant differences ($p < 0.05$) were also observed between species, between sites and between the interactions of site × species, green waste application × species and species × site.

K forms a primary constituent of many clay minerals but its availability is dependent upon the clay mineralogy, the weathering rate and the level of plant uptake. It is also rapidly released from freshly applied organic residues (Rowell, 1994) and is readily leachable. Strong differences in foliar K were observed between sites. This indicates that the primary source of available K appears to be release from mineral non-exchange sites rather than from the applied green waste and this will reflect differences in the mineral composition, release rate and cation exchange capacity of the soil at each site. Hence, in Figure 6, Carnaby site demonstrated poorer foliar K levels than Winterton or Immingham under a zero green waste application.

At Immingham, all foliar K levels were above optimum, suggesting that luxury uptake of K was occurring in excess of tree growth requirements and so tree growth rates would not be limited by K availability. Figure 6 illustrates that sites with sub-optimal available K at a zero green waste application appear to show the greatest improvement in foliar K and benefit most from the higher rates of applied green waste, notably at Carnaby. Foliar K levels may therefore not reflect green waste application rates or treatments in cases where an above optimal supply of K from clay minerals is already available. Sycamore appears to obtain more benefit than alder from improved K availability at higher green waste applications (Figure 6). However, all sites and species have an above optimal supply of K at the highest green waste application rate of 500 t/ha.

K release during green waste decomposition would be rapid and some leaching might be expected. However, K continues to be released in sufficient supply at least four years after green waste application at the higher rates. This suggests either that the K release from the clay material is sufficient to meet tree growth needs or that K from the green waste was adsorbed onto soil cation exchange sites and a residual is still being released to the trees. The results indicate that green waste application at high rates ensure good short term K availability to the establishing trees, especially where K is deficient in the soil.

3.3.2 *Phosphorus*
Foliar phosphorus (P) in sycamore and alder was marginal or deficient up to 2000 and no relationships with green waste treatments could be observed. P levels in 2001 increased from deficient (in 2000) to optimal or marginal on most plots and a significant ($p < 0.05$) interaction between green waste application and foliar P was observed (Figure 7). Foliar P was higher under higher rates of green waste application. Foliar P in sycamore was also in excess of optimal levels for growth for rates of green waste application above 250 t/ha in 2001. Alder demonstrated no improvement and even a slight decline in foliar P at the higher green waste application rates. Significant differences ($p < 0.05$) were also observed between species and sites and the interactions between green waste rate × species and site × incorporation method.

Phosphorus is usually present in organic matter and is released by mineralisation or is present in small

quantities as inorganic phosphate, for example as rock phosphates or held on sesquioxide, calcite or 1:1 clay surfaces. P is rapidly immobilised in soil by adsorption onto sesquioxide surfaces and 1:1 clays or through the formation of inorganic phosphate minerals. The release rate is very slow and dependent upon the pH of the soil water solution, low P availability becoming limiting to growth in acid conditions.

P adsorption and slow P release times may partially explain why significant differences in foliar P between treatments were only becoming evident in 2001. However, the mobility of P is very poor and roots take up phosphorus only from their immediate vicinity, within 2–3 mm radius around the roots (Rowell, 1994). Physical restrictions to root growth due to compaction may also have had a significant detrimental impact on P uptake in the early years of the experiment. This places emphasis on the need to prepare a loose, rootable soil condition before tree planting to ensure a well-distributed rooting system develops. The placement of the P supply close to the newly growing roots may also be a critical factor in determining the success of greenwaste applications.

3.3.3 *Nitrogen*
All sites exhibited poor foliar nitrogen (N) levels during the study, alder being deficient in N (below 2.5% dry weight; Taylor, 1991) from 1998 to 2000 and sycamore deficient (below 2.0%) throughout. Foliar N in alder steadily increased during the study but only reached optimal levels in 2001. Foliar N in 2001 was significantly different ($p < 0.05$) between sites, between species and between the interactions of site × cultivation × species and site × cultivation × green waste rate. Alder had significantly ($p < 0.05$) higher levels of foliar N than sycamore in 2001 (Figure 8). This may reflect improvements in the ability of alder

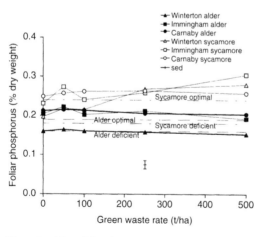

Figure 7. Mean foliar phosphorus results in 2001.

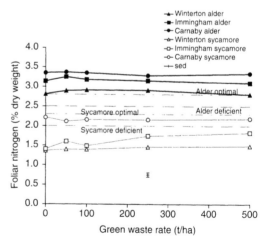

Figure 8. Mean foliar nitrogen results in 2001.

to fix nitrogen as the root system becomes established and more stable, as further root nodulations grow and as populations of N-fixing bacteria associated with those nodulations increase.

N in soil is derived primarily from the fixation of nitrogen gas from the soil atmosphere by micro-organisms to comprise part of the soil organic matter (forming amino groups in proteins) or from fresh applications of plant residues. N is then mineralised from decomposing soil organic matter by bacteria into plant-available form in the soil solution. N is the element usually required in the greatest quantities by the tree and poor N availability is the key nutritional problem identified on restored land, usually due to lack of organic matter (Taylor, 1991).

The 500 t/ha green waste rate with a target N application of 150 kg/ha was expected to supply sufficient N to meet the needs of the newly planted trees. The applied green waste had a low carbon to nitrogen (C:N) ratio of 17:1 (Table 2) which is close to that of natural soil (15:1). The C:N ratio falls below the 25:1 level attributed to a sharp increase in microbial activity and subsequent N immobilisation (Bending et al., 1999). N mineralisation was therefore expected to occur after green waste application and foliar N was expected to increase. The absence of any treatment effect from the applied green waste, particularly in 1998–1999 when most of the N mineralisation would occur is therefore surprising.

The results suggest that the tree rooting in early years was not sufficiently established to take up the available N. The residual effect in 2001 of N on sycamore under high green waste application rates may reflect the slow decomposition of a residual fraction of less-readily mineralised organic components of the green waste.

Other factors which may be of importance in explaining these results include N leaching rates in the different soil types, foliar N being lowest in the sandy Winterton soil and highest in the heavy clay soil at Carnaby. Ground condition factors such as acid pH and the development of anaerobic conditions due to compaction and waterlogging may have had a major impact on N availability. It is more probable, however, that available N was rapidly taken up by the competing ground vegetation cover before the trees were sufficiently established to exploit the nutrient source.

The results indicate that the applied rates of green waste were not suitable to supply non-nitrogen fixing species with available N over the full tree establishment period and repeated applications of fertiliser would be required.

3.3.4 Interactions between nitrogen, phosphorus and potassium

The degree of uptake of P and K by the tree is commonly related to N availability. Where P and K are in plentiful supply (in most normal soils), N is limiting to plant growth and sufficient P and K uptake will occur to meet growth demands. Foliar K in alder is higher than in sycamore (Figure 6), suggesting that there is no limit to the availability of K across the sites and hence growth rate. Poor N availability may, however, be limiting K uptake in the sycamore, which does not have the N-fixing capability of alder. In contrast, foliar P in sycamore is higher than foliar P in alder (Figure 7). This may reflect a relative shortage in availability of P to meet the growth demands resulting from improved N availability in alder. K is available in much larger quantities than P due to supply from mineral exchange sites. The composted green waste addition is the most probable source of P in these soils.

The results therefore suggest that N is the key factor limiting sycamore growth but that as P reserves from organic residues become exhausted, P will limit alder growth. Application of further mineral nitrogen fertiliser is therefore likely to have some benefit to sycamore but will have little impact on the tree height in alder since roots cannot use the applied nitrogen while there is a deficit of phosphate.

The variable foliar nutrition results in earlier years might be caused by different nutrient uptake rates where replacement of dead trees by beat-ups has occurred. The foliar results should also be treated with caution since foliar levels in young trees often fluctuate until the tree reaches a steady growth state, typically after canopy closure at age 10 years. This suggests that further monitoring is required over the next few years to identify the significance of the emerging trends.

3.4 Interactions between foliar nutrition and tree growth performance

The tree growth and foliar nutrition data were examined to identify if any interrelationships could be observed. Figure 9 illustrates the type of scatter diagram which

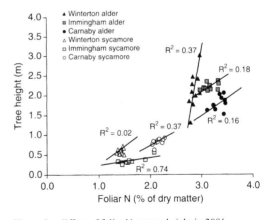

Figure 9. Effect of foliar N on tree height in 2001.

is typical of the evaluation, in this instance showing the results for foliar N. Distinct relationships were observed between sites, which are clustered in different locations on the scatterplot in Figure 9, and a trend between increasing growth with increasing levels of foliar nutrition was evident. There was no effect of incorporation depth or green waste application rate on tree height when evaluating the influence of foliar nutrition. The relationships observed between mean tree height and foliar N, as for the other nutrients, are very weak in most cases (indicated by a low R^2 value which shows a very poor fit of data points to the line predicted by the regression equation).

Tree growth conditions may also fail to reflect foliar nutrition due to other restrictions on tree growth, including mechanical resistance of the soil to root penetration, low soil oxygen levels, periodic stresses induced by waterlogging or droughty conditions and landfill gas migration through the soil.

It is likely that a significant nutrient value component of the applied green waste is not readily available to the establishing trees, perhaps because it may be consumed by other ground vegetation, leaching losses may occur or the roots may not grow rapidly enough or in locations suitable to exploit the nutrient source. This may explain why the relationships between tree foliar nutrient levels and green waste application rates are indistinct.

It remains unclear if the tree growth benefit from green waste identified in 2001 is the result solely of improved nutrition or also of improved rooting conditions resulting from better soil structure, drainage and moisture retention. Further work is required to examine soil physical condition under the treatments. However, it remains apparent from tree growth responses that high rates of applied green waste are beneficial to tree establishment. The degree of improvement remains strongly related to the tree establishment conditions at each individual site.

4 CONCLUSIONS

The aim of applying green waste was to improve moisture retention, to maintain open fissures within the cultivated ground to encourage root penetration and soil structural development, and to act as a longer-term source of nutrient holding (or cation exchange) capacity.

The results of the investigation showed that composted green waste has some considerable potential as an environmentally sustainable and cost effective method of soil improvement for woodland establishment on landfill sites. Tree growth performance in 2001 showed a significant relationship with the applied rate of composted green waste amendment and deep incorporation appeared to give more benefit to growth than shallow mixing.

Strong relationships were also observed between application rates and levels of foliar nutrition, although these interrelationships were very different for the two tree species. Nitrogen was a key factor limiting growth in sycamore, while the N-fixing alder showed better tree growth performance and began to show signs of limited P uptake. This may explain why sycamore appears to show a greater response than alder to composted green waste application.

Successful use of green waste for tree establishment on landfill sites may depend upon:

1. Production of a loose soil condition suitable for good rooting and tree establishment;
2. Incorporation of green waste into the soil in close proximity with the roots to improve nutrient (particularly P) uptake capability by tree roots and to retain open fissures in the soil for further root penetration as the trees grow;
3. Ensuring that the green waste has a low C:N ratio (below 25) so that N mineralisation occurs;
4. Applying high quantities of green waste, at least 4–500 t/ha of green waste, to maximise the nutrient availability;
5. Minimising direct surface applications of green waste mulches for tree establishment or adopting an attentive weed management and control system. However, if the aim is to simply "green" the sites then surfacing mulching or incorporating green waste at a shallow depth in the soil can be very effective;
6. Repeating green waste (or other fertiliser) applications every 3 to 4 years to ensure that trees continue to have available nutrition up to the age of 10 to 15 years. Surface applications would be suitable after Year 4 once trees have become established and are less prone to weed competition.

Further work is required to identify the longer-term ability of the green waste application to improve moisture retention, to maintain open fissures within the cultivated ground to encourage root penetration and soil structural development, and to act as a longer-term source of nutrient holding (or cation exchange) capacity. The impact of the green waste application on tree performance should become more evident as the trees mature and encounter limitations resulting from poor root penetration and restrictions to water and nutrient availability within the landfill soil cover materials.

ACKNOWLEDGEMENTS

This project was funded by Enventure Northern Ltd through the Landfill Tax Credit Scheme under contract to Maslen Environmental Ltd and was conducted by Forest Research. Thanks are due to managers and staff at Waste Recycling Group for supplying the green waste, cultivating the plots to the experimental prescription and allowing access, maintenance and monitoring of the sites. The authors wish to thank the following Forest Research staff: Ian Blair and the team at Wykeham Technical Support Unit for establishing the experimental plots and collecting field data; Environmental Research Laboratory staff for chemical analysis of foliar samples and Alvin Milner and Roger Boswell for the analysis of statistical data.

REFERENCES

Bending, N.A.D., McRae, A.J. & Moffat, A.J. 1999. *Soil-forming Materials: Their Use in Land Reclamation*. The Stationary Office, London.

Bending, N.A.D. & Moffat, A.J. 1997. *Tree Establishment on Landfill Sites: Research and updated guidance*. Forestry Commission, Edinburgh.

Dobson, M.C. & Moffat, A.J. 1993. *The Potential for Woodland Establishment on Landfill Sites*. HMSO, London.

Hodge, S.J. 1995. *Creating and Managing Woodlands Around Towns*. Forestry Commission Handbook 11, HMSO, London.

Moffat, A. & McNeill 1994. *Reclaiming Disturbed Land for Forestry*. Forestry Commission Bulletin 110. HMSO, London.

Rainbow, A. 1998. W*aste not, want not*. Horticulture Week (1 October 1998).

Rowell, D.L. 1994. *Soil Science: Methods and Applications*. Longmans, Harlow.

Taylor, C.M.A. 1991. *Forest fertilisation in Britain*. Forestry Commission Bulletin 95. HMSO, London.

Land Reclamation – Moore, Fox & Elliott (eds)
© *2003 Swets & Zeitlinger, Lisse, ISBN 90 5809 562 2*

The protection of ecology on landfill sites

S.M. Carver
Waste Recycling Group plc, UK

ABSTRACT: Landfill sites, even operational ones, and in particular undeveloped phases of sites, are likely to support a high diversity and abundance of flora and fauna species. The 1981 Wildlife & Countryside Act, The Habitats directive, The Countryside and Rights of Way Act and other wildlife legislation compels the landfill developer and operator to establish the presence or absence of protected species and undertake measures to ensure their conservation.

1 INTRODUCTION

Landfill sites are unique in that they tend to incorporate land previously disturbed either by historic quarrying or mineral extraction activities or waste deposition of some nature. They frequently encompass undeveloped land, reclaimed and restored areas. The nature of these sites means that they provide desirable habitats for wildlife. Because they tend not to be intensively managed, farmed or cropped, they are colonised by a diverse range of species. What could be misconceived to be scruffy and unkempt vegetation and scrub land or sparsely established grassland for example can provide prime habitats for rare or protected species.

This paper explores the optimum ways of safeguarding wildlife on landfill sites whilst ensuring that disruption to the operations of the site activities are kept to a minimum. It assesses the various forces in action to encourage those in charge of land development and specifically in relation to waste management to conserve wildlife.

Although the law provides an incentive to the developer to employ correct wildlife protection measures, it should not be relied upon to be the only mechanism by which ecology is conserved.

2 LAW

Part 1 of the Wildlife and Countryside Act 1981 details the protection afforded to birds, animals and plants.

The law is frequently reviewed and updated so it is important to keep abreast of the latest amendments to legislation, acts and inclusions to the species schedules.

Protection is afforded not only just to certain animals but to their habitat whilst they are using it as well. For example, this is the case for badgers (*Meles meles*) (protected by the exclusive Protection of Badgers Act 1992), water voles (*Arvicola terrestris*) and great crested newts (*Triturus cristatus*) (European protected species), all of which are often encountered on landfill sites.

Subject to certain provisions of the Wildlife and Countryside Act 1981, a person who intentionally damages or destroys or obstructs access to, any structure or place which any wild animal included in Schedule 5 uses for shelter or protection, or disturbs any such animal while it is occupying a structure or place which it uses for that purpose, shall be guilty of an offence. Subject to certain provisions, any person who intentionally picks, uproots or destroys any wild plant included in Schedule 8 or not being an authorised person, intentionally uproots any wild plant not included in that schedule shall be guilty of an offence.

Subject to certain provisions of the Wildlife and Countryside Act 1981, if any person intentionally kills, injures or takes any wild bird; takes, damages or destroys the nest of any wild bird while that nest is in use or being built; or takes or destroys an egg of any wild bird, he shall be guilty of an offence.

The Hedgerows Regulations 1997 makes special provision for the protection of important hedgerows in England and Wales.

Certain offences carry penalties of varying degrees of severity. The potential punishment upon conviction is directly related to the offence committed. The Countryside and Rights of Way Act 2000 (CRoW Act) which came into force in 2001 provides for increased enforcement of the Wildlife and Countryside Act 1981.

In accordance with the EC Habitats Directive 92/43/EEC (the Directive), all development affecting European protected species can only legally proceed with a derogation either by permission from the Member State or a licence issued by The Department for the Environment, Food & Rural Affairs (DEFRA). In order for an appropriate licence to be issued, the application must satisfy the following three main criteria:

1. Preserving public health or safety, or other imperative reasons of overriding public interest including those of a social or economic nature and beneficial consequences of primary importance for the environment.
2. There is no satisfactory alternative.
3. The development will not be detrimental to the population of the species concerned at a favourable conservation status in their natural range.

There is a tendency to presume that the existence of a planning permission is a mitigating circumstance and in some way exempts the developer from complying with wildlife law especially if attached to it there are no conditions relating to ecology. It is however an offence to *recklessly* or intentionally disturb a place or rest or shelter or a nest site of a protected species. This relatively recent alteration to the law removes the ambiguity relating to responsibility. Ignorance of the presence of a protected species or its habitat does not provide an acceptable defence in law for their disturbance or destruction whether it is deliberate or not.

The CRoW Act means that persons committing offences in relation to protected species can be arrested and could be forced to serve custodial sentences of up to 6 months.

3   SURVEYS

Surveys, however intensive, do not always identify the presence of all species. The time of the year during which they are conducted is important in ensuring that all protected species are detected. The timing of surveys is equally significant in that it is imperative that they allow adequate time for a licence to be obtained and any mitigation/translocation to be undertaken with minimal impact on the operation of the site. This avoids deployment of emergency rescue procedures inevitably resulting in less successful mitigation. The extent to which the surveys encroach beyond the boundaries of the development area can directly influence which animal species are identified and the impacts the development may enact. Even though an absence of protected species may be recorded in the area of direct development, it may be vital in providing feeding, hunting, foraging, shelter, hibernation or occasional roost site at certain times during the calendar year.

However, some species and indeed individual animals only become evident at the final stages of site clearance during vegetation and soil stripping operations. This emphasises the advantage of conducting careful site clearance under the supervision of ecologists even if surveys have been carried out. Furthermore, it must not be ruled out that although protected species may be absent at the time of a survey or indeed the granting of a planning permission, they may colonise prior to development if there is a substantial delay to proposed works. Adequate measures must be employed in deterring the reintroduction or intrusion of cleared animals.

It is inevitable that not all developers declare the presence of protected species on their site on all occasions of discovery. The threat of legal proceedings, fines, custodial sentences and the resulting detrimental publicity would not be a sufficient deterrent in disturbing or destroying habitats or their species. More rigorous enforcement of the law brought about by intensive inspections of land development and improved guidance, planning procedures and legislative proposals could raise the standards of wildlife protection.

The protection of wildlife relies to a great extent upon the integrity of developers and landowners and the implementation of proper company environmental policies with high moral standards.

4   HABITAT PROTECTION

The protection of a habitat ensures that the species it is capable of supporting will survive where conditions remain suitable. It could be argued that long-term sustainability should involve a degree of management.

Particularly important during the phased operation of landfill sites is the maintenance of wildlife corridors in ensuring that pathways and migration routes are not obstructed and isolation of species from their habitats or feeding, sheltering, or breeding ground is prevented. Free movement of wildlife can also promote the preservation of genetic diversity.

The receptor site or compensatory habitat must be capable of supporting the translocated species or it must be rendered suitable by appropriate management techniques.

5   MITIGATION AND COMPENSATION (CASE STUDIES)

5.1   *Fauna*

In the case of animals, surveys not only establish the presence or absence but assess the population and extent of range, map the migration routes, detect resting, hibernation, foraging and breeding sites, all of

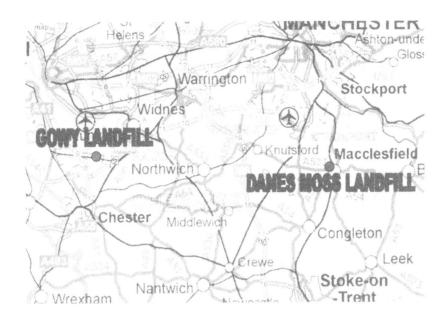

Figure 1.

which can be identified to be some distance from the original sightings. This can provide vital information in determining the optimum compensation and mitigation proposals.

The simplistic approach of adjusting the timing of operations at the site to accommodate nesting activity of birds and avoid their disturbance for example can be the only ecological intervention needed.

5.1.1 Translocation

Water voles were protected on Danes Moss Landfill Site in Macclesfield (Figure 1) by a combination of methods. Displacement of animals from proposed areas of development was encouraged by erecting water vole proof fencing and reducing their food source and rending their habitat undesirable. Physical live trapping of animals along with careful and sensitive stripping of soils was undertaken to expose animals remaining in burrows. Relocation to a favourable receptor site with long term management and monitoring secures the future of the population. Successful breeding has been observed in newly excavated ditches.

Great crested newts were translocated at Gowy Landfill Site in Chester (Figure 1) from areas of proposed development to existing restored habitat. The methodology employed included the amphibian fencing of terrestrial areas and the installation of pitfall traps to capture migrating animals. Enhancement of receptor sites by the creation of a new pond, hibernation areas and future management tailored to amphibians and long term monitoring will promote long term sustainability of the population.

5.2 Flora

Plant species tend to have more static characteristics, however, through seed dispersal colonies can expand into their surroundings and colonise completely new areas depending on the suitability of the soil conditions.

5.2.1 Translocation

Plants may be transplanted by a variety of techniques. Common methods include seed harvesting or the translocation of vegetation turves complete with soil or medium in which they are growing.

Often a condition of a planning permission, and not necessarily for a legal reason, certain plant species, and in particular those of local importance or locally scarce will be required to be protected. Locally scarce Water violet (*Hottonia palustris*) at Gowy Landfill Site has progressively been successfully translocated to a site of Biological Importance within the boundary of the site. The plant occurring in ditches in sedge peat proposed to be lost to landfill development was removed by hand and mechanical methods (a tracked excavator) along with silt and mud deposits and transplanted into re-profiled ditches in safeguarded territory.

Marsh orchids (*Dactylorhiza sp.*) occurring on peat and sand deposits of a peat bog site of special scientific interest at Danes Moss Landfill Site were transplanted as whole turves to restored areas of the site.

5.3 Unintentional rescue

As an incidental result of routine operations in translocation of target species, other wildlife commanding no special protection often benefits. This can occur either by provision or improvement of compensatory habitats where plants and animals are rescued and relocated to safe havens and in the case of plants catalysing the development of a whole new habitat on recently restored areas, for example, where only bare soil existed.

5.4 Habitat preferences of species

Contrary to common belief that most species prefer stable habitats, disturbed or newly restored land of a pioneer nature can attract wildlife and promote colonisation of new species. Areas denuded of vegetative growth are often colonised by animals dispersing from the habitat which they originated from. This is exemplified by young badgers leaving the main sett, young female water voles establishing their own territory or amphibians dispersing to new ponds to avoid cannibalistic tendencies from parents. The loose nature of newly placed soil provides an ideal medium for easy excavation of new chambers by badgers.

Areas undisturbed by human pressures can attract shy elusive animals. Bare stony soils and gravel can provide ideal nesting conditions for the Plover (*Charadriidae*) and infrequently mown unimproved grass for the lapwing (*Vanellus vanellus*) and Skylark (*Alauda arvensis*). Sand martins (*Riparia riparia*) nest in tunnels excavated in sandy or gravely slopes often of the vertical dimensions encountered at mineral extraction or landfill sites. Bare soils and stony banks provide basking sites for butterflies and reptiles.

Bare disturbed soil also provides an ideal niche for the germination of plant seeds. Stressed soils and soil substrates subject to past industrial activity which often exhibit high alkalinity or acidity can support colonies of rare orchids for example. Soils not subject to high concentrations of fertiliser applications associated with modern agricultural practices are capable of yielding a high diversity of wildflowers.

Long unimproved infrequently mown grasslands can, for example, provide habitat for the Brown hare (*Lepus capensis*) which has become increasingly uncommon.

5.5 Succession and diversity

Habitats of varying stages represent the specific points in time of a continually dynamic process. They are able to support complex species structures. This is illustrated by processes of succession encountered on habitats from pioneer to climax stages exemplified by many Landfill Site restoration schemes. Increasing the number of habitat types and stages of succession inevitably promotes an increased species composition. Attempts are often controversially made in the form of specialist management to decelerate the process in order to preserve a particular species composition. This is fundamental to certain species of rare or declining status.

6 COST IMPLICATIONS

Costs associated with wildlife management and conservation of non statutory protected species could be outweighed by the positive publicity received.

The cost of complying with wildlife law is substantial but unquantifiable without being species specific. Conversely, the cost of incurring a fine is likely to be negligible compared with the costs of rigorously adhering to guidance and law. However, the value and consequential impacts of detrimental public relations and delays to operations and development is unmeasurable as is the loss of wildlife and habitats.

7 CONCLUSION

The over-riding driving force for encouraging correct wildlife management is the legal protection that species and their habitats receive. However, the good will and co-operation of landowners and developers seeking to be environmentally responsible is paramount in successful long-term ecological sustainability and conservation.

REFERENCES

Countryside and Rights of Way Act 2000 (HMSO London).
Langton, T., Beckett, C. & Foster, J. (2001) Great Crested Newt Conservation Handbook, Froglife, Suffolk.
Protection of Badgers Act 1992 (The Stationary Office Limited).
Statutory Instrument No. 1160 The Hedgerow Regulations 1997 (DOE).
The Conservation (Natural Habitats, & c.) Regulations 1994 (The Stationary Office Limited).
Wildlife and Countryside Act 1981 (HMSO London).

Land Reclamation – Moore, Fox & Elliott (eds)
© 2003 Swets & Zeitlinger, Lisse, ISBN 90 5809 562 2

Prevention of biorestoration failures along pipeline rights-of-way

M.J. Hann & R.P.C. Morgan
National Soil Resources Institute, Cranfield University at Silsoe, Silsoe, Beds, UK

ABSTRACT: Inappropriate pipeline construction practices and errors in the design and installation of biorestoration on pipeline rights-of-way often give rise to severe erosion. The paper considers experiences of pipeline construction in Europe and South America which have identified a range of problems leading to failures in the biorestoration process. These include:

– Lack of control of the movement, placement, storage and disposal of earthworks.
– The presence of a pipe berm inhibiting the movement of runoff from the right-of-way.
– Poor dimensioning of diverter berms and selection of inappropriate techniques.
– Inappropriate vegetation establishment practices leading to inadequate vegetation cover to avoid severe erosion.
– Poor traffic and grazing management increasing erosion risk.

Examples of inappropriate practices are presented together with the consequent risk to the pipeline and the local environment. Four strategies are proposed for control of erosion on sites which do not meet the performance criteria of Erosion Class 3, the accepted minimum standard for restoration work (Table 1).

1 INTRODUCTION

Advancing technology has led to an increase in the requirement for transporting electricity and communications via cables and liquids and gases in pipelines. These cable/pipe lines often run over large distances installed above, at or below ground level.

There is much evidence throughout the world to show that inappropriate restoration of the right-of-way following pipe/cable installation has given rise to severe erosion.

The consequences of this erosion can be very serious and include infrastructure failures (e.g. roads, bridges and irrigation canals), risks related to exposure and possible damage of pipelines/cables, and effects on local environmental integrity particularly where the rights-of-way are constructed through sites of outstanding natural beauty.

Experiences of pipeline construction and restoration in Europe and South America have identified a range of inappropriate practices on the rights-of-way that have led to failures in the biorestoration process. This work uses the authors' experiences of buried pipeline installations and their restoration in England, Georgia, Azerbaijan and Colombia to identify the types of poor practice which lead to failures in the biorestoration process. It provides recommendations

as to how land-forming operations that should be altered or avoided to prevent erosion. Further it proposes a range of strategies which should reduce erosion along the pipeline corridor and encourage the re-establishment of vegetation thus maximising the soil and environmental protection.

2 CONSTRUCTION PRACTICES

The inappropriate practices that have been found to lead to failures in the biorestoration process and created serious erosion consequences include:

– Lack of control of earthwork movement, placement and storage.
– Poor or inappropriate construction techniques.
– Limited soil management strategies to ensure acceptable vegetation establishment in the necessary time frame.
– Lack of adequate traffic management measures.

These inappropriate practices together with likely consequences of their adoption, as experienced on gas pipelines in Southern England, and oil pipelines in Azerbaijan Georgia and Colombia, are discussed in more detail below.

Pipeline construction produces a large amount of soil, which is surplus to requirement. This includes trench and surface-cut material. The spoil has to be removed from the site as there is limited storage space. The movement of this waste material incurs a high cost and in an effort to minimise costs the contractor may dispose of the spoil close by in inappropriate positions.

Where the pipeline is laid in a trench along a narrow ridge, the top of the ridge is removed and its height lowered to give a wider right-of-way. The width of the right-of-way is further extended by placing the spoil on the side slope. It is not possible to compact this soil so it is left in a weakened state. The combination of steep valley sides and loosely compacted material has resulted in severe gully erosion (Figure 1a).

Where this has occurred in badlands erosion has been very rapid and gullies formed by the collapse of subsurface pipes have started to cut headwards into the ridge, threatening the integrity of the pipe. This is difficult and costly to repair given its position on the slope.

Similarly as shown in Figure 1b, soil is placed to create a level bench, resulting in an embankment between the right-of-way and the hill side slope. The bank material is again poorly compacted, making it vulnerable to gully erosion.

Where the pipeline crosses valley floors with small streams or creeks, the restoration work has filled in the valley bottom, reducing the cross-sectional area of the original channel (Figure 2). The channel is now too small to cope with its original runoff and any extra flow coming from the surrounding right-of-way

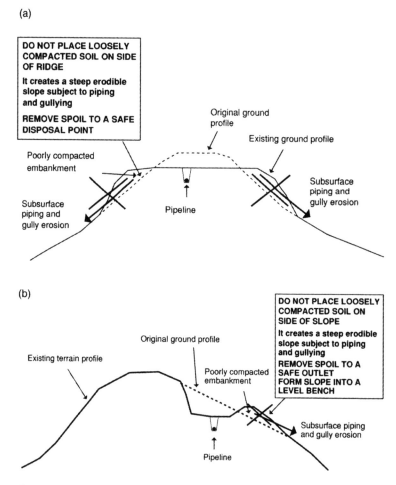

Figure 1.   Examples where loosely-consolidated spoil has led to subsurface and gully erosion: (a) levelling of ridge crest; (b) benching of a hillside.

slopes. In an attempt to adjust its cross section for its discharge, the channel erodes its bed and banks and develops into a gully.

Position and management of stock piles and spoil heaps is sometimes limited. They are often placed in the runoff area with no diverters or local drainage. This produces problems of surface ponding, erosion of the heaps, piles and the surrounding area.

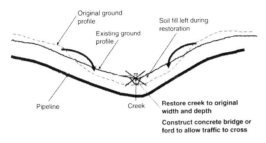

Figure 2. Gully erosion as a result of infilling a creek in valley bottom.

2.2 *Construction techniques*

Several common construction techniques unfortunately do not lend themselves to good soil conservation practice.

A typical cross-section of the pipeline corridor upon a hillside section is shown in Figure 3a. The land has not been restored as a level surface. Instead, a berm (windrow) has been created over the pipeline, which means that, where the pipeline right-of-way lies in a cutting, there is no suitable exit for runoff generated between the pipe berm and the toe of the cut slope. This effect is exacerbated where soil has been placed on top of the cut slope, further raising the bank height, or, in some cases used to form a bank where none previously existed. Gullies are frequently encountered along the cutting toe.

There are also instances on flatter land with a slight cross-fall where the windrow has trapped water and encouraged gully erosion on the up-slope side of the mound (figure 3b). Where the soil is prone to piping, the concentration of water has lead to subsurface erosion either through the mound or along the pipeline.

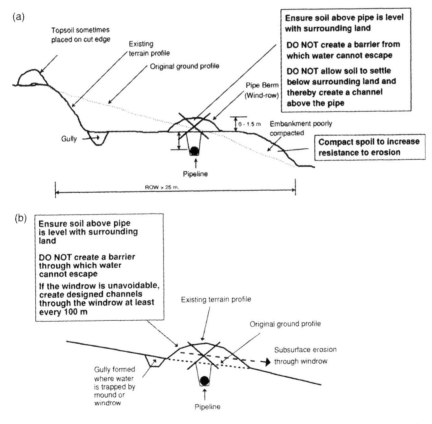

Figure 3. Accelerated erosion as a result of leaving a mound or windrow over the pipe: (a) in cutting; (b) on flat ground with a cross fall.

229

Where the pipeline crosses valley floors with small streams or creeks, the restoration work has used pipe culverts to provide access across the valley bottom. Often the pipe cross section is too small to cope with peak flows of the original channels. In essence the culvert can become a throttling dam in the channel. Flowing water will undermine or in some cases e.g. in the badlands of Azerbaijan and Georgia, will tunnel through the silt/sand soils. In an attempt to adjust the cross section for discharge, the culvert and adjacent channel erodes its bed and banks, developing into a gully.

On some sites outlet channels have been constructed along the corridor. These take runoff from the right-of-way providing an outlet to the waterway network. In several cases these have eroded as they have no reinforcing. Where wicker fences have been installed as check dams, they have invariably been washed out because of a failure to anchor them properly; also they are spaced too far apart allowing overtopping.

2.3 *Erosion control measures*

Once installation is complete erosion control measures are put in place. Whilst many of these are appropriate in theory, they commonly suffer from poor design and faults in workmanship and instead of controlling erosion, often lead to increasing it. Examples of this problem are highlighted below.

Diverter berms constructed from soil are generally too low with virtually no channel on the upslope side. They are invariably set out at too steep a grade usually at between 2 and 10 per cent and sometimes up to 25 per cent compared with the recommended maximum of 0.4 per cent (Hudson, 1981). Further, they are often too short extending only part way across the right-of-way instead of taking runoff on to the surrounding vegetated land. As a result substantial quantities of fast flowing runoff are discharged on the right-of-way itself, leading to the development of gullies.

In many cases where wicker fences are employed as diverters they have failed. They become undermined by the water flow as the body of the fence is not embedded to a sufficient depth. They are also subjected to over turning failure as the support posts are not embedded to a safe depth. Finally the timber in the fences is a useful building material, there is therefore a risk of it being removed by the local population.

Burlap rolls (hessian tubes filled with local soil) are extremely prone to animal damage and weathering. The rolls are also far too easily undermined by water flow resulting in rills and gullies.

2.4 *Soil management and vegetation establishment*

In many cases, particularly in the badlands there has been no care taken to preserve and manage top soil. In fact there is a feeling that there is no discernable top soil. Thus the important, fragile seed bank, organic matter content and structural development are either put at risk or destroyed. Coupled with this is the fact that there are periods of severe moisture stress on certain sections of many pipelines. The result is a very poor growing environment is provided for establishment of vegetation.

Grass seeding is carried out at many sites with varying success. Often the species chosen are inappropriate, the planting rates are far too low and they are sown at a time of drought with no irrigation available. A similar situation exists when plants and shrubs are used with or instead of grasses. The ground cover is therefore not sufficient to provide the necessary protection.

The feasibility of restoring vegetation along a pipeline was evaluated using a simple "screening model" based upon typical climatic and soil conditions (Morgan, 2000). Estimates of vegetation cover in the first year following the completion of construction can be as little as 7 per cent, in low rainfall, low mean temperature regions. This value is estimated to rise to 14 per cent after two years. Field observations show areas along pipelines with in these climatic conditions to have between 2 and 20 per cent cover two years after reinstatement. With such limited cover the rate of erosion has been serious with many rills and gullies developing.

2.5 *Traffic management*

Once completed, a pipeline right-of-way can provide a very convenient form of access, particularly in remote areas which often have relatively poor economies and hence limited roadways. The wheel rutting which occurs as a result of vehicle traffic soon develops into gullie. Vehicles travelling over diverter berms cause their compaction and lowering, leading to more frequent overtopping by runoff. The right-of-way is also used as a grazing area for livestock. This causes some damage to the structures but more importantly, affects the rate of re-vegetation.

3 EROSION PREVENTION

Based upon experience of pipeline installations care must be taken during reinstatement to avoid land-forming operations which will encourage erosion. The following conditions in particular should be avoided.

– Where the right-of-way crosses small rivers or creeks it is critical that the dimensions of the waterway are determined prior to disturbance of the slope and that it is re-established to those dimensions. Pre-construction surveys should be carried out to determine the existence of any gullies etc. downstream. If they are found remediation work will be necessary to control their headward erosion into the right-of-way. Where it is necessary to maintain traffic access it may be necessary to construct a bridge or fording point across the channel, depending upon its depth.

– Where the pipeline is laid on a ridge and the top has to be removed to widen the right-of-way, the spoil must be transported away to a safe disposal point.

4 PROPOSED STRATEGIES

When proposing strategies to mitigate against erosion along rights-of-way it should be stressed that, in the long term, the most cost-effective erosion control is obtained with a dense, self-generating vegetation cover. Since it can take two or more years, depending upon local climate and soil conditions, to establish the generally accepted minimum of 70–75 per cent cover to meet and sustain Erosion Class 3 or better, other control measures are needed in the interim.

It should be stressed that in situations where some structural and revegetation work has already been carried out, albeit with varying degrees of success including, in some cases, enhancing rather than

reducing erosion, the starting point is not an initial heavily compacted bare soil after pipe installation; instead, it is the situation described at the time by field investigation.

The choice of strategies is based upon a detailed analysis of the data gained from each section of the right-of-way. This will give the present situation regarding erosion and vegetation cover, the need to reclaim gullies and how to dispose of runoff from the right-of-way safely.

A range of strategies has been employed and they are summarised below.

4.1 *Do nothing*

There is no need for further action to be taken where the existing control measures and the vegetation cover enable the performance criterion of Erosion Class 3 to be met. The sites should be regularly monitored to ensure that either drought or livestock grazing do not suddenly reduce the vegetation cover.

4.2 *Revegetation*

Establishing sufficient vegetation cover, using local species to ensure self-generation, is the most effective long-term solution to erosion control. The cover

Table 1. Erosion severity classes (Morgan, 2000).

Erosion class	Verbal assessment	Erosion rate (t/ha)	Visual assessment
1	Very slight	<2	No evidence of compaction or crusting of the soil. No wash marks or scour features. No splash pedestals or exposed roots or channels.
2	Slight	2–5	Some crusting of soil surface. Localised wash but no or minor scouring. Rills (channels <1 m^2 in cross-sectional area and <30 cm deep) every 50–100 m. Small splash pedestals where stones or exposed roots protect underlying soil.
3	Moderate	5–10	Wash marks. Discontinuous rills spaced every 20–50 m. Splash pedestals and exposed roots mark level of former surface. Slight risk of pollution problems downstream.
4	High	10–50	Connected and continuous network of rills every 5–10 m or gullies (>1 m^2 in cross-sectional area and >30 cm deep) spaced every 50–100 m. Washing out of seeds and young plants. Reseeding may be required. Danger of pollution and sedimentation problems downstream.
5	Severe	50–100	Continuous network of rills every 2–5 m or gullies every 20 m. Access to site becomes difficult. Revegetation work impaired and remedial measures required. Damage to roads by erosion and sedimentation. Siltation of water bodies.
6	Very severe	100–500	Continuous network of channels with gullies every 5–10 m. Surrounding soil heavily crusted. Integrity of the pipeline threatened by exposure. Severe siltation, pollution and eutrophication problems.
7	Catastrophic	>500	Extensive network of rills and gullies; large gullies (>10 m^2 in cross-sectional area) every 20 m. Most of original surface washed away exposing pipeline. Severe damage from erosion and sedimentation on-site and downstream.

should be uniform avoiding areas in which runoff can concentrate. Plants more than 2 m tall should be avoided as water drip can increase potential erosion. A minimum of 70–75 per cent cover (C = 0.01 in the Universal Soil Loss Equation) is considered sufficient to control erosion in most circumstances. As discussed earlier the likely percentage cover achievable can be estimated for any given region.

Grass seeding may be recommended on sites with less than 10 per cent vegetation cover provided that rainfall is sufficient for grass growth. Prior to seeding the ground should be tilled to a depth of 100 mm. A grass seed mix containing appropriate local species should be drilled in 100 mm rows or broadcast. Seeding should take place either in the spring or autumn, into moist ground and fertiliser applied.

On drier soils, south-facing slopes (in the northern hemisphere) and where rainfall is low relative to evaporation, it will be necessary to increase the soil moisture if establishment is to be achieved. A coarse weave jute mat is recommended where growth is limited. This will act as mulch until sufficient cover is obtained. Whether or not an erosion mat is used, livestock and other traffic should be excluded from the corridor until vegetation cover is established.

Native tree/shrub species should be selected taking account of soil, climate and site aspect. Material should be in the form of whips or sprigs, one to three years old, pre-grown in pots in a nursery. The plants should be spaced at 1 m intervals arranged *en echelon* down slope. The soil around each plant should be loosened and fertiliser added at the time of planting. Where planting is being carried out alongside grass seeding and an erosion mat is being used, planting should take place through holes cut in the mat. In dry areas where no erosion mat is used, the soil around each plant should be earthed up to form a crescent shaped bank on the downslope to provide a basin in which runoff can be harvested.

4.3 Diverter berms

Two types of diverter berm are proposed each formed by digging a channel and throwing the soil on to the downslope side to form an embankment.

- The first is a narrow-based structure (0.8 m) designed to give sufficient depth (0.4 m) to reduce the risk of overtopping. This to be used on all sections of the right-of-way where there will be no traffic; on slopes greater than 20 degrees, the down slope of the berm should be supported by a wooded fence.
- The second is a broad-based structure (4.8 m) with a smaller depth (0.25 m), designed so that it can be driven over with minimal damage. This to be used on those parts of the rights-of-way where vehicle traffic occurs.

The berms should be set at a gentle cross slope grade of approximately 1%.

Berm spacing can be determined using a range of formulae. The recommendations of the study on berm spacing (Morgan et al., 2001) are that it should be performance based and determined in respect of a target soil loss using the Universal Soil Loss Equation.

All diverter berms should discharge to a controlled outfall. Depending upon the nature of the terrain at the side of the right-of-way, the following types of outfall are recommended:

- Stone-lined waterways running down the slope – this should be used where the right-of-way is cut into the surrounding land and the diverter berm ends at the toe of the cutting; the channel should be designed to carry peak runoff expected with a 10 year return period;
- Infiltration pond or soak-away – this should be used where the surrounding land is at the same level or lower than the right-of-way; it can take the form of either an extended berm to enable water to discharge on to surrounding vegetation or a dug-out pond.

4.4 Gully control

Where gullies have developed along the right-of-way, control measures must be introduced. Four types of control measure are recommended, depending on the depth of gully:

- Compacted fill – for use in gullies less than 0.5 m deep;
- Sand bags – for gullies up to 1.5 m deep to infill beneath wicker fences which are still functioning;
- Double wicker fence check dams – for gullies between 0.5 and 1.5 m deep; and
- Gabion check dams – for gullies greater than 1.5 m deep.

In the case of the double wicker fence dam it is important to stress the following aspects:

- the fences must be embedded into the floor of the gully to a depth of 0.2 m and into the gully sides to a distance of 0.5 m to ensure no undermining.
- The fences should be constructed with a central spillway to concentrate the flow in the middle of the channel, to avoid overflow impinging on the down stream gully sides.
- An apron of stone should be placed on the gully floor down stream of the dam.
- The check dams should be spaced according to the "head-to-toe" rule, i.e. the top of one check dam should be at the same altitude as the toe of the next dam down stream.

These aspects are illustrated in Figure 4.

Cross-section

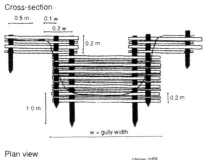

Plan view

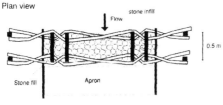

Side view

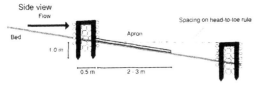

Figure 4. Wicker check dams.

4.5 *Pipe berms/Windrows*

If the pipeline site is either badly eroded or has 5 per cent or less vegetation cover, it is recommended that the slope be regraded, the berm flattened and the soil spread across the right-of-way to form a new lightly-compacted surface. Once the soil has settled, other control measures should be implemented as appropriate to the site.

On slopes of less than 3 degrees windrows can be permitted. The subsoil should not be placed deeper than 0.5 m settled depth. Diverters should be cut through the windrow at not more than 100 m intervals. This will enable any water trapped behind the windrow to escape to a safe outlet.

6 CONCLUSIONS

As indicated in surveys on several pipeline installations, erosion is a problem along sections of the right-of-way.

The problem is largely one inflicted by poor control over construction works, this includes:

- Lack of control of the movement, placement, storage and disposal of earthworks.
- The presence of a pipe berm inhibiting the movement of runoff from the right-of-way.

Also the inadequate design and supervision of installation works for erosion control measures this includes:

- Poor dimensioning of diverter berms and selection of inappropriate techniques.
- Poor management of topsoil and limited effort to maintain the seed bank.
- Inappropriate vegetation establishment practices leading to inadequate vegetation cover to avoid sever erosion.
- Poor traffic and grazing management increasing erosion risk.

Four strategies are proposed for control of erosion on sites, which do not meet the performance criteria of Erosion Class 3:

- Revegetation using grass mixtures and shrubs, ideally supported by a jute erosion mat;
- Diverter berms using narrow-based structures wherever possible and broad-based structures across sections used as roadways; all berms to be at specified grade (1%) and leading to controlled outlets;
- Gully control using compacted fill, double wicker fence dams, or gabions, depending upon gully depth; and
- Regrading of the right-of-way to remove pipe berms, or on land at less than 3% to create breaks in the windrow allowing trapped water to reach a safe outlet.

REFERENCES

Hudson, N.W. (1981). Soil conservation. Batsford, London.

Morgan, R.P.C. (2000). Erosion control: design basis for mountain pipelines. Report to BP International Ltd (second edition). Cranfield University.

Morgan, R.P.C., Hann, M.J. & Gasca, A.H. (2001). The Western Route Export Pipeline, Georgia: recommendations for erosion control. Final Report (revised) to BP Amoco. Cranfield University.

Land Reclamation – Moore, Fox & Elliott (eds)
© 2003 Swets & Zeitlinger, Lisse, ISBN 90 5809 562 2

Reclamation of quarry waste in Morocco: Agadir cement works

C. Harrouni, A. El Alami & F. Gantaoui
Institut Agronomique et Veterinaire Hassan II, Department of Environment and
Natural Resources Management, Agadir, Morocco

H.M. Moore & H.R. Fox
Centre for Land Evaluation and Management, Geography Division,
University of Derby, Kedleston Road, Derby, UK

ABSTRACT: This paper presents interim results of a project for the ecological restoration of an overburden heap resulting from cement works activity in Agadir. Substrate analyses showed that the overburden is low in nutrients but has a high electrical conductivity (2.16 dS/m, calculated on 1/5 extract). In contrast, the local topsoil has a low electrical conductivity, a substantial amount of organic matter (3.55%) and contains adequate amounts of nutrients except for phosphorous (less than 20 ppm) and trace elements. This led to the recommendation that the local topsoil should be scraped and carefully stored as it can subsequently be used for land restoration and re-vegetation if supplied with phosphorous and trace elements. High measured electrical conductivity values for the overburden can be attributed less to sodium chloride salt than to a combination of calcium and magnesium salts which are very high in the substrates of which the overburden is composed. Ecological restoration based on the re-establishment of Argan trees and accompanying plants is described. Local candidate plants are considered the best suited for re-vegetation as they are fully adapted to the harsh site conditions and will help harmonise the overburden with the local environment. However, because of low and unreliable rainfall, irrigation is necessary to ensure initial plant establishment. Appropriate planting and maintenance procedures and management practices are presented.

1 INTRODUCTION

Mineral extraction leads to land degradation in 2 ways: (i) by pit creation at extraction points and (ii) by mound/heap creation of overburden waste. Degradation results from original soil removal and plant disappearance. Such degradation may be damaging to local populations either directly through actual exploitation techniques (explosive blasting, heavy load lorry traffic, dust and noise nuisance...) or indirectly by reducing land productivity and values. In certain instances, especially in developing countries, derelict land may become a no man's land subject to occupation by shantytowns which may in turn lead to conflict situations.

It is therefore highly desirable to restore derelict land with the aim of bringing it back to a reasonable level of self-sustainability. This process may involve the artificial establishment of vegetation which in turn will artificially accelerate the process of plant establishment (Harris et al 1996). Ecological restoration is a fundamental process to maintain biodiversity through

habitat creation thus achieving environmental sustainability (Box 1998).

This paper describes a protocol for the reclamation of a waste overburden at the cement works site owned by Ciments du Maroc near Agadir, southern Morocco (Figure 1). The factory was first established in 1952 with a production capacity of approximately 10,000 metrictons/year. This capacity was increased to 450,000 tonnes in 1976 and now exceeds 1,000,000 tonnes a year.

In Morocco the legal aspects of land restoration are not fully clarified. There is no well defined requirement on industry to restore the land after extraction is completed. The only existing legislation dates back to 1914, stating that the land should be restored back to its original shape. New legislation was prepared in 1994 in the framework of an Environmental Impact Analysis law. However, it remains difficult to apply because (i) it does not make any distinction between different kinds of quarrying, (ii) it does not specify objectives for land reclamation and (iii) it

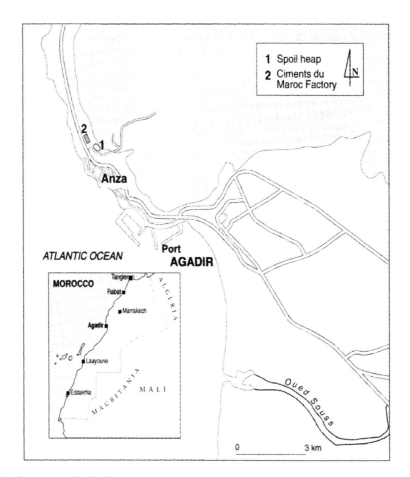

Figure 1. Agadir cement works location map.

does not establish any strategies for the control of restoration.

Due to increasing environmental awareness, Ciments du Maroc has engaged, since 1990, several measures to reduce the impact of their activity on the environment. Their genuine commitment to this cause enabled them to be certified ISO 14001 in August 2001. Ciments du Maroc is a major company in Agadir and their concern about the environment may serve as an example for others. The reclamation of quarry overburden, which is the subject of this study, is among work undertaken by the company in co-operation with IAV Hassan II over the last ten years.

This study represents the first attempt to use wild plant species on a large scale to restore highly disturbed land in a Moroccan context. It is hoped that it will provide valuable information to guide future restoration work on sites with similar environmental conditions in Morocco.

2 SITE DESCRIPTION

2.1 *Local environment*

The Ciments du Maroc factory is located in Anza, a small town 7 km north of Agadir with a population of 30,290 inhabitants in 1994 (RGPH 1994). It is surrounded by dwellings which gives rise to concern about public health and safety.

The overburden heap, which is the subject of this study, is situated on a hillside overlooking the cement works (Figure 1). It reaches an elevation of 160 m above sea level and is visible from the main road that links Agadir to Essaouira in the North. The entire site lies on the Atlantic coast at the western end of the Maritime High Atlas Mountains (Figure 1). The climate is arid but is moderated by the maritime influence. Climate parameters are summarised in Tables 1 to 3. They show that rainfall is low (234 mm per year on average), irregular and mainly concentrated

236

Table 1. Monthly rainfall of Agadir (mm).

	January	February	March	April	May	June	July	August	September	October	November	December
Average	39.42 ± 53.4	27.34 ± 32.6	36.61 ± 42.9	13.57 ± 16.1	3.55 ± 6.2	1.15 ± 3.6	0	0.09 ± 0.3	2.58 ± 4.9	17.82 ± 25.1	35.85 ± 42.4	56.46 ± 57.0
Absolute maximum	174.4	98.2	163.3	51.4	22.5	15	0	1	17	87.9	123.5	184.3

Average values are calculated on measurements made at IAV Hassan II meteorology station located c. 30 km south-east of the study area. Records are from 1982 to 2001.

Table 2. Monthly temperature of Agadir (°C).

	January	February	March	April	May	June	July	August	September	October	November	December
Average	14.36 ± 1.17	15.90 ± 1.36	17.52 ± 1.58	18.16 ± 1.30	19.73 ± 1.18	21.57 ± 0.55	23.60 ± 1.15	23.85 ± 1.71	23.56 ± 2.24	20.83 ± 1.74	18.36 ± 1.24	15.62 ± 1.20
Average maximum	22.31 ± 1.27	23.87 ± 2.03	25.40 ± 2.18	25.76 ± 1.53	27.07 ± 1.50	28.21 ± 0.79	30.12 ± 1.57	30.98 ± 2.51	30.52 ± 2.12	28.56 ± 1.95	26.02 ± 1.48	23.23 ± 1.11
Average minimum	6.30 ± 1.76	8.10 ± 1.37	9.64 ± 1.43	10.52 ± 1.27	12.52 ± 1.16	14.88 ± 0.68	16.39 ± 1.71	16.79 ± 1.72	15.88 ± 1.39	13.22 ± 1.76	10.70 ± 2.01	8.03 ± 2.08
Absolute maximum	28.16 ± 2.93	31.41 ± 2.60	33.16 ± 3.03	35.60 ± 4.54	35.05 ± 6.16	34.02 ± 4.73	38.35 ± 6.46	41.11 ± 6.85	39.94 ± 5.19	35.54 ± 4.18	32.99 ± 3.05	29.30 ± 2.51
Absolute minimum	1.81 ± 1.35	2.51 ± 1.50	4.16 ± 1.54	6.13 ± 2.51	8.14 ± 1.18	10.38 ± 1.04	12.26 ± 1.31	12.28 ± 1.66	11.07 ± 1.48	8.24 ± 1.89	5.58 ± 1.97	2.86 ± 1.87

Average values are calculated on measurements made at IAV Hassan II meteorology station located c. 30 km south-east of the study area. Records are from 1982 to 1997.

Table 3. Monthly relative air humidity of Agadir (°C).

	January	February	March	April	May	June	July	August	September	October	November	December
Average maximum	94.16 ± 7.1	95.84 ± 4.3	95.75 ± 3.0	95.60 ± 6.0	97.14 ± 3.2	97.84 ± 3.71	96.12 ± 4.7	94.70 ± 7.0	97.16 ± 2.1	95.74 ± 3.5	95.49 ± 2.7	95.37 ± 5.2
Average minimum	38.80 ± 11.3	42.00 ± 10.9	41.92 ± 8.2	43.08 ± 6.7	47.23 ± 6.9	51.27 ± 4.2	51.74 ± 5.5	49.10 ± 7.2	48.57 ± 4.1	45.79 ± 8.1	44.57 ± 9.8	42.99 ± 10.4

Average values are calculated on measurements made at IAV Hassan II meteorology station located c. 30 km south-east of the study area. Records are from 1982 to 1997. Max and min values were recorded at 7 am and 2 pm respectively.

Table 4. Natural perennial vegetation on the site.

Plant species	Botanic family	Observation
Argania spinosa	*Sapotaceae*	Endemic small tree
Acacia gummifera	*Mimosoideae*	Endemic small tree
Euphorbia beaumierana	*Euphorbiaceae*	Endemic (cactus like)
Senecio anteuphorbium	*Asteraceae*	Crassulacean shrubby
Withania frutescens	*Solanaceae*	Shrub
Lycium intricatum	" "	Prostrate shrub
Rhus pentaphyllum	*Anacardiaceae*	Erect shrub
Rhus tripartitum	" "	" " "
Asparagus altissimus	*Liliaceae*	Woody climber
Periploca laevigata	*Zygophyllaceae*	Woody climber
Lavandula multifida	*Lamiaceae*	Herbaceous perennial
Lavandula dentata	" "	Herbaceous perennial
Warionia saharae	*Asteraceae*	Endemic erect shrub
Osiris lanceolata	*Santalaceae*	Erect shrub
Suaeda fruticosa	*Chenopodiceae*	Halophytic shrub
	" "	" " "
Atriplex halimus	*Salsolaceae*	" " "
Salsola vermiculata	*Asteraceae*	Milky small shrub
Launea arborea		

between November and March. Significant rainfall events may occur but records show that even the rainiest wet season months may be completely dry (Table 1). Temperatures are moderate (19.4°C annual average) but hot spells occur when south-easterly winds ("*Chergui*") blow, reducing relative air humidity to as low as 30%. December, January and February are the coldest months with absolute minimum temperatures close to 0°C recorded. July, August and September are the hottest months but temperatures higher than 30°C can occur in almost any month of the year (Table 2). Nevertheless, the proximity of the ocean induces fog formation, often accompanied by abundant dew. Thus, relative air humidity is high, especially in summer (Table 3), contributing to the reduction of evapotranspiration and augmenting precipitation totals.

The vegetation cover consists of Argan (*Argania spinosa*) scrub with *Euphorbia beaumierana* (a cactus-like species) as the main accompanying plant. Table 4 shows the main perennial plant species growing naturally in the area. Apart from *E. Beaumierana* and *S. anteuphorbium* all the other plants are selectively grazed by sheep, goats and camels.

Soils locally are thin and stony. They are developed on a calcareous substratum and crusts are frequently found. They are of a serozem type usually unsuitable for agriculture because of its shallowness and stoniness. Topsoil thickness is around 30 cm except in thalwegs where it can be higher. Topsoil physical and chemical properties are described in the section below, in comparison with the substrate of the overburden.

2.2 Overburden characteristics

Cement is made from a mixture of limestone (c.70%) and clay (c.30%) with metal additives. Material unsuitable for cement making (overburden) is stripped off from the limestone and clay quarrying sites and dumped together in mounds. In the study area the first dumping started in 1952 and continued until 1982. The overburden comprises stripped soil, calcareous shale from the limestone and clay quarries and waste products from the actual cement making process. The clay in this overburden is predominantly marl and inter-bedded with the local limestone.

The heap has a low angle top covering 2.5 ha with embankment slopes spreading out over 960 metres. Average slope percentage is about 66%. The heap top has a general west facing slope of 9% (Figure 2). The heap is highly susceptible to erosion. Moreover the heap has a light colour which contrasts with the rest of the surrounding landscape.

Materials forming the overburden are very low in nutrients and coarse textured, consisting of stones and gravel (Tables 5 & 6). The substrates that compose the overburden were analysed and data are compared to local soil where plants grow naturally. Calcareous and marly substrate types were sampled from the overburden and local topsoil samples were taken during strip off operations prior to limestone extraction. All analyses were performed in the IAV Hassan II soil science laboratory using standard analytical methods as reported by Guantaoui (2001). The moisture holding capacity is on average quite low in all samples analysed. However, permeability is lower in the overburden substrates compared to the local topsoil.

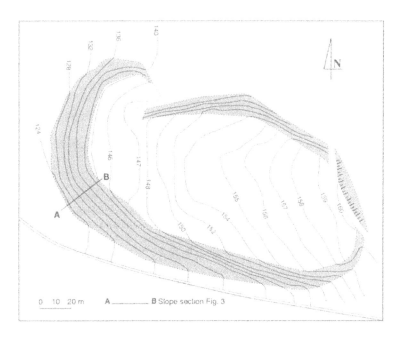

Figure 2. Morphology of the waste heap. Contours in metres.

Table 5. Physical properties of the local topsoil and overburden.

Substrate nature	Gravel >2 mm	Coarse sand 0.2–2 mm	Fine sand 50 μ–0.2 mm	Coarse silt 20–50 μ	Fine silt 2–20 μ	Clay <2 μ
Calcareous	23.60	7.55	3.17	16.58	44.10	28.60
Marly	44.27	3.98	15.28	17.39	49.10	14.25
Topsoil	0	17.03	18.89	34.63	20.75	8.70

	Texture	Holding capacity %	Permeability cm/hour
Calcareous	Moderately stony	34.24	2.52
Marly	Very stony	32.00	3.82
Topsoil	Rather fine	36.61	8.81

Table 6. Chemical properties of the local topsoil and overburden.

Substrate nature	pH (water)	EC DS/m	Total lime (%)	Active lime (%)	Organic matter (%)	NO_3 (ppm)	P_2O_5 (ppm)	K_2O (ppm)
Calcareous	8.6	1.03	47.92	5.60	0.24	1.8	5	252
Marly	9.0	2.16	32.93	7.69	0.72	1.6	38	27.8
Topsoil	7.8	0.58	3.60	2.56	3.55	44.4	18	416

	Cl (ppm)	Na_2O (ppm)	CaO (ppm)	MgO (ppm)	Fe (ppm)	Mn (ppm)	Cu (ppm)	Zn (ppm)
Calcareous	533	1054	1854	690	1.8	1.0	0.4	0.3
Marly	886	795	7073	303	5.4	0.6	0.5	0.9
Topsoil	621	419	632	632	3.4	12	0.5	0.5

The pH in all samples is alkaline but in topsoil it is less than in the overburden. Electrical conductivity is moderate in topsoil but is high in the substrates, with the marly substrate having the highest value. Despite this these values are agronomically acceptable in topsoil. As might be expected total lime is high in the overburden with nearly 50% in the calcareous materials. However, active lime is almost 40% higher in the marly substrate as compared to the calcareous. Organic matter is very low in the overburden but topsoil has a good percentage. Consequently nitrogen content is low in the overburden and rather good in topsoil. This asset is of great importance because soils in such an arid environment are often poor in organic matter. This makes this topsoil very valuable as planting and/or covering material for subsequent plant establishment. However, whilst phosphorous concentrations are very low in all analysed samples, chlorine is present in high proportions.

Exchangeable cations are very significant for such soils given the high electrical conductivity and chlorine content. Calcium is high in all samples but the highest value was found in the marly substrate which was almost four times that of the calcareous substrate, which in turn was nearly 3 times higher than in topsoil. As noted potassium content is low in the substrates while it is rather good in topsoil. Sodium is high in all samples with the calcareous substrate having the highest value followed by the marly one and the topsoil. Magnesium also has high values in calcareous substrate and topsoil but its content is rather good in the marly substrate.

The content of trace elements is in general low in all samples except for Manganese which had a reasonable value in topsoil (12 ppm).

Chlorine and exchangeable cations are the main elements responsible for electrical conductivity which expresses soil salinity. However, as cations other than sodium are present in high proportions in the substrates and topsoil, soil salinity may not be only due to sodium chloride. Sodium adsorption ration (SAR) and exchangeable sodium percentage (ESP) are useful parameters to assess the importance of sodium (Landon 1991). Calculations made from the data presented in Table 5, as described by Landon (1991), showed that SAR and ESP are within acceptable values (Table 7). This suggests that the high electrical conductivity values can be attributed less to sodium chloride salt than to a combination of calcium and magnesium salts. As noted calcium content is particularly high in the substrates composing the overburden (Table 6).

3 OVERBURDEN RESTORATION STRATEGY

As stated earlier the restoration of this overburden will serve as a demonstration for the application of reclamation principles. Wild plants have not previously been used on such a large scale in restoration work in Morocco With respect to the surrounding landscape and to the topographical characteristics, it was decided that ecological restoration should help to harmonise the heap with the local environment and mitigate the visual effects on the landscape.

With regard to topography and in order to reduce erosion risk on the slopes, the embankment was re-shaped in terraces as shown in Figure 3. This was to provide horizontal benches for plant establishment, enabling at the same time rain water accumulation and reducing run-off. The overall tip slope angle was maintained but the embankment was extended. Excavation and backfill slices are shown in Figure 3.

The reclamation option chosen was a combination of afforestation and ecological restoration based on the Argan as the main tree species with some plants of its natural succession typical of the local area. Candidate plants for the site are chosen as follows: *Argania spinosa, Acacia gummifera, Senecio anteuphorbium, Euphorbia beaumierana, Euphorbia Regis Jubae, Withania frutescens, Periploca leavigata, Lycium intricatum, Atriplex halimus,* and *Suaeda fruticosa.* Although *E. Regis Jubae* does not exist in the immediate environment, it occurs as a naturally accompanying species of the Argan in areas to the north and south of Agadir. It is a hardy plant that can withstand the harsh conditions of the site.

Table 7. SAR and ESP in the soil and overburden.

Substrate nature	Sodium adsorption ratio	Exchangeable sodium percentage
Calcareous	0.28	0.85
Marly	0.38	0.78
Topsoil	0.12	1.10

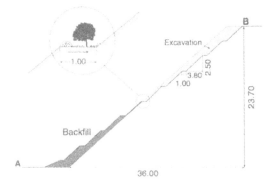

Figure 3. Proposed slope terracing. Overall slope c 66%. All figures in metres.

All these plants were propagated at the IAV Hassan II nursery in Agadir, mostly from seed or cuttings depending on the species. *E. beaumierana* was collected as seedlings from the limestone site prior to topsoil stripping.

Planting took place on June 11th 2002. The Argan trees were planted every 6 m along the terraces. The Acacia were intercalated with Argan plants at 3 m distance. The rest of the plants were randomly planted between Argan and Acacia plants at 1.5 m distance (Figure 3). On the top of the overburden Argan trees were planted on a 6×6 m grid and the rest of the area was rotovated to loosen the substrate with the aim of promoting natural plant colonisation. It is anticipated that the plants established on the embankment will provide seeds for such colonisation in addition to herbaceous annual plants.

In preparation for planting, holes of a circular shape 70 cm in diameter and 70 cm in depth were dug and filled with local topsoil amended with manure, nitrogen, potassium and phosphorous. Manure was added at the rate of 4 kg per plant to improve organic matter content in the soil and hence water holding capacity and fertility. Nitrogen was given as urea at the rate of 0.1 kg per plant, potassium, supplied as potash, at the dose of 0.1 kg per plant and phosphorous, in calcium dihydrogenophosphate form, at the rate of 0.2 kg per plant. Manure, urea, potash and phosphate were incorporated to the soil prior to planting.

Because of the local climatic conditions irrigation is necessary to ensure plant establishment and subsequent growth. The watering regime consists of 32 litres of water dripfed at a rate of 8 litres per hour every 20 days during the dry season (April to September). Following the first irrigation a stone mulch was applied to the ground surface to reduce passive water evaporation.

Plant aftercare consists of monthly monitoring for weeding, staking when required and pruning if necessary. Plant survival and growth are also being monitored. The aftercare will last a maximum of 3 years as by then the plants should be self-supporting. During this period the site will be continually supervised to prevent grazing by sheep and goats.

All outside operations were performed by Ciment du Maroc including waste heap reshaping, topsoil reinstatement, fertiliser application, planting and aftercare, with careful supervision by the authors.

4 INTERIM RESULTS

A full set of plant monitoring results is not yet available at the time of writing. However, observations of Argan and Acacia two months after planting indicated a promisingly substantial amount of biomass production largely due to side shoot development. Indeed the real increase in plant height was generally quite limited though tree species such as Argan and Acacia are relatively slow growing. The average increase in height for the Argan was 2.84 cm $\pm$ 2.5 cm (77 plants) whilst for Acacia the average was 7.51 cm $\pm$ 1.55 cm (60 plants). This represents a 4.3% and 5.8% height increase respectively. It is anticipated that future measurements will provide valuable information on the performance of such wild species in restoration projects.

5 CONCLUSION

Ecological restoration is one aspect of land reclamation. It can be used to harmonise a derelict site with its natural environmental context and to mitigate the visual effects of mineral extraction on the landscape. Moreover, ecological restoration with local plant species may bring the land back to self sustainability, especially in arid lands where climatic conditions make it difficult for exotic plants to grow without engaging costly and tedious aftercare.

The reclamation of the Anza quarry waste heap described in this paper is an important experimental project for such restoration work in Morocco. The use of wild plant species on this scale and on such disturbed land has not previously been attempted in a Moroccan context. For this reason, close monitoring of initial plant establishment is crucial as it will help understand plant behaviour and inform future restoration projects. More research is also required to improve wild plant propagation and determine their nutrient and water requirements. More plants of the local Argan ecosystem need to be studied and included as candidate plants for re-vegetation if they prove to be easy to propagate and handle.

In addition this project is an illustration of the fruitfulness of co-operation between industry and the academic sector for land reclamation. If it is successful, it can be enlarged to other areas and may encourage other companies to begin implementing similar measures for environmental management.

ACKNOWLEDGEMENTS

The authors are grateful to M. Brahim Debbagh for having kept regular weather records since the opening of IAV Hassan II, Agadir. They also acknowledge the genuine involvement of Agadir Cement Works in land reclamation.

REFERENCES

Harris J.A., Birch P. & Palmer J. 1996. *Land restoration and reclamation: principles and practice*. Harlow: Longman.

Box J. 1998. Setting objectives and monitoring for ecological restoration and habitat creation. In H.R. Fox, H.M. Moore & A.D. McIntosh (eds.), *Land reclamation, achieving sustainable benefits*: 7–11. Rotterdam: Balkema.

Guantaoui F. 2001 *Réhabilitation des sites dégradés, Principes et pratique : Cas d'un dépôt de stériles de la Cimenterie d'Agadir.* Mémoire de 3ème Cycle, IAV Hassan II, Agadir.

Landon J.R. (ed.) 1991. *Booker tropical soil manual.* Harlow: Longman.

RGPH (1994) Recensement Général de la Population et de l'Habitat, Direction de la Statistique, Rabat, Morocco.

Land Reclamation – Moore, Fox & Elliott (eds)
© 2003 Swets & Zeitlinger, Lisse, ISBN 90 5809 562 2

Restoration of a magnesian limestone grassland community on former quarry sites

J.D. Riley & D.L. Rimmer
University of Newcastle, Newcastle upon Tyne, UK

D. Brignall
Wardell Armstrong, Newcastle upon Tyne, UK

ABSTRACT: Magnesian limestone grassland is one of the rarest plant communities in the UK. Over the past thirty years increased demand for the underlying stone has put the community at risk. Natural revegetation of former quarries is a slow process, and site restoration is therefore required to save the community from extinction. One of the aims of this project was to develop a strategy for the successful revegetation of magnesian limestone quarry sites with their original grassland community. A field experiment has been established to test various substrates (crushed limestone with and without added soil) and the benefit of sowing seed harvested from a nearby undisturbed grassland site. Results from the first two seasons indicated that the greatest species richness and abundance of desirable species (e.g. *Trifolium pratense, Lotus corniculatus, Rhinanthus minor, Anthyllis vulneraria*) was achieved on those plots that had seed sown onto finely crushed limestone and soil. Depth of the underlying layer of limestone rock did not have a significant effect on species richness.

1 INTRODUCTION

1.1 *Magnesian limestone grassland*

The largest area of magnesian limestone in the UK occurs within the counties of Durham and Tyne and Wear. It forms a triangular area of 600 km² between South Shields, Hartlepool and Darlington (Dunn 1980). Its commercial value lies in its use as bulk stone and hardcore for engineering schemes. The stone has been quarried for centuries; but in the past 30 years in particular, greater demand and altered quarrying techniques (such as the removal of spoil, thus reducing relict habitats) have put the plant and animal communities, which depend upon the soils that develop, at serious risk of extinction. Magnesian limestone grassland is now the rarest calcareous grassland in the UK (Hedley et al. 1997).

Rodwell (1992) designated the community CG8, dominated by *Sesleria albicans* (blue moor grass). The community includes nationally rare species, common species restricted to calcareous soils and species which are more usually found in old pastures (Doody 1980). Common species include *Avenula pratensis* (meadow oatgrass), *Carex flacca* (glaucous sedge), *Festuca ovina* (sheeps fescue), *Galium verum*, and *Plantago lanceolata* (ribwort plantain); while nationally rare

species include *Helianthemum nummularium* (rock rose), *Linum perenne* (perennial flax), *Primula farinosa* (birdseye primrose), and *Sesleria albicans*. The bulk of the sward on primary grassland consists of *S. albicans, F. ovina* and *C. flacca*, while *Cynosurus cristatus* (crested dogstail), *Arrhenatherum elatius (false oatgrass)* and *Dactylis glomerata (cocksfoot)* are frequent members. Although predominately grassland, many forb species can also be present. *Plantago lanceolata, Lotus corniculatus* (birdsfoot trefoil) and *Linum catharticum* (fairy flax) are usually frequent, with occasional *Plantago media* (hoary plantain) and *Viola hirta* (hairy violet).

Rodwell (1992) identified three separate sub-communities – one dominated by *Hypericum pulchrum* and *Carlina vulgaris*, one dominated by *Avenula pubescens*, and one dominated by *Pilosella officinarum*.

1.2 *Quarry reclamation*

Disused quarries now constitute a substantial proportion of the remaining areas of magnesian limestone grassland. Most have revegetated naturally, but this is a slow process. There is considerable interest in accelerating the colonisation and establishment of magnesian limestone grassland.

Surprisingly few revegetation attempts on limestone have been documented in the academic literature (Davis et al. 1985, 1993, Dixon & Hambler 1984, 1993, Hambler et al. 1990, 1995, Cullen et al. 1998). Even less common are records of attempts that have been made to restore a semblance of the original vegetation to the site in question. For example, Dixon & Hambler (1984), and Richardson & Evans (1986), were concerned with establishing examples of vegetation on experimental plots, which were successful and provided an attractive flora. However, the plants concerned were either not usually found on limestone grassland, or only included a few of the species that are typical of such communities.

The only reported attempt at revegetation on magnesian limestone was carried out by Richardson & Evans (1986), although their species mix contained few magnesian limestone grassland species. They carried out field trials on a bare area of Coxhoe Quarry, near Durham. They sowed 18 grass species, including *Agrostis stolonifera*, *Festuca rubra*, *Dactylis glomerata*, *Lolium perenne*, *Arrhenatherum elatius*, as well as species more characteristic of magnesian limestone grassland – *Brachypodium sylvaticum*, *Brizia media*, *Bromus erectus*. They managed to create a verdant sward of grassland, but one that contained few of the typical magnesian limestone species.

Best practice for quarry reclamation schemes depends on the species that are sown and the community being created. Some studies suggest that moderate fertilisation would be desirable (Humphries 1982, Dixon & Hambler 1984), when establishing a sward on bare limestone; others point out that fertilisation, if it is too great, would be undesirable for many limestone grassland species since it allows more competitive species to dominate the community (Janssens et al. 1998). It has also been suggested that for some quarry reclamation schemes 15 cm of subsoil is required, while for others 5–10 cm of topsoil should also be used (Land Use Consultants, 1992a, b). Bishop Middleham quarry in County Durham (NZ 332326) is a working limestone quarry. It is a requirement of the planning authority (Durham County Council) that after quarrying has ended, and the site has been landfilled, an attempt should be made to restore magnesian limestone grassland. The aim of the work reported here was to determine the substrate and seeding requirements for the restoration of magnesian limestone grassland at the Bishop Middleham site after landfilling and capping. It was decided to assess: (i) the optimum depth of limestone to place on the cap, (ii) the need for soil, since some quarries have developed good limestone grassland despite having extremely thin, nutrient-poor soils (Bradshaw et al. 1982; Riley et al., in prep.), (iii) whether the application of seed harvested from a local grassland SSSI was required, and what species would enter the site

naturally from adjacent grassland, as Richardson (1956) suggested would occur.

2 MATERIALS AND METHODS

2.1 Field experiment

The experimental area was at the northern end of Bishop Middleham quarry and was 15 m × 60 m, containing 36 plots. A layer of clay was applied to the whole area to act as an inert base for the experiment and to simulate the cap that would be present once the quarry had been filled. The plots were in six blocks of 5 m × 30 m, and each block contained six plots of 5 m × 5 m. The treatments were arranged in a fully randomized, partially factorial, block design.

The six blocks were made up of two underlying coarse (~75 mm diameter) limestone treatments either 15 cm or 30 cm in depth, each replicated three times. The six treatment plots within each block were combinations of three substrates applied on top of the limestone, with and without added seed application subsequently. The substrates were: (i) control (i.e. bare limestone rock), (ii) 10 cm of limestone "flour" (material with a diameter <3 mm), and (iii) 10 cm of "flour" plus 10 cm of soil. The soil was taken from nearby soil mounds that contained soil removed from the site prior to quarrying. It was a 50:50 mixture of topsoil and subsoil (Table 1). The plots were constructed in such a way that the final surface was level, despite the different depths of limestone and substrate applied.

The seed was sown in mid-September 1999 at a rate of 10 g m^{-2} (100 kg ha^{-1}). It had been harvested from Thrislington NNR (NZ 316327), the best and most extensive example of magnesian limestone grassland in the country. Seed collection was carried out using a hand-held vacuum harvester between mid-July and early September 1999 (Riley et al. 2003). The collected seed was dominated by *Festuca ovina* (50%) and *Briza media* (20%). *Pilosella officinarum*, *Plantago lanceolata*, *Hieracium agg.*, *Leontodon hispidus*, *Carex flacca*, *Lotus corniculatus* and *Rhinanthus minor* each accounted for approximately 3% of the collected seed. All plots were raked by hand before sowing. Fencing was erected around the area in order to prevent rabbit grazing.

Table 1. Mean (±s.d.) phosphate and organic carbon contents of the soils used in the trial.

	pH	Phosphate (mg kg^{-1})	Organic carbon (%)
Topsoil	7.6 (0.3)	24.3 (3.8)	4.3 (0.3)
Subsoil	7.6 (0.2)	6.67 (0.58)	1.7 (0.4)

2.2 Limestone treatments

Sampling was carried out on the limestone treatments prior to sowing, in order to determine whether the flour-amended plots were significantly different in particle size from the controls (bare coarse limestone), because they appeared to be similar on visual inspection. Three 1 kg samples of material were removed from the top 10 cm of each plot. Those plots that were covered with soil were left undisturbed. Each sample was dried for two days at 70°C and then weighed. The material was washed on a 2 mm diameter sieve, dried and weighed again. The material that did not pass through the sieve was converted to a percentage of the total mass.

The proportion of material >2 mm was 33.5% on the control plots and 32.5% on the flour-amended plots. They were not statistically significantly different. The reasons for this similarity were probably a combination of weathering of the coarse limestone to give smaller particle sizes and the washing down of the flour into the underlying limestone.

2.3 Vegetation data collection and analysis

The vegetation data were collected in mid-July 2000 and again in mid-June 2001 by a visual estimate of the percentage cover of every species within each plot. Since the vegetation was sparse, each plot was treated as a single 16 m² quadrat (allowing a 1 m wide buffer zone around each plot in order to avoid plot boundaries).

Since the experimental design was partially factorial, a split-plot ANOVA was performed with depth of limestone rock as one factor, and the six combinations of substrate/seed as six levels of a second factor. The substrate/seed combinations were then compared for significance using the Tukey-Kramer test.

As the residuals of most species could not be transformed to a normal distribution, it was decided to only perform ANOVA on species richness. Canonical Correspondence Analysis (CCA) was carried out instead to construct species-treatment biplots, using CANOCO (ter Braak & Smilauer 1998). Ecological interpretation of the trends displayed in the species-treatment biplots was based upon a comparison of the distribution of species and treatments. Treatments located in a particular sector of the biplots were associated with those species that were located in the same sector.

TABLEMANY (a version of TABLEFIT, Hill 1996) was used to allocate National Vegetation Classification (NVC) plant communities (Rodwell 1992) to the vegetation recorded from each treatment combination. The coefficient of similarity between the sample vegetation and CG8 grassland was noted as a measure of the degree of similarity to this target community.

3 RESULTS

3.1 Species identified

Frequency of occurrence and percentage cover of individual species are shown in Table 2. Those in classes III and IV were usually present on both control and soil treatments. The percentage cover of all

Table 2. Species frequency classes and mean cover (% ± SE) on all treatments in both seasons.

Species	2000	2001
Class IV*		
Agrostis capillaris	4.4 (±0.6)	4.6 (±0.6)
Class III		
Avenula pubescens	2.2 (±0.4)	2.6 (±0.5)
Briza media	5.0 (±1.0)	4.0 (±0.8)
Festuca sp.	6.7 (±1.2)	6.7 (±1.2)
Holcus lanatus	3.3 (±0.7)	3.5 (±0.7)
Rhinanthus minor	5.8 (±1.1)	6.0 (±1.1)
Sanguisorba minor	6.3 (±1.8)	6.7 (±1.9)
Trifolium pratense	2.5 (±0.5)	4.4 (±1.0)
Class II		
Anthyllis vulneraria	1.9 (±0.5)	16.0 (±4.7)
Avenula pratensis	1.5 (±0.4)	2.6 (±0.5)
Cirsium vulgare	2.1 (±0.5)	1.7 (±0.5)
Festuca rubra	1.3 (±0.4)	1.3 (±0.4)
Leontodon hispidus	1.4 (±0.4)	1.4 (±0.4)
Lolium perenne	1.1 (±0.4)	1.1 (±0.4)
Lotus corniculatus	2.8 (±0.7)	3.3 (±0.8)
Matricaria matricarioides	1.7 (±0.4)	1.8 (±0.4)
Papaver rhoeas	1.1 (±0.4)	1.1 (±0.4)
Pilosella officinarum	1.1 (±0.4)	1.1 (±0.4)
Plantago lanceolata	1.8 (±0.5)	1.8 (±0.5)
Poa pratensis	1.8 (±0.4)	1.7 (±0.4)
Rumex sp.	2.2 (±0.5)	2.2 (±0.5)
Valerianella locusta	1.3 (±0.4)	1.3 (±0.4)
Class I		
Aethusa cynapium	0.1 (±0.1)	0.3 (±0.2)
Arabis hirsuta	0.1 (±0.1)	0.3 (±0.2)
Capsella bursa-pastoris	1.7 (±0.9)	1.1 (±0.8)
Carex flacca	1.1 (±0.4)	1.1 (±0.4)
Dactylis glomerata	0.7 (±0.3)	0.7 (±0.3)
Epilobium montanum	0.6 (±0.3)	0.7 (±0.4)
Euphorbia helioscopia	0.4 (±0.2)	0.4 (±0.2)
Hieracium agg.	0.8 (±0.3)	1.5 (±0.6)
Koeleria macrantha	1.0 (±0.3)	1.0 (±0.3)
Phleum pratense	0.7 (±0.3)	0.6 (±0.3)
Poa annua	1.0 (±0.5)	1.0 (±0.5)
Polygonum aviculare	0.8 (±0.4)	1.0 (±0.4)
Ranunculus acris	0.7 (±0.3)	0.7 (±0.3)
Rumex acetosa	1.3 (±0.7)	1.3 (±0.7)
Spergula arvensis	0.8 (±0.5)	0.8 (±0.5)
Taraxacum agg.	0.8 (±0.3)	0.7 (±0.3)
Tussilago farfara	1.3 (±0.7)	1.4 (±0.7)

*IV = present in 60–79% of plots, III = 40–59%,
II = 20–39%, I = <20%.

Table 3. The mean number of species present in each combination of substrate and seed.

Treatment	2000	2001
Control/no seed	2.3[a]	2.7[a]
Flour/no seed	3.3[a]	3.7[a]
Flour + soil/no seed	9.7[b]	9.8[b]
Control/seed	11.3[bc]	11.3[bc]
Flour/seed	14.3[cd]	14.5[cd]
Flour + soil/seed	20.3[e]	20.7[e]

Treatments with the same superscript letter were not significantly different at $p < 0.05$ by the Tukey-Kramer test. SE for all means was 0.9.

species was low, but for *A. vulneraria* it had clearly increased considerably by the second season.

3.2 Treatment effects

There was no effect of depth of underlying limestone on the number of species present on a plot. There were small increases in species number with the addition of flour compared to control (Table 3); but the differences were not statistically significant. This lack of any difference is not surprising, given their similar particle size characteristics (see section 2.2, above).

There were large and significant differences in the number of species present on plots with and without soil and between the sown and unsown plots, but little change from the first season to the second in any of the treatments (Table 3).

On control and flour-amended plots without seed, species such as *Aethusa cynapium*, *Arabis hirsuta*, *Taraxacum agg.*, *Papaver rhoeas* and *Rumex sp.* were most common. On plots with soil but without seed, the number of species increased considerably and included *Tussilago farfara*, *Capsella bursa-pastoris*, *Euphorbia helioscopia*, *Ranunculus acris* and *Poa annua*.

On the control plots with seed *Briza media*, *Dactylis glomerata*, *Carex flacca*, *Holcus lanatus* and *Matricaria matricarioides* were most common. The sown treatment with flour and soil had many species, of which *Lotus corniculatus*, *Anthyllis vulneraria*, *Trifolium pratense*, *Koeleria macrantha*, *Leontodon hispidus*, *Hieracium agg.* and *Pilosella officinarum* were the most common (Figure 1).

There was little substantial difference at the end of the second growing season, except an increase in the percentage cover of *Anthyllis vulneraria* and *Trifolium pratense* (Tables 2 and 4).

Only *Anthyllis vulneraria*, *Trifolium pratense* and *Sanguisorba minor* achieved substantial cover. *A. vulneraria* and *T. pratense* were absent from the control plots; but by the second season achieved considerable cover on the sown soil plots (60–80% and 13–15%

respectively). *S. minor* achieved a cover of 40% on the sown soil plots on 30 cm limestone.

However, the cover of most species remained poor at 5% or less, with most species remaining very small and exhibiting a prolonged juvenile phase, for example, *F. ovina* and *B. media*, which were very common and achieved a moderate cover, but remained less than 5 cm tall.

Rhinanthus minor flowered after the first growing season, and *T. pratense* and *A. vulnereria* flowered in the second growing season. *S. minor* was coming into flower during the final assessment of vegetation. No other sown species had reached the flowering stage after two growing seasons.

A green film developed on some of the control plots, particularly in depressions that were subject to periodic waterlogging. This is probably due to algae and was reported by Dixon & Hambler (1984) after three years in similar conditions.

3.3 Comparison with NVC communities

The main aim of this experiment was to identify treatments that would lead to the rapid establishment of a sward that would develop into *Sesleria albicans – Scabiosa columbaria* (CG8) grassland. All of the communities that had developed on the plots by the end of the second season were very poorly matched (similarity coefficient <50) to any NVC community. None of the unsown plots had a similarity coefficient greater than 5 when compared with CG8 grassland. The sown plots on limestone rock had a mean coefficient of 26 to CG8 grassland, the plots with limestone rock and flour applied had a mean coefficient of 27, and those with limestone rock, flour and soil applied had a mean coefficient of 32. This shows there was little difference between the treatments in terms of their degree of similarity to CG8 grassland. This is unsurprising since the sward was in its establishment phase.

4 DISCUSSION

4.1 Succession

A two-year period was insufficient for succession to occur. Far more important than the effect of time was the effect of the treatment initially applied. The only possible effects of succession over the two years were in the cover of *T. pratense* and *A. vulneraria*. The latter had a very small cover in the first season, but dominated the sown limestone flour + soil treatment plots in the second season. It may take considerably longer before more general effects of time become apparent. The prolonged juvenile phase exhibited by most of the plants is common during the early stages of restoration schemes, until fertility increases (Park 1982), while

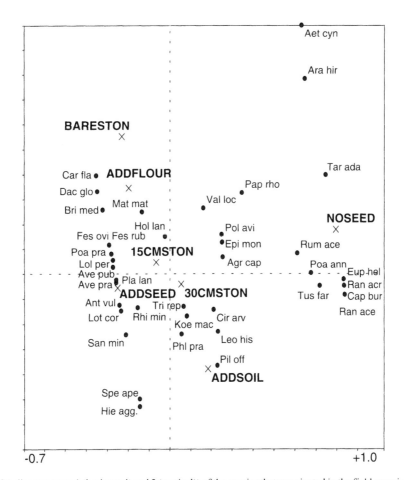

Figure 1. CCA diagram (axes 1 (horizontal) and 2 (vertical)) of the species that germinated in the field experiment and their relationship to treatments in the first growing season.

Table 4. The mean cover (%) on sown plots during the first two seasons.

Species	Flour		Flour + soil	
	2000	2001	2000	2001
Anthyllis vulneraria	5.8	21.6	5.8	74.2
Trifolium pratense	5.0	8.3	5.8	14.1

numerous other factors e.g. drought, limestone dust cover and temperature can delay growth on limestone (Humphries 1982; Park 1982). Davis et al. (1985) and Davis & Coppeard (1989) also reported slow establishment of their sown species during the first few years, while Richardson & Evans (1986) observed very poor germination of harvested species on limestone, although those plants of *B. media*, *B. erectus* and *B. sylvaticum* that did germinate were "strong and healthy individuals" (Richardson & Evans 1986).

However, considerable quantities of fertiliser were applied in that experiment.

4.2 Seed source and dispersal

After two seasons, all unsown plots had developed sparse vegetation, frequently of small individuals. A two-year period clearly proved insufficient for the ingress of substantial numbers of desirable migrant species (particularly orchids) from surrounding vegetation that Richardson (1956) suggested would happen in considerable quantity. A far more diverse sward, including numerous forbs, was achieved on those plots that had been sown. Clearly sowing is vital for the rapid development of a sward, but will nonetheless take longer than two growing seasons to develop an acceptable cover of most species. However, within two years, seed applied at a rate of 100 kg ha^{-1} led to a community that included numerous important magnesian limestone grassland species, such as *S. minor*, *F. ovina*,

A. pratense, A. pubescens, A. vulneraria, T. repens, L. corniculatus, R. minor, B. media, P. lanceolata and *P. officinarum*, but lacked the most characteristic, such as *S. albicans* and *S. columbaria*. Therefore, a seed collection targeted at those species that were absent from the collected seed will be necessary.

The presence of *P. officinarum, L. hispidus* and *K. macrantha* on unsown plots could mean that these species were entering the site from elsewhere (or from the soil seed bank in the case of *P. officinarum*) or that some seed had blown onto these plots from adjacent sown ones. *Poa annua, Rumex sp., Taraxacum agg., Papaver rhoeas, Epilobium montanum* and *Polygonum aviculare* all appeared to be entering the site from elsewhere, since they were present on unsown plots without soil. *Tussilago farfara, Euphorbia helioscopia, Ranunculus acris* and *Capsella bursa-pastoris* all appeared to be either emerging from the soil seed bank, or entering the site from elsewhere and establishing mainly on the soil plots.

4.3   Soil nutrient status and available niches

It is clear that a 15-cm depth of coarse limestone rock is adequate as a basic substrate for the restoration of these sites.

Since the soil used in this experiment had a better nutrient status than the limestone (Table 2); Booy (1975) reported the nutrient status of limestone from Thrislington Quarry as $2.9 \, mg \, kg^{-1}$ nitrate, $0.6 \, mg \, kg^{-1}$ phosphate and $4.4 \, mg \, kg^{-1}$ potassium), and probably a better structure and water-holding capacity, it is unsurprising that those plots to which soil had been applied provided more niches for establishment, and thus developed a more species-rich sward. Nonetheless, despite the minimal nutrient status, limestone without soil was sufficient for some species to establish.

The strong performance of *Trifolium pratense* and *Anthyllis vulneraria* on the more fertile soil plots, plus the association of *L. corniculatus* with this treatment concurs with Richardson & Evans (1986) who recorded an increase in the growth of legumes when fertiliser was added to magnesian limestone. The good performance of *Anthyllis vuleraria* on the limestone rock + flour treatments reflects the fact that legumes are among the first colonisers of quarries (Bradshaw et al. 1982). It remains unclear why this species should be absent from the bare rock alone treatment.

If the fertility on the sown soil plots is still too low for many species to grow well, the dominance of these two legumes may increase fertility sufficiently in the future that other species will increase their growth rate, without the need for additional fertilisers. However, factors other than fertility may be restricting the growth of most species. The association of *C. flacca* and *F. ovina* with the bare limestone rock/flour treatments suggests that, for some species, these substrates may provide more favourable conditions than soil, even with minimal fertility. However, since growth was poor for most species on all treatments, species associations with either soil or limestone could easily change as growth improves.

Despite precautions, rabbits did enter the site, but the vegetation was probably insufficiently developed to be of much interest to them as a food source. Localised enrichment via their dung (Davis et al. 1993) is possible; but no areas of enrichment were noticed.

In addition to fertility, one should also consider the effect of species-species interactions on growth. For example, *Rhinanthus minor* is a hemi-parasite and may be negatively affecting the performance of its hosts (Davies & Graves 1998), and thus affecting species distribution in the field.

4.4   Practical implications

At this early stage in development, the species richness and cover were greatest when plots with 10 cm soil and 10 cm limestone flour on either 15 or 30 cm limestone rock were sown with the harvested seed mix, and the resulting sward contained numerous desirable species. However, the greatest species richness is not necessarily the ultimate target since that could include a sward with many highly competitive and few desirable species. It is much too early to say which substrate combination is best for the creation of *Sesleria albicans – Scabiosa columbaria* grassland, since they all result in vegetation that matches poorly to the community. However, the sowing of harvested seed will be essential for any rapid revegetation scheme. Despite the presence of numerous undesirable weed species, their cover remained small, and weeding is probably not required until after two years, if it is to be considered at all. Those species that flowered can be expected to increase their cover of the plots in the next season.

It is likely that a second seed sowing, focusing on species missed during the first seed collection, coupled with more time would lead to development of a better sward. Except for *T. pratense, S. minor* and (particularly) *A. vulneraria*, some form of fertilisation of both soil and limestone plots could also be beneficial in order for plants to develop past the juvenile stage. However, if fertiliser is used, the rate should be kept low and consist of slow release granules, in order that strong competitors such *as H. lanatus* (which was present in the harvested seed) are not encouraged at the expense of desirable species such as *S. albicans* or *H. nummularium*, that may enter the sward either naturally or by sowing at a later stage. Alternatively, the legumes were the most common species, and it may be that they will increase nitrogen levels sufficiently to enable other species to improve growth, without the aid of fertiliser.

ACKNOWLEDGEMENTS

The authors would like to thank Dr. Roger Smith, University of Newcastle, who co-supervised this project, for his contribution to the preparation of this paper. The authors extend thanks to W & M Thompson (Quarries) Ltd for permission to establish, and assistance with the construction of, the field experiment at Bishop Middleham Quarry. Thanks also to English Nature and Lafarge Aggregates for permitting the collection of seed from Thrislington NNR. The work was financed by a CASE postgraduate studentship (to JDR) from the Ministry of Agriculture, Fisheries and Food (subsequently Department of the Environment, Farming and Rural Affairs), with additional contributions from W & M Thompson (Quarries) Ltd.

REFERENCES

Bradshaw, A.D., Marrs, R.H. & Roberts, R.D. 1982. Succession. In B.N.K. Davis (ed.), *Ecology of Quarries*: 47–53. Cambridge: Institute of Terrestrial Ecology.

Booy, M.E. 1975. Botanical studies in land reclamation after magnesian limestone quarrying. MSc thesis, University of Newcastle upon Tyne.

Cullen, W.R., Wheater, C.P. & Dunleavy, P.J. 1998. Establishment of a species-rich vegetation on reclaimed limestone quarry faces in Derbyshire, UK. *Biological Conservation* 84: 25–33.

Davies, D.M. & Graves, J.D. 1998. Interactions between arbuscular mycorrhizal fungi and the hemiparisitic angiosperm *Rhinanthus minor* during co-infection of a host. *New Phytologist* 139: 555–563.

Davis, B.N.K. & Coppeard, R.P. 1989. Soil conditions and grassland establishment for amenity and wildlife on a restored landfill site. In G.P. Buckley (ed.), *Biological Habitat Reconstruction*: 221–231. London: Bellhaven Press.

Davis, B.N.K., Lakhani, K.H., Brown, M.C. & Park, D.G. 1985. Early seral communities in a limestone quarry: an experimental study of treatment effects on cover and richness of vegetation. *Journal of Applied Ecology* 22: 473–490.

Davis, B.N.K., Lakhani, K.H. & Brown, M.C. 1993. Experiments on the effects of fertilizer and rabbit grazing treatments upon the vegetation of a limestone quarry floor. *Journal of Applied Ecology* 30: 615–628.

Dixon, J.M. & Hambler, D.J. 1984. An experimental approach to the reclamation of a limestone quarry floor: the first three years. *Environmental Conservation* 11: 19–28.

Dixon, J.M. & Hambler, D.J. 1993. Wildlife and reclamation ecology: rabbit middens on seeded limestone quarry spoil. *Environmental Conservation* 20: 65–73.

Doody, J.P. 1980. Grassland. In T.C. Dunn (ed.), *The Magnesian Limestone of Durham County*: 44–50. Durham: Durham County Conservation Trust.

Dunn T.C. (ed.) 1980. *The Magnesian Limestone of Durham County.* Durham: Durham County Conservation Trust.

Hambler, D.J., Dixon, J.M. & Cotton, D.E. 1990. The relative potentials of six grass cultivars for rehabilitation and stabilisation of a limestone quarry spoil bank. *Environmental Conservation* 17: 149–156.

Hambler, D.J., Dixon, J.M. & Hale, W.H.G. 1995. Ten years in the rehabilitation of spoil: appearance, plant colonists and the dominant herbivore. *Environmental Conservation* 22: 323–334.

Hedley, S., Clifton, S. & Mullinger, S. 1997. *The Durham magnesian limestone natural area.* Peterborough: English Nature.

Hill, M.O. 1996. *TABLEFIT – version 1.0, for identification of vegetation types.* Huntingdon: Institute of Terrestrial Ecology.

Humphries, R.N. 1982. The establishment of vegetation on quarry materials: physical and chemical constraints. In B.N.K. Davis (ed.), *Ecology of Quarries*: 55–61. Cambridge: Institute of Terrestrial Ecology.

Janssens, F., Peeters, A., Tallowin, J.R.B., Bakker, J.P., Bekker, R.M., Fillat F. & Oomes, M.J.M. 1998. Relationship between soil chemical factors and grassland diversity. *Plant and Soil* 202: 69–78.

Land Use Consultants. 1992a. *Amenity reclamation of mineral workings. Annual report.* London: HMSO.

Land Use Consultants 1992b. *The use of land for amenity purposes.* London: HMSO.

Park, D.G. 1982. Seedling demography in quarry habitats. In B.N.K. Davis (ed.), *Ecology of Quarries*: 32–41. Cambridge: Institute of Terrestrial Ecology.

Richardson, J.A. 1956. The ecology and physiology of plants growing on spoil heaps, in clay pits and quarries in the coal measure and magnesian limestone areas of County Durham. PhD thesis, University of Durham.

Richardson, J.A. & Evans, M.E. 1986. Restoration of grassland after magnesian limestone quarrying. *Journal of Applied Ecology* 23: 317–332.

Riley, J.D., Craft I.W. & Rimmer D.L. 2003. Restoration of magnesian limestone grassland: optimizing the time for seed collection by vacuum harvesting. *Restoration Ecology* (submitted).

Rodwell, J.S. (ed.) 1992. *British Plant Communities Vol. III. Grasslands and Montane Communities.* Cambridge: Cambridge University Press.

ter Braak, C.J.F. & P. Smilauer. 1998. *CANOCO reference manual and users guide to CANOCO for Windows: software for canonical community ordination (ver. 4).* Ithaca: Microcomputer Power.

Reclamation Research:
Achievements and Challenges

Land Reclamation – Moore, Fox & Elliott (eds)
© *2003 Swets & Zeitlinger, Lisse, ISBN 90 5809 562 2*

Passive *in situ* remediation of acidic mine waste leachates: progress and prospects

P.L. Younger
Hydrogeochemical Engineering Research & Outreach (HERO), School of Civil Engineering and Geosciences, University of Newcastle, Newcastle Upon Tyne, UK

ABSTRACT: The reclamation of former mining sites is a major challenge in many parts of the world. In relation to the restoration of spoil heaps (mine waste rock piles) and similar bodies of opencast backfill, key challenges include (i) the establishment of stable slopes and minimization of other geotechnical hazards (ii) developing and maintaining a healthy vegetative cover (iii) managing the hydrological behaviour of the restored ground. Significant advances have been made over the past four decades in relation to all four of these objectives. One of the most recalcitrant problems is the ongoing generation and release of acidic leachates, which typically emerge at the toes of (otherwise restored) spoil heaps in the form of springs and seepage areas. Such features are testament to the presence of a "perched" groundwater circulation system within the spoil, and their acidity reflects the continued penetration of oxygen to zones within the heaps which contain reactive pyrite (and other iron sulphide minerals). Two obvious strategies for dealing with this problem are disruption of the perched groundwater system and/or exclusion of oxygen entry. These strategies are now being pursued with considerable success where spoil is being reclaimed for the first time, by the installation of two types of physical barrier (dry covers and water covers). However, where a spoil heap has already been revegetated some decades ago, the destruction of an established sward or woodland in order to retro-fit a dry cover or water cover is rarely an attractive option for dealing with the "secondary dereliction" represented by ongoing toe seepages of acidic leachates. More attractive by far are passive treatment techniques, in which the polluted water is forced to flow through reactive media which serve to neutralize its acidity and remove toxic metals from solution. A brief historical review of the development of such systems reveals a general progression from using limestone as the key neutralizing agent, through a combined use of limestone and compost, to systems in which almost all of the neutralization is achieved by means of bacterial sulphate reduction in the saturated compost media of subsurface-flow bioreactors. In almost all cases, these passive treatment systems include an aerobic, surface flow wetland as the final "polishing" step in the treatment process. Such wetlands combine treatment functions (efficient removal of metals from the now-neutralized waters down to low residual concentrations, and re-oxygenating the water prior to discharge to receiving watercourses) with amenity value (attractive areas for recreational walking, bird-watching etc) and ecological value.

1 INTRODUCTION

1.1 *Abundant mine wastes*

The reclamation of former mining sites is a major challenge in many parts of the world. Three principal forms of mining waste typically require rehabilitation before the land they occupy can be safely re-used. These include spoil heaps ("mine waste rock piles" in international parlance), bodies of opencast backfill, and redundant tailings dams. In this paper, for the sake of brevity, the focus is on the first two, albeit many of the comments made are applicable also to the reclamation of tailings dams. The growing need for adequate reclamation strategies for mine wastes is underlined by the fact that some 70% of all material moved by mining worldwide at the present time is mineral waste (overburden, interburden and mineral processing residues) (Hartman 1987). While some types of mining operation (particularly opencast coal workings) are predicated on progressive restoration of waste rock masses during the course of mining, where the ratio of overburden to ore falls into single figures the double-handling of material which progressive reclamation entails can be financially unattractive. This is particularly so, for instance, in cases where low grade ores are worked by open-pit methods (e.g. the copper porphyry deposits of Chile): virtually all the material extracted from the pit will be processed outside of the pit, even though only a few percent of the total mass will end up as saleable product. Transporting the remainder some distance back to the pit is generally not undertaken in the absence of binding legal conditions. In the case of deep mines, return of mineral wastes to the subsurface is rarely feasible without major re-design of the mine haulage system, which would usually be on such a scale as to render the mine uneconomic. (An interesting exception to this is currently under development at the Boulby Potash Mine in Cleveland, UK, where slurried processing wastes are being returned to old voids deep beneath the North Sea.)

1.2 Drivers for mine waste reclamation

There are three principal reasons why it is usually desirable to reclaim large surface depositories of mining wastes:

(i) *Protection of human health*. The most usual driver in this regard is minimizing health risks associated with wind-blown dust arising from bare spoil. These dusts can pose a substantial silicosis risk to nearby residents even where the particles themselves are not chemically toxic. Less common drivers relate to the dramatic hazards to life and limb associated with geotechnically unstable spoil heaps. These hazards were most infamously manifest in the Aberfan disaster of Friday October 21st 1966, when a substantial part of a large, steep-sided colliery spoil heap failed catastrophically, and slid down a mountainside into the mining village of Aberfan (near Merthyr Tydfil in South Wales), where it engulfed several homes and part of Pantglas Junior School, claiming 144 lives, 116 of whom were schoolchildren (McLean & Johnes 2000).

(ii) *Landscape improvement*. Many members of society deem it desirable to replace the stark, "badlands" topography of weathered, unvegetated spoil with attractive, biologically productive landforms, which may then be useful for agricultural and/or forestry. It is worth noting, however, that significant examples are known in which local residents actually object to the re-shaping of former spoil heaps, on the grounds that it removes local landmarks and removes some of the most striking reminders of a heritage of which many people are proud. Such objections have actually culminated in the granting of conservation status to certain spoil heaps, including the Kilton spoil heap in the Cleveland Iron Orefield and several of the "pink bings" of the Lothian oil shale fields of central Scotland.

(iii) *As a contribution to minimizing water pollution associated with polluted runoff/spoil leachates*. This is the particular topic of this paper. Of the many thousands of former spoil heap sites in the UK, only around a hundred or so are known to give rise to serious pollution of adjoining watercourses. For instance, in a breakdown of water pollution problems associated with abandoned mine-sites in Scotland, Younger (2001) noted that only 2% of the total volume of polluted drainage leaving such sites was sourced from the spoil heaps (as opposed to flooded mine workings). Of course this volumetric breakdown is not the entire story, for spoil heaps, being generally shallower than mine voids, tend to be far more exposed to the Earth's atmosphere than are the deeper portions of flooded voids, and the sulphide minerals which they contain are therefore more likely to be subjected to oxidation, leading to the release of acidity. This means that mine waste depositories "punch above their weight" in the degree to which they release pollutants: many spoil heap leachates are acidic, whereas many deep mine discharges revert to being alkaline in the long-term. Thus in terms of the loadings (tonnes/day) of contaminants released by the two types of source, we obtain the following breakdown for the Scottish example (Younger 2001, 2002a):

- flooded mine voids: 72%
- old tailings deposits/waste rock piles: 28%

Acidic mine waste leachates commonly exhibit pH values in the range 3–4.5. In extreme circumstances, pH can drop as low as 1.6, at which level it is typically buffered by equilibrium with one or other members of the jarosite family of minerals (Saaltink et al. 2002). Typical metal contaminants associated with acidic mine waste leachates include (in order of typical decreasing abundance): Fe, Al, Mn, Zn, Cu, Ni, Cd. With the exception of Mn (which is problematic only where the receiving watercourse drains towards a public water supply intake) all of the above metals are highly toxic to aquatic invertebrates and fish when present at the elevated concentrations typical of mine leachates. Non-metal contaminants are dominated by sulphate (SO_4), which although barely toxic to aquatic fauna even at very high concentrations, does contribute substantially to the overall salinity of the water, which may render it unfit for other uses (such as for irrigation and domestic supply). Arsenic (a metalloid which is mobile as an oxyanion at near-neutral pH) is rarely a significant contaminant in *acidic* mine waste leachates, but can be present at toxic concentrations in some circum-neutral metal mine spoil leachates (such as are found locally in Cornwall). For further information on the discussion on the typical concentrations and ecological impacts of these contaminants, the reader is referred to Kelly (1988), Jarvis & Younger (1997) and Younger (2002c).

1.3 Spoil reclamation and leachate remediation

Activities involved in the reclamation of spoil include (i) re-shaping and re-grading to establish stable slopes, thus minimizing geotechnical hazards (ii) developing and maintaining a healthy vegetative cover, and (iii) managing the hydrological behaviour of the restored ground. The motivations for each of these activities correspond roughly to their respectively numbered "drivers" listed in Section 1.2. Nevertheless, all three of them make some contribution to the alleviation of acidic leachate pollution, for instance: (i) Where spoil erosion is minimized by re-shaping, the formation of deep gullies (which often allow the introduction of atmospheric oxygen to the spoil, promoting acidity release) is prevented. (ii) Unreclaimed spoil often absorbs all of the rainfall which it receives, and returns it to the environment in a polluted state. Indeed, the reason why many spoils remain bare in the absence of deliberate reclamation efforts is often not so much to do with the toxicity of mine wastes as with the fact that spoil fragments (and thus the pores between them) are so coarse that all incoming rainfall immediately soaks into the spoil, effectively leading to water shortage at the spoil surface, precluding plant colonization (Jenkins, pers comm.). Thus the establishment of a dense vegetal cover, and the associated development over time of a humus layer (soil A zone), results in greater transpirational losses from the spoil than typically occurred in the pre-reclamation state. (iii) Efforts to further manipulate the hydrological behaviour of mine wastes are almost always focused on pollution minimization, and will be discussed in detail below.

2 PASSIVE PREVENTION OF POLLUTANT RELEASE

2.1 Logic behind techniques employed

The European Commission's current research project PIRAMID (which concerns the prevention and/or treatment of acidic mine drainage by passive means; see www.piramid.org) defines "passive prevention of pollutant release" as "...the surface or subsurface installation of physical barriers (requiring little or no long-term maintenance) which inhibit pollution-generating chemical reactions (for instance, by permanently altering redox and/or moisture dynamics), and/or directly prevent the migration of existing polluted waters...". Examples of such approaches include the use of wet or dry covers on waste rock piles, the grouting of permeable pathways to prevent rapid migration of contaminants and the use of ground solidification techniques.

While "high-tech" methods such as the injection of bactericides into spoil heaps have been investigated in the past for the minimisation of acidity release (this

release is bacterially mediated), these methods are very expensive and have never been found to work for more than 6 months before repeated treatment is necessary (for a review, see Younger et al. 2002a). The focus of attention in recent years has thus been on simpler methods which effectively exclude oxygen from pyrite surfaces (thus preventing bacterial acid-generating activity), such as the use of dry covers and water covers to (e.g. Gustafsson et al. 1999).

2.2 Dry covers

The access of both atmospheric oxygen and moisture to pyrite surfaces within spoil heaps can be severely limited by the installation of low-permeability covers, now generally known as "dry covers" (Gustafsson et al. 1999). Most so-called "dry covers" used in the mining industry are variations on the clay cap technology commonly used in many contaminated land remediation schemes, although PFA-based materials have also been used to form dry covers in some cases (Younger et al. 2002a). Clay caps are usually at least 0.5 m thick, and are generally compacted such that they retain permeabilities less than 10^{-4} m/d at all points. The application of the term "dry cover" to clay caps is in some ways misleading, since one of the main reasons simple clay caps prevent oxygen ingress to the underlying spoil (at least in cold, humid regions) is that their pore networks are almost always saturated with water (or nearly so). Diffusion of oxygen occurs far less readily in water-filled pores than in air-filled pores (Nicholson et al. 1989).

In cold, humid regions, a single-layer clay cap may be all that is required to achieve minimal ingress of water and oxygen to underlying mine wastes. However, in temperate regions, adequate dry covers will generally also include a coarse grained "capillary break" layer between the clay cap and the underlying spoil, which serves to prevent upward migration of pore-waters in response to seasonal surface layer desiccation. In arid tropical areas, dry covers need to be further engineered to incorporate high permeability layers above the clay cap, in order to allow for storage and later release of storm waters arising from high-intensity convective storms (e.g. Durham et al. 2000). Such storage-and-release covers prevent the erosion which would occur if the bare surface of the clay cap were exposed at the surface.

2.3 Water covers

As in clay-based covers in cold, humid regions, the use of water covers to inhibit the leaching of acidity from mine wastes is predicated on the slower rate of oxygen diffusion through water than through air. The design of water covers is deceptively simple: all that is required is that some form of impoundment is arranged to ensure that a minimum depth of water is maintained above the surface of the flooded waste materials at all times. Younger et al. (2002a) recommended a minimum water depth of one metre, because where water covers are shallower than one metre (i) they are usually so well-mixed that the oxygen content at the sediment surface is little less than at the water surface; indeed, on these grounds, there are reasons to advocate water depths of several metres if this is consistent with other site constraints (access, safety etc). (ii) the surface of the underlying tailings is prone to agitation by wind-blown waves of the magnitude which can develop in impoundments with areas up to several hundred hectares (i.e. of sizes appropriate to mine waste management situations). Re-suspension of tailings can result, leading to complications in water quality management.

In practice the topographic relief and other constraints of a given site (e.g. availability of runoff, intensity of evaporation) may make a 1 m minimum depth difficult to attain/maintain. Compromise solutions inevitably develop, with water covers as shallow as 0.3 m being developed in some cases. While re-suspension of tailings can be an issue with such shallow covers, observations of oxygen penetration into waste sediments below them show that they are still very effective in minimizing sulphide oxidation (e.g. Li et al. 2000).

In general, relatively few mine sites have topography ideally suited to the inexpensive development and low-cost long-term maintenance of water covers. They are therefore less common in practice than dry covers, but are indubitably effective where conditions permit their use.

3 PASSIVE TREATMENT OF ACIDIC LEACHATES

3.1 Acidic vs neutralized leachates

Before proceeding, it is worth emphasizing that although this paper focuses on the treatment of acidic leachates, once a given acidic leachate has been rendered neutral, techniques which are suitable for the primary treatment of naturally neutral leachates may then be employed to complete the overall treatment of the water. However, conciseness precludes an extensive review here of passive treatment technologies applicable to neutralized leachates. The reader seeking such information is directed to the recent major synthesis of Younger et al. (2002a).

3.2 Active vs passive treatment

The emergence of passive mine water treatment technologies over the last decade or so, first in the USA

(Hedin et al. 1994) then in the UK (Younger 1997, 2000) has been one of the most significant developments in mining environmental engineering in living memory. Until the early 1990s, the only "proven technologies" for the treatment of polluted mine waters were variants of conventional wastewater treatment technologies, involving extensive use of electricity for pumping, mixing etc, and the ongoing addition of neutralizing agents, oxidants, flocculants and other reagents (see Younger et al. 2002a for a recent review). While there is nothing intrinsically wrong with such technologies, they are characterized by relatively high operating expenditures (opex). On active mine sites, this opex might represent no more than mean marginal additions to existing contracts covering power supply and processing reagent delivery. However, on abandoned mine sites, active treatment has the disadvantage of being the only process requiring the perpetuation of such costs and associated site manning and security measures.

Passive treatment technologies were developed specifically to deliver solutions which have low opex requirements, albeit the initial capital expenditure (capex) may equal or exceed that of an equivalent active treatment system. With low opex requirements, a passive treatment system can run unattended for long periods of time, with only occasional maintenance being required. Thus the PIRAMID project defines "passive treatment" as an improvement in water quality which is "achieved using constructed (or appropriated, natural) gravity-flow systems, in which all treatment processes use only naturally-available energy sources (such as topographical gradient, microbial metabolic energy, photosynthesis and chemical energy), and which require only infrequent (albeit regular) maintenance to operate successfully over their design lives". Passive treatment technologies currently include wetland-type systems, subsurface reactive barriers, and a burgeoning array of gravity-flow geochemical reactors.

3.3 Geochemical principles of passive treatment approaches for acidic leachates

The most obvious natural process which serves to neutralize acidic waters is carbonate dissolution, the process which is responsible for the spectacular caves found in most of the limestone regions of the world. While carbonate dissolution is not sufficiently rapid to be attractive in active treatment applications, for passive treatment systems it has an ideal combination of reactivity (yielding as much as 5 meq/l of alkalinity in 14 hours retention time; Hedin et al. 1994) and convenience (it is far easier and far less hazardous to manipulate than typical "active" neutralizing agents). In the early days of passive treatment, limestone was almost exclusively used to generate alkalinity in acidic waters, and it continues to make an important contribution to overall neutralization in most modern passive treatment plants (Younger et al. 2002a). It is important to note that the carbonate mineral of choice for passive treatment applications is calcite, for dolomite dissolves more slowly (and also, apparently, incongruently under conditions found in passive treatment systems) and other common carbonates such as ankerite and siderite release sufficient iron into solution that they effectively have no net-neutralization capacity.

A second natural neutralization process widely harnessed in passive treatment is dissimilatory bacterial sulphate reduction. In this process, which is only significantly operative in the complete absence of oxygen, the SO_4 which is ubiquitous in mine leachates is reduced by certain (widespread) strains of bacteria. Their action releases bicarbonate alkalinity (HCO_3) into the water, which acts to neutralize acidity, and also generates dissolved sulphide, which reacts with certain contaminant metals (especially Fe^{2+}, Zn, Cu and Ni) to form sulphide minerals. These are highly insoluble as long as anoxic conditions are retained. Other metals, which do not readily form sulphides, may also be removed under sulphate-reducing conditions, by hydrolysis (e.g. Al forms $Al(OH)_3$ as pH rises above 4.5) or carbonate precipitation (e.g. $MnCO_3$ precipitates under anoxic conditions at near-neutral pH).

3.4 Passive unit treatment processes for acidic leachates: typology and evolution

In this section, the varieties of unit processes currently used for treating acidic mine waste leachates are described, more-or-less in order of evolution. For detailed referencing of the original descriptions of these unit processes, the reader is referred to Younger et al. (2002a).

The earliest passive unit process used for acidic leachate treatment harnessed calcite dissolution in the form of an "anoxic limestone drain" (ALD) in which the acidic water is forced to flow through a buried bed of limestone gravel (Fig. 1). With a retention time in excess of 10 hours, waters with less than about 3 meq/l total acidity would be rendered neutral and

thus suitable for removal of residual iron etc in simple surface-flow aerobic reed beds.

The simplicity of ALDs belies some of the complications associated with their practical implementation, for where the polluted water contained dissolved concentrations of more than 1 mg/l of oxygen, aluminium and/or ferric iron (Fe^{3+}), then the drain would tend to clog and cease to function within 6 months. While a substantial number of mine waters in the USA were found to be amenable to ALD treatment (typically waters with pH around 5; Kleinmann, pers. comm.) this has so far not proven to be the case in Europe (Younger 2000).

Even where ALD treatment is technically feasible, if the water contains much more than about 3 meq/l of total acidity, one ALD will not be sufficient to neutralize the water completely, and a second phase of alkalinity generation will be necessary. As the water leaving the aerobic reed beds (Fig. 1) will now contain more than 1 mg/l of dissolved oxygen (DO), a second ALD is not feasible. This observation motivated engineers to develop "successive alkalinity producing systems" (Kepler & McCleary 1994), in which the effluent from an aerobic reed bed in a setting such as Figure 1 is fed through a saturated compost layer, which removes DO and converts any dissolved Fe^{3+} back to the Fe^{2+} form, which means that the water can then pass through a subsequent "ALD" layer without causing clogging. The resultant arrangement is a vertical-flow wetland-type system with a bed off compost overlying a bed of limestone gravel (Fig. 2). Such units are days generally referred to as "Reducing and Alkalinity Producing Systems" (RAPS).

Parallel with the developments of ALD technology and the emergence of the RAPS concept, observations of some established wetland systems receiving acidic waters led to the deduction that bacterial sulphate reduction was a significant source of alkalinity. Attempts to mimic this effect led to the creation of so-called "compost wetlands" (Fig. 3) in which water flows through and over thick substrates of organic matter which promote dissimilatory bacterial sulphate reduction (Hedin et al. 1994; Younger et al. 1997; Jarvis and Younger 1999). Suitable organic substrates are many and varied. Compost made from the manure of large herbivorous mammals are invariably

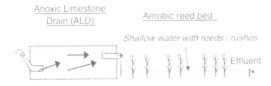

Figure 1.   Schematic cross-section of a typical application of an anoxic limestone drain, neutralizing acidic leachate which is then treated in an aerobic wetland.

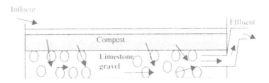

Figure 2.   Schematic cross-section of a typical RAPS unit, used to treat acidic leachates which initially contain more than 1 mg/l of DO, Fe^{3+} or Al (after Younger 2000).

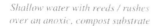
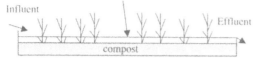

Figure 3. Schematic cross-section of a compost wetland, used to treat acidic leachates where there is insufficient head for installation of a RAPS unit.

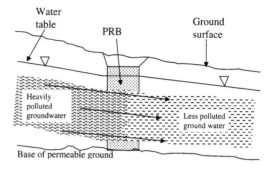

Figure 4. Schematic cross-section of a permeable reactive barrier (PRB) treating contaminated mine site leachate within a groundwater flow system (after Younger 2000).

suitable, with horse manure and straw, cow byre screenings and even llama droppings all performing well in field trials. In addition, composted municipal waste, bark mulch and spent mushroom compost have all been used in full-scale applications. The similarities between compost wetlands and the compost layer of typical RAPS units is more than coincidental, and the two unit processes may in some ways be regarded as variations on the same theme. Because of the fact that all the leachate is forced to move through the compost substrate in a RAPS unit, whereas there is substantial scope for the development of by-pass flows in a compost wetland, the former are far more efficient. RAPS require typically less than half (sometimes only 20%) of the land area an equivalent compost wetland to occupy in order to achieve the same level of treatment. Hence RAPS would always be preferred over compost wetlands where the site geometry permits the construction of either type of system.

As with ALDs, it is common practice for RAPS and compost wetlands to be followed downstream by aerobic reed beds, which serve not only to remove residual dissolved iron but also to raise the DO of the effluent and diminish any BOD which it has acquired during its passage through the compost.

All of the unit processes described so far are appropriate for circumstances in which the polluted leachate emerges at the ground surface in the form of a spring or seepage zone. Where the leachate flows through the subsurface, reaching its target without ever surfacing first, then alternative passive treatment strategies are called for. Using the same principles as ALDs and RAPS, the *in situ* treatment of subsurface leachates requires the installation of permeable reactive barriers (PRBs) (Fig. 4).

In addition to organic matter (Benner et al. 1997; Amos & Younger 2003), as is also used in RAPS and compost wetlands, a range of other media have been tested for incorporation in mine leachate treatment PRBs. These include zero-valent iron, zeolites, carbonate minerals and blast furnace slags (Younger et al. 2002a). All of these substrates are characterised by distinct reactive tendencies in relation to acidic mine waters. For instance, zero-valent iron leads to the establishment of highly reducing conditions,

which can be useful in trapping redox-sensitive metals such as uranium. Zeolites trap metals by sorptive processes. Carbonates and blast furnace slag tend to raise the pH, favouring both sorption and hydrolysis of metals to form (hydr)oxide solids.

The kinetics of these various processes differ amongst themselves, but are all sufficiently rapid that they are amenable to deployment in passive treatment reactors in which waters are typically resident for a few hours to a few days.

In terms of the current state-of-the-art, key areas of active unit process development include:

(i) modifications of RAPS systems to cope with elevated Al concentrations (Younger 2002b).
(ii) investigation of sewage sludge filter cake, paper mill wastes and other organic media as alternatives to composted animal wastes in RAPS and compost wetland substrates (see Section 4.2.10).
(iii) the use of thoroughly mixed limestone-compost substrates in RAPS instead of the two layer arrangement shown in Figure 2 (Younger et al. 2002b).
(iv) polishing of the effluents of RAPS and/or compost wetlands for removal of residual metals to very low concentrations, by using bone charcoal as a sorbent (see Section 4.2.1).

Some of these innovations are currently being evaluated at full-scale on UK sites, as will become evident in the following section.

4 UK EXPERIENCES 1994–2002

4.1 Overview

Excepting the USA where the technology first emerged (Hedin et al. 1994), the UK has more full-scale passive mine water treatment systems than any other country (Younger 2000). While the majority of

Figure 5. Locations of passive *in situ* remediation systems treating acidic mine spoil leachates in the UK. Details for each site are given in the text. Key: A = Aspatria; B = Benhar Bing; Bo = Bowden Close; D = Deerplay; Do = Dodworth; N = Nailstone; O = Oatlands; Q = Quaking Houses; R = Renishaw Park; S = Shilbottle.

these systems are treating water from flooded mine voids (which are predominantly of neutral pH, and are in any case beyond the scope of this paper), a substantial canon of experience has been building up in relation to the passive *in situ* remediation of acidic mine waste leachates. Figure 5 shows the locations of systems in this category constructed to date in the UK. In the following paragraphs, summaries of experiences gained in creating these systems are presented, in roughly chronological order of the commissioning of each system. The manner in which the uptake of the technology has proceeded, in terms of both process selection and scale of system installed, can therefore be clearly observed, paving the way for a closing discussion of likely future trends.

4.2 Summaries of UK passive systems treating acidic mine waste leachates

4.2.1 Dodworth, South Yorkshire
The leachate at Dodworth emanates from the former spoil heap and buried coal washery lagoons of Dodworth Colliery (1862–1984), ownership of which passed from British Coal to Barnsley Metropolitan Borough Council (BMBC) as part of a package of land acquired for re-development to industrial units. The leachate from the spoil heap is generally of low pH (3–4.5), and contains up to 150 mg/l Al, 3 mg/l Zn, 70 mg/l Fe and 2.2 mg/l Ni. The first passive system at Dodworth (and in the UK) was installed by BMBC in the summer of 1993 (Bannister 1997). This was a simple pilot-scale system comprising an ALD feeding a 50 m^2 aerobic reed bed. Although this proved reasonably successful, regulatory pressure at the time favoured a shift in strategy towards a semi-active approach involving alkali dosing and sedimentation followed by two aerobic reed beds. Subsequent monitoring revealed that the lack of a passive alkalinity generating process, and a simple lack of sufficient area for treatment, were hindering the attainment of desired effluent concentrations of a range of ecotoxic metals (Fe, Al, Cu and Ni). At the time of writing BMBC are considering the possibility of installing a RAPS unit at the site, with possible downstream polishing of Ni, Cu, Cd and Zn concentrations using bone charcoal sorption reactors.

4.2.2 Quaking Houses, County Durham
Acidic leachate, containing up to 35 mg/l Al, 40 mg/l Fe, 5 mg/l Zn and a pH often around 4.3, has been seeping from the spoil heap of the former Morrison Busty Colliery, County Durham, since about 1985. In the absence of any identifiable "problem owner" the residents of the former coal mining village of Quaking Houses, whose local stream was badly affected by the leachate, developed a partnership with the University of Newcastle to develop an innovative, low-cost solution to the pollution problem. This led in February 1995 to the installation of Europe's first compost wetland at the site (Younger et al. 1997). This system, which was of pilot-scale, was manually installed and was monitored for 18 months, yielding an impressive 11 g/d/m^2 removal rate for total acidity. Encouraged by these findings, philanthropic funding was obtained (most notably from the Northumbrian Water Kick-Start Fund) and a full-scale compost wetland was commissioned in November 1997 (Jarvis & Younger 1999). Removal rates since then have averaged around 10 g/d/m^2 total acidity, and the local stream has been entirely restored to its pre-contamination condition. Following installation of the wetland, investigative works on the spoil heap identified a pyrite oxidation "hot spot" which is responsible for generating most of the acidity entering the wetland. Localised installation of a simple boulder clay cap in the vicinity of this hot spot has successfully reduced acidity generation by as much as 80%. A thorough description of the Quaking Houses project has recently been published by CL:AIRE (2002).

4.2.3 Nailstone, Leicestershire

Nailstone Colliery was one of the last working deep mines in the Leicestershire Coalfield. After first merging with Bagworth Colliery in the 1980s, all underground working in the area ceased in 1991. After demolition of surface buildings and sealing of shafts and 1-in-4 conveyor decline at the Nailstone site, a substantial area of acid-generating spoil remained, together with an extensive former coal stocking area. The site drainage was complex, including open channels and french drains beneath the coal stocking area (which led the water to some sedimentation ponds), and surface ditches and erosional gullies draining the outer slopes of the spoil heaps into the River Sence. In the mid-1990s the site was acquired by the property management division of a firm (later bought by Viridor Ltd, who now own the site) which was interested in developing an integrated site remediation and regional waste management facility. Amongst the first steps taken by the firm was the installation in the summer and autumn of 1997 of three simple passive treatment facilities. The first of these was essentially a linear aerobic wetland, which serve to strip suspended aluminium from previously acidic water, which had become neutralized before discharging at the ground surface. This greatly improved the River Sence in a matter of days. A number of small surface flow compost wetlands were also installed, (totalling about $2500\,m^2$ in area), which successfully lowered dissolved iron concentrations from 46 to 15 mg/l, dissolved aluminium concentrations from 47 to 9 mg/l, and raised pH from 4.9 to 5.6.

4.2.4 Benhar Bing, West Lothian.

The spoil heap ("bing" in Scottish dialect) associated with the former Benhar Ironstone Mine in central Scotland has been a persistent source of acidic leachate pollution to headwaters of the River Almond since the early 1900s. Improvements in water quality elsewhere in the catchment finally made clean-up of the Benhar Bing seepage essential to the further improvement of the local environment. Reclamation of the bing was initiated in 1996, with sewage sludge amendments being used to develop a soil-forming material which has successfully supported the establishment of woodland cover (birch and alder) on the surface of the bing. In 1997, leachate which continued to seep from the toe of the bing was diverted into a settling pond and a 0.4 ha compost wetland, which has a 0.5 m-thick substrate of mushroom compost (Heal & Salt 1999). Monitoring of the performance of this compost wetland revealed removal efficiencies on the order of 33% for total acidity, and 20–40% for Fe, Al, Mn and SO_4. Effluent from the wetland is still acidic (pH 2.8, total acidity around 6 meq/l) with significant Fe (120 mg/l) and Al (44 mg/l) (Heal & Salt 1999). These concentrations are very high when compared with those achieved in other compost wetland systems (e.g. Quaking Houses and Nailstone). This may be due to the fact that a design water depth of 0.3 was implemented at Benhar Bing, which is substantially deeper than in the more successful compost wetlands (where water depths rarely exceed 0.1 m). Deep water in a compost wetland hinders thorough exchange of solutes between the substrate and the entire water column, which is necessary for satisfactory functioning of such systems (e.g. Younger et al. 2002a).

4.2.5 Oatlands, Cumbria

Acidic drainage from the former Oatlands Colliery site in west Cumbria has long affected the River Keekle (a tributary of the River Ehen, which is a major salmonid fishery). For some years the precise source of the contamination (i.e. mine voids versus spoil leachate) was unclear, but site investigations finally revealed that spoil leachates are the source (Warner 1997). These contained as much as 85 mg/l Fe, with a pH typically around 4.0. Alongside drainage improvements on the former colliery site, designed to minimize leachate generation, an aerobic wetland was installed in September 1998. Although this wetland was soon successfully removing around 70% of the iron, the lack of a passive alkalinity-generating process within the system meant that pH also dropped, reaching 3.1 at the final effluent from the wetland (Younger 2000). Four years after establishment, however, treatment performance has improved (presumably by spontaneous development of an anoxic substrate) sufficient to ensure satisfactory water quality conditions in the River Keekle.

4.2.6 Renishaw Park, Derbyshire

Following the surface reclamation of the Renishaw Park Colliery spoil heap, acidic spoil leachate continued to drain from the toe of the heap at a part of the site which remained in Coal Authority ownership, closely adjoining a local watercourse. The spoil seepage was causing vivid ochre staining in a local watercourse, which not only degraded the ecology of the watercourse but also detracted from visual amenity in a site undergoing re-development as an up-market golf course. In November 1999, Europe's first permeable reactive barrier to treat acidic spoil leachate was installed at the site. Although this has been highly effective (the staining of the stream has ceased) flow leaves the PRB in a diffuse manner and has been impossible to monitor effectively, so that precise performance data have proven unobtainable at this site.

4.2.7 Bowden Close, County Durham

Three discrete spoil leachate discharges on the former Bowden Close Colliery site (near Crook in County Durham) together form one of the most prolific sources of pollution encountered to date in the former

Durham Coalfield. The leachates are generally strongly acidic (pH 4) with up to 80 mg/l Al, 100 mg/l Fe and elevated concentrations of Mn and Zn. Pilot-scale field tests undertaken by the University of Newcastle and Durham County Council in 1999–2000 demonstrated the feasibility of passively treating the Bowden Close leachates using RAPS technology (Younger 2002b). Based on these findings, a full-scale passive treatment system is currently under construction at the site. This incorporates two RAPS units, specifically designed to optimize Al^{3+} removal, and a polishing aerobic reed/rush wetland. With significant logistical and financial support from CL:AIRE and the BOC Foundation, this new system is set to become one of the most extensively and intensively monitored examples of its type anywhere in the world, and is thus confidently expected to yield important insights into (*inter alia*) the principal treatment mechanisms operative within RAPS (microbiological vs physicochemical etc). Results will be disseminated by CL:AIRE, who have adopted the project as their Technology Demonstration Project No 5.

4.2.8 Deerplay spoil heap, Lancashire

As part of its national rolling programme of remediating the most serious sources of mine water pollution in the UK, the Coal Authority implemented a major pump-and-treat solution on the site of the former Deerplay Colliery, high on the Lancashire-Yorkshire border. Although there was no polluting discharge from deep mine workings on this site, it was an opportune location at which to sink a well and pump deep mine water into a treatment system, causing enough drawdown to "switch off" a previously uncontrolled mine water discharge in a steep-sided valley (Black Clough) on the other flank of Deerplay Moor (in which a treatment system could not readily be constructed). As part of the site works at Deerplay, the Coal Authority decided to capture and treat acidic spoil leachate which had long caused localised degradation of the nearby stream. To this end, in summer 2001 the Authority installed a RAPS-type alkalinity generating unit (in which the compost-based bioreactor was separated in space from the limestone gravel bed) followed by an aerobic wetland. At the time of writing, extended periods of wet weather have hindered the site works necessary to complete the commissioning of this passive treatment system. However, early monitoring data (Jarvis & England 2002) show strong alkalinity generation (2.4 meq/l being added to the water) coupled with a sharp rise in pH (from 3.3 to 6.7) and substantial removal of total aluminium (from 14 down to 0.5 mg/l).

4.2.9 Shilbottle, Northumberland

From the early 19th Century until October 1982, Shilbottle Colliery worked a single seam of coal in the vicinity of Alnwick, Northumberland. Much pyritic shale was removed from the colliery during working, and deposited at surface to form a large spoil heap (locally termed the "Brass Heap" on account of the visible, brassy-coloured pyrite present in the shale). The extreme contaminant concentrations in the leachates emanating from the Brass Heap (pH 3–4; ≤1200 mg/l Fe; ≤680 mg/l Al; ≤300 mg/l Mn) qualify the Shilbottle site for the dubious distinction of being the worst example of its genre in England. Discharge of this leachate to the adjoining Tyelaw Burn has long resulted in spectacular pollution of the stream, and has also affected water quality to such an extent that it at times necessitates great efforts to maintain compliance with drinking water limits for manganese in a public supply abstraction from the River Coquet downstream. Mass-balance studies by the University of Newcastle demonstrated that around two-thirds of the total loading of contaminants leaving the spoil heap as leachate were flowing via subsurface pathways directly into the Tyelaw Burn, thus completely by-passing treatment which might have been afforded by aerobic reed-beds which were installed in the mid-1990s during the first phase of spoil reclamation. After a period of field and lab characterisation (Amos and Younger, 2003), a solution for this site was developed which involved diversion of the Tyelaw Burn and the installation of a PRB along the toe of the spoil heap, which feeds water into three oxidation ponds, whence it can flow via the existing reed-beds back to the Tyelaw Burn. The PRB contains a thoroughly mixed substrate comprising 50% limestone (or blast furnace slag) gravel, 25% green waste compost and 25% horse manure and straw. The bulk of the work on the site was completed in 2002, and the system is scheduled for final commissioning in Spring 2003 (Younger et al. 2002b). Early monitoring during commissioning suggests substantial alkalinity generation and metal attenuation in the PRB and ponds. The site has now been adopted as CL:AIRE Technology Demonstration Project 13.

4.2.10 Aspatria, Cumbria

Intermittent surface runoff combined with baseflow seepages from the former colliery spoil heaps of Aspatria, in north-west Cumbria, have caused significant degradation of the local stream for many years. The pollution loadings emanating from the site where exacerbated when a failed attempt to establish a conifer plantation on some of the more acid-generating portions of the spoil led to the deep ploughing (and thus accelerated oxidation) of some shales which are very rich in pyrite. While a RAPS unit would be a logical choice for these highly acidic waters, the lack of topographic relief between the point of leachate emergence and the banks of the receiving watercourse means that insufficient hydraulic head is available for this option to be implemented. On the other hand, a large area of flat land is available between the two. Therefore one of the world's largest-ever compost wetlands is planned for installation at the site. Because of its large size, however, the acquisition of sufficiently large volumes of reactive material for the substrate is an important issue. Accordingly, a pilot system has been installed to field-test some alternative substrates. Three cells, each 15 m by 4 m in size, were commissioned in December 2002. In one of these the substrate is paper waste; in another the substrate is sewage sludge filter cake, and in the third a 50:50 mixture of both materials has been installed. At the time of writing performance data are not yet available for this pilot system.

5 PROGRESS AND PROGNOSIS

In reading through the case studies presented above, which are believed to offer comprehensive coverage of all full-scale passive systems treating acidic spoil leachates in the UK at the time of writing, certain trends in design concept development are apparent. Some of the earliest attempts at passive treatment of acidic spoil leachates were made using simple aerobic reed beds (e.g. Dodworth in 1993, Shilbottle in 1995 and Oatlands in 1997). Such aerobic reed beds are now considered inappropriate for such applications, unless they are preceded by a compost wetland or RAPS. They are considered inappropriate because they tend to result in a *lowering* of pH (due to hydrolysis of Fe^{3+}, which is an acid-generating reaction; Younger et al. 2002a), albeit they do remove a reasonable amount of iron from solution by bacterially-catalysed oxidation.

Beginning with the Quaking Houses pilot wetland in February 1995, deliberate introduction of bacterial sulphate reduction as an alkalinity generating process began to find favour. In 1997 it was implemented at full scale at Quaking Houses, Nailstone and Benhar Bing, with further pilot-scale work on novel substrates commencing at Aspatria in 2002. The RAPS variant of the bacterial sulphate reduction process was more recently implemented (following a successful application to acidic deep mine waters in South Wales in 1998; Younger 1998), first at pilot-scale (at Bowden Close in 1999) then at full-scale (at Deerplay in 2001), with the full-scale system at Bowden Close being scheduled for completion in the spring of 2003. PRB systems based on bacterial sulphate reduction were also introduced for the first time in 1999 (at Renishaw Park, on a small scale) with the first large-scale PRB application being installed at Shilbottle in 2002–3. These PRB applications differ significantly from the "in-aquifer" PRB applications at Canadian metal mine sites, such as that described by Benner

et al. (1997), in that the reactive material is emplaced within the contaminated spoil itself and discharges water downstream not to any aquifer, but to surface water features.

This progression in technology development and implementation reflects both a growing sophistication in treatment process understanding (e.g. it is now clear that simple aerobic reed beds are not an adequate solution for acidic leachates) and a growth in confidence in the process (on the part of both designers and regulators), which is manifest in increasing innovation in system design and a willingness to break new ground in full-scale installations (of both RAPS and PRBs).

This is not to pretend that no problems have been encountered in the implementation of the technologies described in this paper. However, the main problems experienced to date have related to rather mundane issues common to many engineering projects (such as provisions for flood water management, erosion control and other aspects of hydraulic design) rather than biogeochemical process failures.

Looking to the future, there are clearly substantial grounds for optimism that the remaining 90 or so seriously polluted acidic spoil leachates in the UK (and many more elsewhere in the world) can be either minimized at source (using dry cover techniques) or passively treated at relatively modest cost (none of the systems described above cost more than £100K to construct, and most cost less than half this figure). There are, nevertheless, significant challenges remaining, including:

- the need to establish robust methods for predicting the lifetimes of reactive components (especially organic substrates) in compost wetlands, RAPS and PRBs.
- the development of strategies for the management of sulphide-rich spent substrates, which must inevitably be faced at some stage in the future for all such systems.
- The desirability of a robust understanding of the key microbial/geochemical processes operative in these systems, which should provide crucial clues into possible treatment failure scenarios. (The guiding principle here should be that, even though a treatment system is demonstrably working, it isn't necessarily working for the reasons we had in mind at the time of design.)

In order to encourage targeted research into these and other issues, CL:AIRE has recently announced the establishment of a new national research facility, named CoSTaR (Coal mine Sites for Targeted Remediation Research), which incorporates three of the sites discussed in this paper (namely, Bowden Close, Quaking Houses and Shilbottle). Funding has recently been confirmed for research at these three

CoSTaR sites within the framework of a new "LINK" research project ("ASURE") co-funded by the UK government (Department of Trade and Industry), two mining companies (Rio Tinto and Scottish Coal) and two consultancy/property management firms (IMC Ltd and Parkhill Estates Ltd). ASURE will run from April 2003 to March 2006, with a total budget of £920K, and involves research teams at Newcastle University (biogeochemistry and engineering specialists) and the University of Wales Bangor (microbiological specialists). ASURE has three principal objectives:

– To identify processes and rates of biogeochemical transformations of sulphur, iron and carbon in compost-based reductive passive mine water treatment systems, paying particular attention to their effects on pH and alkalinity.
– To characterise indigenous microorganisms engaged in S, Fe and C cycling in the compost-based passive systems, identifying both spatial and seasonal variations in organism distribution and metabolic activity.
– To advance engineering design concepts for compost-based passive systems, building on the findings of research undertaken in pursuit of objectives A and B, by integration of experimental results with treatment performance metrics, using conceptual and mathematical modelling techniques to assess the degree of consistency between concepts and data.

By the time these objectives are realised, the technology is likely to have spread to a range of other sites in the UK and beyond, and the time will be ripe to begin to rise to some of the challenges posed by the need to sustain these biologically-based treatment systems as both functional units and landscape features of ecological value for decades to come.

ACKNOWLEDGEMENTS

I am grateful to Paul Beck from CL:AIRE and to Dr Howard Fox of the University of Derby for their encouragement to prepare this paper. Many present and former members of the HERO research team contributed to the development of the concepts summarized in this paper; special mention must be given to Catherine Gandy and Adam Jarvis. The latter, in his present post with IMC Ltd, has also been instrumental (along with his colleague David Laine) in making available information on the Oatlands, Renishaw Park, Deerplay and Aspatria sites, which was useful in drafting Section 4. Dave Jenkins (formerly of the Soil Science unit at the University of Wales, Bangor) gave me insight into the reasons for a persistent lack of vegetal cover on the long-established, non-toxic, spoil heaps of Bethesda, Gwynedd. Funding from the European Commission (PIRAMID project;

EVK1-CT-1999-000021), EPSRC (GR/S07247/01) and NERC (GST022060) has also been key to assembling some of the ideas presented here. The development of the ASURE proposal (LINK project Biorem 4, EPSRC grant number GR/S07247/01) also provided a stimulating focus for the synthesis of some of the ideas presented here, with inputs from Prof Andy Aplin (Newcastle University), Dr Barrie Johnson (University of Wales, Bangor), John Thompson (Scottish Coal and Parkhill Estates) and Dr Chris Cross (Rio Tinto). Nevertheless, the views expressed in this paper are not to be interpreted as representing the views of any of the organisations named above.

REFERENCES

Amos, P.W. & Younger, P.L. 2003. Substrate characterisation for a subsurface reactive barrier to treat colliery spoil leachate. *Water Research* 37: 108 – 120.
Bannister, A.F. 1997. Lagoon and reed-bed treatment of colliery shale tip water at Dodworth, South Yorkshire. In Younger, P.L. (editor) *Minewater Treatment Using Wetlands. Proceedings of a National Conference held 5th September 1997, at the University of Newcastle, UK.* 105–122. London: Chartered Institution of Water and Environmental Management.
Benner, S.G. Blowes, D.W. & Ptaceck, C.J. 1997. A full-scale porous reactive wall for prevention of acid mine drainage. *Ground Water Monitoring and Restoration* (Fall 1997): 99–107.
CL:AIRE 2002. A constructed wetland to treat acid mine drainage from colliery spoils at Quaking Houses, County Durham. *CL:AIRE Case Study Bulletin CSB2*, 4pp. London: Contaminated Land: Applications in Real Environments (CL:AIRE).
Durham, A.J.P. Wilson, G.W. & Currey, N. 2000. Field performance of two low infiltration cover systems in a semi arid environment. *Proceedings of the Fifth International Conference on Acid Rock Drainage, (ICARD 2000), Denver, Colorado, May 21–24, 2000.* 2: 1319–1326.
Gustafsson, H.E. Lundgren, T. Lindvall, T. Lindahl, L-E. Eriksson, N. Jönsson, H. Broman, P.G. & Göransson, T. 1999. The Swedish acid mine drainage experience: research development and practice. In Azcue, J.M., (editor), *Environmental impacts of mining activities. Emphasis on mitigation and remedial measures.* 203–228. Heidelberg: Springer.
Hartman, H.L. 1987. *Introductory mining engineering.* New York: Wiley.
Heal, K.V. & Salt, C.A. 1999. Treatment of acidic metal-rich drainage from reclaimed ironstone mine spoil. *Water Science & Technology* 39: 141–148.
Hedin, R.S. Nairn, R.W. & Kleinmann, R.L.P. 1994. Passive treatment of polluted coal mine drainage. *Bureau of Mines Information Circular 9389.* Washington DC: United States Department of the Interior.
Jarvis, A.P. & England, A. 2002. Operational and treatment performance of an unique Reducing and Alkalinity Producing System (RAPS) for acidic leachate remediation in

Lancashire, UK. In Merkel, B.J., Planer-Friedrich, B., & Wolkersdorfer, C. (eds) *Uranium in the aquatic environment. (Proceedings of the International Conference Uranium Mining and Hydrogeology III and the International Mine Water Association Symposium, held in Freiberg, Germany, 15–21 September 2002).* 25–40. Berlin: Springer-Verlag.

Li, M. Catalan, L.J.J. & St-Germain, P. 2000. Rates of oxygen consumption by sulphidic tailings under shallow water covers – field measurements and modelling. *Proceedings of the Fifth International Conference on Acid Rock Drainage, (ICARD 2000), Denver, Colorado, May 21–24, 2000.* 2: 913–920.

Jarvis, A.P. & Younger, P.L. 1997. Dominating chemical factors in mine water induced impoverishment of the invertebrate fauna of two streams in the Durham Coalfield, UK. *Chemistry and Ecology* 13: 249–270.

Jarvis, A.P. & Younger, P.L. 1999. Design, construction and performance of a full-scale wetland for mine spoil drainage treatment, Quaking Houses, UK. *Journal of the Chartered Institution of Water and Environmental Management* 13: 313–318.

Kelly, M.G. 1988 *Mining and the freshwater environment.* Elsevier Applied Science, London.

Kepler, D.A. & McCleary, E.C. 1994. Successive Alkalinity Producing Systems (SAPS) for the Treatment of Acidic Mine Drainage. *Proceedings of the International Land Reclamation and Mine Drainage Conference and the 3rd International Conference on the Abatement of Acidic Drainage. (Pittsburgh, PA; April 1994). Volume 1: Mine Drainage.* 195–204.

McLean, I. & Johnes, M. 2000. *Aberfan. Government and Disasters.* Welsh Academic Press, Cardiff. 250pp.

Nicholson, R.V. Gillham, R.W. Cherry, J.A. & Reardon, E.J. 1989. Reduction of acid generation in mine tailings through the use of moisture-retaining cover layers as oxygen barriers. *Canadian Geotechnical Journal* 26: 1–8.

Saaltink, M.W., Domenech, C. Ayora, C. & Carrera, J. 2002. Modelling the oxidation of sulphides in an unsaturated soil. In Younger, P.L. & Robins, N.S. (eds) *Mine Water Hydrogeology and Geochemistry. Geological Society, London, Special Publications* 198: 187–204.

Warner, I. 1997. Oatlands Colliery – West Cumbria. In Bird, L., (Editor), *Proceedings of the UK Environment Agency Conference on "Abandoned Mines: Problems and Solutions". Held at Tapton Hall, University of Sheffield, 20th–21st March 1997.* 40–51.

Younger, P.L. (Editor). 1997. *Minewater Treatment Using Wetlands. Proceedings of a National Conference held 5th September 1997, at the University of Newcastle, UK.* London: Chartered Institution of Water and Environmental Management.

Younger, P.L. 1998. Design, construction and initial operation of full-scale compost-based passive systems for treatment of coal mine drainage and spoil leachate in the UK. In *Proceedings of the International Mine Water Association Symposium on "Mine Water and Environmental Impacts", Johannesburg, South Africa, 7th–13th September 1998.* 2: 413–424.

Younger, P.L. 2000. Holistic remedial strategies for short- and long-term water pollution from abandoned mines. *Transactions of the Institution of Mining and Metallurgy (Section A: Mining Technology)* 109: A210–A218.

Younger, P.L. 2001. Mine water pollution in Scotland: nature, extent and preventative strategies. *Science of the Total Environment* 265: 309–326.

Younger, P.L. 2002a. Mine waste or mine voids: which is the most important long-term source of polluted mine drainage? *United Nations Environment Programme, Mineral ResourcesForum (http://www.mineralresourcesforum.org/), "Current Feature" paper Nov–Dec 2002.* 12pp.

Younger, P.L. 2002b. A reducing and alkalinity-producing system (RAPS) for the passive treatment of acidic, aluminium-rich leachates emanating from revegetated colliery spoil materials at Bowden Close, County Durham. In *Proceedings of the CL:AIRE Annual Project Conference, 11th April 2002, Imperial College, London.* (ISBN 0-9541673-1-7). Paper 7, 21pp.

Younger, P.L. 2002c. Mine water pollution from *Kernow* to *Kwazulu-Natal*: geochemical remedial options and their selection in practice. (Scott Simpson Lecture 2002). *Geoscience in Southwest England (Proceedings of the Ussher Society)* 10: 255–266.

Younger, P.L. Curtis, T.P. Jarvis, A.P. & Pennell, R. 1997. Effective passive treatment of aluminium-rich, acidic colliery spoil drainage using a compost wetland at Quaking Houses, County Durham. *Journal of the Chartered Institution of Water and Environmental Management* 11: 200–208.

Younger, P.L. Banwart, S.A. & Hedin, R.S. 2002a. *Mine Water: Hydrology, Pollution, Remediation.* Dordrecht: Kluwer Academic Publishers. 464pp.

Younger, P.L. Jayaweera, A. Elliot, A. Wood, R. Amos, P. Daugherty, A.J. Martin, A. Bowden, L. Aplin, A.C. & Johnson, D.B. 2002b. Passive treatment of acidic mine waters in subsurface-flow systems: exploring RAPS and permeable reactive barriers. *In* Nuttall, C.A. (editor), 2002, *Mine water treatment: a decade of progress. Proceedings of a Conference held in Newcastle Upon Tyne, Nov 11–13th 2002.*

Land Reclamation – Moore, Fox & Elliott (eds)
© 2003 Swets & Zeitlinger, Lisse, ISBN 90 5809 562 2

The spirit and purpose of land restoration

John F. Handley
Centre for Urban and Regional Ecology (CURE), School of Planning and Landscape, University of Manchester

ABSTRACT: This paper explores difficult and perhaps hazardous territory, the inter-disciplinary field between the theory and practice of land restoration on the one hand and insights which can be gained from environmental philosophy on the other. It is argued that here, as in related disciplines, the development of an appropriate environmental framework will assist research and development to continue in a manner which commands respect and confidence. Drawing on the literature, a typology from environmental ethics is proposed and applied to the field of land restoration and management. The paper emphasises that each of us, whether "environmental scientist" or "environmental design professional" will have their own philosophical position, either explicit or implicit. It suggests that a degree of self awareness may be helpful in placing our individual contribution and activity within a greater whole.

1 INTRODUCTION

"Shall we not learn from life its laws, dynamics, balances? Learn to base our needs not on death, destruction, waste, but on renewal. In wisdom and in gentleness learn to walk again with Eden's angles? Learn at least to shape a civilisation in harmony with the earth."

Ansel Adams and Nancy Newhall
This is the American Earth, 1960

This arresting quotation from Ansel Adams and Nancy Newhall beautifully illustrates their perspective on the land and its stewardship. It will be familiar to the land reclamation community, not least because it was used to introduce *The Restoration of Land*, the standard text by Bradshaw and Chadwick (1980), that synthesised research findings and experience from the second half of the twentieth century. The unwritten implication is that here we find a "land ethic" that can guide and shape land reclamation activity in theory and practice.

In related disciplines, such as medicine and genomics, scientific discovery, allied to technical progress, has posed challenging questions of immediate interest to researchers, practitioners and society at large. Here, the development of an appropriate ethical framework is seen to be an essential requirement for research and development to continue in a manner which commands respect and confidence. This paper will argue that scientific and technical progress has been no less striking in the recycling of land and

reconstruction of ecosystems, and that, if anything, this will intensify in the 21st century. A land ethic is needed to help avoid the despoliation that too often accompanied past economic activity, and to ensure that, in future, land restoration achieves its full potential. Even where land restoration activity takes place outside such a framework, we should at least be aware of the value system which guides our actions.

Land Reclamation 2003 brings together two broad groups of people – scientists and technologists, on the one hand, who seek to understand environmental processes, identify potential threats and pose solutions; on the other is an array of planners, engineers, managers and designers, dubbed the "environmental design professionals" by Thompson, (2000), who seek to make these solutions work in practice. As shown in Figure 1 there are four interlocking areas of ethics of potential interest to the environmental design professional – personal ethics, business ethics, professional ethics and environmental ethics. The paper will focus on environmental ethics and, in exploring the principles of land restoration from this perspective, it will seek to connect both communities of interest – the scientist and the practitioner.

2 THE SCOPE OF ENVIRONMENTAL ETHICS

In broad terms environmental ethics is the moral discourse which explores the relationship between humankind and the natural environment. As Taylor

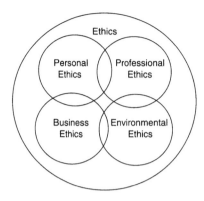

Figure 1. Ethical concerns for the environmental design professional. Source: after Woolley and Whittaker (1995).

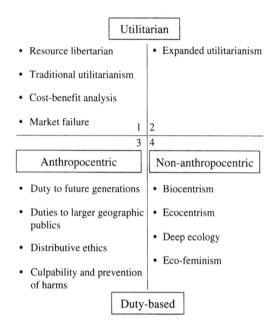

Figure 2. Categorising theories of environmental ethics. Note: the concepts and theories contained in each quadrant are meant to be illustrative, not exhaustive. Moreover, no attempt is made within each quadrant to array the theories along the two axes. Source: adapted from Beatley (1994).

(1986, p. 3) notes, it is *"concerned with the moral relations that hold between humans and the natural world. The ethical principles governing those relations determine our duties, obligations and responsibilities with regard to the Earth's natural environment and all the animals and plants that inhabit it."* Environmental ethics implies choices and decisions, both about what we do and how we do it. Moreover, as Beatley (1994) emphasises this is not an optional question, even those who have little or no regard for environmental impacts on the natural environment have, in a sense, a set of environmental ethics – they simply treat the environment, and considerations of it, as being of limited value.

If we are to think about "our position" on what, for some, may be unfamiliar ground we need some kind of a map. Beatley (1994) and Merchant (1992) both provide helpful organising frameworks in which the primary gradient moves from a human centred position (anthropocentric/egocentric) to one that is nature centred (biocentric/ecocentric). The strong anthropocentric position has been clearly stated by Baxter (1974) in his book *People or Penguins*:

"My criteria are oriented to people, not penguins. Damage to penguins, or sugar pines, or geological marvels is, without more, simply irrelevant. One must go further, by my criteria, and say: Penguins are important because people enjoy seeing them walk about rocks ... In short, my observations about environmental problems will be people-oriented, as are my criteria. I have no interest in preserving penguins for their own sake."

Those who take the opposite position argue that nature, or elements of nature, have inherent worth and intrinsic value, irrespective of the value they may hold instrumentally to human beings. This involves an extension of moral philosophy to consider the interest of living things (as in Paul Taylor's biocentric

approach) which is discussed below. The ecocentric, or earth centred position goes further to recognise the interconnectedness of humankind and nature through a shared "community of interest" as first expressed by Leopold (1949):

"The land ethic simply enlarges the boundaries of the community to include soils, water, plants and animals, or collectively: the land...a land ethic changes the role of Homo sapiens from conqueror of the land community to plain member and citizen of it. It implies respect for his fellow members, and also respect for the community as such."
                                        Aldo Leopold (1949)
                                        A Sand County Almanac

The second axis, drawn out by Timothy Beatley op cit, concerns the extent and nature of moral obligation. On the on hand we have the utilitarian (teleological position) in which "the ultimate criterion or standard of what is morally right is the non moral value brought into being by our actions" (Frankena, 1973). This contrasts with deontological or duty-based theories in which there are other considerations that may make an action or rule right or obligatory besides the goodness or badness of its consequences. The resulting theoretical framework is shown in Figure 2.

266

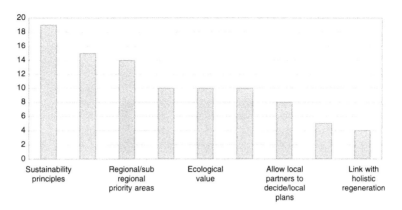

Figure 3. Reclamation priorities as identified by a questionnaire of practitioners in Northwest England (by number of responses). Source: Northwest Land Reclamation Review Steering Group (2001).

3 ETHICAL PERSPECTIVES AND PRACTICE OF LAND RESTORATION

The paper will concentrate on three quadrants of the matrix in Figure 2. First we should recognise the problem of derelict and despoiled land as a symptom of market failure. Is this purely an historic problem and to what extent can dereliction and contamination be prevented in the future?

The next quadrant is also human centred but takes on board duties to future generations and larger geographic publics. This is the domain of sustainable development which has provided the main organising principle for land restoration policy and practice in the United Kingdom since at least the publication of *This Common Inheritance*, the government's environmental white paper in 1990.

Today sustainable development, and the principles which underpin it, is firmly established as a key driver of land restoration activity, as evidenced by a recent survey of land reclamationists (Fig. 3) who were invited to identify regional priorities for land reclamation (Northwest Land Reclamation Review Steering Group, May 2001). The influence of different notions of sustainability and sustainable development will be explored in the context of land restoration. Of special assistance here is an excellent review by Ian Thompson (2000) which examined the ethics of sustainability, from the perspective of the environmental design professional.

One strand of government policy has prioritised the reclamation of brownfield land for residential, commercial and industrial use. Land recycling for "hard" end-use may relieve pressure for development on greenfield sites, as well as contributing to urban regeneration by stimulating development and improving environmental quality. The twin objectives of minimising

the use of scarce land resources for development and maximising the efficiency with which previously developed land is recycled both contribute to the principles of sustainable urban development.

However, there is often a mismatch between supply and demand which may inhibit land recycling and make governmental targets for reuse of brownfield land difficult to achieve (Handley 2001). The factors at work here include:

1. the physical and chemical condition of the land demands a significant development premium;
2. land recycling is much more effective in regions with high demand; and
3. the geographic distribution of land within a region may not be conducive to land recycling: a significant proportion of derelict land is in rural locations, often with Green Belt status, and derelict land tends to be concentrated in older industrial areas where social deprivation is high and the private sector is loathe to invest.

More importantly, we need to recognise the landscape, wildlife and cultural values which may have accrued through time. Here we begin to cross the boundary into the fourth quadrant of Beatley's matrix, from the human centred to the nature or earth centred realm. This will require "extending the boundaries".

4 EXTENDING THE BOUNDARIES

As we have seen biocentric theories extend the boundaries of moral significance to include life forms other than humans. The biocentric position is particularly well expressed by Paul Taylor in his book *Respect for Nature* (Taylor 1986). He identifies four beliefs that form the core of the biocentric outlook:

(a) *"The belief that humans are members of the Earth's Community of Life in the same sense and on the same terms in which other living things are members of that Community.*

(b) *The belief that the human species, along with all other species, are integral elements in a system of interdependence such that the survival of each living thing, as well as its chances of faring well or poorly, is determined not only by the physical conditions of its environment but also by its relations to other living things.*

(c) *The belief that all organisms are teleological centres of life in the sense that each is a unique individual pursuing its own good in its own way.*

(d) *The belief that humans are not inherently superior to other living things."*

From this position, Taylor develops a series of principles to guide human action with respect to nature. Of particular relevance here are the principles of minimum wrong, distributive justice and restitutive justice. Where human interests, expressed for example through the development process, are in conflict with nature that calls into question the very case for development. Where the human interest is overriding we should seek to minimise damage, for example through environmental impact assessment (EIA) and to avoid, mitigate or compensate for that damage. EIA may also be helpful in the land reclamation process itself as a means of identifying and maximising the potentially beneficial effects of a project (Jay and Handley 2001) besides pointing up the potentially adverse or conflicting side-effects of a project.

The principle of restitutive justice applies wherever the basic interests of nature have been overridden. This principle is of particular relevance to restoration ecology because of our growing capability to recreate opportunities for nature (Gilbert and Anderson 1998) through habitat creation. The wide range of habitat creation activity in the UK is well illustrated (Fig. 4) by the survey of Jones (1990). Box (1998) has provided helpful guidance on setting objectives and monitoring for ecological restoration and habitat creation.

Gilbert and Anderson recognised that techniques developed to compensate for serious habitat loss could be used as an excuse to destroy further habitat. They emphasise that which *cannot* be achieved by habitat creation "*is the subtlety, complexity and biodiversity of an ecosystem that has evolved over time*" (Gilbert and Anderson 1998). The dangers inherent in

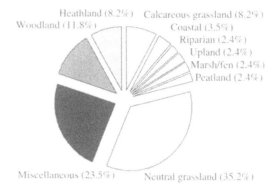

Figure 4. Habitat creation activity in the UK, as assessed by a questionnaire survey. Source: Jones (1990).

a less well considered approach have been clearly expressed by the philosopher Robert Elliot:

"That environmental restoration is flourishing and that its technologies are becoming increasingly adept may be welcome developments, provided the possibility of adequate restoration is not appealed to in justification of the destruction or degradation of natural areas. They are welcome developments only to the extent that they provide some basis for hoping that natural areas that are already degraded, destroyed and drained of natural value, may be restored, eventually to be reclaimed by nature. Such developments should provide no excuse and no justification for those who want to degrade or destroy the earth's shrinking, relatively natural, areas"

Robert Elliot (1997)
Faking Nature – the ethics of environmental restoration

Robert Elliot's worries about "faking nature" would have struck a cord with Aldo Leopold, one of the seminal figures of the modern environmental movement. As shown by the earlier quotation, Leopold stresses that as human beings we belong to a community of living things and that our community is something we must nourish and support if we are sustain the nourishment and support it provides in return (Pratt *et al.* 2000). The viewpoint he expressed has been described as "ecocentric" and the insights it brings to restoration ecology have been well expressed by Bradshaw in his essay entitled *Restoration: an acid test for ecology* (Bradshaw 1987) which advocates an ecologically grounded approach to the rebuilding of ecosystems. Subsequent developments by landscape ecologists such as Forman (1995) seek to combine scientific understanding of the ecological processes with the meaningful engagement of local people. Humankind is seen to be within nature, not external to it, and the human community should be fully engaged in the process of rebuilding damaged landscapes. We (Handley *et al.* 1998) explored such an "ecological approach" to landscape restoration in the preceding Conference of Land Reclamationists with particular reference to Bold Moss, St Helens. Subsequent experience within a large-scale "experimental" programme (Groundwork's Changing Places) has confirmed the viability and validity of that approach (Morgan 2002) and it has been codified in Ecoregen.com, a website and manual which provides a "toolkit for community led regeneration of derelict land".

The Status Report which set out the case for "Changing Places" (Handley 1996) also highlighted the need for long-term stewardship of restored land. The ecological approach to land restoration minimises the aftercare requirements by working with the grain of natural recovery but sustained management input is critical for long-term success. The Status Report identified a management deficit within the post-industrial landscape and advocated the establishment of a land restoration trust (analogous to The National Trust for Places of Historic Interest or Natural Beauty in England, Wales and Northern Ireland). It is pleasing that, as this paper goes to press, the Office of the Deputy Prime Minister has announced the government's intent that English Partnerships, Groundwork, the Forestry Commission and the Environment Agency will create a Land Restoration Trust, which will work with local communities, to restore and manage brownfield land that is suitable only for use as public greenspace (ODPM 2003).

5 CONCLUSION

In this paper we have explored difficult and perhaps hazardous territory; the interdisciplinary field between land restoration (theory and practice) and environmental philosophy. There are traps here for the unwary, particularly "the naturalistic fallacy". Facts, according to the naturalistic fallacy, can never yield values: philosophers insist that one cannot deduce from how things *are* to how they *ought* to be. Taylor, for example, in expounding his biocentric position comes close "but avoids with skill" the naturalistic fallacy (Pratt *et al.* 2000) by explaining that "the key step in the move from the scientific facts to the conclusion that we ought to respect nature is the denial of human superiority". The "ecocentric" position of Leopold is perhaps easier to defend because of the clear inter-dependence of human welfare and the ecological health of natural systems, whether at the level of the ecosystem or the biosphere as a holistic entity (Midgley 2001). The important point however, is to recognise that as scientists and "environmental design professionals" we each have our own philosophical position, even if this is implicit rather than explicit, and a degree of self awareness may be helpful in understanding our modus operandi.

REFERENCES

Baxter, W.F. 1974. *People or Penguins: The case for optimal pollution.* New York: Columbia University Press.

Beatley, T. 1994. Environmental ethics and the field of planning: Alternative theories and middle-range principles. In H. Thomas. *Values and Planning.* Aldershot: Avebury.

Box, J. 1998. Setting objectives and monitoring for ecological restoration and habitat creation. In H.R. Fox, H.M. Moore & A.D McIntosh. *Land Reclamation: achieving sustainable benefits.* Rotterdam: Balkema.

Bradshaw, A.D. 1987. Restoration: an acid test for ecology. In W.R. Jordan III, M.E. Gilpin & J.D. Aber. *Restoration Ecology: a synthetic approach to ecological research.* Cambridge: Cambridge University Press.

Bradshaw, A.D. & Chadwick, M.J. 1980. *The Restoration of Land.* Oxford: Blackwell Scientific Publications.

Elliot, R. 1997. *Faking Nature: the ethics of environmental restoration.* London and New York: Routledge.

Forman, R. 1995. *Landscape Mosaics: the ecology of landscapes and regions.* New York: Springer.

Frankena, W. 1973. *Ethics.* Englewood Cliffs, New Jersey: Prentice Hall.

Gilbert, O.L. & Anderson, P. 1998. *Habitat Creation and Repair.* Oxford: Oxford University Press.

Handley, J.F. 1996. *The Post-industrial Landscape – A resource for the community, a resource for the nation?* Birmingham: Groundwork Foundation.

Handley, J.F. 2001. Derelict and Despoiled Land – Problems and Potential. In C. Miller. *Planning and Environmental Protection.* Oxford and Portland: Hart Publishing.

Handley, J., Griffiths, E.J., Hill, S.L. & Howe, J.M. 1998. Land restoration using an ecologically informed and participative approach. In H.R. Fox, H.M. Moore & A.D. McIntosh. *Land Reclamation: achieving sustainable benefits.* Rotterdam: Balkema.

Her Majesty's Government. 1990. *This Common Inheritance: Britains Environmental Strategy (CM 1200).* London: HMSO.

Jay, S. & Handley, J. 2001. The application of environmental impact assessment to land reclamation practice. *Journal of Environmental Planning and Management* 44(6): 765–782.

Jones, G.H. 1990. Learning from experience. *Landscape Design* 193: 40–44.

Leopold, A. 1949. *A Sand Country Almanac.* New York and Oxford: Oxford University Press.

Merchant, C. 1992. *Radical Ecology: The search for a liveable world.* London and New York: Routledge.

Midgley, M. 2001. Individualism and the concept of Gaia. In K. Booth, T. Dunne & M. Cox. *How might we live? Global ethics in a new century.* Cambridge: Cambridge University Press.

Morgan, P. 2002. Changing Places – going soft on derelict land. *Ecos* 22(3/4): 4.

Northwest Land Reclamation Review Steering Group. 2001. *Reclaim the Northwest!* Warrington: North West Development Agency.

Office of the Deputy Prime Minister. 2003. *Sustainable communities: building the future*. London: ODPM.

Pratt, V., Howarth, J. & Brady, E. 2000. *Environment and Philosophy*. London and New York: Routledge.

Taylor, P. 1986. *Respect for Nature – A theory of environmental ethics*. Princeton, New Jersey: Princeton University Press.

Thompson, I.H. 2000. Ethics of sustainability. In J.F. Benson & M.H. Roe. *Landscape and Sustainability*. London: Spon Press.

Woolley, H. & Whittaker, C. 1995. Ethical practices in the landscape profession: a research note. *Landscape Research*. 20(3): 147–151.

Land Reclamation – Moore, Fox & Elliott (eds)
© 2003 Swets & Zeitlinger, Lisse, ISBN 90 5809 562 2

Some mechanisms responsible for the alteration of soil hydraulic properties by root activity

V.B. Powis, W.R. Whalley & N.R.A. Bird
Silsoe Research Institute, Wrest Park, Silsoe, Bedford, UK

P.B. Leeds-Harrsion
National Soils Resources Institute, Cranfield University, Silsoe, Bedford, UK

ABSTRACT: When roots grow through soil they modify their physical environment in a number of ways. In the agricultural context the extraction of water by roots can cause many soils to shrink and crack. When roots explore compacted soils, such as those on degraded sites, they need to compress the soil further to make a channels for the roots. The activity of roots has both an immediate impact on the soil's hydraulic status as well as consequences for the longer term. In this paper we consider both of these aspects. We describe the results of laboratory experiments designed to demonstrate the ways in which roots can change the hydraulic properties of soil. We show that different species can have different effects on the hydraulic conductivity of soil. As roots decayed the hydraulic conductivity of the soil can change, but the direction of this change depends upon the water status of soil during root growth. In this paper we discuss the possible mechanisms that could explain these findings. To conclude we consider the implications of our results for the restoration of compacted and polluted sites. We consider our results to be important in the context of a better understanding of how to improve the physical structure of soil.

1   INTRODUCTION

The importance of soil structure and hydraulic properties on rooting and plant growth is widely documented. Poor soil conditions, such as high density or excess or limiting water, can limit root and plant growth and subsequently reduce crop yields. However, the effect of root growth on the soil, especially soil hydraulic properties has received less attention until recently. Early work in this area recognised that there was an effect of rooting on soil properties (e.g. Martin & Craggs 1949) and subsequent studies have begun to identify the nature of the effects.

In agriculture interest in the use of zero-tillage or biological tillage, where any change to the soil structure and properties is achieved through the choice of crops within a rotation has increased. The use of plant activity to improve the soil physical status is often called "biological tillage" (e.g. Cresswell & Kirkegaard 1995). The recent interest in biological tillage is partly due to the recognition of soil damage and erosion associated with the continuous use of conventional tillage (White 1997). In some cases mechanical tillage has limited effects in reducing compaction by heavy machinery and can result in roots clumping into narrow channels. Here the use of a crop to penetrate the subsoil and improve soil conditions for subsequent root growth may be and effective solution (Hairiah & van Noordwijk 1989).

Reclaimed land often has structural and hydraulic properties that are different from most indigenous soils. This can be due to the replacement of the subsoil and topsoil over minespoil or landfill. A compact layer is often required to isolate the material beneath the subsoil and limit excess water infiltration (Bengtsson *et al* 1994). However the inclusion of such a layer and the replacement of the subsoil may cause poor drainage of the soil and hence limit root growth and distribution (Fairley 1985). The yields of different species have been shown to vary with soil depth, clay content and hydraulic properties on reclaimed sites (Merrill, Ries & Power 1998).

If a plant species can be chosen for its ability to grow in specific soil conditions and alter the hydraulic properties to reach some predetermined expectation there may be greater scope for managing reclaimed

273

land by "biological tillage". In this paper we describe two laboratory experiments which explore the effects that different plants have on soil hydraulic properties and also the effect of soil water status during growth. We also describe that mechanistic basis for our observations.

2 MATERIALS AND METHODS

2.1 Experiment 1: effect of soil water status

2.1.1 Soil types
Synthetic soils were used throughout this work, obtained by mixing sand (grade 65 obtained from Hepworth Minerals, Redhill, Surrey) and kaolinite (obtained from ECC, St. Austell, Cornwall). The soils were mixed by hand in small quantities then stored in airtight containers until required. Two different soil ratios were made ~4 sand : 1 kaolin as used by Barley (1954) and 10 sand : 1 kaolin. The soils were packed in permeameters (see 2.1.3) to a bulk density of $1.6 \, Mg \, m^{-3}$. The soils were packed in thin layers to minimize differences in density through the profile.

The equilibrium water content at potentials between 0 and $-18 \, kPa$ was measured in $-1 \, kPa$ increments using a tension plate. Samples of the two different soil mixtures were packed into metal rings to a bulk density of $1.6 \, Mg \, m^{-3}$ and placed onto a tension table. The samples were gradually wetted up until saturated. Once saturated one sample was removed to determine the water content by weighing the sample and then re-weighing after oven drying at 105°C for 48 hours. The water table connected to the tension table was then lowered in $-1 \, kPa$ increments and a sample removed at each increment for water content determination.

The four part sand to one part kaolinite soil was used for the experiment requiring saturated conditions to enable a comparison with Barley's work in 1954. However, to obtain unsaturated conditions with approximately 20% of the saturated water filled pores becoming air-filled required a water table depth greater than 1.2 m, which was not practical. The second soil mixture of ten parts sand to one part kaolinite allowed unsaturated conditions to be achieved with a water table depth of 0.6 m.

2.1.2 Plant materials and growth conditions
For this experiment maize (*Zea mays* L. Toledo) was used. The seeds were pre-germinated by placing them on wet filter paper and leaving in a dark room at 25°C for three days. The germinated seeds were then planted at a depth of 5 mm beneath the soil surface. Fifteen seeds were planted into each permeameter. The plants were grown in a controlled temperature environment with a bank of growth lights above the

permeameters. The lighting created a 16-hour daylight period with temperatures between 25–28°C. During the 8-hour night period a temperature of 20°C was maintained. A layer of white plastic beads was placed onto the soil surface to minimize evaporation. A nutrient solution was used throughout the work for growth and measurement of hydraulic conductivity. The length of the growth period and foliage harvesting time differed between the two methods. For the saturated conditions the plants were grown for 10 days and the foliage was harvested after the measurement of saturated hydraulic conductivity. For the unsaturated method the plants were grown for 21 days and the foliage harvested prior to the measurement because of the increased size of the plants.

2.1.3 Root growth environment and permeameter design
Constant head permeameters were constructed from plastic tubing 75 mm diameter and 325 mm in length. Figure 1 shows a schematic of the permeameter design. A series of holes were made in the side of each permeameter at 25 mm, 115 mm, 220 mm and 255 mm depth from the surface. Manometer tubes were connected using right-angle joints into the upper three holes. Plastic wool was inserted into the interface between the tubes and the permeameter to prevent soil particles moving into the manometer tubes during measurements. A porous cup 52.5 mm long and 22 mm external diameter was inserted into the lowest hole. This was connected to the nutrient solution reservoir which could be raised and lowered to alter the saturation conditions.

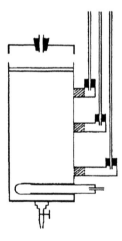

Figure 1. A schematic of the permeameters used in this experiment. Manometer tubes and a porous cup were inserted into the side of the permeameter. A sealed base unit included a tap to allow water movement, with a screen to limit soil migration. Water inlet was either through a fitted lid or a plastic cylinder inserted into the top.

Figure 2. Plants growing the permeameter tubes which are shown schematically in figure 1.

The base consisted of a screw-threaded unit with a 20 mm hole cut into the centre to allow a tap to be fitted. A nylon retaining screen was attached to the inside of the base and the unit was sealed onto the bottom of the permeameter to create a watertight fitting. The upper unit for the saturated conditions consisted of a plastic cylinder 75 mm diameter and 200 mm high, which was inserted into the top of the permeameter. A hole was made in the side of the cylinder for use as a water inlet point during measurement. The top of the cylinder was sealed with tape during measurement to create watertight conditions. For the unsaturated conditions a modification to the permeameter was included as the foliage was harvested prior to measurement. This allowed a screw-fitted unit similar to the base to be used during measurement, with a water inlet point in the centre of the unit and a water tight seal around the edge. During the experiment the plants were grown in the permeameter cylinders (Figure 2).

2.1.4 *Measurement of hydraulic conductivity*
Before the initial measurement of hydraulic conductivity the permeameters were wetted up gradually from below by raising the level of the nutrient solution. This was done to minimize air entrapment within the sample. Once saturated the lids were inserted and the Mariotte bottle connected to allow a constant head of 750 mm to be maintained. Figure 3 shows the system used to measure hydraulic conductivity.

For the saturated soil conditions the conductivity was measured using the method suggested by Barley (1954). Measurements were made over a 150 hour period, during which a peak value was attained. For the modified method it was found that the peak value occurred within a 4–6 hour period from the start and that the value obtained was stable over 150 hours. Therefore, two replicate measurements were made

over a ten hour period for the modified method at each time point. The difference in time scale to reach a peak value is probably due to the different soil types and associated pore size distributions. For both methods the first reading was taken after 30 minutes to ensure the water was moving through the system. The two methods were compared side-by-side with measurements of saturated hydraulic conductivity made at five time periods, before growth, after growth, and 1, 2 and 3 months after the foliage was removed. Figure 3 shows the permeameters being used to measure saturated hydraulic conductivity.

2.2 *Experiment 2: effect of plant species*
The method for this experiment was the same as the method used in the previous chapter for unsaturated soil conditions where plants were grown in unsaturated soil for 21 days. The hydraulic conductivity was measured before and after growth and during decay. For this experiment three plant species were used and a measure of the root volume and length were made. The water release curve before and after the experiment was also measured.

2.2.1 *Soil*
The synthetic soil used for all the experiments in this chapter consisted of 10 parts sand (grade 65 obtained from Hepworth Minerals, Redhill Surrey) and kaolinite (From ECC, St. Austell, Cornwall). This mixture is the same as that used for the unsaturated growth method in experiment 1. In all the experiments presented it was packed to a dry bulk density of $1.6 \, Mg \, m^{-3}$.

2.2.2 *Soil shrinkage and cracking*
To examine the possibility of shrinkage of the soil during drying and water extraction by the roots, the standard shrinkage curve of the soil was measured. The synthetic soil was wetted up to saturation and packed into troughs, measuring 140 mm long with an internal radius of 11.25 mm. Half of the samples were

weighed and placed into a constant temperature environment set at 20°C. The remaining samples were placed into an oven at 105°C after weighing. All samples were weighed daily over a 14 day period. At the end of this period there was no shrinkage or change in appearance despite loss of water.

2.2.3 Measurement of K_{sat}

The permeameters were fixed into the frame below a bank of growth lights and connected to a reservoir of nutrient solution which was gradually raised to saturate the soil. The measurement of saturated hydraulic conductivity follows the procedure described for experiment 1.

2.2.4 Plant species and growth conditions

Three species were used for this work – maize (*Zea mays* L. Toledo), wheat (*Triticum aestivum* L. Hussar) and pea (*Pisum sativum* L. cv. Cossack). The seeds were obtained from Zeneca Seeds UK Ltd, Kings Lynn, Norfolk. These species were chosen based on the differences between their rooting systems. Pea is a dicotyledonous species with a dominant tap root, maize and wheat are both monocotyledonous with the rooting system of maize dominated by large nodal roots and that of wheat dominated by fine seminal roots (Arnon 1972, Gregory 1988). The seeds were pre-germinated by placing them on wet filter paper and leaving them in a dark room at 25°C for three days. The germinated seeds were then planted at a depth of 5 mm from the soil surface. Fifteen seeds were planted into each permeameter. The layout of the species was randomised within each block of permeameters (See Figure 1).

The growth conditions and nutrient solution were the same as for the previous experiment to provide a non-limiting environment during the 21 day growth period. This consisted of a 16 hour daylight period with temperatures between 25–28°C. Any possible edge effects due to reduced light were minimized by shielding the sides with aluminium foil as shown in Figure 2. A temperature of 20°C was maintained for the 8 hour night period. A 10 mm layer of white plastic beads was placed onto the soil surface to minimize evaporation.

2.2.5 Water release characteristic before and after growth and decay

A further comparison of the hydraulic properties was made by comparing the moisture characteristic curve of the soil before and after the experiment. The water release characteristic before the experiment was measured on replicate samples packed to a density of 1.6 Mg m^{-3}. These were placed on a tension table and the water table was lowered in −1 kPa intervals. At each potential one sample was removed and the water content determined by oven drying.

After the final K_{sat} measurement had been made, one permeameter for each species was removed. The base of the permeameter was removed and a small layer of soil removed. A layer of Plaster of Paris was then inserted into the base and allowed to dry. This provided a good contact to be maintained between the permeameter and the porous plate of a tension table.

In order to measure the water content at various depths, small holes were drilled into the side of each permeameter for the insertion of the TDR probes. A total of three probes per column were used. The probes consisted of three brass wires, 70 mm long and 3 mm diameter, attached to the probe head at 10 mm spacing. The transit time of the pulse was then measured using a Tektronix 1502 cable tester (Tektronix, Beaverton, OR) with a standard coaxial cable connecting the pins to the cable tester. The permeameters were placed onto tension tables and the water table raised until standing water remained on the plate. The transit time for the first probe was monitored until a stable value was reached. The values for the other probes were then measured. After measurement for all probes the water table was lowered in −1 kPa increments and measurements were taken when a stable value was obtained.

2.2.6 Root volume and length – direct measurement and estimation from foliage weight

Following the measurement of K_{sat} after growth the foliage was harvested from each permeameter and the dry weight determined. The foliage was oven dried at 48°C for 48 hours then weighed. One permeameter per species was destructively sampled to obtain measurements of root length, volume and distribution. These values were used to obtain ratios between the foliage and the root length and root volume, following the method of (Barley 1954). This ratio was then applied to the measured foliage dry weights for the remaining permeameters to obtain an estimate of the root length and volume. The root system was excavated from the permeameter and the root volume measured. This was done by immersing the root system in a cylinder of water and measuring the volume of displaced water. The root length was then measured with each root measured by hand and the length of laterals roots included in the total length. The depth of the main root branches and lateral growth was noted to provide a guide to the depth distribution of the roots.

3 RESULTS

3.1 Experiment 1: effect of soil water status

3.1.1 Temporal changes in hydraulic conductivity following the growth of maize

Figure 4 shows the changes to hydraulic conductivity following root growth and decay for soil which was

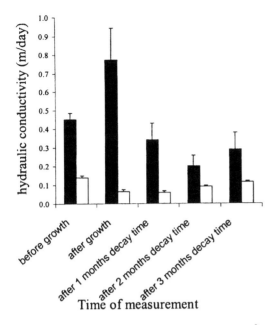

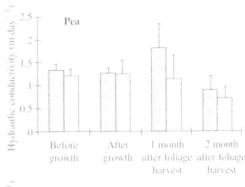

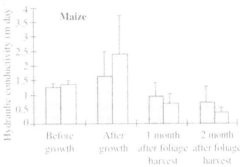

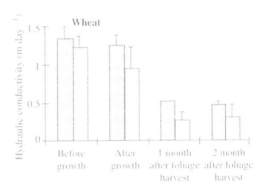

Figure 4. Mean K_{sat} over the entire permeameter for five measurement periods using different experimental conditions: (■) unsaturated growth and decay in a 10 part sand : 1 part kaolin soil, (□) saturated throughout in a 4 part sand : 1 part kaolin soil. The error bars represent the standard error of the mean from replicate permeameters.

Figure 5. Mean saturated hydraulic conductivity for the upper (▨) and lower (□) sections of permeameters, grouped by specie.

either saturated or unsaturated during root growth. The error bars represent the standard error of the mean. The unsaturated growth environment has larger error bars indicating increased variability. There is a significant difference between the methods for the baseline value, which is consistent with the different soil types used with lower K_{sat} values for the soil with the higher clay content. There are significant differences ($p < 0.05$) in K_{sat} following plant growth for both soil saturation states in comparison with the initial immediately after growth and in the initial decay period (1 month, see Figure 4). After 2 months decay the differences in K_{sat} between the 2 treatments were not significant.

3.2 Experiment 2: effect of plant species

3.2.1 Temporal changes to saturated hydraulic conductivity following the growth of maize, wheat or pea in unsaturated soil

Figure 5 shows the mean saturated hydraulic conductivity at two different depths in the permeameter for maize wheat or pea. Measurements were taken immediately after growth and during the period of root decay. The conductivity of soil rooted by maize increased after growth and then decreased as the roots decayed. The growth of wheat and pea had little effect on hydraulic conductivity immediately after the end of the growth period. However, as for maize, when the roots of these two plant species decayed the hydraulic conductivity decreased. The growth of pea roots had the least effect on soil hydraulic conductivity. The effect of wheat and maize was similar apart from the hydraulic conductivity measured immediately after the end of the root growth period.

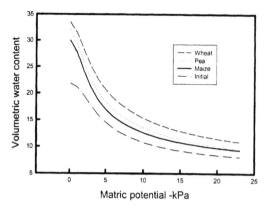

Figure 6. Water retention data for soil following the growth of wheat, pea or maize in comparison with the initial condition.

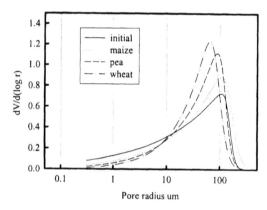

Figure 7. Pore size distributions calculate from the water retention data shown in figure 6.

3.2.2 *Water retention charactersitic of rooted and non-rooted soil.*

Figure 6 shows the water retention characterisic of the rooted soil for wheat, maize and pea in comparison with the initial condition. The bulk porosity of the soil, indicated from the water content at 0 kPa, appears to be higher following root growth. The pore size distributions calculated from the water retention data, shown in Figure 7, show that the pore size distribution is skewed towards smaller size following root activity.

3.2.3 *Relationship between root volume and hydraulic conductivity*

We did not find a strong evidence of a relationship between root volume or root length and the hydraulic

conductivity of the soil at any of the stages of growth and decay that we considered in this work.

4 DISCUSSION

The experimental results that are described in this paper illustrate that the changes in soil bulk hydraulic conductivity following root growth are complex and depend upon both the crop species and soil conditions during growth. When the soil was saturated during root growth hydraulic conductivity decreased immediately after root growth, but then increased during the period of root decay. This result for maize is in agreement with the findings of Barley (1954) also for maize. It appears that it may be possible to reduce the hydraulic conductivity of a saturated soil by growing maize, however, this effect is temporary and the hydraulic conductivity can be expected to return to its original value as the roots decay. Nevertheless this effect can be large, Barley observed a 9 fold reduction in hydraulic conductivity and in this work the reduction was 2 fold. This reduction is assumed to be due to compression of soil by roots reducing pore size. Analysis of pore size distributions in the soil from experiment 2 would support this hypothesis.

When maize was grown in unsaturated soil the hydraulic conductivity increased. This result in experiment 1 was repeated in experiment 2. This is an important finding because it demonstrates that the effect of root activity on soil hydraulic properties depends not only on the plant species, but also on the soil conditions during root growth. It is important to note that the different effects of maize on soil hydraulic properties are repeatable under the different soil conditions.

When plants were grown in unsaturated soil, only maize increased the hydraulic conductivity of the soil. For pea and wheat the effect of plant growth on hydraulic conductivity was initially neutral, but during the decay period hydraulic conductivity declined. The decay period for maize also resulted in a reduction in hydraulic conductivity. This reduction in hydraulic conductivity during root decay is likely to be due to microbial clogging of the pore network.

The data presented relate to how root activity can change the soil's saturated hydraulic conductivity. From these data we can infer some generalisations. Firstly, when plants grow in water-saturated soil the value of hydraulic conductivity is reduced. Secondly, when roots which have grown in saturated soil decay the hydraulic conductivity tends to increase. Thirdly, when plants are grown in unsaturated soil the effect on conductivity is smaller than when the soil is saturated. Finally, when roots grown in unsaturated soil decay the hydraulic conductivity deceases. These generalisations may apply only to the laboratory

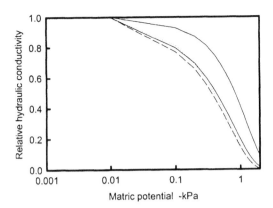

Figure 8. The values of calculated relative hydraulic conductivity estimated for maize, wheat and pea (the closely grouped lines) and also for the initial condition (the upper line).

model systems that we used for this work (Figures 1 to 3). It will be important to establish if these ideas can be applied to the field environment.

During much of the time soil is unsaturated. By fitting the van Genuchten equation to the water retention characteristic in figure 6 it is possible to predict the relative unsaturated hydraulic conductivity (van Genuchten 1980). These data are shown in Figure 8 and they are of great interest because they show that root activity has a large effect on the hydraulic conductivity of unsaturated soil. There is very little data on this aspect of root activity available in the literature. The data in figure 8 suggests that root activity can have a large effect on hydraulic conductivity of unsaturated soil even if any change in the value of the hydraulic conductivity of the saturated soil is hard to detect. Interestingly in this data there is very little difference in the estimate of relative conductivity between the different species.

The results reported in this paper relate to special soil conditions where there is little initial structure and negligible organic matter. In wet soil under these conditions root growth tends to rearrange particles by displacement as they push through the soil. Under these conditions the porosity of soil adjacent to the root can increase due to plastic deformation of soil. Here we can see that the hydraulic conductivity of soil reduces immediately after root growth (Figure 4). When the soil is unsaturated it may not deform plastically and the root may need to overcome a mechanical impedance that is due to an effective stress as well as the over burden stress. Under these circumstances the root may need to fracture the soil and create a macropore structure, which would be consistent with an increase in hydraulic conductivity. This was observed

for maize (Figures 4 and 5) but not for pea or wheat (Figure 5).

In the field environment root growth will lead to the development of soil structure through repeated wetting and drying. Here the hydraulic properties of the soil may be very different from those reported in this paper. However, knowledge of how vegetation changes the hydraulic properties of a newly established sites is important, because it is under these circumstances that the soil is likely to be most sensitive to damage by rain and water runoff.

5 CONCLUSIONS

The experimental work reported in this paper show that laboratory experimentation can give repeatable results. We found that the effect of plant growth on soil hydraulic conductivity depends on the soil condition during the growth period. When the soil was saturated the hydraulic conductivity reduced following root growth in this work the reduction was by a factor of two. During root decay hydraulic conductivity recovered towards its original value. When the plants were grown in unsaturated soil the effect of root activity on hydraulic conductivity was neutral. When roots had been growing in aerated soil hydraulic conductivity always decreased as the roots decayed.

ACKNOWLEDGEMENTS

The work reported in this paper was supported in part by BBSRC grant 204/BRE13684.

REFERENCES

Arnon, I. 1972. *Crop production in dry regions Volume II: Plant Science Monographs* (general editor N. Polunin) Leonard Hill. London.

Barley, K. P. 1954. Effects of root growth and decay on the permeability of a synthetic sandy loam. *Soil Sci.:* 205–210.

Bengtsson, L., Bendz, D., Hogland, W., Rosqvist, H. & Akesson, M. 1994. Water balance for landfills of different age *J. of Hyd.* 158: 203–217.

Cresswell, H. P. & Kirkegaard, J. A. 1995. Subsoil amelioration by plant roots – the process & the evidence *Aust. J. Soil Res.* 33: 221–239.

Fairley, R. I. 1985. Grass root production in restored soil following opencast mining In A. H. Fitter, D. Atkinson, D. J. Read & M. B. Usher (eds) *Ecological interactions in soil: plants, microbes and animals* British Ecological Society Special Publication No. 4.

Gregory, P. J. 1988. Growth & functioning of plant roots In A. Wild (ed) *Russell's Soil Conditions & Plant Growth:* Chapter 4, Harlow Longman.

Hairiah, K. & van Noordwijk, M. 1989. Root distribution of leguminous cover crops in the humid tropics and effects on a subsequent maize crop In J. van der Heide *Proc. Symp. Nutrition Management for Food Crops in Tropical Farming Systems:* Malang, Indonesia.

Martin, J. P. & Craggs, B. A. 1946. Influence of temperature and moisture on the soil-aggregating effect of organic residues *J. Am. Soc. Agron.* 38: 332–339.

Merrill, S. D., Ries, R. E. & Power, J. F. 1998. Subsoil characteristics and landscape position affect productivity of reconstructed mine soils *Soil Sci. Soc. Am. J.* 62: 263–271.

van Genuchten, M. Th. 1980. A closed-form equation for predicting the hydraulic conductivity of unsaturated soil. *Soil Sci. Soc. Am. J.* 44: 892–898.

White, R. E. 1997. *Principles and practice of soil science* Third Edition Abingdon: Blackwell Science.

Land Reclamation – Moore, Fox & Elliott (eds)
© *2003 Swets & Zeitlinger, Lisse, ISBN 90 5809 562 2*

Phytoremediation of a contaminated dredged canal sediment

A. Royle & N.M. Dickinson
School of Biological and Earth Sciences, John Moores University, England

P.D. Putwain & R. King
School of Biological Sciences, University of Liverpool, England

E. Gray-Jones
Environment and Regeneration Department, Warrington Borough Council, Warrington, England

ABSTRACT: Sediments in a derelict section of canal in Warrrington consist of a wet, black, odorous and oily mud containing a wide range of elevated contaminants including Cu, Zn, Ni, As, Pb, Cd, Cr, mineral oils, TPHs, PAHs and sulphides. Recognition of the high level of contamination and associated costs of disposal have been a constraint for restoration of the canal, which has been out of use for 50 years or more. We are investigating the feasibility of using *in situ* phytoremediation as a low-cost alternative to cart and dump. Our objectives are to demonstrate that metals can be rendered immobile and non-hazardous in soils and biomass, whilst plant roots and developing biota optimise conditions for the natural attenuation of organics. In the present paper, the rationale of this approach to reclamation and the first year of establishment of the project are described.

1 INTRODUCTION

1.1 *Phytoremediation*

Phytoremediation is receiving considerable attention as a low-cost treatment technology for land and groundwater contaminated with heavy metals (Chaney et al. 1997, Glass 1999, Pulford et al. 2002, Vangronsveld et al. 2000) and organic compounds (Campanella et al. 2002, Carman et al. 1998, Meagher 2000, Susarla et al. 2002). One strategy is to use fast-growing trees, particularly *Salix* and *Populus* species, to remove, stabilise or enhance the volatalisation of polluting chemicals (Greger & Landberg 2000, Jones et al. 1999, Pulford et al. 2002, Vervaeke et al. 2002). There is a real possibility that plants can be used to reclaim contaminated land and restore sustainable and healthy soils (Dickinson 2000, Kearney & Herbert 1999).

1.2 *Contaminated canal sediment*

Environmental improvement of a derelict section of canal in Warrington was prohibitively expensive due to the cost of disposal of some 40,000 t of contaminated sediment. From an earlier survey it was known that the canal sediment consisted of a wet, black, odorous and oily mud (up to 1.5–1.7 m depth) containing a wide range of elevated contaminants including Cu, Zn, Ni, As, Pb, Cd, Cr, phenols, mineral oils and S. The proportion of bioavailable metals in this type of sediment may be as much as 40% of total, even after 60 years without disturbance. This may present a considerable hazard of dispersal to the wider environment if wet, reduced sediments are disturbed or dredged for land disposal.

The 2 km section of canal at Woolston was excavated in 1821 to improve navigation along the river Mersey by cutting out a large, looping section and hence straightening its course. A weir just after the upstream end of the canal marks the limit of the tidal reaches of the river and hence the downstream end of the canal is under tidal influence. Water flow into and out of the canal was controlled by lock gates and an aqueduct took overflow water across the river to feed into another nearby canal. Historical industrial development along the Woolston New Cut Canal appears to have been limited to a chemical works (active for at least 40 years, maybe as much as 70 or 80 years), a gunpowder mill, a tannery (only operational for perhaps a decade), an abattoir, a brick works and a piping and tubing works. These industries will have discharged into the canal and left their mark in the composition of the sediment. The original sailboats and barges which used the canal gave way to oil power and this is undoubtedly another source of pollution via leaks and spillages. Adjoining sections of the river Mersey have supported various industrial and engineering works including many metal works, more tanneries and a gas works.

Rail and road transport came to outdo canal transportation and the canal was allowed to fall into disrepair. It has not been used since the late 1940's. Perhaps associated with the demolition of the aqueduct in 1978, the canal appears to have ceased to flow. Water levels gradually dropped and the standing water and wet sediment has now been colonized by vegetation, dominantly *Typha latifolia* (Reedmace), with *Salix atrocinerea* (Sallow) at the edges.

Modern developments along the canal include housing estates, an industrial estate and a green waste recycling facility.

1.3 *Aims of the research project*

The project described in this paper is a case study of the feasibility of using phytoremediation *in situ* in the contaminated sediment as a low-cost alternative to cart and dump. By modeling this ecosystem, the objectives are to investigate whether metals can be rendered immobile and non-hazardous in soils and biomass, whilst plant roots and developing biota optimise conditions for the natural attenuation of organics as well as transforming the sediment into a soil.

2 METHODS

2.1 *Site preparation*

Established vegetation was cleared from a 150 m section of the New Cut Canal south bank and from shallow sediment on that side. The south bank is the shallow side of the canal, the deep side being the north side with the towpath running along the bank.

A raised platform (3.5 m wide) above existing water level was then created through dredging and transfer of sediment from the towpath side to the shallow side of the canal (Figure 1). The platform was divided into 6 experimental blocks. Twelve short-rotation coppice species, hybrids and clones of willows, poplars and alders (Table 1), were then planted on the raised platform in double rows (0.5 m × 0.5 m) of 6 plants of each clone, with 1 m between rows. *Salix* and *Populus* were planted as pegs, and *Alnus* was planted as 50–70 cm rooted stock (which was pruned back after establishment). Control plots were also included in the design. Species were randomly allocated within each block.

Three sediment samples (0–15 cm) were taken with an auger between each row, and then bulked for each

Figure 1. Planting platform of dredged sediment being prepared on the shallow side of the canal.

Table 1. Tree species and clones planted on the raised platform of sediment within the canal.

Salix viminalis "Jorunn" *
Salix viminalis × *schwerinii* "Tora" *
Salix caprea × *cinerea* × *viminalis* "Calodendron" *
Salix viminalis × *burjatica* "Ashton Stott" *
Salix viminalis × *caprea* "Sericans"
Salix fragilis *
Salix atrocinerea #
Poplus deltoides × *nigra* "Ghoy"
Poplus trichocarpa "Trichobel" *
Alnus glutinosa
Allnus incana
Alnus cordata

* Recommended for SRC use in FC Information Note 17.
# From cuttings of trees that had naturally colonized the canal.

of the 6 blocks. The samples were thoroughly mixed and then evenly divided into four. One set was sent to each of three national UKAS accredited laboratories for analysis and the forth set was kept in house.

Five boreholes were established to the side of the canal to monitor groundwater in case of contamination. Due to the canal having a clay liner, any contaminants which may leach from the sediment bank should remain within the bounds of the canal and not cause contamination. Establishment of growth of the trees was monitored during the first year, and invasive plants were controlled by hand weeding.

3 CURRENT PROGRESS

3.1 *Contamination levels*

Results of sediment analyses showed considerable variation between the three laboratories (Table 2). Part of the explanation for these differences concern

Table 2. Mean values ($\mu g\,g^{-1}$) for sediment contaminants, as provided by separate UKAS accredited laboratories. Values are means of 7 samples. Shaded values are those exceeding standard thresholds of contamination.

Determinand	Laboratory		
	A	B	C
Sulphate (% SO4/SO3)	0.53	6022 (total)	0.97
Sulphide	1078	354	84.0
Arsenic	349	682*	416
Boron	69.0*	3.5	1.6
Cadmium	12.7	13.6	18.1*
Chromium	790*	1471*	977*
Copper	567*	1076*	736*
Lead	1221*	2150*	1443*
Nickel	67	78	783.1
Mercury	3.8	7.5*	4.8
Zinc	3631	5835*	4286
Cyanide	103*	10.6	23.2
Total PAH	216*	141	121
TPH	7636*	5671*	2207* (C6-C40)

* Significantly different to results of other laboratories.

different analytical methodologies – the details of which were not automatically provided in full detail by any of the laboratories. Whilst the three laboratories were UKAS accredited, standard methods for sample preparation, extraction and analysis vary (Dickinson et al. 2000). Although there is a general consensus in the determinands that exceed existing thresholds, significant differences existed between the laboratories for most determinands. The only exception was for sulphide where data were very variable between samples which probably masked differences between laboratories.

3.2 Tree establishment

Tree establishment was very successful, with only 25% mortality in the first year. The least successful species was *S. atrocinerea* (mortality 67%). Although this species had naturally colonized the canal, it appeared to be a difficult species to root from cuttings. Survival of all other species was good.

3.3 Further research

During the course of the project we will be studying various aspects of the functioning of the ecosystem. At present a pot experiment is underway to assess the effects of various amendments on the rate of remediation. Soil fauna will be studied and its composition monitored and related to soil health. Laboratory experiments will determine the changes in the sediment and its contaminants during drying and oxidation. The bioavailability of the metals in the sediment will also be monitored over the course of the project. We will look at the transfer of foliar metals to herbivorous animals in a laboratory environment as a model for food chain transfer.

4 DISCUSSION

The sediment of the canal, whilst waterlogged, provides an anoxic, reduced environment. Metals will be in reduced forms and breakdown of organic contaminants will be minimal. After dredging, dewatering and exposure to air, the sediment will start to oxidise. Oxidation will induce a host of chemical and biological changes in the sediment and the contaminants contained within it. Sediments deposited on land following dredging have been the subject of previous studies in Belgium (Tack et al. 1996, Tack et al. 1998, 1999, Tack & Verloo 1999) and the UK (Stephens et al. 2001). Most metals show a redistribution from residual to mobile phases during drying and oxidation that is also associated with decreasing sulphide/sulphate ratio. Metal bioavailability may be particularly high in the early stages, but long-term prediction of metal migration is uncertain.

In the present project, the sediment was retained within the canal and it is likely that migration of metals to the wider environment will be controlled to a large extent by the clay liner of the canal. However, disturbance of the bottom sediment and leaching from the dredged sediment may affect contaminant concentrations in the water column of the canal. Hence, before this can be seriously considered as a treatment technology, it is important to demonstrate that contaminants are not quickly dispersed to the wider environment. Another potential source of dispersion is through food chains (Vandecasteele et al. 2002, 2003) and this will be assessed over the next two years.

5 CONCLUSION

Manipulating the processes of contaminant dispersion or immobilisation offers a real possibility of treating contaminated sediment without removal, whilst contributing to a healthy, sustainable, non-hazardous landscape of high ecological and amenity value. If successful, this will provide a realistic, generic and transferable methodology with wide application for cost-effective *in situ* reclamation of contaminated sediments.

REFERENCES

Campanella, B. F., Bock, C. & Schroder, P. 2002. Phytoremediation to increase the degradation of PCBs and PCDD/Fs. Potential and limitations. *Environmental Science and Pollution Research International* 9: 73–85.

Carman, E. P., Crossman, T. L. & Gatliff, E. G. 1998 Phytoremediation of No. 2 fuel oil-contaminated soil. *J. Soil Contam.* 7: 455–466.

Chaney, R. L., Malik, M., Li, Y. M., Brown, S. L., Angle, J. S. & Baker, A. J. M. 1997. Phytoremediation of soil metals. *Current Opinion in Biotechnology* 8: 279–284.

Dickinson, N. M., 2000, Strategies for sustainable woodlands on contaminated soils. *Chemosphere* 41: 259–263.

Dickinson, N. M., MacKay, J. M., Goodman, A. & Putwain, P. 2000. Planting Trees on Contaminated Soils: Issues and Guidelines. *Land Contamination and Reclamation* 8: 87–101.

Glass, D. 1999. International activities in phytoremediation. In: A. Leeson & B.C. Alleman (eds.), *Phytoremediation and innovative strategies for specialized remedial applications.* Columbus: Battelle Press. 95–100.

Greger, M. & Landberg, T. 2000. Use of willows in phytoextraction. *International Journal of Phytoremediation* 1: 115–123.

Jones, S. A., Lee, R. W. & Kuniansky, E. L. 1999. Phytoremediation of trichloroethene (TCE) using cottonwood trees. In: A. Leeson & Alleman, B. C. (eds.), *Phytoremediation and innovative strategies for specialized remedial applications.* Columbus: Battelle Press. 101–108.

Kearney, T. & Herbert, S. 1999. Sustainable remediation of land contamination. In: A. Leeson & B.C. Alleman (eds.), *Phytoremediation and innovative strategies for specialized remedial applications* Columbus: Battelle Press. 283–288.

Meagher, R. B. 2000. Phytoremediation of toxic elemental and organic pollutants. *Current Opinion in Plant Biology* 3: 153–162.

Pulford, I. D., Riddell-Black, D. & Stewart, C. 2002. Heavy metal uptake by willow clones from sewage sludge-treated soil: the potential for phytoremediation. *International Journal of Phytoremediation* 4: 59–72.

Royle, A. 2002. "Site History: The Woolston New Cut Canal". *Internal publication.*

Stephens, S. R., Allowy, B. J., Parker, A., Carter, J. E. & Hodson, M. E. 2001. Changes in the leachability of metals from dredged canal sediments during drying and oxidation. *Environ. Pollut.* 114: 407–413.

Susarla, S., Medina, V. F. & McCutcheon, S. C. 2002. Phytoremediation: An ecological solution to organic chemical contamination. *Ecological Engineering* 18: 647–658.

Tack, F. M., Callewaert, O. W. J. J. & Verloo, M. G. 1996. Metal solubility as a function of pH in contaminated dredged sediment affected by oxidation. *Environmental Pollution.* 91: 199–208.

Tack, F. M., Singh, S. P. & Verloo, M. G. 1998. Heavy metal concentrations in consecutive saturation extracts of dredged sediment-derived surface soils. *Environmental Pollution.* 103: 109–115.

Tack, F. M., Singh, S. P. & Verloo, M. G. 1999. Leaching behaviour of Cd, Cu, Pb and Zn in surface soils derived from dredged sediments. *Environ. Pollut.* 106: 107–114.

Tack, F. M. G. & Verloo, M. G. 1999. Single extractions versus sequential extraction for the estimation of heavy metal fractions in reduced and oxidised dredged sediments. *Chemical Speciation and Bioavailability* 11: 43–50.

Vandecasteele, B., De Vos, B. & Tack, F. M. G. 2002. Cadmium and zinc uptake by volunteer willow species and elder rooting in polluted dredged sediment disposal sites. *The Science of The Total Environment* 299: 191–205.

Vandecasteele, B., Lauriks, R., De Vos, B. & Tack, F. M. G., 2003, Cd and Zn concentration in hybrid poplar foliage and leaf beetles grown on polluted sediment-derived soils. *Environ. Monit. Assess.* in press.

Vangronsveld, J., Ruttens, A., Mench, M., Boisson, J., Lepp, N. W., Edwards, R., Penny, C. & van der Lelie, D., 2000. In situ inactivation and phytoremediation of metal- and metalloid-contaminated soils: field experiments. In: D.L. Wise & D.J. Trantolo (eds.), *Bioremediation of Contaminated Soils.* New York: Marcel Dekker, Inc. 859–885.

Vervaeke, P., Luyssaert, S., Mertens, J., Meers, E., Tack, F. M. G. & Lust, N. 2002. Phytoremediation prospects of willow stands in contaminated sediment: a field trial. *Environ. Pollut.* Reviewed manuscript July 2002.

Land Reclamation – Moore, Fox & Elliott (eds)
© 2003 Swets & Zeitlinger, Lisse, ISBN 90 5809 562 2

CL:AIRE – five years on

P. Beck
Chief Executive, CL:AIRE

ABSTRACT: CL:AIRE was established to develop and provide confidence in sustainable and cost-effective methods for remediating contaminated land. CL:AIRE's objectives involve: developing demonstration projects on contaminated sites, developing a research strategy, disseminating information, preparing educational materials and securing funding. A total of 43 project applications have been received of which 24 have been approved as CL:AIRE projects. CL:AIRE has developed a research strategy to identify remediation topics that it believes should be addressed through field demonstrations of research and technology. Disseminating information that is of interest to the contaminated land community and education are key roles for CL:AIRE.

Social and corporate responsibility, combined with UK and European legislation have created a growing interest in the consideration of alternative remediation methods to landfilling. CL:AIRE is helping to increase the confidence in the use of alternative methods by disseminating impartial and balanced assessment of demonstrations and research to stakeholders in the industry.

1 INTRODUCTION

CL:AIRE (Contaminated Land: Applications in Real Environments) is a public/private partnership, which was incorporated in March 1999. Its purpose is to demonstrate research and technologies to develop and provide confidence in sustainable and cost-effective methods for remediating contaminated land. Included within the technical remit are innovative methods for site characterization and monitoring.

This paper will discuss the development of the organization heading into its fifth year and reflect upon where it now stands and how it can go forward.

The two main aims of CL:AIRE which are embodied in its mission statement are to (i) improve the quality of the UK environment and (ii) improve the UK economy through cost effective treatment and market opportunities.

2 ORGANIZATION

CL:AIRE is a not-for-profit company, limited by guarantee. It is also a registered charity and an environmental body registered with ENTRUST. The guarantee is provided by the six organizations which make up the legal membership and include: Soil and Groundwater Technology Association (SAGTA), English Partnerships, Environment Agency, Scottish Environment Protection Agency, Welsh Development Agency

and Department of Environment (NI). In addition to forming CL:AIRE, the members along with the following group of companies, provided core funding support to ensure the economic viability of CL:AIRE for a period of three years: Department of the Environment, Food and Rural Affairs (DEFRA); UK Coal; Scottish Enterprise; Shanks; Scottish Power; Unilever and Viridor Waste Management.

The Board of Trustees which provides the overall direction for the company has representation from the six member organizations and from the following stakeholders: small and medium-sized business, research councils, local authorities and the insurance industry. Day to day operations are carried out by a four-person Management Team headed by the Chief Executive.

The Management Team is supported by two advisory groups, a User Group, and a Technology and Research Group. The User Group is comprised of individuals with expertise in the characterization and cleanup of contaminated land, representing site owners, developers, local authorities and environmental consultants. The Technology and Research Group consists of individuals with experience in the practical aspects of research development and includes academics, industry practitioners and regulators. The advisory groups provide advice to CL:AIRE and review CL:AIRE working documents. They participate in the selection of sites and projects, and help to formulate a research strategy for CL:AIRE. The

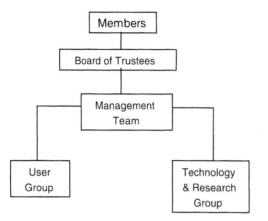

Figure 1. Plan showing the structure of CL:AIRE.

reporting structure of the organization is illustrated in Figure 1.

From the outset the organization has operated from an ethos which has emphasized transparency and strong corporate governance. It was recognized early on that CL:AIRE held a unique position in the contaminated land industry by providing independent assessment function to all stakeholders with an emphasis on the practical and applied aspects of remediation. Consequently, a high value has been placed on establishing and maintaining the organization's independence.

3 OBJECTIVES

The objects and powers of CL:AIRE are legally defined in its Memorandum of Association. Its operational objectives are to:

- establish a network of characterised, contaminated sites throughout the UK
- demonstrate the application of remediation technologies which may offer improved site investigation, monitoring or remediation solutions
- develop a strategy for remediation technology research in the UK
- disseminate information related to research and technology demonstrations on contaminated land
- prepare and provide educational materials for the environment industry, students and the general public
- procure sources of funding to support its activities

4 ACTIVITIES

The activities of CL:AIRE are carried out to meet its objectives. Whilst the organization's objectives have remained constant from the beginning, the activities have evolved to reflect a changing market and the need to modify approaches where first attempts may have been unsatisfactory.

4.1 Sites

It was expected that CL:AIRE would develop a portfolio of well characterized, secure sites throughout the UK which would be made available for technology demonstrators and researchers to carry out their field trials. While more than 50 sites have been put forward, only a handful are considered to be well characterized.

The difficulty of obtaining well characterized sites is tied to the economics and future use of the site. Because of the pressure on site owners to generate immediate returns on their investment, those sites that have been well characterized are usually turned over to meet requirements for immediate disposal or re-development, consequently they are not available for CL:AIRE projects. However there are opportunities to use sites which are already undergoing full scale remediation, to provide a one-off technology demonstration project.

Within the existing CL:AIRE site portfolio, a number of special sites of national interest have been developed. These are large well characterized sites available for demonstration and research.

The SIReN site (Site for Innovative Research into Monitored Natural Attenuation) is an active petrochemical facility which has been made available for research into monitored natural attenuation (Lethbridge et al. 2002). SIReN is a joint initiative involving Shell, the Environment Agency, AEA Technology and CL:AIRE.

CoSTaR (Coal Mine Sites for Targeted Remediation Research) which includes six closely spaced, former coal mining sites, has been established between the University of Newcastle, the Coal Authority, the County Councils of Durham and Northumberland, and CL:AIRE. The facility will serve to investigate and find practical, low cost, low energy solutions to land contaminated by coal mining activities (Younger et al. 1997, Jarvis & Younger 1997, Batty & Younger 2002, Nuttall 2002). A similar research site is being developed to address contamination from metal mining operations.

4.2 Projects

CL:AIRE has established a comprehensive application process to encourage the submission of field demonstration projects. The process starts with discussions with potential applicants. Projects are developed as either technology demonstrations, or research projects. Applicants then submit their application

Table 1.	Technology demonstration projects.	
Project no.	Technology type	Project status
TDP1	Low temperature thermal desorption	Completed
TDP2	Soil washing	Completed
TDP3	Zero valent iron permeable reactive barrier	Completed
TDP4	Slurry phase bioremediation	In progress
TDP5	Wetlands	In progress
TDP6	Solid phase biopile	Completed
TDP7	Solid phase biopile	Cancelled
TDP8	Stabilisation/solidification	Completed
TDP9	Air sparging	Completed
TDP10	Low temperature thermal desorption	Completed
TDP11	Soil washing	Completed
TDP12	Solid phase biopile	In progress
TDP13	Biotic permeable reactive barrier	In progress
TDP14	Soil washing	Completed
TDP15	Windrow composting	In progress

Table 2.	Research projects.	
Project no.	Technology type	Project status
RP1	Characterising adsorption	In progress
RP2	Characterising bioactivity	Completed
RP3	Characterising natural attenuation in a petroleum contaminated chalk aquifer	In progress
RP4	Statistical method to optimise investigation to characterise contaminated land	Not commenced
RP5	Bonemeal phosphates to immobilise metal contamination on soil	In progress
RP6	Phytoextraction of metals: investigation of hyperaccumulation and field testing	In progress
RP7	Oxidation and bioremediation of PCBs	Not commenced
RP8	Geophysical methods for site characterisation	Not commenced
RP9	Phytotoxicity indicators	In progress

which is evaluated by the Technology and Research Group (TRG). Projects may be approved, conditionally approved or rejected. Approved projects are sent to the Board for ratification. Conditional approval is given to those projects, which in the opinion of the TRG, have merit but may raise issues regarding the project design or work plan. The TRG provides comments and feedback to the applicant to help improve the quality of the scientific and technical aspects of the project and its deliverables. If the applicant modifies the application to address the concerns of the TRG, the project is approved.

More than 70 individuals/organizations have expressed an interest in participating in a CL:AIRE research or technology demonstration project. This interest has generated, as of January 2003, 43 project applications of which 24 have been approved by the TRG and ratified by the CL:AIRE Board. One project application has been conditionally approved and four are with the TRG for consideration. CL:AIRE has met its targets of one project in Year 1, 5 projects in Year 2 and 10 projects in Year 3. Technology demonstration and research projects are listed in Tables 1 and 2.

4.3 Research strategy

CL:AIRE has developed a research strategy to identify remediation topics that it believes should be addressed through field demonstrations of research and technology. The strategy was initially developed through discussions with key UK researchers, consultants, contaminated land owners and regulators during 2000. The strategy is very much influenced by the need to develop practical solutions. It undergoes annual revision and is ratified by the CL:AIRE Board. The strategy document is distributed to researchers, funders and environmental practitioners, and is posted on the CL:AIRE website.

4.4 Dissemination of information

Disseminating information that is of interest to the contaminated land community is a key role for CL:AIRE. With constant feedback from stakeholders, CL:AIRE has developed a number of vehicles for disseminating information. It constantly receives requests for information and has become a first port of call for contaminated land enquiries. The following are key features of the CL:AIRE dissemination process:

- Database
- Newsletter
- Website
- Project reports
- Factsheets
- Technical, Research, Case Study and Guidance Bulletins
- Conferences and Workshops

CL:AIRE has developed a database of subscribers who have a direct interest in the management and remediation of contaminated land. From an initial base of 300 subscribers in February 1999, the number has grown to more than 4650. Subscribers can register with CL:AIRE by contacting the CL:AIRE office or through the website. There is no cost to become a subscriber and to register with CL:AIRE, and it ensures free distribution for most CL:AIRE publications.

CL:AIRE view is a quarterly newsletter which has been published since December 1999. The newsletter provides a forum for presenting the news and views from key stakeholders in the contaminated land industry in the UK and internationally. Back issues of *CL:AIRE view* are available on the CL:AIRE website.

The CL:AIRE website at www.claire.co.uk was launched in September 1999 and activity has shown a steady increase growing from an average of 2,300 hits and 300 visits per month to more than 100,000 hits and 2,500 visits at the end of 2002. The website contains information about CL:AIRE, allows downloading of CL:AIRE documents, provides links to other organizations and provides summaries of CL:AIRE projects. An ongoing exercise is to post all contaminated land projects being undertaken by research organizations in the UK. There are more than 120 universities and research organizations carrying out contaminated land research.

CL:AIRE publishes the results of its projects in the form of Technology Demonstration Project (TDP) reports and Research Project (RP) reports. The first TDP (TDP3) report on permeable reactive barriers was published in November 2001. Other TDP reports and 1 RP report (RP1) are in preparation for publication in 2003. TDP1 will report on a low temperature thermal desorption trial while TDP 2 describes the full scale cleanup of a gas works using soil washing. Both reports are expected to be published in early 2003. Other reports will be published later in the year including: TDP4, TDP6, TDP8, TDP9, TDP10, and TDP11. Two-page Factsheets which summarize each project report are prepared and distributed to database subscribers. Factsheets are currently available for TDP2 and TDP3.

CL:AIRE is developing a series of four-page Technology (TB), Research (RB) and Case Study Bulletins (CSB) and eight-page Guidance Bulletins (GB) to report on innovative techniques/products, practical aspects of research, useful case studies and guidance. The purpose of the Technical, Case Study and Guidance Bulletins is to raise the level of best practice in the UK. The purpose of the Research Bulletins is to try to improve the uptake of applied research to the benefit of both the research community and environmental practitioners. The first bulletins, TB2 and CSB1 were published in March 2002 and distributed to CL:AIRE subscribers. TB1 and CSB2 were published and distributed toward the end of 2002.

2002 proved to be a busy year for CL:AIRE special events. The first Annual Project Conference was held in April and showcased 13 CL:AIRE projects. The first technology workshop on bioremediation was held in June, and a conference on DNAPL research was held in September. Events were well attended and feedback indicates that they offer good value to delegates. It is CL:AIRE's policy to keep fees as low as possible for these events to attract those who are on restricted budgets. Other events have been planned for 2003. The Annual Project Conference was held in March and a CL:AIRE Workshop on wetlands for treating acid mine drainage is offered at the Land Reclamation Conference held in Runcorn in May 2003.

4.5 *Education*

The preparation of educational material has always been a key objective and it is a statutory requirement that charities provide some form of education function.

Initially, CL:AIRE considered developing educational materials for secondary school children, but early contact with the Department for Education and Skills was not encouraging, and indications suggested that active participation in trying to develop materials such as teacher training packs for the national curriculum was beyond the resources of CL:AIRE. As a result, a conscious decision was made to direct more of the education objective toward the contaminated land stakeholders themselves where feedback continues to demonstrate a real need. The forms of information dissemination described above help deliver the education objective.

In addition, CL:AIRE has undertaken video filming of field demonstration projects for the purpose of developing educational videos for the public, students and environmental industry practitioners.

4.6 *Funding*

CL:AIRE is constantly searching for new sources of income to meet expenses for:

- delivering CL:AIRE activities
- financing CL:AIRE projects

Whilst CL:AIRE was not set up originally with a source of project funds, it has been able to lever more than £1M of public and private funding for technology demonstration and research projects.

5 THE CHANGING LANDSCAPE OF CONTAMINATED LAND

The contaminated land market is constantly changing. There has been an observable increase in the interest in the use of alternatives to landfilling. Whilst the majority of contaminated soil is still taken to landfill, social and corporate responsibility, and UK and European legislation have created a growing interest in the consideration of alternatives. Environmentally aware contaminated land owners are requiring their consultants to assess and cost alternative remedial methods alongside landfilling. Some remedial techniques such as bioremediation and soil vapour extraction have now become so common place that they are now part of mainstream remediation and under the right conditions, offer cost-effective alternatives to landfilling.

Government policy and regulatory issues are key drivers for change and have had a major influence on the contaminated land industry since the inception of CL:AIRE. The Office of the Deputy Prime Minister (ODPM), requires that redevelopment agencies reclaim 1,400 hectares of brownfield land each year (Rolls, pers. comm.). As the former Department of the Environment, Transportation and the Regions (DETR), ODPM developed policy that requires 60% of all new and refurbished housing to be developed on Brownfield land.

Contaminated site owners, environmental practitioners and local authorities are becoming familiar with the new contaminated land regime, introduced by Part IIA of the Environment Protection Act 1990. Regulations came into force in 2000 in England and Scotland and in 2001 in Wales. Part IIA defines contaminated land and embodies the principle of polluter pays. It takes a risk-based approach to the management of contaminated land. Local authorities are required to identify contaminated land within their jurisdiction. There is statutory guidance associated with Part IIA which includes derivation of soil and water clean up standards (Department of the Environment, Transport and the Regions, 2000).

The Pollution Prevention and Control Regulations 2000 implement EC Directive 96/61/EC on integrated pollution prevention and control (IPPC). They address land contamination from current licensed industrial activities. Operators are required to establish baseline environmental conditions which must be met on closure and on surrender of their permit.

Other European legislation has implications for the UK. The recent Landfill Directive 1999/31/EC which came into effect in England and Wales through new Landfill Regulations in July 2002, identifies three classes of landfill: hazardous, non-hazardous and inert. Waste acceptance criteria for each class will determine the type of waste that can go to each landfill. The new regulations end the practice of co-disposal of hazardous and municipal waste. Waste that cannot meet waste acceptance criteria will require pre-treatment. These new regulations are expected to increase the disposal costs of hazardous waste, and will add to the cost of remediation for those wastes which require pre-treatment.

The Water Framework Directive 2000/60/EC is another key piece of European legislation which will place the management of water resources within the context of a watershed or drainage basin. Regulators will be required to undertake inventories of their groundwater and surface water resources and develop appropriate policies to ensure compliance with statutory water quality objectives.

The licensing of remedial works is a major hurdle to the use of alternative technologies. The process of applying for a license to carry out remediation can be time consuming and licenses are costly to maintain. A streamlining of the process was one of the recommendations of Lord Rogers' Urban Task Force. The Remediation Permit, a recent initiative led by Phil Kirby, OBE of SecondSite Property Holdings Ltd (formerly Lattice Property Holdings Ltd) aims to simplify the permitting of remedial technologies in order to reduce barriers to the regeneration of Brownfield lands (Remediation Permit Working Group 2002).

6 WHERE DOES CL:AIRE STAND HEADING INTO YEAR 5?

Within this changing landscape brought about by legislation and market conditions, the increasing demand for the outputs from CL:AIRE demonstrate a growing interest in the technical aspects of contaminated land management. As a consequence, CL:AIRE is developing into a major player in the contaminated land scene. One of the reasons for this is its independence. Because CL:AIRE does not have obligations to any party, it is seen as impartial and has considerable credibility and is trusted by most stakeholders in the industry. The following summary provides highlights CL:AIRE's progress. CL:AIRE:

- provides sites for research and technology demonstrations
- develops projects to advance knowledge and uptake of sustainable remediation methods
- provides comment on research proposals and letters of support for projects which are within CL:AIRE's remit
- identifies research needs in its research strategy which is circulated annually
- has been responsible for directing more than £1 M of private and public funding toward research and technology demonstration projects despite not having a budget for project funding
- is a member of FIRST Faraday, a centre of excellence for the remediation of the polluted environment, and is associated with several contaminated land networks including PRB network at Queen's University, Belfast and STARNET at Cambridge University
- is active with other research programmes. Contaminated land researchers applying through the LINK programme and FIRST Faraday are encouraged to direct their projects through CL:AIRE
- is increasingly involved in project steering committees and other initiatives involving contaminated

land. CL:AIRE was part of the Remediation Permit Working Group
- is recognized as a key source of information related to contaminated land and maintains an active programme of information dissemination
- adds value to a number of stakeholder organizations by providing impartial information or opinions. An assessment by English Partnerships using an independent consultant confirmed that CL:AIRE was providing a useful service to the contaminated land industry. The Office of the Deputy Prime Minister (ODPM) requires that Regional Development Agencies in England and Wales involve CL:AIRE in major regional regeneration schemes that include remediation.

7 WHAT LIES AHEAD?

In the near term, CL:AIRE will operate in the same way as it has for the past four years and continue the delivery of its objectives. CL:AIRE will continue to establish sites of national interest for research and to develop high quality demonstration projects. Disseminating the results of projects and other information will remain a key activity. The CL:AIRE website will be expanded with information about research and technology demonstration projects and important publications. CL:AIRE will host three conference/workshop events annually.

Because of CL:AIRE's unique position and experience as facilitator and independent assessor of technologies, there are opportunities for CL:AIRE to be involved at a more strategic level with a variety of research and technology demonstration projects as well as having input into the development of UK contaminated land policy.

At this point in CL:AIRE's development, it is necessary to achieve stable funding to ensure that CL:AIRE can continue to deliver its objectives. CL:AIRE has developed a modest revenue stream, but this is not sufficient to cover annual expenses. The initial tranche of core funding which was provided after CL:AIRE was incorporated has been paid out. Whilst several of the original supporters have continued or have expressed an interest in continuing to support the organization, additional sources of funding will need to be found within the next year.

All in all, CL:AIRE has made considerable progress in the past four years in a market that has grown and continues to undergo change. The feedback that we get from our client base tells us that we have an important role to play in improving the knowledge base within the contaminated land industry and we look forward to the challenges that lie ahead.

REFERENCES

Batty, L.C., & Younger, P.L. 2002, Critical role of macrophytes in achieving low iron concentrations in mine water treatment wetlands. *Environmental Science & Technology*, 36: 3997–4002.

Council of the European Community. 1996. Council Directive of 24 September 1996 concerning integrated pollution prevention and control (96/61/EC). OJ L257, 26–40.

Council of the European Community. 1999. Council Directive of 26 April 1999 on the Landfill of Waste (1999/31/EC). L182, 1–19.

Council of the European Community. 2000. Council Directive of 23 October 2000 establishing a framework for community action in the field of water policy (2000/60/EC). L327, 1–72.

Department of the Environment, Transport and the Regions 2000. *Contaminated Land*. DETR Circular 02/2000.

Jarvis, A.P., & Younger, P.L. 1997, Dominating chemical factors in mine water induced impoverishment of the invertebrate fauna of two streams in the Durham Coalfield, UK. *Chemistry and Ecology*, 13: 249–270.

Lethbridge, G., Scott, P., Neaville, C., Earle, R., Hart, A., Macnaughton, S., & Swannell., S. 2002. SIReN: The Site for Innovative Research into Monitored Natural Attenuation. In CL:AIRE (editor), *Proceedings of the CL:AIRE Annual Project Conference April 11th 2002.*

Nuttall, C.A. 2002. Testing and performance of a newly constructed full-scale passive treatment system at Whittle Colliery in Northumberland. In Nuttall, C.A. (editor), *Mine water treatment: a decade of progress. Proceedings of a Conference held in Newcastle Upon Tyne, Nov 11–13th 2002.* 16–24.

Remediation Permit Working Group (Chair Phil Kirby OBE) 2002. *The Remediation Permit: Towards a single regeneration licence.*

Rolls, M. 2002. Office of the Deputy Prime Minister, Eland House, London

Younger, P.L., Curtis, T.P., Jarvis, A.P., & Pennell, R. 1997, Effective passive treatment of aluminium-rich, acidic colliery spoil drainage using a compost wetland at Quaking Houses, County Durham. *Journal of the Chartered Institution of Water and Environmental Management*, 11: 200–208.

Soil Restoration

Land Reclamation – Moore, Fox & Elliott (eds)
© *2003 Swets & Zeitlinger, Lisse, ISBN 90 5809 562 2*

Use of biopellets in establishing species-rich grassland on reclaimed colliery spoil

J.P. Newell Price
Land Research Associates, Durham, United Kingdom

E.A. Allchin
Entec UK Ltd., Gosforth, Newcastle-upon-Tyne, United Kingdom

ABSTRACT: This paper presents the results of a field trial into the benefits of biopellets as an amendment to colliery spoil at the former Hawthorn Colliery and Coke Works, Murton, Co Durham. It considers the relative benefits of various biopellet application rates to the growth of a wild flower seed mix and also investigates the impact of the colliery spoil and the biopellets on water quality. The biopellets provide sufficient nutrients for the growth of a good cover of wild flowers. Soil water sulphate concentrations were consistently above the discharge limit of 2000 mg/l but were related to the depth of colliery spoil through which the water was moving. Nitrogen concentrations in both soil water and foliar samples were correlated with the rates of biopellet application. Leaching in drainage water was a problem throughout the first winter at the higher application rates. The best vegetation covers were achieved on the biopellet plots.

1 INTRODUCTION

The intended end use for the One North East owned former Hawthorn Colliery and Cokeworks site at Murton, Co Durham was woodland and native grassland, with the possibility of future industrial development in the central zone. According to a feasibility report by Entec UK Ltd. (1999), the reclamation would have required the importation of over 360,000 m³ of soil which would be difficult to source, and costly and unacceptable from a logistical and environmental point of view.

An alternative solution was proposed involving the use of a thermally dried sewage sludge product, ("biopellets") as an amendment to on site soil-forming materials. One North East accepted this initiative but, before the method could be used across the whole site, the Environment Agency asked for an investigation into the possible impacts on surface and groundwater quality. Consequently a trial was established with the aim of determining the biopellet application rate that would provide sufficient nutrients to sustain vegetation cover and species diversity over both the short and long term, while not compromising on and off site water quality.

2 TRIAL DESIGN

The trial was sited on a 0.8 ha area in the southern central area of the site and was contoured such that the southern 0.6 ha drained to the south and the northern part drained to five separate sumps in the north.

The soil-forming material chosen for the trial was a colliery spoil from the south of the site. This is a poor plant growth medium, which is coarse textured and stony. It is free draining when loose tipped, but can become compacted when handled wet or left bare. It was considered to be the on-site material that would be most likely to transfer potentially toxic elements (PTEs) to drainage water.

During March through to May 2000, the colliery spoil was loose tipped onto the trial area. Operations were regularly interrupted due to wet soil conditions, and earthmoving was discontinued until conditions were sufficiently dry. As a result, trial establishment was delayed, but a trial area with uniform physical soil conditions was achieved.

The whole area was stone picked to remove material greater than 100 mm, twenty 20 m × 20 m plots marked out in four rows (Fig. 1) and soil samples taken from each for analysis from the upper 25 cm (Table 1).

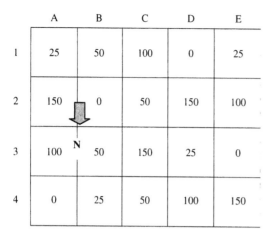

	A	B	C	D	E
1	25	50	100	0	25
2	150	0	50	150	100
3	100	50	150	25	0
4	0	25	50	100	150

Figure 1. Plot layout at the Hawthorn Reclamation Site. The plot biopellet application rate (t/ha) is given in the centre of each square.

Table 1. Colliery spoil analyses results: mean values per subsequent biopellet treatment and means for all plots.

Determinand	Biopellet application rate (t/ha)					All plots
	0	25	50	100	150	
pH	5.6	5.7	6.0	5.6	6.0	5.8
Total N (%)	0.4	0.3	0.4	0.3	0.3	0.4
P index	0	0	0	0	0	0
K index	2+	2+	3	2+	3	2+
Mg index	5	5	4	5	4	5
Zn (mg/kg)	267	139	99	111	145	152
Cu (mg/kg)	94	59	70	72	89	77
Ni (mg/kg)	52	48	53	51	56	52
Cd (mg/kg)	0.5	0.2	0.2	0.1	0.2	0.2
Pb (mg/kg)	9	23	27	9	15	17
Hg (mg/kg)	0.3	0.3	0.3	0.2	0.4	0.3
Cr (mg/kg)	107	107	101	108	118	108
Mo (mg/kg)	4.7	7.7	6.4	5.8	6.3	6
Se (mg/kg)	0.3	0.4	0.5	0.4	0.3	0.4
As (mg/kg)	27	30	44	27	33	32
F (mg/kg)	18	17	18	15	15	17

Table 2. Nutrient and metal analysis of the biopellets used in the trial.

Determinand	Analysis	Additions at 50 t/ha
	(%)	(kg/ha)
N	3.3	1650
P_2O_5	2.8	1400
K_2O	0.1	65
Dry Matter	90.7	45,350
	(mg/kg)	
Zn	380.0	19.0
Cu	275.0	13.8
Ni	33.0	1.7
Cd	<1.0	<0.05
Pb	70.0	3.5
Hg	0.7	0.04
Cr	43.0	2.2
Mo	4.0	0.2
Se	2.9	0.2
As	2.4	0.1
F	470.0	23.5

Table 3. Species composition of the original seed mixture.

Sown species
Ox-eye Daisy – *Leucanthemum vulgare*
Common Flax – *Linum usitatissimum*
Bird's-foot Trefoil – *Lotus corniculatus*
Greater Bird's-foot Trefoil – *Lotus pedunculatus*
Ragged Robin – *Lychnis flos-cuculi*
Sainfoin – *Onobrychis viciifolia*
Ribwort Plantain – *Plantago lanceolata*
Selfheal – *Prunella vulgaris*
Meadow Buttercup – *Ranunculus acris*
Cowslip – *Primula veris*
Meadow Crane's-bill – *Geranium pratense*
Common Sorrel – *Rumex acetosa*
Lady's Bedstraw – *Galium verum*
Common knapweed – *Centaurea nigra*
Pignut – *Conopodium majus*
Salad Burnet – *Sanguisorba minor*
Small Scabious – *Scabiosa columbaria*
Common Vetch – *Vicia sativa*
Yarrow – *Achillea millefolium*
(% composition by number/weight of seeds is not known)

Zinc and nickel were found to be about double the average concentration, chromium about three times the normal level and copper about four times the mean level in UK topsoils. The contents of mercury and molybdenum were also elevated above the levels in natural soils.

Biopellets were then spread according to the rates and layout in Figure 1. Their analysis, and the amount of nutrients added at each application rate, is set out in Table 2.

The whole area was sown with the wild flower seed mix shown in Table 3 on May 22nd 2000.

Five sumps on the northern boundary of the trial area were installed to collect drainage water and, in order to sample water from both within and below the root zone, PTFE suction samplers were installed at depths of 25 cm and 90 cm in each plot.

3 MONITORING

3.1 *Sumps*

Drainage water was collected from the sumps and sent for analysis on eight occasions between May 2000 and April 2001. Water samples were analysed for pH, sulphate-S, nitrate-N, ammoniacal-N, phosphorus, phosphate, BOD and COD, zinc, copper, nickel, cadmium, lead, mercury and chromium.

3.2 *Suction samplers*

Water was extracted from the suction samplers on nine occasions between July 2000 and May 2002. Ten composite samples, made up from four replicates of each of the five treatments and from two depths, were analysed for the same parameters as the sump water samples.

3.3 *Vegetation*

Germination was assessed on 19th July 2000 in terms of the number of seedlings and the estimated percentage vegetation cover within a 2.25 m² area. Four assessments were made for each treatment.

Monitoring visits were undertaken on three subsequent occasions in order to evaluate the success of the trial with regard to vegetation cover, species composition and species abundance.

Because the vegetation cover was patchy in October 2000, the abundance of each species was estimated in only one 1.44 m² quadrat in each plot. It was important to ascertain which of the sown species had failed to establish and how many species had colonised naturally, so a complete species list was made for each plot, Entec UK Ltd. (2000).

In 2001 and 2002 a more detailed estimate of species abundance was made using a conventional sampling techniques of two 1 m² quadrats in each plot. The percentage ground cover of vegetation was assessed by eye and the cover of each species estimated to the nearest 5%.

Foliar samples were taken in 2001 and analysed for total nitrogen, total sulphur, phosphorus, potassium, magnesium, calcium, manganese, zinc, copper, nickel, cadmium, lead, mercury and chromium.

4 WATER QUALITY

4.1 *Sump water*

Figures 2–7 present the results of the analyses of sump water. Plots 4A, 4B, 4C, 4D and 4E correspond to biopellet applications 0, 25, 50, 100 and 150 t/ha. Table 4 presents the discharge limits, as defined by the Environment Agency for water leaving the Hawthorn site. These discharge limits relate to the provisions of the Water Resources Act (1991).

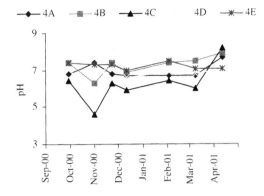

Figure 2.   Trends in sump water pH.

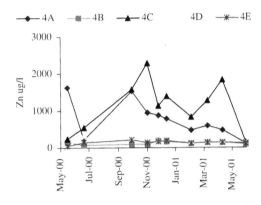

Figure 3.   Trend in sump water zinc concentrations (μg/l).

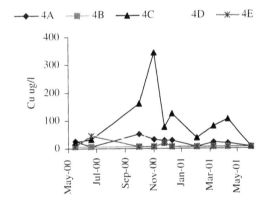

Figure 4.   Trends in sump water copper concentrations (μg/l).

4.1.1 *pH*

Throughout the monitoring period soil water pH remained fairly stable across all plots. Only on two plots did the pH ever fall below the discharge minimum

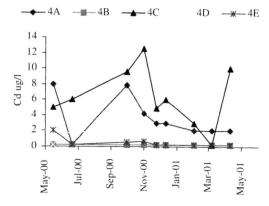

Figure 5. Trend in sump water cadmium concentrations (μg/l).

Table 4. Discharge limits for the Hawthorn site, as defined by the Environment Agency.

Determinand	Discharge limit(s)
BOD	40 mg/l
pH	<6 and >9
NH_4-N	5 mg/l
Dissolved Cr	500 μg/l
Dissolved Ni	400 μg/l
Dissolved Cu	80 μg/l
Dissolved Pb	500 μg/l
Total Zn	2000 μg/l
Sulphate	2000 μg/l
Cadmium	0.5 μg/l or detection limit
Mercury	detection limit

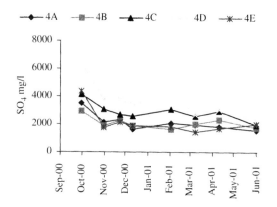

Figure 6. Trend in sump water sulphate concentrations (mg/l).

of 6. On plot 4D the pH fell to 5.5 in October 2000. On plot 4C, the pH fell to 4.6 in the same month and to 5.9 in December 2000. All other plots maintained a pH of between 6.3 and 8.2. By April 2001 the soil water pH on all plots was above 7.0.

4.1.2 PTEs

The first samples were taken on May 15th 2000, before the biopellets were applied. Mercury was detected in the sump water across all five plots. Cadmium was also detected in the sump water from plots 4A, 4C and 4E; and levels of nickel were above the discharge limit for plots 4A and 4C.

From these results, it might be assumed that the colliery spoil must have elevated levels of these three elements. However, the spoil samples taken in May 2000 did not show elevated levels of total cadmium in the top 25 cm on any of the twenty plots analysed (see Table 1).

Since the biopellets were applied, levels of mercury have dropped to below the detection limit, apart from on one occasion (2nd February 2001), when mercury was detected in water from sumps 4D and 4C.

Levels of cadmium and nickel have been recorded well above discharge limits in the water draining from plots 4A, 4C and 4D. On plot 4C, levels of copper were also measured above discharge limits on five separate occasions, while the discharge limit for zinc was exceeded on October 31st 2000. Plot 4C regularly recorded the highest levels of dissolved copper, nickel, zinc and cadmium. The control plot (4A) had the second highest levels of these metals. Figure 5 shows that cadmium was detected in the water draining from plots 4A and 4C right to the end of the monitoring period.

There is no correlation between biopellet application rate and the level of any of the PTEs found in the drainage water. Indeed, levels of cadmium and nickel in the water draining the control plot (Plot 4A: no biopellets applied) were only exceeded by levels on plot 4C.

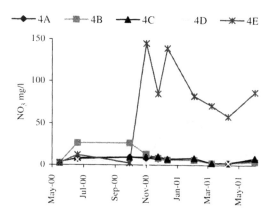

Figure 7. Trends in sump water nitrate concentrations (mg/l).

4.1.3 Sulphate

At the beginning of the monitoring period, the sulphate discharge limit of 2000 mg/l was exceeded in the waters draining from all five plots. Sulphate concentrations then declined during the monitoring period, such that by the end of May 2001, only plot 4D was discharging water with dissolved sulphate above 2000 mg/l (Fig. 6).

4.1.4 Nitrate & ammoniacal nitrogen

Levels of ammoniacal nitrogen remained below the discharge limit across all plots for the whole period (Fig. 7).

As far as nitrate-N is concerned, the monitoring period can be roughly divided into two parts. In the first period, up to October 2000, dissolved nitrate-N remained below 50 mg/l. The highest concentration of 26 mg/l was measured in water draining plot 4B (25 t/ha biopellets) on 19th June and 29th September 2000.

In the second period, nitrate-N levels on plots 4A, 4B and 4C remained at low levels, while on plots 4D (100 t/ha biopellets) and 4E (150 t/ha biopellets), levels increased to up to 54 mg/l and 145 mg/l, respectively.

4.1.5 Sump water BOD

The highest measure of biological oxygen demand was prior to biopellet spreading, when BOD ranged from below the detection limit to 16 mg/l. This is well below the discharge limit of 40 mg/l. Since then, levels have varied within this range, but have not exceeded 16 mg/l.

4.2 Soil water

Water withdrawn from the suction samplers from each of four replicates from each treatment were collected and tested as a composite sample for pH, PTEs, sulphate-S, nitrate-N, ammoniacal-N, phosphorus, phosphate, BOD and COD. Water was collected from 25 cm (within the root zone) and from 90 cm (below the root zone) for each treatment. The results for zinc, nickel, sulphate and nitrate are presented in Figures 8–14.

4.2.1 pH

Throughout the monitoring period, soil water pH ranged from 6.4 to 8.2.

4.2.2 Zinc

Total zinc levels remained below half the discharge limit throughout the monitoring period. The variations in zinc level showed no consistent temporal or spatial variations. However, zinc levels at 90 cm were generally below those found within the root zone.

4.2.3 Copper

Within the root zone, only the control plot produced copper concentrations above the discharge limit of

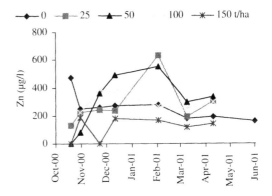

Figure 8. Soil water zinc concentrations at 25 cm depth.

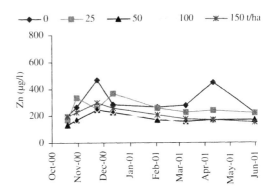

Figure 9. Soil water zinc concentrations at 90 cm depth.

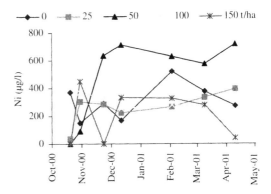

Figure 10. Soil water nickel concentrations at 25 cm depth.

80 µg/l. Water with a dissolved copper content of 250 µg/l was collected in October 2000. All other plots and all other dates produced water that was well below the discharge limit. The same pattern was observed at soil water from 90 cm depth, although in October 2000

297

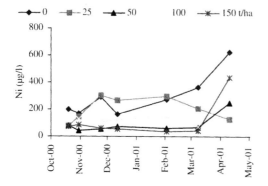

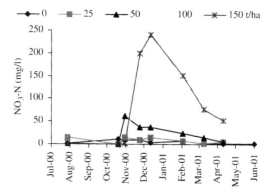

Figure 11.   Soil water nickel concentrations at 90 cm depth.

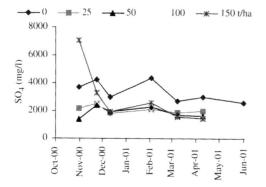

Figure 12.   Soil water sulphate concentrations at 25 cm depth.

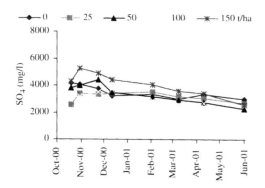

Figure 13.   Soil water sulphate concentrations at 90 cm depth.

Figure 14.   Soil water nitrate concentrations at 25 cm depth.

From November 2000 onwards, dissolved nickel contents in water collected from these plots ranged from 574 to 722 µg/l. Beyond the root zone, within the same period, nickel levels were reduced to below 100 µg/l. The only other treatment to produce water with dissolved nickel above the discharge limit was the control in February and April 2001.

4.2.5   Cadmium
Within the root zone, all plots at one time or another produced cadmium concentrations above the detection limit. Beyond the root zone, only the control produced water in which dissolved cadmium could be detected.

4.2.6   Mercury
Mercury was detected on at least one occasion in water from the control plots, the 25 t/ha plots, the 100 t/ha plots and the 150 t/ha plots. However, there was no relationship between biopellet application rate and dissolved mercury levels.

4.2.7   Sulphate
Sulphate levels showed the same temporal trend as in the sumps. Sulphate concentrations were well above the discharge limit of 2000 mg/l at the start of the monitoring period and declined over time.

Within the root zone, the biopellets seem to have had the effect of slightly suppressing soil water sulphate concentrations. By March 2001, sulphate concentrations were below the discharge limit.

Beyond the root zone, at 90 cm depth, sulphate concentrations were still well above the discharge limit by the end of the monitoring period.

4.2.8   BOD
Within the root zone, the highest BODs were found in water extracted from the 100 t/ha and 150 t/ha plots.

the 50 t/ha treatment also produced dissolved copper levels of over 80 µg/l.

4.2.4   Nickel
Within the root zone, the only treatment that consistently produced nickel concentrations above the discharge limit of 400 µg/l was the 50 t/ha treatment.

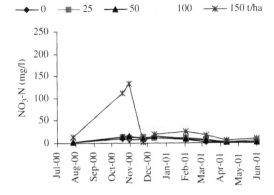

Figure 15.   Soil water nitrate concentrations at 90 cm depth.

Beyond the root zone, the 150 t/ha plots also produced high BOD water. However, the control plot also yielded high BOD ground water over the winter of 2000-01, so the high BOD at this depth may not necessarily have been related to the biopellets.

4.2.9   Ammoniacal-N & nitrate-N

The highest ammoniacal-nitrogen concentrations were measured from the 100 and 150 t/ha plots. However, levels did not exceed the discharge limit of 5 mg/l.

The highest nitrate water was extracted from the 100 and 150 t/ha plots. Within the root zone, soil water nitrate concentrations were very well correlated with biopellet application rate ($R^2 = 0.91$).

There is a high risk of nitrate leaching from the 100 and 150 t/ha treatments, particularly during the late autumn and winter months. Other biopellet treatments seem to provide sufficient nitrogen for plant growth without compromising water quality.

4.2.10   Phosphorus and phosphate

Orthophosphate was not detected in the water from any of the treatments. Phosphorus was detected from all treatments, but did not correlate with bio-pellet application rate ($R^2 = -0.03$).

4.3   Summary of sump and suction sampler results

The results have shown that PTE concentrations were not related to biopellet application rate. High sulphate concentrations were related to the depth of colliery spoil, while the biopellets seem to have had the effect of suppressing sulphur concentrations.

Soil water nitrate concentrations were well correlated with biopellet application rate. There is a high risk of winter nitrate leaching from biopellet application rates of 100 and 150 t/ha.

Table 5.   Degree of germination by mid July 2000.

Biopellet treatment (t/ha)	Mean vegetation cover (%)	Mean no. of seedlings
0	1.0	18
25	2.8	85
50	6.4	106
100	2.4	31
150	0.7	5

5   VEGETATION GROWTH AND DIVERSITY

5.1   Germination

Table 5 presents the results of the germination assessment.

The 50 t/ha treatment produced the best results in terms of germination. The poor incorporation of the biopellets on the 150 t/ha plots provided a hostile seedbed. Where pellets were poorly incorporated the vegetation cover was less than 0.1% with only two or three seedlings emerging.

By mid August, vegetation covers on the biopellet plots ranged from 30 to 50%. Vegetation covers on the control plots varied from 1 to 15%. There was no evidence that relatively high PTE levels recorded on some of the plots were having any effect on the vigour or growth of the vegetation.

5.2   Ground cover and vegetation height

The total ground cover for each treatment, and the contribution of sown and unsown species to this cover in 2001 and 2002, are illustrated in Figures 16–18. Plot means were used to calculate the overall treatment mean, while the cover of sown and unsown species was obtained by summing the estimated cover of individual species.

In October 2000, vegetation covers were highest on the 50 t/ha plots and lowest on the control plots. The 150 t/ha treatment also had low vegetation covers of less than 20%, probably due to the poor incorporation of the biopellets. At this stage, the vegetation was dominated by seed mix species, although some ruderals had started to invade (see section 5.3). Five months after the seed mix had been sown, it was clear that the biopellets were having a positive effect on germination and growth.

By June 2001, mean vegetation covers on the 25 to 100 t/ha treatments ranged from 66 to 82%. By contrast, most of the control plots had less than 1% vegetation cover. On the biopellet plots, seed mix plants still dominated the naturally colonising species (Fig. 17).

In July 2002, total vegetation covers were similar to those found in June 2001. However, on the 100 t/ha

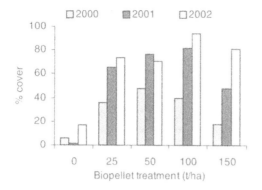

Figure 16. Mean total vegetation covers for each treatment from 2000 to 2002.

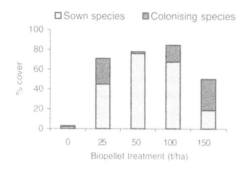

Figure 17. Mean vegetation covers for sown and colonizing species in June 2001.

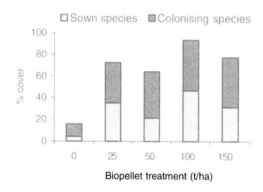

Figure 18. Mean vegetation covers for sown and colonizing species in July 2002.

biopellet plots, total covers had increased to a mean figure of 94%.

In the first two years of the trial, vegetation covers on the 150 t/ha plots had been suppressed due to the poor incorporation of the biopellets. However, by July 2002 mean vegetation covers had increased to 81%,

Table 6. Vegetation height in the experimental plots, with the range of heights recorded in parentheses.

| | Mean vegetation height (cm) | |
Treatment	2001	2002
0	0	8 (2–40)
25	52 (30–77)	27 (0–90)
50	39 (10–68)	29 (0–90)
100	46 (10–76)	37 (6–84)
150	44 (0–89)	45 (4–90)

mainly due to increases in the naturally colonising species (Fig. 18). Indeed, the naturally colonising species increased across all plots between 2001 and 2002 and served to suppress the cover of sown species in most plots.

Table 6 shows the mean vegetation heights recorded in 2001 and 2002, measured at four random spots in each plot. In 2001, the mean sward height was greatest in the 25 t/ha plots and lowest in the 50 t/ha treatment. In contrast, the vegetation heights recorded in 2002 increased with the application rate. However, all treatments showed a wide range of variation in vegetation height.

5.3 Species diversity

In October 2000, plant diversity was very low on the control plots with only 4 to 5 species on most plots. The majority of species were represented in the original seed mixture, in particular yarrow, ox-eye daisy, ribwort and sainfoin. The few naturally colonising species included sow thistle (*Sonchus sp*) and coltsfoot (*Tussilago farfara*). Of the biopellet plots, species diversity was highest on the 25 to 100 t/ha plots, where the number of species per plot ranged from 6 to 18. The 150 t/ha plots were more sparsely vegetated and species diversity ranged from 4 to 12.

The most frequently recorded species were yarrow, ox-eye daisy, ribwort plantain, sainfoin, salad burnet and lady's bedstraw. Other species recorded at lower frequency included common knapweed and bird's-foot trefoil. Species recorded more rarely included selfheal, common vetch, greater bird's-foot trefoil, meadow crane's-bill, ragged robin, common flax and meadow buttercup.

Of the newly colonising species, the most common were creeping bent (*Agrostis stolonifera*), prickly sow thistle lesser trefoil (*Trifolium dubium*) mayweed (*Tripleurospermum maritimum*) and coltsfoot.

Between October 2000 and June 2001, there was a decline in the total number of species for all treatments (see Figure 19). This was mainly due to a decline in the number of the rarer sown species, such as greater

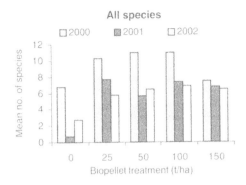

Figure 19. Number of species in the experimental plots.

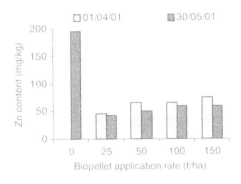

Figure 20. Foliar zinc content at various biopellet rates.

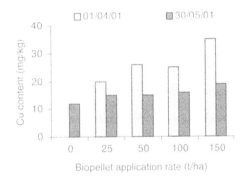

Figure 21. Foliar copper content at various biopellet rates.

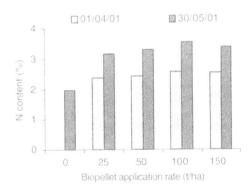

Figure 22. Foliar nitrogen content at various biopellet rates.

bird's-foot trefoil, meadow crane's bill, ragged robin and common flax. At the same time, colonizing species such as creeping bent, annual meadow grass and scented mayweed were becoming more established. However, yarrow and ox-eye daisy still dominated the vegetation.

In 2001, the only treatment where naturally colonizing species were more abundant than sown species was the 150 t/ha biopellet treatment. The most common sown species on these plots were yarrow (9% cover) and ribwort plantain (2% cover), while the dominant colonising species were scented mayweed (13% cover) and prickly sow-thistle (2% cover). A high influx of these ruderal species could prevent the establishment of seed mixture species, especially on these high nutrient plots.

By 2002, yarrow, ox-eye daisy and lady's bed straw were still the most common sown species across all treatments. Covers of these three species ranged from 5 to 20%. However, the main trend was the continued increase in the cover of creeping bent and other colonising species, such as Yorkshire fog (*Holcus lanatus*), common couch (*Elymus repens*), coltsfoot and rosebay willowherb (*Chamerion angustifolium*). The cover of creeping bent ranged from 5 to 70%. The change was particularly apparent in the 150 t/ha plots, where the scented mayweed and prickly sow-thistle had been completely replaced by creeping bent, as well as false oat grass (*Arrhenatherum elatius*) and rough meadow-grass (*Poa trivialis*) in places.

Common flax, cowslip, common sorrel and pignut were included in the seed mix but not found on any of the monitoring visits and failed to establish in any of the experimental plots.

6   PLANT UPTAKE OF NUTRIENTS & PTES

Samples of ribwort plantain were taken on 1st April and 30th May 2001 from each of the five treatments and tested for nitrogen, phosphorus, potassium, calcium, magnesium, sulphur, zinc, copper, lead and nickel. Some of the results are presented in Figures 20–23. A number of relationships could be discerned.

Of the major nutrients, both nitrogen ($R^2 = 0.93$) and phosphate ($R^2 = 0.93$) were significantly correlated with biopellet application rate. This was reflected in the differing species compositions between plots. For example, American willowherb (*Epilobium ciliatum*)

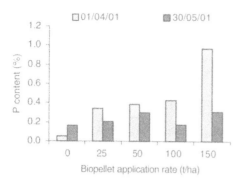

Figure 23. Foliar phosphorus content at various biopellet rates.

and rosebay willowherb were only found in plots with biopellet applications of 50 t/ha and above.

Sulphur, potassium, calcium and magnesium showed no significant correlation with biopellet application rate.

Of the minor nutrients both copper and zinc were consistently correlated with biopellet application rate ($R^2 > 0.90$). However, for copper the correlation was only significant once the high soil copper concentrations on the 50 t/ha plots were taken into account.

No relationship was found between biopellet application rate and any of the PTEs, nickel, chromium, cadmium, lead and manganese. The concentration of PTEs in plant tissue was more related to PTE soil concentrations than biopellet application rate.

7 DISCUSSION

7.1 *Water quality*

The suction sampler and sump investigations showed that PTEs have leached from both biopellet and control plots. Similarly, high sulphate water was draining from all plots and seems to be related to the depth of colliery shales through which the water is moving. Water sulphate concentrations declined over time, but are still sufficiently high to discount the use of colliery shales in the final restoration.

The biopellets did affect soil water nitrate concentrations. At the highest biopellet application rates of 100 and 150 t/ha, soil water nitrate concentrations increased to 241 mg/l in December 2000, while groundwater nitrate concentrations increased to 139 mg/l. The risk of nitrate leaching is such that biopellet application rates were limited to 50 t/ha for the final restoration.

7.2 *Vegetation*

It is clear that the application of biopellets has a positive effect on plant germination and growth. Plots with

biopellets displayed much greater plant coverage and diversity than plots that had received no biopellets.

Most wildflowers included in the original seed mixture germinated in a relatively short space of time, which is encouraging, as some species in seed mixtures require winter chilling prior to germination. The species included in the seed mix that failed to germinate throughout the monitoring period tend to be less robust or have more specific germination requirements than those that did establish. For example, cowslip prefers short turf, base rich soils and does not thrive under shade, Stace (1991). Pignut seldom colonises new sites and tends to be associated with the later stages of succession. Small scabious, like cowslip, is a plant of short, dry limestone turf, is dependent on mycorrhiza to thrive and is not a colonist of new sites (Grime *et al.* 1990). Common sorrel is largely a damp meadow plant, and does not have a persistent seed bank (Grime *et al.* 1990).

The highest species diversity was found on the 25, 50 and 100 t/ha plots, while vegetation covers and the number of seed mix species were lower on the 150 t/ha plots. This could have been due to the difficulty of incorporating biopellets into the soil surface prior to seed sowing. Pellets lying on the surface may have acted as a mulch and prevented the germination of seeds below.

In 2001, the 50 t/ha plots produced the most consistent set of results in terms of plant cover and species diversity. The dominant sown species were yarrow and ox-eye daisy, with lady's bedstraw, sainfoin, ribwort plantain and salad burnet also present.

By 2002, the competitive sown species yarrow, ox-eye daisy and lady's-bedstraw were surviving well, but the most frequent and abundant species by 2002 was creeping bent, reaching the highest mean cover in the 100 tonnes category (48.12%).

As expected, a range of colonising plants invaded the plots from adjacent habitats. In 2001, ruderal additions such as prickly sow-thistle, scented mayweed and annual meadow grass were commonly found alongside creeping bent. By 2002, the competitive tall grasses and other wind-sown species such as thistles and willowherbs were the prominent colonisers and the previous ruderal additions such as prickly sow-thistle and annual meadow grass had disappeared. Any increase in cover of some of the more competitive naturally-colonising species such as creeping bent could be detrimental to the establishment and spread of the plants from the seed mixture.

7.3 *Uptake of nutrients and metals*

The uptake of nitrogen, phosphate, copper and zinc were consistently correlated with biopellet application rate ($R^2 > 0.90$). At the higher application rates, the increased uptake of nitrogen did have an effect on the size of the plants, but there was no discernible disbenefit from the increased uptake of either copper or zinc.

The increased uptake of nitrogen and phosphate on the 100 and 150 t/ha plots is an indication of a high nutrient environment, which will not favour the establishment of species rich grassland (Gilbert & Anderson 1998).

7.4 *Optimum biopellet application rate*

The results have shown that the optimal biopellet application rate at the Hawthorn Reclamation Site is around 50 t/ha. It is the one rate that does not compromise off site water quality, while still optimizing the chances of a successful reclamation in both the short and long term. It produced the best results in terms of germination, first year establishment and vegetation cover. This rate is less likely to encourage invasion by competitive tall grasses, such as creeping bent, than the higher application rates By year three, it also produced some of the best vegetation cover and species diversity. The risk of nitrate leaching is minimal, which is not the case for the higher application rates of 100 and 150 t/ha.

Total nitrogen additions of 1650 kg/ha and organic matter additions of 25–30 t/ha meet the guidelines set out by Bradshaw (1983) and DETR (1999). The 25 t/ha rate supplies only half these quantities. So, while the first and second year results from the 25 t/ha rate have been encouraging, it is doubtful whether the vegetation would continue to thrive in the long term.

8 CONCLUSIONS

Investigations into the quality of runoff water, plant growth and species diversity have shown that the optimal biopellet application rate at the Hawthorn Reclamation Site is around 50 t/ha.

The study into water quality demonstrated that high levels of sulphate and some PTEs in the drainage water were due to levels in the colliery spoil rather than the biopellets. However, the higher biopellet application rates of 100 and 150 t/ha did result in nitrate leaching during the winter months.

All the biopellet treatments, apart from the 150 t/ha rate, produced consistently encouraging results in terms of the successful establishment of vegetation, good vegetation covers and species diversity. The non-establishment of some of the species in the seed mix was of some concern and could lead to an alternative approach in the future.

Rather than sowing a wildflower seed mix as a new crop, the restoration site could be sown with a single grass species as a matrix for weed control and soil stabilisation (Jones 1991). This will serve as a nurse crop for the gradual introduction of other species. In the right conditions, suitable species will introduce themselves within 2 or 3 years, with the advantage that they will be local ecotypes. Then small patches could be treated with the desired specialist species, inoculating the soil with seed and fungal partners together (Merryweather 1991).

The problem with most reclamation sites is the lack of original topsoil along with the associated fungal partners of a number of desired species. Mycorrhizal inoculation techniques could be used in order to enable the introduction of plants such as small scabious and pignut, which could form an important and valued part of a new species-rich grassland habitat. There is also the added advantage that, if the intended pre-reclamation soil conditions (as viewed on the landscape architect's plans) are not produced in all areas, species can be introduced according to post restoration soil conditions.

ACKNOWLEDGEMENTS

The authors would like to acknowledge the financial contribution of One North East to this project.

REFERENCES

Bradshaw, A.D. 1983. The reconstruction of ecosystems. *Journal of Applied Ecology* 20: 1–18.
Department of the Environment 1991. Water Resources Act 1991 (c. 57). ISBN 0105457914. London: HMSO.
Department of the Environment, Transport and the Regions. 1999. *Soil-forming Materials: Their Use in Land Reclamation*. Bending, N.A.D, McRae, S.G. & Moffat A.J. London: HMSO.
Entec UK Ltd. 1999. *One North East Regional Development Agency: Sewage Pellet Proposal. Stage 1. Assimilation of Information.* (unpubl.). Entec UK, Gosforth.
Entec UK Ltd. 2000. *Hawthorn biopellet trial. Report on the first monitoring visit.* (unpubl.). Entec UK, Gosforth.
Gilbert, O.L. & Anderson, P. 1998. *Habitat Creation and Repair.* Oxford: Oxford Universuty Press.
Grime, J.P., Hodgson, G.J. & Hunt R. 1990. *The Abridged Comparative Plant Ecology.* London: Chapman & Hall.
Jones, A. (2001) We plough the fields, but what do we scatter? *British Wildlife* 12(4): 229–235.
Merryweather, J. 2001. Meet the Glomales – the ecology of mycorrhiza. *British Wildlife* 13(2): 86–93.
Stace, C. 1991. *New Flora of the British Isles.* Cambridge: Cambridge University Press.

Land Reclamation – Moore, Fox & Elliott (eds)
© *2003 Swets & Zeitlinger, Lisse, ISBN 90 5809 562 2*

Geotextiles in land reclamation: applications for erosion control and vegetation establishment

R.J. Rickson

National Soil Resources Institute, Cranfield University at Silsoe, Silsoe, Bedfordshire, UK

ABSTRACT: Land disturbance can incur damaging environmental processes such as soil erosion, which must be addressed in land reclamation projects. Newly reclaimed sites may suffer from over-steepened slopes, exposure of highly erodible soils, and lack of vegetation cover, contributing to high erosion risk. Vegetation can reduce soil erosion rates by 90% compared to unvegetated sites, but establishing an effective vegetation cover may take several seasons. Geotextiles begin to control soil erosion immediately after installation. Effective geotextiles mimic the erosion-control properties of vegetation (% cover of the soil surface, roughness imparted to flow, water retention and improved soil infiltration). Research since the 1980s has evaluated the effectiveness of these products under different environmental scenarios. In addition to their erosion control effectiveness, geotextiles can enhance vegetation establishment and growth on reclaimed sites, by modifying local microclimates, in terms of temperature and soil moisture content, which affect the rates of seedling emergence and subsequent growth.

1 INTRODUCTION

Artificially engineered slopes can be highly susceptible to the forces of soil erosion by water and wind. Some of the highest erosion rates in the literature have been recorded on construction sites such as highway cuttings or embankments (for example, 480 t/ha /yr^{-1} (Diseker & Richardson 1962)), and from industrial sites such as mine spoils. The consequences of high erosion rates include difficulties in establishing vegetation and concerns over declining water quality when eroded sediments reach waterbodies. This paper considers the use of geotextiles (or erosion control mats) for the control of sediment production and for the establishment of vegetation on newly constructed slopes. To set the scene, the reasons why erosion risk is so high on engineered slopes will be reviewed, as well as the impacts of allowing erosion rates to continue unchecked. Finally, the potential for using geotextiles in land reclamation projects is considered, sourcing the currently available data and literature on this technique.

2 THE PROBLEM OF SOIL EROSION

Land disturbance during slope engineering or construction activities changes the inherent properties or "capability" of the land resource. Land capability affects which ultimate land use can be sustained on the reclaimed site, and is determined by the land characteristics of slope gradient, soil depth, texture and permeability and erosion status (Klingebiel & Montgomery 1966), all of which may be affected by the slope modifications resulting from land disturbance. Declines in land capability are reflected in higher erosion risk on such slopes, because of changes in the soil, slope and land cover characteristics of the site.

2.1 *Changes in slope*

Landscape engineering may change slope gradients, lengths and profiles (both across- and down-slope). An increase in slope gradient relative to that present pre-disturbance is often favoured by land engineers because of the reduced land take (and thus economic cost) afforded by steeper slopes. Such slopes have higher erosion risk however, because any overland flow generated is of higher velocity and greater volume because hydraulic gradient is greater on steeper slopes and time of concentration of flow is shortened. Any increase in flow velocity increases the kinetic energy and hence erosive power of the flow, such that detachment and transport rates are greater on steeper slopes. Greater velocity and volume of flow also increase the transport capacity of the flow, so that more detached material can be transported off-site. It has been shown that detaching power of runoff varies with v^3, and transporting power varies with v^5 (Morgan 1995).

Over-lengthened slopes may be a result of continuous, systematic, long term deposition of mine spoil or other waste materials. Greater volumes of flow may be generated, as effective catchment area over which runoff can generate is increased on longer slopes. Any overland flow generated will also accelerate over the longer distance, so that flow velocity is increased, with consequent effect on flow erosivity (see above).

Slope form across- and down-slope (concave versus convex versus straight) is also modified during land reclamation and engineering works. It has been shown that convex slopes generate more soil erosion that straight ones, which in turn generate soil loss more than concave slopes. This is because for convex slopes, the steepest section of slope is at the bottom, where maximum overland flow velocity and volume occur, such that erosion is greatest there. Conversely, for concave slopes, the maximum slope steepness occurs at the top of the slope, where effective catchment area for runoff generation is limited in size, and any flow that is generated has travelled only a short distance down slope, such that flow velocity is limited (Meyer & Romkens 1976).

2.2 *Changes in soil properties*

During engineering works or landscaping, soil is often highly disturbed because of constant re-working and trafficking by engineering plant. Even if topsoil is removed prior to earthworks, to be stored in storage heap, the disturbance to its structure can be significant and maybe even irreversible. Sub-soil is often left exposed on site, with weak aggregate stability because of low levels of organic matter and stabilising base minerals.

Any disturbance to the soil profile will adversely affect aggregation. The stability of aggregates is the most significant factor affecting soil erodibility (the susceptibility of soil to the forces of detachment and transport by erosive agents) (Bryan 1968). When aggregates are less stable, the soil is more prone to the processes of capping and sealing, as particles are detached and then redeposited on the soil surface, blocking any micro-pores running through the profile (Farres 1978, Simmons 1998). This affects infiltration rates, so that for any given rainfall event, more (potentially erosive) runoff is generated. Vegetation establishment is also difficult as emergence by seedlings through the cap can be impeded.

Another consequence of aggregate breakdown is the increase in bulk density as smaller aggregates and individual particles are able to pack more tightly, so

increasing bulk density and reducing porosity. Soil disturbance by engineering plant can result in compaction (Spoor & Godwin 1979). Again this adversely affects infiltration rates and capacity, which changes slope hydrology so that more runoff is generated. Despite these consequences, one common practice is to "finish" slopes post-earthworks, producing a compacted, smooth slope profile (devoid of vegetation), in an attempt to control erosion. The theory behind this practice is that compacted slopes have higher shear strength, and are thus able to resist the shear forces of overland flow. This is difficult to justify however, because (a) this practice actually increases the volume, velocity and thus kinetic energy of flow and (b) the soil susceptibility to raindrop impact and subsequent rainsplash erosion is not reduced.

Any increase in bulk density reduces soil porosity and thus the availability of water, air and nutrients to establishing or existing vegetation on site.

2.3 *Removal of vegetation*

The lack of vegetation cover on newly constructed and engineered slopes means that there is no protective cover mitigating the processes of raindrop detachment and runoff transport of eroded particles (Morgan 1995). Removal of vegetation by land disturbance and slope engineering increases the susceptibility of soil to erosive forces by as much as three orders of magnitude. This is evidenced by comparing soil losses from bare slopes to those from vegetated slopes under the same environmental conditions of rainfall, soil type and slope gradient and length. Wischmeier & Smith (1978) quantified this comparison with the use of C factor values in the USDA developed Universal Soil Loss Equation. Assigning the soil loss measured from the bare soil a C factor of 1.0; the equivalent vegetated slope had a C factor value of 0.004. This illustrates the protective effect of vegetation once established, and conversely, the susceptibility of slopes to erosion when vegetation is removed.

3 THE NEED FOR EROSION CONTROL

Having established that engineered slopes have greater erosion potential, the question arises as to what are the consequences of allowing accelerated erosion rates to continue? Where is the justification for the time and expense involved in implementing erosion control measures?

Soil erosion is a natural process, whereby any loss in soil is balanced by the soil formation rate. However, when erosion rates exceed the rate of soil formation, there is a discrepancy, and the soil resource is lost (sometimes irreversibly over time). Soil depths decline, reducing (a) available rooting media, (b) available nutrient pool, and (c) available water capacity. This will limit vegetation establishment on such sites.

Also, soil erosion is a selective process, with preferential movement of seeds, fertilisers, nutrients and medium textured soil particles (especially silts), as evidenced in enrichment ratios, showing eroded soil is much more fertile than the uneroded in-situ soil. There are significant costs to the land manager of re-seeding slopes that have been eroded, clearing out ditches that have silted up, and even greater costs associated with reclaiming erosion features such as gullies.

Once eroded, the soil particles and aggregates carried in the overland flow become pollutants, or even contaminants, depending on whether environmentally damaging chemicals are adsorbed onto the eroded particle surface. Sediment affects water quality by increasing turbidity, reducing light penetration through the flow, increasing temperatures and reducing available oxygen for micro- and macro-organisms. Micro-organisms are adversely affected by even low concentrations of sediment, and this may then have consequences for other predator species further up the aquatic food chain. At very high levels of sediment concentration ($>$20,000 mg/l), sediment affects higher orders of species, clogging fish gills and destroying spawning grounds for salmon (Mason 1991, Sharpley & Smith 1990).

These impacts are not only the concern for environmental organisations or regulators. Increasingly there is an ethos that the "polluter should pay" – which is common parlance in the United States. Land managers in some States are compelled by local bye-laws to have an erosion control plan in place before any land disturbance work is permitted by the local authority (Burrell & Karimi 2002). Inspectors ensure that the plans are put into practice, and substantial fines of up to $25,000 can be imposed for any contravention of the regulations.

Whether such coercive tactics will be *de riguer* in the UK and Europe will be seen in the future, but certainly responsibility is being shifted onto the land manager to ensure sediment production from engineered slopes does not become an environmental concern.

A less coercive approach to justify erosion control practices would be to compare the costs of damage or impact by erosion (the "do nothing scenario") with the likely costs of remediation.

4 MITIGATION OF SOIL EROSION

The factors affecting erosion can be summarised in the following equation:

$$A = R.K.LS.C.P \qquad (1)$$

Where

A = annual soil loss
R = rainfall erosivity factor
K = soil erodibility factor
LS = slope gradient and length factor
C = land cover factor
P = land protection factor (Wischmeier & Smith 1978).

Land managers are able to modify all of these factors, except rainfall erosivity, although to different degrees of practicability and success.

Soil erodibility can be reduced by the use of supplements such as organic matter (which is positively correlated with aggregate stability, up to about 10% organic matter content), compost and textural supplements. However, these effects on erodibility are usually long term, and do not address the immediate problem of high erosion risk during, or immediately following completion of engineering works.

The influence of slope on erosion risk has been noted above, and this factor can be taken into account by landscape architects when planning the final slope morphology of the site. Artificially over-lengthened, oversteepened convex slopes should be avoided, and slope breaks should be incorporated to reduce erosion risks. In practice this may be difficult to achieve, especially if land take has to be minimised. The use of traditional soil conservation practices such as terraces, contour berms and cut off drains may have potential for reducing slope length and gradient on artificially created slopes (Hudson 1981).

Managing the vegetation cover is probably the most effective, self-regenerating and practicable approach to mitigation of erosion risk. In addition to the aesthetic value of vegetation, the ability of plants to control soil erosion is well-documented (Coppin & Richards 1990, Morgan & Rickson 1995).

Plant canopies intercept incoming rainfall, breaking the fall of individual raindrops and reducing their impact velocity at the ground surface. Interception can also change the drop size distribution of incoming rainfall, making median drop sizes smaller. This has a direct effect on raindrop kinetic energy ($0.5\,m\,v^2$) and thus erosivity. Plant canopies can also store incoming rainfall, slowly releasing it to the ground, so reducing the effective rainfall intensity (mm/hr), which in turn is strongly related to rainfall erosivity. These effects are evidenced by an exponential reduction in soil loss with increasing canopy cover (Laflen & Colvin 1981).

Plant stems are also effective in reducing erosion rates. Depending on the density and type of stems, a roughness is imparted to overland flow, which sets up turbulence and eddying reducing flow velocity and kinetic energy for detachment and transport (Rickson & Morgan 1988). This effect can be expressed by the

Mannings n value, which has been quantified for different vegetation species, stem densities and plant heights (Morgan 1986). The resultant reduction in flow velocity encourages deposition of transported sediment, as flow transport capacity declines rapidly with reduction in flow velocity. Thus the stems appear to have a filtering effect – which is used to good effect in vegetative filter strips.

Finally, the root component of vegetation is important in controlling erosion. Laterally spreading roots are especially effective in increasing infiltration rates and capacity of the soil (Dissmeyer & Foster 1985). This reduces production of overland flow, so reducing erosion risk. A dense, fibrous root network also increases the cohesiveness of the soil, so that the energy required to detach soil particles by rainsplash and runoff is much higher than for an unrooted soil. Well-rooted soil also enhances microbiological activity which can affect erodibility and soil fertility to encourage good vegetation growth in the following season. Organic matter levels are increased in well rooted soil, improving soil resistance to erosion via higher aggregate stability and improved water holding capacity of the soil, so reducing runoff generation potential.

The benefits of vegetation in reducing soil erosion are clear from the discussion above, and from the evidence of increased erosion rates on slopes devoid of plant cover. However, these beneficial effects will only occur once the vegetation has established. Depending on the site's environmental conditions, this make take one or more seasons to achieve, and it is during this window of high erosion risk that slopes are susceptible to erosive forces. This critical period may be lengthened by the difficulties in establishing vegetation as mentioned above (thin soils, poor water holding capacity, poor nutrient source and potentially high levels of phytotoxic compounds). When erosion does occur during this time, any seeds sown will be removed by the erosion process (along with any particulate fertilisers and/or nutrients being used by the site contractor).

Thus, there is a clear need for immediate protection of exposed, engineered sites, to minimise the time the slopes are susceptible to soil erosion processes. Techniques such as mulching and hydroseeding have been used to differing effects, but this paper seeks to review the use of geotextiles in providing immediate and effective control of soil erosion and sediment production.

5 THE USE OF GEOTEXTILES IN LAND RECLAMATION

Geotextiles are "permeable textiles used in conjunction with soil, foundation, rock, earth or any geotechnical engineering related material, as an integral part of a man made project" (John 1987). Whilst they are used in numerous applications (filtration, separation, slope stabilization and drainage) it is the applications of erosion control and vegetation management that are considered in this paper. Some examples of geotextiles used for these purposes are given in Table 1.

Table 1. Examples of geotextiles used for soil erosion control and vegetation establishment.

Product	Material	Weight (g·m²)
Geojute	100% jute woven	500
Soil saver	100% jute woven	500
Antiwash	100% jute woven	500
Erosamat	100% jute woven	500
Grassmat	Jute non-woven reinforced with polyethylene mesh	950
Geomat	100% coir woven	700
Greenways Culture quilt	Straw blanket with jute mesh	650
BonTerra SK	50% straw 50% coir blanket with jute mesh	250
BonTerra K	Coir non-woven fibres held in polypropylene mesh	250
Enkamat	Nylon 3D mat	265
Tensarmat	Polypropylene 3D mat	450

5.1 Geotextiles for erosion control

There are many erosion control geotextiles available on the market today. They can be classed into natural and synthetic products, and come in the form of 2d and 3d mats, sheets, grids or webs. The natural products comprise materials such as jute, coir, sisal, paper, straw and wood chips. The lifespan of the products is determined by the material used and the environmental conditions at the site (notably ambient temperature and moisture regime). For example, under temperate conditions the jute and coir mats of similar weight/ unit area last approximately 2 and 5 years, respectively. The natural products are thus temporary, and their erosion control effectiveness declines as they degrade. However, as the natural geotextiles break down under UV and microbiological attack, vegetation should be establishing concurrently on the geotextile protected slopes.

Most synthetic geotextile products are made of nylon, polypropylene or polyester, and have much longer lifespans (commonly 20 years and more). Here the products are designed to control erosion both in the short term, and then work with the establishing vegetation to control subsequent erosion. The interactions between geotextile and plant erosion control have been investigated elsewhere (Selenje 1994).

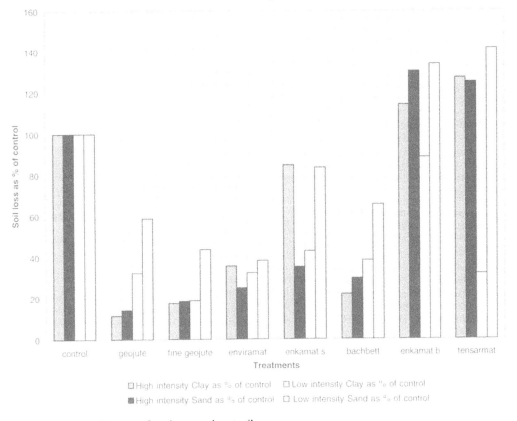

Figure 1. Relative performance of erosion control geotextiles.

Until recently, there were very few studies which quantified geotextile performance for erosion control. Even fewer studies attempted to explain how the geotextiles worked in controlling erosion. Specification and selection of products (if at all) was based on intuition or qualitative experience, rather than on hard scientific, objective data.

A research programme was initiated at Cranfield University in the mid 1980s to rectify this situation and to fill gaps in the knowledge of geotextiles for erosion control. This research programme is still ongoing and has produced some unique insights into how geotextiles work at controlling erosion. This knowledge is essential for end users as it helps in the selection of the most effective product. The information is also useful to manufacturers in that it identifies what physical properties are required when developing new, more effective products.

5.1.1 Geotextile performance

Geotextiles mimic vegetation in controlling erosion; the canopy, stem and root components of vegetation (see above) are all evident in erosion control geotextiles.

Many researchers have tested the effectiveness of different geotextiles in controlling erosion (e.g. Armstrong & Wall 1992, Cancelli et al. 1990, Cazzuffi et al. 1991, Fifield & Malnor 1990, Godfrey & McFalls 1992, Sutherland & Ziegler 1996, Rickson & Vella 1992). This extensive work has shown that geotextile performance is highly variable, depending on the product tested. Even for the same product performance is affected by the erosion processes operating on the site (whether they are detachment or transport limited (Morgan, 1986)), rainfall intensity, soil type and slope. This makes it very difficult to prescribe any one product as being the "best" available, as this depends on site specific conditions. This can be illustrated in Figure 1, which shows the relative performance of the same selected erosion control geotextiles on different soil types and under different simulated rain storms (slope gradient and length were constant). The soil loss is expressed as a ratio of that lost from the bare soil slope (where no geotextile is present), and it can be seen that relative performance is highly variable.

Despite this useful, comparative data few studies go beyond simple quantification of soil losses observed - data is often descriptive rather than explanatory. However, Rickson (2000a) attempted to set up hypotheses to explain how the different geotextiles appear to perform differently under identical conditions of rainfall, soil and slope. This experimental programme involved different slope lengths and gradients, different soil types and different rainfall events. The results of these tests are reported elsewhere. Overall, these tests and a review of other geotextile testing programmes worldwide enabled Rickson (2000a) to identify the most important properties of effective erosion control geotextiles. These were:

5.1.1.1   Area of the geotextile (% cover)
This property is important in determining rainsplash erosion rates in particular, as reflected in the strength of the correlation between percentage area of geotextile and soil loss results where only rainsplash erosion was simulated ($r = -0.917$; significant at $p < 0.01$; Rickson, 2000a). The soil is protected from the incoming raindrops, so raindrop energy is transferred to, and dissipated by the geotextile, rather than being expended on soil detachment. Higher areal coverage by the geotextile also intercepts any splash particles that may have been detached in the apertures between the geotextile components (such as in the openings created by the warp and weft yarns for the woven products). Interception of splashed material by the geotextile results in very limited transport distances, so rainsplash erosion is "transport limited" in this case. The buried geotextiles have low percentage area, so can neither absorb raindrop energies, nor intercept any splashed soil particles.

5.1.1.2   Water holding capacity of the geotextile
It was thought that products with high water holding capacities would be able to absorb rainfall and overland flow, so that these agents of erosion would be no longer able to detach or transport material. However, mean total runoff volumes measured do not support this hypothesis. There is a very weak relationship between water holding capacity and runoff volume ($r = +0.3575$ – not significant). Clearly the geotextiles do not control soil loss by reducing runoff volume.

Another link between water holding capacity and erosion rate has to be found to explain the significant correlation between these two variables. The significance of water holding capacity for erosion control may lie in the effect this parameter has on weight of the geotextile (see below). Products with high water holding capacity increase their weight when wet. Heavier products have good contact between the geotextile and the soil beneath, which was identified by Reynolds (1976) as being the most important factor

affecting the ability of geotextiles to control erosion. The natural fibre products have high water holding capacities and, when wet, can become five to six times heavier than their dry weight.

The synthetic products have very low water holding capacities, so they do not absorb water and do not become heavy with any additional weight. This may explain their poor erosion control performance.

5.1.1.3   Geotextile induced roughness to the flow
Whilst the geotextile treatments tested are unable to control runoff *volumes*, runoff *velocity* is reduced by some products. Roughness is imparted to overland flow by the fibres and yarns of the geotextiles, so reducing flow velocity, increasing effective flow depth and increasing Manning's roughness coefficient (n) values. Any reduction in runoff velocity will have dramatic consequences on flow transport capacity, which varies with the fifth power of runoff velocity (Morgan 1986). Hence, even if material can be detached by raindrop impact, this material cannot be transported by the flow, so that the erosion rate is controlled, or "transport limited".

The buried, synthetic products have low values of Geotextile Induced Roughness, because (a) they were installed beneath the surface overland flow and (b) their smooth, thin fibres have little effect on flow regime, even when exposed.

5.1.1.4   Weight of geotextile when wet
This property reflects the ability of the different products to absorb water and become heavier as discussed above. When products become wet and heavy, the "drapability" or "huggability" between the geotextile and the soil surface is enhanced, which avoids the undermining of the geotextile by overland flow.

5.1.1.5   Depth of flow ponded by the geotextile
The ability of the products to reduce runoff velocity will increase the effective depth of overland flow. This can have effects on detachment rates by rainsplash. Increased flow depths can act as a buffer between incoming raindrops and the soil beneath. The water cushions the impact of raindrops and hence the ability of those raindrops to detach soil.

The buried synthetic products are unable to increase depth of flow on the slope. Thus, the cushioning effect of a layer of water intercepting raindrops was not observed for these products, and rates of rainsplash detachment were consequently higher than for the other geotextile products.

5.2   *Geotextiles for vegetation establishment*

Due to limited empirical data, until recently the effect of geotextiles on plant growth could only be extrapolated from the literature and experiments involving mulches (for example, Foster, Johnson & Moldenhauer

1982, Sprague & Triplett 1986, Laflen & Colvin 1981, Rickson 1995). From this work, it was thought that geotextiles might help to establish vegetation by creating more stable, non-eroding conditions by controlling erosion processes (Kill & Foote 1971). Controlling soil erosion is essential on newly seeded slopes. The sections above have illustrated the effectiveness of geotextiles in achieving this aim. However, these products are also beneficial in creating site conditions conducive to vegetation establishment. This is an essential part of most land reclamation schemes.

5.2.1 *Geotextiles effects on micro-climate*
Geotextiles are similar to mulches in their impact on microclimate. In temperate conditions, geotextiles tend to increase night time temperatures and suppress daytime maxima compared with bare soil plots (Fifield, 1987). At night time, soil covers tend to reduce heat loss to the atmosphere because of their insulation properties, whereas in the day they keep temperatures relative cool because of limited solar insolation through the cover. Cooler daytime temperatures also result from relatively higher soil moisture contents under blankets because of reduced evaporation rates (which in turn is related to lower insolation levels).

Overall, geotextiles will reduce diurnal fluctuation in temperature on a site, which may or may not be conducive to vegetation growth. Certainly frost risk is reduced under geotextiles, and excessively high daytime times are also suppressed under the geotextile cover.

Soil moisture contents of soil protected by a geotextile are higher than bare soil, because of differential evaporation rates, and the water retention capacity of the geotextile. Natural products have higher water holding capacities compared to synthetic products (Dudeck et al. 1970, Rickson 2000a). However, whilst this may appear to be an advantage (especially where water is limiting for plant growth), maintaining relatively high soil moisture content encourages only shallow root development. When the geotextile then degrades, the shallow roots are unable to tap water reserves at depth, and after a lush growth in the first 2 years after sowing, the vegetation dies back, and sometimes the slope needs complete reseeding.

The degree to which different products affect microclimate will depend on the environmental conditions on the site and on the type and composition of the individual product (Reynolds 1976).

5.2.2 *Geotextile effects on soil properties*
Very little data is available on the effect of geotextile installation on soil properties in either the short or long term. Manufacturers have claimed that natural products add organic matter to the soil as they degrade, and that the geotextile fibres are a source of soil nutrients as they break down. There is little data

in the scientific literature to support these claims, and in any case, whether these nutrients are available to any establishing plants has not been tested. The effect (beneficial or adverse) of geotextiles on soil properties requires a long term study, and at present no such study is underway.

5.2.3 *Experimental data*
Some experimental data on geotextile effects on vegetation establishment are available. Rickson (2000b) reports on the initial emergence and final vegetation cover achieved by a number of natural and synthetic geotextiles. The results reported here are restricted to one soil type and one seed mix.

Seedling emergence was first observed 3 days after seeding. The mean number of germinated seeds for each treatment, recorded on Day 5 is shown in Figure 2. All the geotextile treatments had greater germination rates than the bare soil control plots, although the difference was only significant for the straw blanket products, BonTerra SK and Greenway's Culture Quilt (p = >0.01).

The differences in germination and seedling emergence are undoubtedly linked to the higher soil moisture contents and temperature differences generated by the products. The high percentage cover products, Grassmat, BonTerra K and most noticeably the straw based BonTerra SK and Greenway's Culture Quilt produced the highest rates of seed germination. This is because they retained more soil moisture by restricting evaporation losses. The blankets also insulated against heat losses from the soil trays at night, thereby maintaining optimum temperatures for germination. The jute woven products, with their relatively low percentage cover did not increase the germination rates to such an extent, but did increase rates compared to the control.

The poor germination rates for the bare soil and synthetic products is related to the watering of the plots. The frequent irrigation disturbed the seeds on the soil surface. On the bare soil plots, seeds floated on the ponded water and were deposited towards the edge of the tray. The surface laid synthetic products are not as drapable as the natural materials, whose close geotextile/soil contact prevents such seed disturbance. Although the seed germination rates were slightly higher for these synthetic plots compared to the control, vegetation was denser towards the tray edges.

Only the Grassmat product inhibited vegetation emergence. The heavy, needled punched, non-woven jute material was penetrated by only a few grass seedlings. Other seedlings, which germinated beneath the mat, either died, or survived by growing to the side of the tray where they emerged through the edge of the mat.

After Day 10, the vegetation was assessed visually on a percentage cover basis. The results are shown

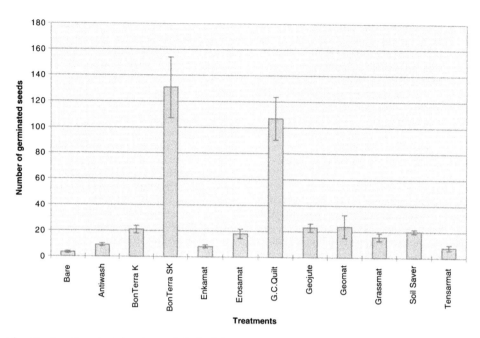

Figure 2.   Number of germinated seeds recorded after 5 days (error bars = one standard error of the mean).

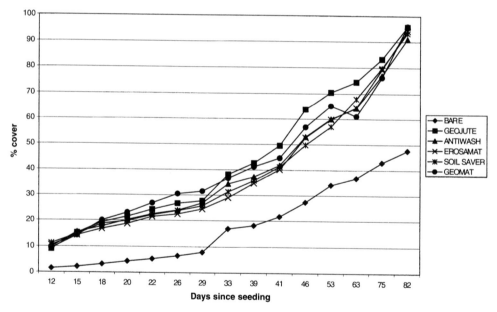

Figure 3.   Vegetation establishment: Percentage cover for the jute and coir woven products.

graphically in Figures 3 and 4. The twelve treatments have been split into two graphs for clarity. The bare soil control result has been plotted on both graphs.

The figures show that all the products had a relatively steady increase in vegetation cover over the course of the trial. All the woven jute products helped promote the growth of the seeded clover, which produced a much denser cover than the dicotyleous species found on the control plots. The Grassmat trays had a relatively dense vegetation growth, but only around

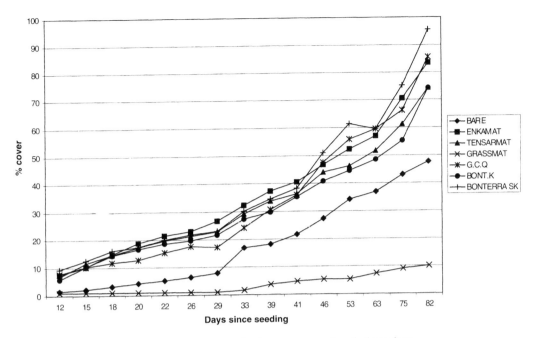

Figure 4. Vegetation establishment: Percentage cover for the non-woven and synthetic products.

the outside of the mat, where the seedlings did not have to penetrate through the thick non-woven mat.

All the geotextiles tested, with the notable exception of the jute non-woven mat increased significantly the amount of vegetation growth compared to the bare soil plots (p = >0.01). The final vegetation covers generated by the four different $500\,g\,m^2$ jute woven products and the $700\,g\,m^2$ coir mesh are not significantly different. These products tend to produce better vegetation coverage in the initial 30–40 days of the trials, with the other treatments eventually matching this performance at a later date.

6 CONCLUSION AND RECOMMENDATIONS

Geotextiles are used for soil erosion control and vegetation establishment on newly constructed, bare soil slopes. Whilst there is a significant range in effectiveness for erosion control, those products with high percentage cover, high water holding capacity, roughness imparted to the flow, wet weight and ability to pond overland flow appear to give most effective and reliable control of erosion.

From some preliminary findings, it is clear that geotextiles can aid vegetation establishment on newly engineered sites by modifying the conditions under which vegetation germinates, emerges and grows. This will reduce the time taken for the initial revegetation phase of land reclamation to be completed.

Very few studies have been carried out on other slope forming materials, and the assumption that geotextiles perform in the same way and to the same extent on these materials as they do on agriculturally based soil needs to be tested.

REFERENCES

Armstrong, J.J. & Wall, G.J. 1992. Comparative evaluation of the effectiveness of erosion control materials. In *The environment is our future, Proceedings of the IECA Annual Conference* 23: 77–92. Nevada, US.

Bryan, R.B. 1968. The development, use and efficiency of indices of soil erodibility. *Geoderma,* 2: 5–26.

Burrell, C.R. & Karimi, H. 2002. Implementation of an effective erosion control and storm water management enforcement programme in Washington, D.C. In *Adventures in erosion education, Proceedings of the IECA Annual Conference* 33: 227–235. Nevada, US.

Cancelli, A., Monti, R. & Rimoldi, P. 1990. Comparative study of geosynthetics for erosion control. In Den Hoedt (ed.), *Geotextiles, geomembranes and related products*: 403–408. Rotterdam, Balkema.

Cazzuffi, D., Monti, R. & Rimoldi, P. 1991. Geosynthetics subjected to different conditions of rain and runoff in erosion control application: a laboratory investigation. In *Erosion control: a global perspective, Proceedings of the IECA Annual Conference* 22: 193–208. Florida, US.

Coppin, N.J. & Richards, I.R. 1990. *Use of vegetation in civil engineering.* CIRIA. Butterworths.

Diseker, E.G. & Richardson, E.C. 1962. Erosion rates and control methods on highway cuts. *Trans. Am. Soc. Agr. Engrs.* 5: 153–155.

Dissmeyer, G.E. & Foster, G.R. 1985. Modifying the Universal Soil Los Equation for forest land. In El Swaify, S.A., Moldenhauer W.C. & Lo, A. (eds) *Soil erosion and conservation.* 480–495. Ankeny, IA: Soil Conservation Society of America.

Dudeck A.E., Swanson, N.P., Mielke, L.N. & Dedrick, A.R. 1970. Mulches for grass establishment on fill slopes. *Agronomy Journal* 62: 810–812.

Farres, P. 1978. The role of time and aggregate size in the crusting process. *Earth Surface Processes and Landforms* 3: 243–254.

Fifield J.S., Malnor, L.K., Richter, B. & Dezman, L.E. 1987. *Field testing erosion control products to control sediment and establish dryland grasses under arid conditions.* HydroDynamics Incorporated, Parker, CO, USA.

Fifield, J.S. & Malnor, L.K. 1990. Erosion control materials vs. a semiarid environment, what has been learned from three years of testing? In *Erosion control: technology in transition, Proceedings of the IECA Annual Conference,* 21: 235–248. Washington DC, US.

Foster, G.R., Johnson, C.B. & Moldenhauer W.C. 1982. Hydraulics of failure of unanchored cornstalk and wheat straw mulches for erosion control. *Transactions of the American Society of Agricultural Engineers* 25: 940–947.

Godfrey, S. & McFalls, J. 1992. Field testing program for slope erosion control products, flexible channel lining products, temporary and permanent erosion control products. In *The environment is our future, Proceedings of the IECA Annual Conference,* 23: 335–339. Nevada, US.

Hudson, N.W. 1981. *Soil conservation.* London: Batsford.

John, N.W.M. 1987. *Geotextiles.* Glasgow: Blackie & Son.

Kill, D.L. & Foote, L.E. 1971. Comparisons of long and short fibred mulches. *Transactions of the American Society of Agricultural Engineers* 14: 942–944.

Klingebiel, A.A. & Montgomery, P.H. 1966. Land capability classification. *USDA Soil Conservation Service Agricultural handbook,* 210.

Laflen, J.M & Colvin, T.S. 1981. Effect of crop residue on soil loss from continuous row cropping. *Transactions of the American Agricultural Engineers* 24: 605–609.

Mason, C.F. 1991. *Biology of freshwater pollution.* Longman Scientific and Technical, USA.

Meyer, L.D. & Romkens, M.J.M. 1976. Erosion and sediment control on reshaped land. In *Proceedings of the Third Interagency sediment conference, PB-245-100, Water Resources Council, Washington, D.C.:* 2–65; 2–76.

Morgan, R.P.C. 1986. *Soil erosion and conservation.* 2nd edition. Longmans UK Ltd.

Morgan, R.P.C. 1995. *Soil erosion and conservation.* 3rd edition. Longmans UK Ltd.

Morgan, R.P.C. & Rickson, R.J. (eds) 1995. *Slope stabilisation and erosion control: a bioengineering approach.* E & FN Spon, London.

Reynolds, K.C. 1976. Synthetic meshes for soil conservation use on black earths. *Soil Conservation Journal of NSW* 34: 145–159.

Rickson, R.J. & Vella, P. 1992. Experiments on the role of natural and synthetic geotextiles for the control of soil erosion. In *Proceedings of the Congress "Geosintetico per le costruzioni in terra – Il controllo dell'erosione".* Bologna, Italy.

Rickson, R.J. 2000a. *The use of geotextiles for soil erosion control.* Unpublished Ph.D. thesis, Cranfield University, Cranfield, U.K.

Rickson, 2000b. The use of geotextiles for vegetation management. In *Vegetation management in changing landscapes. Aspects of applied biology* 58: 107–114.

Rickson, R.J. 1995. Simulated vegetation and geotextiles. In Morgan, R.P.C. & Rickson, R.J. (eds), *Slope stabilisation and erosion control: a bioengineering approach.* 95–132. E & FN Spon London.

Rickson, R.J. & Morgan, R.P.C. 1988. Approaches to modeling the effects of vegetation on soil erosion by water. In Morgan, R.P.C. & Rickson, R.J. (eds), *Erosion assessment and modeling.* C.E.C., DG VI, EUR 10860 EN: 237–254.

Selenje, M.J. 1994. *Effectiveness of composite measures of erosion.* Unpublished M.Sc. thesis, Cranfield University, Cranfield, UK. MS 94/2112.

Sharpley, A. & Smith, S. 1990. Phosphorous transport in agricultural runoff: The role of soil erosion. In Boardman, J., Foster, I. & Dearing, J. (eds), *Soil erosion in agricultural land.* John Wiley & Sons.

Simmons, R.W. 1998. *The effect of polyacrilamide based soil conditioners on structural sealing at a sub-process level.* Unpublished Ph.D. thesis, University of Kent at Canterbury.

Spoor, G. & Godwin, R.J. 1979. Soil deformation and shear strength characteristics of some clay soils at different moisture content. *Journal of Soil Science* 30: 483–498.

Sprague, M.A. & Triplett, G.B. 1986. Tillage management for a permanent agriculture. In M.A. Sprague & G.B. Triplett (eds) *No tillage and surface tillage agriculture.* Chichester: Wiley.

Sutherland, R.A. & Ziegler, A.D. 1996. Geotextile effectiveness in reducing interrill runoff and sediment flux. In *Erosion Control Technology ... Bringing it home. Proceedings of the IECA Annual Conference* 27: 393–406.

Wischmeier, W.H. & Smith, D.D. 1978. Predicting rainfall erosion losses. *USDA Agricultural Research Service Handbook* 537.

Land Reclamation – Moore, Fox & Elliott (eds)
© 2003 Swets & Zeitlinger, Lisse, ISBN 90 5809 562 2

Soil functioning in natural and planted woodlands on slate waste

J. Williamson, D. Jones, E. Rowe & J. Healey
Institute of Environmental Science, University of Wales Bangor, UK

R. Bardgett
Institute of Environmental and Natural Sciences, University of Lancaster, UK

P. Hobbs
Institute of Grassland and Environmental Research, N. Wyke, Okehampton, UK

ABSTRACT: Quarries and minesites are examples of extreme disturbance. Frequently, materials that form soil are scarce and ecological restoration is easier to achieve than productive land. Establishment of soil microbial function is critical to ecological restoration and a key objective of the study. An organic fertilizer containing a mix of sewage and paper sludges was designed to promote soil functioning during the revegetation of slate waste. We compared soil formation under naturally established birch trees (*Betula pubescens*) on slate waste with that of container-grown seedling birches of local provenance in slate waste amended with either organic or mineral fertilizer. Hypotheses that an organic nutrient source would lead to more rapid establishment of microbial communities and nutrient cycling than mineral fertilizer, and result in a substrate biochemically comparable to that under naturally established revegetation, were supported.

1 INTRODUCTION

Quarries and minesites are examples of extreme disturbance; soil-forming material is often scarce at these sites and ecological restoration is the only viable option. Ecological restoration is "*the process of assisting the recovery and management of ecological integrity. Ecological integrity includes a critical range of variability in biodiversity, ecological processes and structure, regional and historical context and sustainable cultural practice*" (SER 1996). The key to restoration success is the early establishment of soil microbial function (Allen et al. 1999) and this was the primary goal of our study.

We report on ecological restoration at Penrhyn Quarry, Europe's largest slate quarry, which lies adjacent to a National Park and two designated Sites of Special Scientific Interest, in Wales. The blocky nature of the slate waste tips (slate is extremely resistant to weathering) and the lack of topsoil make the re-vegetation of these waste tips by natural means very slow; one hundred year-old waste tips are only sparsely colonized by trees and woody shrubs. Major constraints to plant establishment here were assumed to be low nutrient availability, low water-holding capacity and grazing pressure from sheep and rabbits.

The early benefit of organic matter on nutrient and water availability on plant establishment is well documented but effects below-ground are rather less well presented in the literature. This paper describes changes in soil quality indicators during the early establishment phase of trees planted into slate waste ameliorated with either organic or mineral amendments and compares these planted systems with soils under trees that had naturally established on slate waste. In addition, phospholipid fatty acid profiling was employed to study changes in the composition of microbial communities. The following hypotheses were tested that:

– theoretical C:N considerations can be used to design substrates from organic wastes to establish trees
– organic amendments accelerate nutrient cycling recovery
– organic amendments create substrates biochemically comparable to those of naturally established vegetation.

2 MATERIALS AND METHODS

2.1 Materials

Naturally established groves of similar-sized birch trees growing in undisturbed slate waste were selected and ranked according to tree height (range 20–1000 cm). Six replicate groves were selected for each tree height. On the same location, additional areas of naturally established heathland, herbs (predominantly woodsage – *Teucrium scorodonia*) and bare slate waste were identified for soil sampling, also replicated six times. The slate waste had been undisturbed for approximately 15 years.

Nearby, seedling birch trees (local provenance) were planted into either slate waste, slate waste amended with polyacrylamide gel or boulder clay (treatments addressing water-holding capacity), using pocket planting, to which either no fertilizer, mineral NPK fertilizer or organic fertilizer was added. (Pocket planting allows the *in situ* substrate to be improved as a growth medium through the addition of new material to provide the seedling with sufficient water and nutrients during the early years of establishment: this encourages roots to grow beyond the pocket into finer materials deeper within hard rock waste tips.) The mineral fertilizer was a controlled-release product, Osmocote® Plus (15% N, 4.4% P, 8.3% K) + MgO + trace elements, with an estimated 16–18 month release, applied at 550 kg N per ha. The organic fertilizer was a mix of digested sewage cake and papermill sludge. The sewage and paper was mixed using a cement mixer to achieve a target C:N ratio in the range of 15:1 to 20:1; this was achieved by mixing equal proportions (wet weight basis) of each product. The sewage-paper mix was applied at a rate to provide the same amount of mineral N in the first year as the Osmocote® treatment, based on data from WRc (1985). There were nine replicate plots for each fertilizer treatment.

Soil samples were taken one year after planting, from a depth of 0–5 cm. Soil from the naturally established trees was sampled at the same time and to the same depth.

2.2 *Soil analytical methods*

Soil total C and N were determined combustion in a high-frequency induction furnace (Leco, Michigan, USA). Microbial N was determined by chloroform fumigation extraction followed by colorimetric determination of ninhydrin-reactive N (Joergensen & Brookes 1990). Soil respiration was derived from CO_2 evolution measured by infrared gas analyzer. Soil microbial diversity was derived from phospholipid fatty acid (PLFA) signatures (Frostegard et al. 1993). Relative abundance was calculated for 14 indicator PLFAs, specific to soil bacteria, fungi and actinomycetes, within the entire microbial biomass (Frostegard & Baath 1996).

3 RESULTS AND DISCUSSION

Fertilizer application increased tree basal area, mineral fertilizer by five-fold and organic fertilizer by

Table 1. Effects of fertilizer treatment after one year on the soil microbial biomass (mg N kg^{-1}), soil respiration (mg C kg^{-1} h^{-1}), microbial diversity (from Simpson's Index) and tree growth (mm^2) under planted birch and compared with naturally established (woodland) birch trees growing in slate waste.

| Planted trees: | Fertilizer treatment | | | |
	No fertilizer	Organic fertilizer	Mineral fertilizer	Woodland
Microbial biomass N	21 a†	135b	29a	137b
Soil respiration	0.40a	3.29b	0.44a	2.63b
Microbial diversity	4.3a	6.8b	5.3a	7.2b
Tree growth	5a	41c	26b	na

† values labelled with the same letter were not different i.e. $P > 0.05$.

eight-fold, compared with unfertilized trees (Table 1). The organic fertilizer (sewage-paper sludge mix) greatly accelerated nutrient cycling as indicated by soil microbial biomass, soil respiration and microbial diversity (Table 1). Indeed, the values of these soil biological variables found in the organic amended slate were very similar to those exhibited by those soils sampled from naturally established birch on slate waste. The mineral fertilizer, however, had no effect on these variables, compared with the unfertilized treatment. An active microbial biomass is important for maintaining the nutrient capital of the system in the longer term and provides a strong argument for adding organic matter with some relatively plant-available N and P (Sopper 1993).

Microbial and respiratory quotients, expressing the proportion of microbial C relative to soil organic C and the proportion of CO_2-C respired relative to microbial C (respectively), have been often used as indicators of how efficiently the soil microbial biomass utilises C resources in developing ecosystems (Insam & Domsch 1988). In general, microbial communities undergoing disturbance will tend to have a lower microbial quotient and higher respiratory quotient than communities that are stable and have attained steady state. Microbial quotients generally vary between 1 and 4 (Lynch & Panting 1980) with values of <1 depicting a very impoverished soil or new substrate (Williamson & Johnson 1991). The microbial quotient of "soil" under birch trees planted for one year was lower than found in the soils of naturally established trees, on average, but the range was very much greater in the planted trees, with both the maximum and minimum values falling outside the accepted range (Table 2). High respiratory quotients,

Table 2. Comparing soil quality indicators (mean, with the range in brackets) in planted (n = 27) and naturally established (woodland; n = 24) birch trees growing in slate waste. Microbial quotient is the percentage of soil total C that is microbial biomass C; respiratory quotient is the amount of CO_2C respired hourly by the microbial biomass, expressed as a percentage of the microbial biomass C; microbial biomass C:N is the ratio of C to N in the biomass.

	Planted trees	Woodland trees
Microbial quotient	2.5 (0.3–8.6)	3.1 (1.3–4.4)
Respiratory quotient	3.5 (0.3–5.9)	0.3 (0.2–1.0)
Microbial biomass C:N	3 (1–5)	11 (6–14)

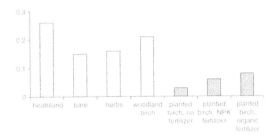

Figure 1. Soil microbial PLFA of planted and naturally established (woodland) birch trees, heathland and herbs growing on slate waste: ratio of fungal to bacterial PLFA $(nmol\,g^{-1})$.

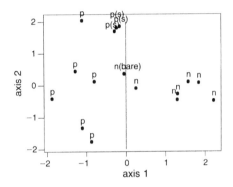

Figure 2. Soil microbial PLFA variation in natural and planted vegetation. Plot of coordinates derived from detrended correspondence analysis (Canoco). p denotes planted systems; n denotes natural systems. N(bare) denotes undisturbed bare slate waste. (s) denotes sewage-paper sludge amendment.

generally taken as >1, indicate that the microbial community is respiring C rather than immobilizing C into the biomass for growth and reproduction. High respiratory quotients therefore suggest some stressor is present or the community is rather wasteful of resources and therefore has not attained steady state. Soils from planted trees were shown to exhibit far higher respiratory quotients than those soils from naturally established trees (Table 2) and again, the range was greater in the soils of planted trees.

The ratio of C to N of the microbial biomass has been used to indicate changes in the relative proportions of microbial groups within the community. The C:N ratio often quoted for microbial biomass lies somewhere between 6 and 7. Biomass dominated by fungi has a higher C:N ratio, up to 11, whereas biomass dominated by bacteria has a lower C:N ratio, down to 2 (Anderson & Domsch 1980). Soils from planted trees exhibited a low microbial C:N ratio characteristic of a bacterial dominated community; even the maximum value obtained was still below the average accepted value. Conversely, the microbial C:N ratio of soils from naturally established trees was very high, characteristic of a fungal dominated community.

Just as above-ground biomass, namely plants, undergo successional changes then so does the below-ground biomass. Primary microbial colonizers in soil are usually bacteria; they are able to metabolize simple sugars only, are quick to respond to new nutrient sources, tend to have higher respiratory quotients than late-successional colonizers. Fungi tend to colonize substrates after the bacterial flush, are capable of metabolizing complex organic substrates, are slower to respond to new nutrient supply and have a slower respiratory quotient compared with primary colonizers (Bowen 1980).

So far, much of the information about differences in the soil microbial communities of planted and natral trees has been inferred from the soil quality indicators presented in Table 2. Using PLFAs as biomarkers for specific groups of microbes we were able to directly measure the composition of the soil

microbial communities. Soil microbial PLFA profiles revealed that soils from naturally established trees had a far greater fungal biomass than soils of planted trees (Fig. 1). This was true of all the naturally established plant systems sampled at the locality. Therefore PLFA profiling supported inferences that were made from the soil quality indicators. Analysis of PLFA in planted and natural systems also revealed the ratio of Gram + to Gram − bacteria was far greater in the latter (data not shown).

Plotting the coordinates derived from detrended correspondence analysis neatly separated the planted and natural systems using the PLFA data, with axis 1 accounting for >50% of the variation (Fig. 2). The planted system that most closely aligned to the natural system plots is the sewage-paper sludge amended treatment, shown as p(s) in Fig. 2.

4 CONCLUSIONS

Substrates can be designed from organic wastes to successfully establish native woodland tree species, using theoretical C:N considerations. In this study, digested sewage cake and papermill sludge was mixed to produce an organic fertilizer with a gross C:N ratio in the range of 15–20; this organic fertilizer outperformed mineral NPK fertilizer with a similar mineral-N content in terms of tree growth, in the first year.

This organic fertilizer accelerated soil nutrient cycling recovery in mineral slate waste as evidenced by increases in microbial biomass, soil respiration and microbial diversity, compared with the other treatments. In fact, all three variables were very similar to those found in naturally established woodland soils on slate waste. This strongly suggests that the created "soil" exhibits the key constituents of a functional soil.

Whilst the created soil in the planted system was biochemically very similar to the woodland soil, its microbial composition was quite different. Soils from planted trees were characterized by lower microbial C:N ratio, higher respiratory quotient and a greater proportion of bacterial PLFA, especially Gram + bacteria, compared with natural woodland systems. These differences collectively point to a much younger soil. However, the addition of an organic fertilizer appeared to increase the proportion of fungal PLFA resulting in a PLFA profile closer to natural systems than the other treatments, indicating the rate of soil development on slate waste may be accelerated.

ACKNOWLEDGEMENTS

This study was funded by the European Commision Life Environment programme, Alfred McAlpine Slate Ltd., The Slate Ecology Company Ltd., amd Pizarras-Villar del Rey.

REFERENCES

Allen, M.F., Allen, E.B., Zink, T.A., Harney, S., Yoshida, L.A., Siguenza, C., Edwards, F., Hinkson, C., Rillig, M., Bainbridge, D., Doljanin, C. & MacAller, R. 1999. Soil microorganisms. In L.R. Walker (ed), *Ecosystems of Disturbed Ground*: 521–544. Amsterdam: Elsevier.

Anderson, J.P.E. & Domsch, K.H. 1980. Quantities of plant nutrients in the microbial biomass of selected soils. *Soil Science 130*: 211–216.

Bowen, G.D. 1980. Misconceptions, concepts and approaches in rhizosphere biology. In D.C. Ellwood, J.N. Hedger, M.J. Latham, J.M. Lynch & J.H. Slater (eds), *Contemporary Microbial Ecology*: 283–304. London: Academic Press.

Frostegard, A., Baath, E. & Tunlid, A. 1993. Shifts in the structure of soil microbial communities in limed forests as revealed by phospholipid fatty acid analysis. *Soil Biology & Biochemistry 25*: 723–730.

Frostegard, A. & Baath, E. 1996. The use of phospholipid fatty acid to estimate bacterial and fungal biomass in soil. *Biology & Fertility of Soils 22*: 59–65.

Insam, H. & Domsch, K.H. 1988. Relationship between soil organic C and microbial biomass on chronosequences of reclamation sites. *Microbial Ecology 15*: 177–188.

Joergensen, R.G. & Brookes, P.C. 1990. Ninhydrin-reactive nitrogen measurements of microbial biomass in 0.5 M K₂SO₄ soil extracts. *Soil Biology & Biochemistry 22*: 1023–1027.

Lynch, J.M. & Panting, L.M. 1980. Cultivation and the soil biomass. *Soil Biology & Biochemistry 12*: 29–33.

Society for Ecological Restoration. 1996. www.ser.org/

Sopper, W.E. 1993. *Municipal sludge use in land reclamation*. Boca Raton: Lewis.

Williamson, J.C. & Johnson, D.B. 1991. Microbiology of soils at opencast coal sites. II. Population transformations occurring following land restoration and the influence of ryegrass/fertilizer amendments. *Journal of Soil Science 42*: 9–15.

WRc. 1985. *The agricultural value of sewage sludge, A farmers' guide*. Marlow, UK: WRc.

Land Reclamation – Moore, Fox & Elliott (eds)
© 2003 Swets & Zeitlinger, Lisse, ISBN 90 5809 562 2

The development of intuitive software ("ROOTS") to provide a decision support system for the production of custom built specifications for the creation of woodland on disturbed land

N.A.D. Bending
Progressive Restoration Limited, Herefordshire, UK

C.R.H. Robinson
Forestry Commission, Land Regeneration Unit, Staffordshire, UK

J.A. Evans & C.J. Pile
Muddy Boots Software Limited, Herefordshire, UK

ABSTRACT: English Partnerships and the Forestry Commission, two of the principal land management organisations in the United Kingdom, have formed a partnership in order to establish a framework for social and economic regeneration in disadvantaged areas by the creation of woodland on disturbed land. The success of woodland planting on disturbed land in the UK has been mixed in the past. The reasons for this are manifold but include the failure of recommendations from research to be implemented and the neglect of results from site investigations in the course of compiling specifications. To overcome these difficulties, English Partnerships and the Forestry Commission have supported the development of a PC-based decision support system software ("ROOTS") capable of producing custom-built specifications for the planting and establishment of trees in new woodlands on individual sites. The software provides extensive practical and technical support to users to assist in decision making and provides a valuable educational resource, delivering best practice to users by stealth at the most appropriate point of delivery.

1 INTRODUCTION

1.1 Background

English Partnerships and the Forestry Commission, two of the principal land management organisations in the United Kingdom, have formed a partnership in order to establish a framework for social and economic regeneration in disadvantaged areas, particularly coalfields, by the creation of woodland, in all its facets, on disturbed land. The objectives of the two organisations are compatible and roles of the two agencies are complementary. Regional Development Agencies working under the umbrella of English Partnerships provide funds and project management for reclamation schemes while the Forestry Commission offers long-term ownership and management of the newly created woodlands.

The success of planting on disturbed land in the UK has been mixed in the past and the causes of failure have been the subject of investigation by the Forestry Commission's Research Agency (Moffat &

Laing 2002). The study revealed that woodland establishment on brownfield land is generally of a disappointing quality, due to a combination of poor restoration and low standards of aftercare, including woodland establishment and maintenance. This is despite a legislative framework which requires mineral and landfill operators to bring the land to a state "reasonably fit" for planting for the purpose of forestry (Town and Country (Minerals) Planning Act 1981). The research identified serious weaknesses in the expertise of those charged with reclaiming sites to woodland and a presumption towards overly ambitious schemes featuring exclusively native species, seemingly at the behest of local authorities and without due regard to their suitability. Also apparent was a failure to recognize the significance of soil properties, most notably compaction and waterlogging, on plant survival and growth and to address these issues prior to planting. Great reliance appeared to be placed on the plethora of planting accessories available including tree shelters and guards, planting compost, slow release mineral fertilizers, mulches and inoculation with soil mycorrhizae. The use of whips and standard trees in lieu of smaller, more robust planting stock was commonplace in order to meet a desire for an "instant impact" to the planting. Rarely was a commitment to inspection and maintenance identified. Although the motivation of some operators to deliver schemes of a high quality appeared weak, the failure of many schemes could not be attributed to a lack of funding but to a lack of coherent site investigation and misguided specification choice. The research called into question the impact of several modern keynote publications by the Forestry Commission, Department of the Environment, Food and Rural Affairs (DEFRA) and the British Trust for Conservation Volunteers (BTCV) on the industry and also the effectiveness of agencies involved in disseminating guidance on "good practice" (Moffat and McNeill, 1994; Bending, Moffat & McRae, 1999; Agate, 2000).

In an attempt to overcome these difficulties in March 2000 English Partnerships and the Forestry Commission commissioned the development of a PC-based decision support system software ("ROOTS"), capable of producing custom-built specifications for the planting and establishment of trees in new woodlands on disturbed land. The aim of this software program is to aid the specifier in the decision making process (without relegating the need for professional judgment) while at the same time producing a completed set of contract documents for use in tendering exercises. The software requires that the user systematically collects and collates information on site attributes and soil properties and then considers these in producing detailed design proposals, identifying the need for soil amendments and plant protection measures, compiling tree and shrub planting mixes and ultimately the specification itself. To the user the software delivers relevant information extracted from research conducted largely in the UK and which is contained in a wide array of publications and periodicals representing "good practice". This is packaged in a concise format which can be readily assimilated by users. The originality of the approach is to target the specifier and deliver "good practice" at what is, to all intents and purposes, the point of sale by stealth, making it virtually impossible to ignore.

2   STRUCTURE AND FUNCTION

2.1   Overview

The software contains four inter-relating components. The first is a stand alone guidance document for site investigation and proforma for data entry which permits site and soil properties to be characterized and from which a constraints report can be generated as an output. The second component is a series of what have been coined "ready reckoners" which provide essential tools for practitioners to address commonly occurring issues. These include determination of the most cost effective means of providing plant protection, assessment of soil quality, calculation of rates of application for fertilizers and liming materials and the suitability of waste materials as amendments. The tools include functions to compare costs of alternative approaches and provide summary statements of information which can be generated as an output for the purpose of audit. The third component is a set of tools which allow the user to compile tree and shrub planting mixes. The suitability of an extensive list of native and non-native species is assessed by the software with data on site and soil properties interrogated in the background. The tools include a function to refine selection by expression of preferences to a range of available options and the ability to specify the type and age of planting stock required. Grass, wildflower and legume mixes can be compiled using a similar procedure. The final component is a tool which enables users to create custom built specifications. The software contains a comprehensive list of specification clauses covering hardworks and softworks commonly included in specifications. These are grouped into a series of main and sub-headings which are available for the user to select and can be edited by the user to meet their particular requirements.

A fuller, but not exhaustive, description of each of the components is provided below, along with examples to demonstrate the functionality available.

3   SITE CHARACTERISATION

3.1   Site sub-division

The software requires that the user systematically collects and collates information on site attributes and soil properties. In order to permit this the software includes a stand alone guidance document to enable data to be collected during site survey in a format which can be easily inputted without the need for reinterpretation. A combination of desktop and site survey information is required to complete the proforma which contains approximately 100 items.

The pro-forma can accommodate a variety of "typical" soil profiles including natural topsoil over natural subsoil in a reinstated soil profile (though no division of upper and lower subsoil is made), natural topsoil or subsoil reinstated over soil-forming material (SFM), weathered SFM over unweathered SFM or a profile comprised entirely of SFM.

The user has the opportunity to sub-divide a site into up to six individual sub-sites where substantial differences in site attributes or soil properties occur and guidance is offered as to the basis for subdivision. For example, within a former deep mine complex it may be possible to sub-divide areas of greenfield land earmarked for tipping but never used, steeply sloping benched tip sides and capped lagoons. Differences in the construction of the soil profile may provide a basis for discrimination, with areas boasting a soil cover split from those containing only exposed soil. Sub-division is intended to allow differences in land capability and importantly limitations on the operation of equipment to be recognized, so that these can be accommodated in the design.

3.2 *General information*

The general information required for entry includes the mineral planning authority, national grid reference, site type, altitude, rainfall, exposure, proximity of occupied property, occurrence of groundwater source protection zones or sites and proximity of sites with designated conservation status, etc. Extensive support is offered to the user to complete this section quickly and accurately. For example, pop up lists of all mineral planning authorities in England, Scotland and Wales and a classification of site types by mineral deposit are presented. The potential of the World Wide Web to provide information has been recognized with hyperlinks to relevant organizations. For example, to the Meteorological Office to obtain statistics on annual rainfall, the Environment Agency to access Groundwater Vulnerability Maps and English Nature, Countryside Council for Wales or Scottish Natural Heritage (as appropriate) to identify sites with designated conservation status. Assistance is provided to enable the user to complete the proforma and this includes narrative indicating why the factor is potentially important to decision making. For example, the likely restriction the presence of an aquifer may have on the use of organic amendments or the need to consider proximity of properties in operational matters to minimize nuisance.

Entries are generally based on selection from a number (usually three to five) of pre-determined choices with, where appropriate, the user being guided to enter the range within which a value falls, rather than the absolute value itself. For example, an annual rainfall total of 1230 mm would be entered as a range of 1001–1500 mm. Where the user is required to enter data the validity of entries is automatically assessed and warnings appear to advise of invalid entries. For example, the omission of a prefix from a National Grid Reference.

3.3 *Detailed information*

A combination of walkover survey and intrusive investigation is required to provide the detailed information required. The former includes the presence of shafts, gas venting and buried and overhead services together with assessment of drainage provision, surface condition, ground stability, vegetation cover, evidence of vegetation failure and the presence of noxious weeds, etc.

Guidance is offered to the user on how to record features and the approach is pragmatic with qualitative rather than quantitative information gathered. The focus is entirely towards completing the proforma, with information which is not relevant to this ignored. Entries are once again based on selection from a number (usually three to five) of fixed choices or multiple selections from pop up lists.

The intrusive investigation includes field based descriptions of soil physical properties supported by laboratory analysis to obtain information on soil chemical properties. The assistance offered to users in this section is enhanced to provide explanation of the significance of soil factors on plant growth. Evidence suggests that understanding of these aspects by practitioners is generally poor due either to the absence of training in soil science or for some, the passage of time since contact with the subject. The soil physical properties include texture, stone content, stone size, compaction, characteristic of failure and presence and type of foreign objects, etc. The assessment of soil physical properties adopt soil science conventions where appropriate but the presumption is once again towards the use of descriptive terms and the adoption of measures which can be easily interpreted for decision making. For example, the use of a scale measuring compaction based on the terms "unconsolidated", "loosely consolidated", "firmly consolidated", "moderately compacted", "highly compacted" and "not yet *in situ*" are used rather than direct measures of bulk density, though the relationships between quantitative measures and qualitative terms are addressed so as to allow extrapolation where appropriate.

In order to ensure that data on soil chemical properties is interpreted correctly, given the effects of laboratory method on results and common confusion surrounding reporting unit, both are specified in stand alone guidance which can issued to laboratories. The methods used are generally consistent with those of the relevant British Standard (BS) (British Standards Institute 1994). These are widely available at relatively low cost. The soil chemical properties include for natural topsoils and subsoils, pH, electrical conductivity, organic matter content, total nitrogen, "available" phosphorus, potassium and magnesium together with a suite of potentially toxic elements (heavy metals only), etc. This list is expanded for spoils to include the degree of weathering, the presence of precipitated iron in watercourses, total sulphur, total sulphate and acid neutralizing capacity. Entries are based on selection from predetermined ranges, with the user selecting that in which in the majority of values fall. For example, the choices offered for soil pH are as follows; less than 3.5, 3.5–5.5, 5.6–7.0, 7.1–8.0, 8.1–9.0 and greater than 9.0. In the event of the user identifying variation in pH which can be defined spatially, the presumption is to create a number of subsites.

Additional information is collected on the threats to newly planted trees presented by humans, wildlife and domesticated animals.

3.4 *Constraints report*

A constraints report is created in the background by the software, based on interpretation of the data entered. Responses considered as potentially likely to represent a constraint have been identified for each item within the software, and these are highlighted in the report. These are classified as "moderate" or "severe". For example, entries for pH of less than 3.5 and greater than 9.0 are regarded as severe constraints and entries of 3.5–5.5 and 8.1–9.0 as moderate constraints. The constraints report contains guidance on why the factor is important (drawn from the data entry front end), detailed explanation of the likely causes of occurrence of the problem and its effect on plant survival and growth, along with identification of possible solutions and exceptions i.e. circumstances where the factor is unlikely to emerge as a problem. The information presented seeks to be intelligible to those with limited scientific and technical knowledge. The information presented is not generalized and the occurrence of constraints as they apply to individual sub-sites is presented. For example, different narrative for pH is prepared for severe acidity, moderate acidity, severe alkalinity and moderate alkalinity. The constraints reports produced are bespoke and the user can opt to switch the output to include any or all of the elements identified above. In addition, users with extensive knowledge of the software can select a summary report only.

4 READY RECKONERS

4.1 *Overview*

The "ready reckoners" provide the cornerstone of the software package and are designed to offer practitioners with problem solving tools for issues which commonly occur in reclaiming disturbed land. For example, the most cost effective means of providing protection to newly planted trees and calculation of application rates of organic amendments. Evidence suggests that decision making in these areas is often poorly informed and sometimes muddled by the opposing viewpoints of interest groups. Common errors include provision for plant protection based on perceived, rather than actual, risk and poor choice of organic amendments and application rates set above those strictly required by the vegetation. Notwithstanding this, the opportunities for recycling waste materials are highlighted wherever appropriate and the potential cost savings illustrated.

The ready reckoners offer a stepwise approach which is transparent and leads users logically through the decision making process. The accompanying narrative provides a valuable educational resource. Relevant information contained within the site characterisation proforma is captured in the background and automatically entered into each ready reckoner. A brief description of the ready reckoners currently available is given below.

4.2 *Cost effective plant protection*

This ready reckoner allows the user to compare the relative costs of protecting newly planted trees individually or by perimeter fencing. The user may select an individual compartment, compartments, sub-site or whole site as the basis for comparison. The types of fence and tree shelters or guards appropriate for use on the basis of the combination of risks presented by humans, wildlife and domesticated animals are automatically identified to the user. The user is able to select a fence type (from a classification of 36) and up to three tree shelters or guards (from a classification of 49) and is offered an outline specification of each for review. For example, the specification for fences includes details of overall height, woodwork, timber treatment, timber dimensions, timber spacing, netting, wire and means of attachment. A cost is assigned to each fence (per m) and tree shelter or guard (each) based on the cost of materials and labour involved in construction/erection. The costs include adjustments for variation in timber costs and labour rates geographically within the UK and these are calculated automatically.

In order to increase refinement the user is able, having selected a fence type and tree shelter or guard, to manipulate the standard specification presented. For example, for fences to use Forestry Standards Council (FSC) certified timber, use non tannalised timber, increase or decrease woodwork dimensions, extend or reduce strainer and post intervals and use alternative heavyweight or lightweight netting, etc. Only options which are relevant to a particular fence type or tree shelter and guard are offered to the user and the basic cost is adjusted, upwards or downwards as selections are made.

A function is provided for the user to select up to three tree shelters or guards for use and to assign these to individual species contained within planting mixes (see 5.0). Recommendations on the suitability of individual products are given to assist selection.

Important considerations include response to shelter microclimate, top opening diameter, overall height and netting mesh size.

Total costs of protecting newly planted trees individually or by perimeter fencing are finally calculated and a short statement summarising the information is prepared.

4.3 *Acid base accounting*

This ready reckoner allows the user to calculate the acid generating potential of spoil containing iron sulphides, most notably iron pyrites and to compare this to its acid neutralizing capacity. Acid base accounting is a predictive tool which is widely employed to quantify what (if any) requirement exists for the addition of neutralizing material to raise or maintain soil pH. Use is made of total sulphur in the ready reckoner rather than iron pyrites directly, to determine the maximum potential to generate acidity. Sulphur occurring as sulphate is subtracted from the total to improve accuracy. This conversion requires use of molecular weights and detailed explanations of the assumptions involved are presented.

In order to accurately gauge lime requirement, the user is prompted to enter the rootable thickness of spoil to be cultivated in which neutralizing material is to be added. A series of automatic calculations are made to identify the total lime requirement and a corrected value, after allowance is made for the assorted carbonate bases contained within the spoil. These provide innate neutralizing capacity. The final lime requirement is presented, expressed in terms of tonnes of ground limestone per hectare. The user may include a factor of safety in the calculation, the default for which is set at 10% but can be adjusted.

A function is provided for the user to compare the quantities of other liming materials that could be used as alternatives to ground limestone to deal with acid generation. Details on a wide range of premium, reject and waste products offering a worthwhile neutralizing capacity are contained in a library. The comparative costs of these are calculated automatically using typical neutralising values (NV) and standard costs. The user may, in addition, enter these details for a specific material from a known supplier and likewise compare costs.

4.4 *Soil supply and quality*

This ready reckoner allows the user to compare the physical and chemical properties of soils and soil-forming materials available for use in restoration either *in situ*, in storage on site or via importation.

The user is able to select one of several recognised standards as the basis for evaluation (Bending *et al.* BSI 1994). Additional refinement is afforded by factoring the proposed soil use and position in the restored profile into the computation. Information on recent soil use and management history, including the application of fertilizers, organic wastes, neutralising material and land use is required to improve quality assurance.

Stand alone guidance documents are available for each standard and can be issued to the laboratory to ensure compliance with the methods and reporting specified. The user is able to enter data for up to 30 individual samples of either topsoil, subsoil or soil-forming material (but not a combination of these) into a profoma for between 40 and 48 items, depending on the standard selected, and data may be entered by row or column according to the preference of the user. Automatic entries are made where appropriate. For example, ADAS indices based on the values reported in mg/l for phosphorus, potassium and magnesium and textural classification based on particle size distributions (UK system).

A report is produced confirming whether samples pass or fail for each of the soil properties included within the standard applied. The failed samples are listed and guidance similar to that produced for the site characterisation is presented to the user, explaining the significance of the problem and how this might possibly be overcome (see 2.0). Where samples reach the required standard but the result still possibly presents a constraint, this is highlighted. For example, an electrical conductivity value of 2,700 uS/cm would meet the standard for general purpose topsoil but despite this, sensitive plants might still be affected by salinity.

4.5 *Organic amendment supply and quality*

This ready reckoner allows the user to assess the quality of a wide range of biosolids including liquids, cakes, dried granules and pellets, composts and alkaline stabilized materials in order to determine their suitability for use in reclamation and calculate application rates.

The user is required to enter data on the waste producer, waste processing facility or treatment works and certificates of analysis supplied before providing more detailed information. The proforma lists 36 items including dry solid and organic matter content, plant nutrient content, concentrations of potentially toxic elements, occurrence of pathogenic micro-organisms, presence of viable weed seed and other properties that may affect use such as dust and odour. Automatic entries, such as expressing the percentages of plant nutrients as oxides are made where appropriate. A report is automatically created that identifies and explains the significance of properties which are "abnormal" and may limit or prevent use. For example, high concentrations of nitrate, excessive concentrations of one or more potentially toxic elements, the presence of pathogens, occurrence of root fragments, strong odour or a major component of dust.

A function is provided for the user to calculate an appropriate application rate for a biosolid product and to enable this, information on the recipient soil properties is entered automatically. The user makes a selection from a list of proposed land uses, on the basis of which information the software automatically sets target application rates for the quantity of organic matter, total nitrogen and total phosphate to be applied (all expressed on a per hectare basis). The user is required to adjust the application rate to achieve these inputs with the important caveat that the values set should not fall short or be exceeded by more than 33%. The user is required to obtain the "best fit" possible within these confines.

Once the optimum application rate is set the software automatically calculates the quantities of potentially toxic elements (heavy metals) to be applied and compares these to the maximum statutory loadings under the relevant legislation (Department of the Environment 1989; DEFRA 2002), warning where these are exceeded. In addition, the theoretical increase in the concentration of these elements is calculated and in the event that the concentration after application exceeds that permitted by the above regulations, the user is warned of this danger.

4.6 *Supplementary nitrogen requirement*

This ready reckoner allows the user to calculate supplementary nitrogen requirement to permit the application of carbon rich organic waste materials in restoration. The addition of organic amendments with carbon to nitrogen ratios in excess of 25:1 carries the risk of short to medium term immobilization of nitrogen causing deficiency in plants. The method provides a predictive tool which at its simplest compares the amount of nitrogen which can be expected to be utilised by micro-organisms in the course of decomposition, with the combined amount of nitrogen concurrently supplied by the amendment itself and that present within the recipient soil. The quantity of nitrogen which may be required to be added to redress any shortfall is calculated.

For the purpose of the calculation the user may choose to enter details for a specific material from a known supplier, or alternatively use data held for a wide range of products including wood residues, industrial and food processing wastes in a library. Critical factors influencing decomposition are identified within the ready reckoner including fragment size, existing carbon to nitrogen ratio, target carbon to nitrogen ratio and the period of time to be used for the calculation. The process is simplified by extensive use of defaults to guide the user with detailed explanations

highlighting the circumstances where the user should adjust these within preset parameters.

The user is required to enter a proposed application rate per hectare of the material from which a series of automatic calculations are made to determine the net amount of nitrogen required to prevent immobilization. A correction can be applied to the result to allow for the ability of the soil to "cushion" the effect and a factor of safety can be added.

A function is included for the user to determine the quantities of mineral fertilizer and nitrogen rich organic amendments, especially biosolids, required to meet the need identified. The comparative costs of these may be calculated from library information or using details for a specific material of known origin.

5 SPECIES SELECTION

5.1 *Overview*

The software contains provision for the user to compile tree and shrub planting mixes, together with wildflower and legume mixes. Information provided during the site characterisation is linked to a database which contains details of the preferences and sensitivities (rather than tolerance) of individual plant species to important soil chemical and physical properties. For example, for trees and shrubs these include final altitude, exposure, coastal influence, atmospheric pollution, soil texture and soil pH. The user is able to select species that are "generally suited" or, to improve refinement, able to confine the selection to those species regarded as "well suited". Information on the soil and site requirements of plant species is based on primary sources describing their natural distribution (Clapham, Tutin & Warburg 1981). The software uses soil information for the uppermost layer in the restored soil profile but where this is less than 250 mm in thickness for woodland or 150 mm thickness for grassland, the system automatically defers to properties for the underlying layer.

The species selection is not linked to the National Vegetation Classification (NVC) and where sites or subsites possess a natural undisturbed soil cover, users are encouraged to employ the Forestry Commission's Ecological Site Classification (ESC) software in order to create woodland planting mixes (Pyatt, Ray & Fletcher 2001). Conversely, ESC should not be used for determining planting on disturbed sites given the common fate of overly ambitious schemes (see 1.0).

5.2 *Tree and shrub planting mixes*

Tree and shrub planting mixes can be compiled by the user and assigned to an individual compartment or selection of compartments. The user is required to indicate a stocking density. Different plant spacings may be used for trees and shrubs contained within a single mix to allow core and edge planting. Alternatively, separate mixes can be compiled and applied to a sub-compartment.

Functions are available which allow the user to continually refine species selection. The inclusion only of broadleaved or conifers or both can be specified. Species included can be restricted to those which are nationally native or regionally native (the latter are identified in the background via a linkage to the mineral planning authority). Additional options available are to confine choice only to shrubs or trees or both, to identify a limiting height or to base selection on growth rate, suitability for coppice or suitability for wetland planting.

The user can opt either for only species which meet the criteria or for all species to be shown. Suitable species are coloured green and unsuitable species red and an option is presented to obtain further information on each species to identify the basis on which they are rejected.

The user is able to select species for inclusion within one or more planting mixes and to assign a percentage to each. Plant numbers are automatically calculated on the basis of the stocking density, percentage and information held in the background on the size of compartments.

A function is provided for the user to select a specification for the supply of each chosen species. The user is presented with a drop down list of the age and type of plants ordinarily expected to be available from nurseries which are appropriate to individual species. These include cuttings, bare rooted seedling, transplants and cell grown plants. All height/grading classes are listed and these are not generalized but relate specifically to individual species. Defaults are set for each species and these are determined on the basis of obtaining the most robust planting stock most cost effectively. For example, the default for *Alnus glutinosa* (Common alder) is a 1 + 0 30–40 cm seedling, *Quercus robur* (English oak) is a 1u1 30–40 cm transplant and *Betula pendula* (Silver birch) is a 1 yr cell grown plant (CGP) 30–40 cm. The defaults are recommendations only and can be easily adjusted by the user. Additional functions allow the volume of plugs to be specified for CGPs and the minimum girth at the root collar to be set for those species selected.

5.3 *Grass, legume and wildflower seed mixes*

Grass, legume and wildflower seed mixes can be compiled by the user and assigned to an individual compartment or selection of compartments, area of open ground or areas of open ground.

The user can compile seed mixes containing any or all of the three components listed, except legumes and

wildflower, in any combination given these are contradictory. Functions are available which allow the user to select species on the basis of seed availability, allowing for the possible difficulties of obtaining seed in sufficient quantities of some species at reasonable cost. The default is for species with good or moderate availability to be shown. Species selection can also be refined by restricting choice to species compatible with newly planted trees and these are species which are generally sensitive to herbicides which are least harmful to the environment. Other options include confining selection to species suitable for use in creating wildflower meadows, those possessing agronomic value or naturalistic appearance. Additional information required from the user, over and above that automatically drawn from the site characterisation, includes proposed use of organic amendment and assessment of soil moisture conditions.

The system of displaying information on species suitability is as previously described (see 5.2). The user is able to print a hard copy of the listing of species and use this to identify an "off the shelf" seed mix available commercially from a reputable supplier, or alternatively proceed using the software to create a bespoke mix. Where the latter option is selected the user is required to select a sowing rate from within a predetermined range though a default is set. The sowing rates recommended are dependent on the components contained within the seed mix. For example, the range for a grass seed mix only is 200–400 kgs/ha, with a default of 250 kgs/ha, compared to that for a wildflower only mix which is 6–10 kgs/ha with a default of 8 kgs/ha. Sowing rates are expressed on a per hectare and per square metre basis. Where wildflowers or legumes are included within a grass seed mix, the two components are listed and totaled separately. The user is able to select species for inclusion from within a seed mix and to assign a percentage to each allowing the quantity of seed required to be calculated on the basis of information on size of compartments and areas held in the background. The range of wildflowers included is confined to those recommended by Landlife (Lickorish, Luscombe & Scott 1997).

6 SPECIFICATION

6.1 *Principles*

The software provides the facility for the user to create custom built specifications for site reclamation and contains a comprehensive list of specification clauses covering hardworks and softworks. These are method, rather than performance, based and seek to ensure that works are carried out within the correct legislative and regulatory framework, that materials are properly specified and inspected, that suitable equipment is used under appropriate conditions, that high standards of workmanship are consistently achieved and that protection is afforded to the site. Reference is made to up to date British Standards where appropriate but the aim is to present specifications that stand alone and are clear and transparent. Practices which do not conform to the requirements of the specification but which are nonetheless sometimes employed for the sake of expediency, are identified and explicitly prohibited. Practical courses of action to be followed by the employer in the event of any failure of the contractor to comply with the requirements of the specification are clearly set out. While this approach may appear overly prescriptive, experience has shown that well constructed method specifications do provide an "even playing" surface for contractors to price confidently and competitively. However, they also significantly reduce opportunities for claims resulting from poor or mis-specifying. The procedure involved in creating a specification is described below.

6.2 *Constructing a specification*

The procedure involved in creating a specification is desktop PC-based and two stage. The first involves selection of the specification clause and the second involves reviewing and editing the clause to meet the detailed requirements of the user. Guidance is provided to support the decision making at both levels. The specification produced is genuinely custom built and is devoid of any superfluous content which is so often the cause of ambiguity.

The user is presented with a classification of specification clauses based on a main heading, within which are a selection of sub-headings representing the individual clauses. For example, under the main heading of "source of supply of trees and shrubs" are four sub-headings: "in house", "nominated nursery", "local approved nurseries" and "any national supplier". The user is offered guidance setting out concisely the advantages and disadvantages of each of the choices available. For example, the benefits of locally approved nurseries could include good availability of plants grown from seed of local provenance, plants grown under climatic conditions comparable to those of the planting site and employment of local labour. The disadvantage maybe a slight increase in cost compared to that that could be achieved if any national supplier were permitted. The user is prompted to make a selection which results in the narrative for that clause being placed in a working document.

The user is then required to review and edit the content of the clause offered. The clauses contain prompts for the user to enter information and defaults which can be adjusted using preset options. For example,

where the contractor will be required to collect and plant trees grown "in house", the user is requested to enter (using a blanks system) the address of the distribution point, the contact name and opening hours together with any weight, height or width restrictions that affect vehicular access. In addition, the user is requested to indicate whether the contractor will be responsible for the loading of stock onto vehicles and the number of days notice required before collection. Other fixed aspects covered include the issue and signing of advice notes and the return and condition of re-useable packaging materials which can be revised or deleted. The specification clauses require the user to consider in detail (but not to the point of insignificant minutiae) all aspects which will have a bearing on the cost and quality of completed work.

The process continues with the user identifying headings and making choices within subheadings to build the completed specification. However, the process is not a random "pick and mix" and the software works intuitively to highlight appropriate choices. For example, if the user has exclusively selected cell grown plants in compiling tree and shrub planting mixes, the specification clauses describing the quality, packaging and labeling of these plants are highlighted over that for bare rooted plants and that for a combination of cell grown and bare rooted trees. Likewise, if during the site characterisation the presence of noxious weed species is indicated, the specification clauses describing the control of these are highlighted to the user.

The software contains extensive provision for maintenance with the user able to include specification clauses for a wide array of items ranging from hand weeding, chemical weeding, vegetation mowing, top dressing (with inorganic and organic fertilizer) and foliar analysis, etc. Users are encouraged to opt for a flexible programme of maintenance based on routine monitoring which allows effort to be tailored on the basis of need. Strict adherence to a fixed progress of maintenance or worse, the complete failure to make any provision for weeding, has been a blight to successful woodland establishment (see 1.0).

The items selected by the user in the course of constructing the specification are listed in a bill of quantities which is created automatically. This prevents the possibility of omission of items. The software creates a general and detailed specification which may be incorporated in any form of contract. However, the option will be available for the user to generate a set of preliminaries based on the Forestry Commission's standard form of contract for use.

6.3 *Auditing*

The achievement of good practice and attainment of high standards is dependent not only on the production of rigorous specifications but their implementation. In recognition of this fact an auditing tool is being produced simultaneously which synchronizes the PC-based specification with a hand held device. Each specification clause is deconstructed into a series of measurable parameters to provide the basis on which to judge compliance to the specification. For example, the list for a stock proof fence includes alignment, overall heights, the diameter and length of strainers, struts and intermediate posts (individually listed), timber treatment, spacing of posts and strainers, additional strainers at changes of gradient, gauge of netting and wire, number of staples, gap between bottom wire and gap between netting and top wire, etc. The details presented are not generalised but consistent to the selections made within the specification. For example, the precise length and diameter of the strainer specified is given. The user is required to complete a series of tick boxes to indicate acceptance or rejection. A facility is available to produce short annotated notes for the guidance of the contractor in addressing defects. The audits produced provide the basis for payment on completed work.

The device also provides a platform for use in monitoring. The user is able to produce schedules for maintenance from lists of items contained within the bill of quantities on a compartment by compartment basis. Instructions for maintenance issued previously can also be called up to check whether these works have been completed satisfactorily.

7 CONCLUSIONS

The software will provide a valuable tool to practitioners involved in land reclamation where the planting and establishment of trees for the creation of new woodland is the central objective. The strength of the product is to bring site and soil properties to the fore in all aspects of the decision making process but most critically in the preparation of specifications, where it has so often in the past remained neglected. The software provides a valuable educational resource and an effective mechanism for disseminating the results of future research and updated guidance on good practice. The translation of the specification from desktop PC to handheld device provides a powerful, labour saving, auditing tool. While the product has been designed to be used offline, the application and content data can be updated virtually instantaneously via the World Wide Web. A beta version of product is due to be completed in June 2003 and will be available for testing shortly afterwards.

ACKNOWLEDGEMENTS

This paper represents the views of the first author which are not necessarily those of the Forestry Commission or English Partnerships. The author wishes to thank Dr. Andy Moffat and colleagues at the Forestry Commission Research Agency for providing a technical review of the contents of the software.

REFERENCES

Agate, E. (ed.). 2000. *Tree planting and aftercare: A practical handbook.* Wallingford: BTCV.

Bending, N.A.D., Moffat, A.J. & McRae, S.G. 1999. *Soil-forming materials: Their use in land reclamation.* London. The Stationery Office.

British Standards Institute. 1994. Specification for topsoil. *BS 3882:1994.* London: British Standards Institute

Clapham, A.R., Tutin, T.G. & Warburg, E.F. 1981. *Excursion flora of the British Isles* (third edition). Cambridge: Cambridge University Press.

Department of the Environment. 1989. *The sludge (use in agriculture) regulations 1989.* Statutory Instrument 1056. London: HMSO.

Department of the Environment, Food and Rural Affairs. 2002. *The sludge (use in agriculture) (amendment) (England and Wales) regulations 2002.* Statutory Instrument (to be announced). London: HMSO.

Lickorish, S., Luscombe, G. & Scott, R. 1997. *Wildflowers work: A technical guide to creating and managing wildflower landscapes* (second edition). Liverpool: Landlife.

Moffat, A.J. & McNeill, J. 1994. *Reclaiming disturbed land for forestry.* Forestry Commission Bulletin 110. London: HMSO.

Moffat, A.J. & Laing, J. 2002. An audit of woodland performance on brownfield land in England. *Unpublished research report.* Farnham: Forestry Commission Research Agency.

Pyatt, G., Ray, D. & Fletcher, J. 2001. *An ecological site classification for forestry in Great Britain.* Forestry Commission Bulletin 124. Edinburgh: Forestry Commission.

Case Studies

Land Reclamation – Moore, Fox & Elliott (eds)
© 2003 Swets & Zeitlinger, Lisse, ISBN 90 5809 562 2

A comparison of limestone quarries and their potential for restoration and after-use in Libya and the UK

I.D. Rotherham, F. Spode, S. Elbah & D. Fraser
Centre for Environmental Conservation and Outdoor Leisure, Sheffield Hallam University, UK

ABSTRACT: The environmental impacts of quarries and their minimisation are widely known in the UK. Despite the extensive literature there is little documentation for North African countries. The history and cultural background of limestone extraction and processing in Libya and the UK are compared. The importance of quarried limestone to North Africa is described, and the uses of former quarries noted. This is considered in terms of the extreme environmental conditions that characterise the North African Mediterranean coast.

Potential re-use of Libyan quarry sites for housing, farming, horticulture and nature conservation is discussed and compared with the UK. Environmental sensitivity and pressure for land and extensive planning controls in the UK contrast with lack of controls other than legal consents to operate in Libya. The different approaches to planning and management of these industries are discussed, with possible transfer of procedures and controls from the UK to Libya.

1 INTRODUCTION

The overall aim of this research was to increase understanding of issues and opportunities relating to the restoration of limestone quarries in Libya. This was achieved by means of a comparative study of the histories of both exploitation of limestone and of quarry after-use or restoration in the two contrasting situations of the UK and Libya.

The research specifically aimed to:

1. Review the planning process for opening or extending limestone quarries in both Libya and the UK.
2. Compare and contrast the Environmental Impact Assessment procedures of the two countries.
3. Consider the history of limestone exploitation in Libya and examine aspects of practice transferable to the UK.
4. Assess after-use of defunct limestone quarries in the UK and potential application in Libya.
5. Adapt ideas on quarry restoration from the UK to the cultural and environmental conditions in Libya.
6. Assess the methodologies for addressing environmental problems of mineral extraction in Libya with recommendations for improved practice.

The after-use of post-industrial sites is now a major consideration in western economies such as the UK and the USA (Land Use Consultants 1992 and 1996).

In particular, not only are adverse affects considered important, but the potential after-use for recreational and conservation functions are increasingly highlighted (Land Use Consultants 1996, Moffatt & McNeill 1994). Gunn *et al.* (1997) have considered the potential for more effective restoration of former limestone quarry sites in the UK, and aspects of this research could be usefully transferred to applications in North Africa.

These issues become increasingly relevant to emerging economies, as communities become more wealthy and have increased expectations of both environmental and leisure quality. Furthermore, as industrial and post-industrial landscapes become more widespread, their relevance and potential will be increasingly considered (Gilbert & Anderson 1998). So far, there has been little work on this in North Africa generally, or Libya in particular.

This research relates to issues of sustainability, environmental management, and of tourism and leisure management. North African countries such as Libya increasingly look to tourism as a long-term economic contributor to growth. In this scenario, the adoption of better and more acceptable environmental management practice will be important.

This paper presents the finding of the comparison of planning and Environmental Impact Assessment (EIA) procedures, and the comparison of restoration issues and transferability.

2 METHODOLOGY

A critical review of the relevant literature and methodologies was carried out, followed by a detailed evaluation of environmental impact assessment methodologies and procedures (in both Libya and the UK). Appropriate models for planning and impact assessment in Libya, and in the UK were considered, assessed and critically evaluated. These models were then further developed as the final phase of research.

Information on the broad background to the research topic was gathered from both Libya and the UK, although it is recognised that detailed information on the Libyan situation is hard to find. Many of the literature sources used were translated from Arabic originals.

Individual case study sites were identified in Libya and the UK, and contact made with the appropriate personnel. Relevant information and data were then collected. Information from the core case studies and from other locations (50–100 sites in each country) was collated on map bases and in suitable computer databases, for evaluation and presentation. Information on nature conservation and on leisure/recreation uses was gathered. The information was then rigorously interrogated and the research findings evaluated.

The degree to which the current systems applied in the UK are transferable to Libya was considered, and the standard models of planning procedures in the UK and in Libya critically compared and contrasted. This was in the context of Environmental Audit and Environmental Management Systems (EMS). Impact assessment models used elsewhere in the emerging economies of the world, were investigated.

Based on the above, ideas are presented to help integrate appropriate procedures and processes of site after-use and rehabilitation to the Libyan scenario. These are based on established practice at case-study locations in the UK, and focus in particular, on the potential for habitation, leisure, farming and nature conservation as after-uses. The application of these ideas is set in the context of the emerging economic regeneration of Libya.

The appropriateness of methodologies and the potential benefits of transfer of ideas from the UK and other countries are discussed.

3 THE HISTORY OF LIMESTONE QUARRYING IN LIBYA

For Libya as a developing country seeking a more stable and affluent lifestyle, limestone for construction work is a vital resource. Not only this, but the utilisation of limestone resources in Libya has been important throughout the region's long history. Now for the benefit of its people, the Government seeks to exploit important natural resources within its boundaries. These include limestone.

Settlers from the Greek and Roman periods built cities of native limestone blocks excavated and cut to size. This is evident in the remains of cities such as Cyrene, Susah and Leptis-Magna. Following this period of active building on a grand scale, the local people influenced by cultural factors such as the coming of Islam to the country, changed their building styles. The local vernacular building methods used superficial stone, for rubble-built construction, bound by lime mortar and with roof timbers made from Date Palms. This type of development did not necessitate large-scale quarrying of limestone. For nearly fifteen centuries the quarries that did exist were on a relatively small scale of 2–4 ha.

From the time of the Italian occupation of 1911, and because of a lack of political and economic stability, and associated low levels of economic activity, construction of infrastructure such as roadways *etc.* was simple in both design and execution. All materials used (such as aggregate, sand and paving stone) were of local origin and production.

The native peoples tended to build houses for themselves, or for domestic animals, of blocks of limestone together with branches of trees. They divided the date tree stems to use them as pillars to link and tie their constructions together. The most important available building material was limestone. The usage of limestone became more diverse, sometimes as blocks, and other times as powder (in cement and mortar).

4 THE LIBYAN CEMENT INDUSTRY

The contemporary Libyan Cement Industry originated in the late 1960s, with increased development activity in Libya leading to more demand for construction materials, especially cement. The process of development along with political circumstances created the desire for self-sufficient production of essential raw materials. The manufacture of cement is in every sense a basic industry, and the establishment of cement manufacturing units is one of the first initial steps taken in a modern industrialised economy (Libyan Industrial Research Centre Report 1992). Indeed the industrialised production of cement in Libya is very recent in origin, the first cement works being in 1968, at Al Khums, 150 km east of Tripoli. From then on the industry has grown steadily. By 1998–2001 it had a capacity of 8.5 million tonnes with an additional 2 million tonnes, through new installations established by the Libyan Cement Company. Another 3 million tonnes capacity is planned in the five years (2002–2007) (Libyan Industrial Research Centre Annual Report 1998). Further initiatives are planned for Libya to become an exporter of cement.

5 THE PLANNING PROCESS

EIA and EMS for operations such as quarrying and associated industrial activities are often imposed in part through a planning system. It is therefore important to compare the planning procedures in the UK and in Libya. It is necessary to be clear about the differences between EIA, EMS, and the planning process. Whilst these vary in detail from country to country, and from region to region, they also reflect the particular economic and political system and historical context of each. EIA is a process applied to evaluate and mitigate adverse impacts of developments or industrial processes on the environmental resource. This may include aspects of economic and social impact too, and may be part of a bigger process. The EIA may be voluntary and part of an organisation's good practice, or it may be compulsory. The degree to which an EIA is mandatory will vary from country to country.

An environmental management system (EMS) may relate to an environmental audit (based on financial auditing models) and can include EIA in its process. Audit and Management Systems provide a framework for good practice and good housekeeping in terms of the processes, procedures and monitoring of an organisation, a factory, an industrial process or other activity. The Audit includes assessments and monitoring with a review process designed to improve standards and to meet targets. Formal audits are accredited and meet stringent national and international criteria. Many commercial organisations in Europe and the USA now require bodies with whom they trade to confirm compliance to an appropriate environmental management system. Furthermore audits are usually undertaken or monitored by an approved external auditor.

In contrast to the above, the Planning Process is essentially a procedure that may vary from country to country, or region to region, for the granting or withholding of certain permissions for activities that require consent. As with some EIAs these are basically statutory tools. The planning processes applied in Libya and the UK are radically different and they reflect the contrasting histories of the two countries and their people. As this research shows they also reflect the needs, issues and requirements of local communities and the different impacts of quarrying in the very different environments of North Africa and the North-western Atlantic environment of the UK. The planning process to grant consent, and the requirement to restore (or allow appropriate after-use) also reflect contrasting

issues of land-use, land area and population density in the two countries.

The planning process may require an EIA before a decision is reached and might specify the need for appropriate monitoring and environmental management systems once work is underway.

For planning consent for a quarry in Libya the developer must obtain:

1. Permits from the Electricity Company to confirm that there are no underground electricity cables now or planned.
2. Permits from the regional municipality to confirm that there are no plans to establish new settlements on the site; and no drainage or water pipelines now or planned.
3. Permits from the Agriculture Department stating that the proposed site is unsuitable for farming.
4. Permits from the Transportation Department indicating any routes, motorways or railways planned.

All these permits are prior conditions for permission. (*The Libyan Mining and Quarrying Procedures Handbook* 1985).

The developer also makes a commitment to refill the quarry holes if not required by the government, or for an agreed private purpose. They are also required to reinstate plantations, maintain and report on artefacts or ruins of heritage value, and are responsible for any environmental impacts from operations such as blasting.

The requirements for planning consent for a quarry in the UK.

The UK is considered to be at the cutting edge of processes to control and regulate quarrying, and of techniques and procedures to restore or reclaim post-extraction sites. The Libyan procedures and development contracts lack specific environmental conditions, and it was in part this issue that the research intended to address. The process in the UK is based around the following:

1. Strategic Planning and Resource Planning Context developed by government (National, Regional and Local), by employed officers with policy approved by elected, political members.
2. Proposals are developed by individual minerals companies or quarry operators. These are submitted to the local authority planning system whichever level is the "Minerals Planning Authority". These may then go through a Planning Application process and quite probably a Public Inquiry, chaired by an inspector appointed by the relevant Government Minister.

During the process there is full opportunity for other Government Agencies, Voluntary Conservation Bodies, Local Communities, and individuals to comment on the proposals. The officers for a Planning

Application, or for a Public Enquiry the Inspector, will make a recommendation to accept perhaps with conditions, or to reject. The final decision is political made by the elected members. There may also be various rights of appeal by the developer if a decision is negative. The application may require an EIA from the applicant.

6 ENVIRONMENTAL PLANNING AND IMPACT ASSESSMENT

International policies for environmental impact assessment and protection aim to preserve ecological balance, to safeguard natural and human resources, to prevent and reduce the pollution, and to reconcile the needs of development and of the environment. These are drawn together by a wide range of international and national strategies and legislation to promote environmental improvement. This is often implemented through environmental impact assessment (EIA), and in the context of sustainable development, in accordance with the principles of the 1992 Rio Conference (Brundtland (Ed.) 1987).

The EIA process has been developing since the 1960s when it was first given formalised status through the USA's NEPA (*National Environmental Policy Act* USA 1969). This required EIA for federally funded or supported projects, which were likely to have environmental effects (Weston 1997). It is one way to mitigate adverse impacts of developments. The EIA has been defined by a number of authors e.g. Wood (1995), Canter (1996), Weston (1997), Wathern (1992), Glasson *et al.* (1994), Jorissen & Coenen (1992) and Munn (1979). This can be considered in the context of particular countries and organisations. Gilpin (1995) provides a broad overview of the international approaches and issues, and across the spectrum of international environmental policy. However, in many cases the social, economic and environmental issues relating to emerging economies around the world are given relatively scant treatment. Barrow (1997) presents case studies with some relevance to North Africa generally, and to Libya specifically. However, there is no literature derived from case studies based in Libya itself.

Environmental Impact Assessment (EIA) is the evaluation of effects arising from a major project (or other action) significantly affecting the natural and anthropogenic environment (Wood 1995). Canter (1996) defined EIA as the systematic identification and evaluation of potential impacts (effects) of proposed projects, plans, programmes, or legislative actions relative to the physical, chemical, biological, cultural, and socio-economic components of the total environment. EIA is also defined as a process for identifying likely consequences of activities for the biogeophysical environment, and for human health and welfare. This is for particular activities and also to collate information to inform the decisions of those sanctioning development proposals (Weston, 1997). Wathern (1992) suggests that EIA is a process with the ultimate objective of providing decision-makers with an indication of the likely consequences of their actions. Glasson *et al.* (1994) argue that EIA is a systematic process for the examination of environmental impacts of development and the emphasis, compared with many other mechanisms for environmental protection, is on prevention. EIA is described by Jorissen & Coenen (1992) as an instrument of preventative environmental management.

The UK DoE (Department of the Environment) (1989) defined the EA as: "*The term 'environmental assessment' describes a technique and a process by which information about the environmental effects of a project is collected, both by the developer and from other sources, and taken into account by the planning authority in forming their judgements on whether the development should go ahead.*"

EIA is a technical exercise, the object of which is to provide decision-makers and the public with an account of the implications of proposed courses of action before a decision, is taken.

The above definitions provide a broad indication of the objectives of EIA, and illustrate differing concepts of EIA. Finally, they emphasise the role of EIA in informing the decision-making process for example, the UN Economic Commission for Europe (UNECE) in 1991 declared that the EIA is: "*An assessment on whether the development should go ahead*".

Canter (2000) suggests that, the following four themes may characterise the EIA process in the United States in the coming decade: (1) usage of the World Wide Web; (2) continued introduction of new topical and policy themes into the process; (3) greater emphasis on implementation of mitigation measures; and (4) recognition that numerous permit applications represent "targeted" impact studies. Greater confidence in these predictions are associated with the first years of the decade, therefore a continuing trend during the coming decade is expected to be associated with new topical and policy themes being introduced into the EIA process. In addition, application of the EIA process to the lifecycle analysis of manufactured products, as well as the construction, operation, and decommissioning phases of projects, is also expected to receive greater emphasis.

6.1 *Environmental impact assessment in Africa*

Most African states are considered "developing countries". Due largely to their economies, social and political structures, EIA processes are poorly implemented. For these and other reasons, the EIA process in the whole of Africa will take a long time to be accepted. Even if accepted, it will take a long time to approach the same level as that in the western world. Ofori (1991) in Gilpin (1995) states that: *"Although the use of impact assessment is spreading, Africa generally lags behind in the adoption of environmental impact analysis. South Africa, Egypt (considered earlier, with Mediterranean states), Nigeria and Ghana have made most progress with environmental impact assessment."*

6.2 EIA in North Africa

With the exception of Egypt, the EIA has only recently been introduced in North Africa. What little progress there has been is a result of ongoing demands from Southern European Countries and through agreements at conferences of the North African and the Southern European countries. Tunisia, Morocco and Algeria have all made progress in developing EIA and planning procedures. However for success, moves to this end and imposition of environmental measures depend on the necessary infrastructures for training and monitoring. These are often lacking.

Egypt has made more progress with impact assessment than most African countries. Since the 1980s the Egyptian Government has been training impact assessors, and an Egyptian Environmental Affairs Agency has been established. Egypt has provisions to compel developers to deposit an indemnifying bond to fund mitigation of adverse impacts after completion. The requirement to pay can be waived on some projects (Barrow 1997).

During the 1990s, isolated by UN sanctions, Libya made little progress in its EIA regulations and procedures. However, the Libyan Government has made modest attempts to encourage people and voluntary groups to establish organisations to address environmental issues. This is under the Libyan Government Supervision Organisation called The Environmental Common Institution (ECI). Prior to UN economic sanctions, the EIA procedures in Libya are variable and concentrated on Oil Companies' project sites. Training centres were established but related mostly to oil production and potential impacts. There are serious issues of lack of awareness and of limited education in relation to environmental risks. Since the sanctions were suspended in 1999, the Libyan Government has engaged in moves to redress the imbalance and to establish appropriate processes and procedures. With regard to land restoration and reclamation in Libya, sanctions were a key factor preventing ideas and technologies being adopted by Libyan industry, restoration and reclamation being largely neglected.

There is now progress in the restoration and reclamation of quarries. This is still limited especially in the private sector. With large projects such as limestone or clay extraction quarries for cement manufacture, the EIA procedures are still minimal, and restoration is restricted to tree planting to prevent soil erosion on surrounding slopes.

Economic factors encourage the government to overlook restoration of many former quarries associated with cement manufacture, and environmental or amenity issues are frequently neglected. However, ideas on derelict land improvement and re-use of former quarries now reach the Libyan Government from countries such as the UK. New environmental departments at two Libyan Universities have been opened under Libyan Research Centre supervision with the specific aim of collecting environmental information to disseminate to local Universities, Research Centres, and Domestic Environmental Organisations. This was in part a result of the present research

7 RESTORATION AND AFTER-USE

In addressing quarry impacts on the environment and local people, it is necessary to consider avoidance or mitigation of adverse effects, and also site restoration to positive use. Such restoration in both Libya and the UK will normally be "conditioned" when consent for quarry working is given. Restoration may be for agriculture or amenity and conservation. It may involve the re-contouring and re-vegetation of all of the worked land or selected areas. It may include the "naturalisation" of former extraction surfaces – rock faces *etc.* by means of positive restoration tools such as restoration blasting. Restoration will generally involve the site being brought back to a condition of little adverse landscape impact or even positive impact, and perhaps a positive impact on wildlife *etc*. Once again the cultural and historical context of the industry and the site will influence this.

Reclamation can be re-use and making available for future land utilisation. This will involve the removal of any contamination, the making safe of dangerous structures and the bringing of the site into a condition fit for a future use. Finally, after reclamation and/or restoration there is the after-use of the site and the location. The need for restoration and reclamation is influenced by the potential after-use. Likewise the possible after-use will be determined by the condition of the site and the extent of reclamation and restoration.

7.1 The incorporation of Libyan traditions, and Roman and Greek ideas in contemporary quarry restoration

It is useful to compare natural mountain topography and dale-sides with the steep, sharp, worked slopes

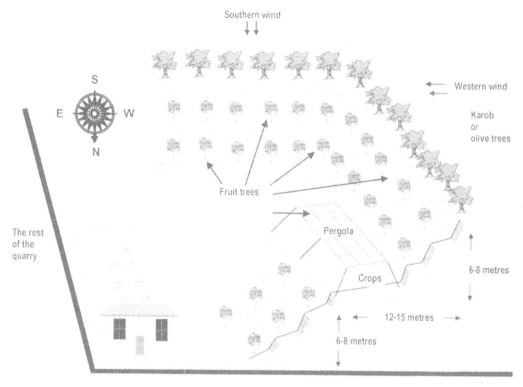

Figure 1. Representation of potential after-use of Libyan quarry site.

and the faces of a recently abandoned quarry. There is particular interest in adapting these landforms to support people, livestock and vegetation. An arid country, Libya has sunshine all year round, and either hot southern winds, or cold western winds. This encourages communities to maximise environmental protection within the landscape. Both natural and artificial structures (such as quarries) are utilised to protect buildings (for habitation, horticulture, agriculture and domestic animals). These traditions extend back to prehistoric times. This sheltering of dwellings, other buildings, agricultural structures for animals, pergolas for climbing plants, and fruit tree plantations is a useful function for restored quarries. A generalised diagram to illustrate this presently under-appreciated use is given below (Figure 1).

This situation contrasts with the UK where environmental priorities and pressures are very different. With a less extreme climate than Libya the domestic use of former quarry sites is not a priority. With a larger land-take for Libyan quarries compared to the UK, they have the potential for the development of entire villages within them. Most UK quarries are deeper and are often filled prior to final landscaping in the upper levels.

Interestingly, both Greeks and Romans followed these principles, establishing their cities in large, topographically protected areas. The present-day Libyan Bedouins also utilise the landscape in this way. In combining Libyan Bedouin approaches with the modern process of restoration, another and distinctive type of quarry restoration becomes possible. This approach reflects that of earlier civilisations, but places this environmental function into a modern, contemporary context. Furthermore, and very importantly, this method of site restoration blends with the natural landscape, and reflects the needs and requirements of local people and the climatic constraints of this extreme environment.

Advocating this new approach highlights the need to recognise the differences in landscape, environment, and culture between the two countries – Libya and the UK. These differences must be considered both in terms of the aims and objectives of restoration and indeed throughout the process itself. The background to the UK situation is around two thousand years of increasingly intensive human activity and increasing population. There is now a high population density and a potential shortage of land and a serious degree of vulnerability in for example, the ecological resource. The backdrop to this is a landscape of largely fertile soils, and usually plentiful water. These factors have encouraged and allowed people to exploit almost every available area of land. Furthermore, the demand for limestone for building and other industrial and construction purposes has been huge. Conservation restoration of disused quarries is clearly a favourable option. This can use techniques such as restoration blasting to help develop "naturalness" and to benefit nature conservation interest.

In contrast with the UK, countries such as Libya, with arid conditions and severe environmental constraints have human activity essentially restricted to areas where water is available. Since much of the countryside is still relatively virgin and uncultivated, and lacking vegetation, the gathering of surface boulders for benching and covering the walls of the quarry faces is easy and cheap. The effectiveness of this approach to restoration depends on the residual engineering factors such as the final shape and the height of the quarry faces, the width of the benches as well as the relief of the quarry floor. These will determine what must be applied and the precise way it should be done.

7.2 An evaluation of site after-use in Libya and the UK

Quarry restoration and uses were assessed at one hundred and forty-eight quarry sites in the two case study countries (Libya and the UK). These have a total of a hundred and ninety-seven restoration sub-sections.

7.2.1 UK quarries
Data were gathered for seventy-eight quarries: (Limestone = 73; Clay = 0; Sandstone = 5) from Minerals Planning officers of UK counties known to have limestone deposits, together with some from managers of large aggregates companies known to operate limestone quarries, sourced from the *Directory of Mining and Quarrying, UK* (1998).

Restoration and after-use are summarised as follows:

1. (31) sites left to re-colonise naturally.
2. (10) sites restored to agricultural end-uses, i.e. grain crops and animal feed.
3. (3) sites restored to low-level restoration, for caravan camping park.
4. (7) sites restored to landscape screening function with tree planting.
5. (14) sites restored to nature conservation/amenity.
6. (8) sites restored to water areas, some as lakes for recreation, and others for water storage.
7. (6) sites restored to wildlife and forest areas for conservation and amenity.
8. (4) sites restored by re-grading quarry faces to reduce visual impacts, to stabilise loose faces, and prevent rock fall.

9. (4) sites converted without restoration to commercial uses such as car storage, aggregate, materials stockpiles, and manufacture or storage of construction industry materials.
10. (1) site converted to grazing pasture.
11. (6) sites back-filled with industrial and domestic waste.
12. (7) sites with future restoration plans but not yet implemented because they are still active.
13. (11) sites with no information about their restoration plans.

7.2.2 Libyan quarries
Data were gathered for seventy quarries (Limestone = 64; Sandstone = 0); (Mixed limestone & clay = 6) with a total of eighty-five restoration sub-sections, directly from the Libyan Cement Company and Libyan Industrial Research Centre.

The site restoration and after-uses are summarised below:

1. (30) sites left to re-colonise naturally.
2. (23) sites restored to agriculture and horticulture uses (a very important demand from the local Bedouin people).
3. No sites received low-level restoration.
4. (4) sites restored for landscape benefit with tree planting by local people and local government bodies; and some associated residential areas.
5. (2) sites restored to nature conservation and amenity (these were close to tourist areas).
6. (5) sites converted to water bodies for recreation and for watering livestock.
7. (3) sites restored to wildlife and forest areas (woodland).
8. (4) sites restored by re-grading quarry faces to reduce risks of collapse.
9. (8) sites converted to commercial and industrial uses such as cattle markets, agricultural and horticultural produce markets and industrial units, for construction and building materials.
10. (4) sites converted to grazing pasture (shallow quarries with faces less than 8 m high).
11. (2) sites converted to domestic waste dump and recycling units.

The most frequent after-use of quarries in both countries was being left to re-colonise naturally or at least without restoration intervention. Clearly in many cases in the UK abandoned limestone quarries may acquire considerable nature conservation interest, although in the absence of effective management this may be ephemeral.

Agricultural and horticultural restoration appears to relate to the culture of the people and their lifestyle. In the UK study sites less than 10% were restored to agriculture with uses for specific kinds of crops and animal stocking. In Libya in both finished and unfinished sites, agricultural and horticultural use was more frequent (27%) with some associated domestic usage too. It seems restoration reflects people's activities and interests. They look to gain benefit from their mineral exploited lands. They also reflect the contrasting environmental and especially climatic circumstances. In the UK more sites are used for landfill or restored to nature conservation. In Libya, a greater proportion of sites are used for horticulture and/or agricultural use.

In both situations approximately the same proportions of sites are left to follow passive ecological succession – i.e. abandoned and not restored. Of course these may be turned to another use at a later date, and many will have "accidental" nature conservation value, often greater than that of landscaped sites.

8 DISCUSSION

Environmental impact assessment within the planning processes of Libya and the UK show major differences in both evolution and application. Approaches in the UK have elements transferable to Libya, but with the proviso that they will require a degree of cultural integration. This is further demonstrated by the specific examples of limestone quarries and for cement manufacturing. Much work refers to situations in the UK or Western Europe and in detail does not easily transfer, but the principles are relevant and transferable to Libya. It is in detail where they would fail in the Libyan context due to differences in environmental, cultural and political conditions. In terms of detailed restoration approaches, those advocated for arid areas in the USA (by the Society for Ecological Restoration and others) are more relevant to North Africa (Jordan *et al.* 1987).

Comparison of the planning procedures in the UK and suggest major differences in the procedures. There is a "bottom-up" approach in Libya, where the developer seeks permissions from various agencies of Government before submitting the application for approval at ministerial level. No environmental impact assessment is required. In the UK the procedures are more "top-down", where government departments prescribe the various stages. Under this system the Developer must adhere to the various stages, including the submission of an environmental impact assessment statement and an after-use proposal prior to lodging the application with the relevant "Minerals Planning Agency". In both cases, if conditions are met then permissions are granted. In both countries consultation with the local communities is facilitated. In Libya the decision-makers have government strategies and regulations to follow but above all they have to deal with the local tribes (the land owners) by explaining the consequent benefits from the proposed project and written guarantees for helping them to restore the site according to their wishes.

In the UK local opinion is considered during the application when environmental impact may only be resolved through the payment of compensation. Furthermore, the ultimate consent may be tightly controlled through the application of a number of conditions, all of which have to be adhered to otherwise the permission may be withdrawn.

Environmental Impact Assessment has developed as a distinct element of the planning process in western countries. The applications of EIA vary in each country. In Libya there is a growing desire to implement greater environmental protection and to allow the quality of life to rise. Here the processes of environmental planning, EIA, and environmental management systems, were started by western oil companies. These multi-national organisations brought with them a culture of environmental responsibility and of response to relevant national and international environmental legislation and agreements or controls. However, whilst this is to be welcomed in principle, its impact has so far been very limited. There now needs to be substantial and rapid progress to incorporate these procedures and cultures into mainstream industrial operations in Libya. In particular this research has identified the need to make rapid progress to implement appropriate policies for limestone extraction and cement manufacture. The operation of such systems in the UK industry can provide a template for such legislative systems. This is because even though the current planning procedures start from such different positions, they basically achieve the same goals in permitting or controlling the development and exploitation process. It is now necessary to integrate these approaches more fundamentally into the Libyan situation (Table 1).

Cultural differences and developments over a long period of time have markedly influenced the systems adopted in the two countries. However, there are aspects of the systems employed in the UK and other western countries that can be used if modified to bring them into sympathy with the politics and culture of Libya.

Environmental conditions relating to topography, climate and ecology impose further constraints on the application of restoration techniques. These methods need to be carefully adapted to local conditions. If done this will permit the beneficial use of former limestone quarries. One major conclusion is that cultural history and environment of a country impact on both potential after-use of former industrial sites such as quarries, and also on local people's attitudes to them. For quarries in Libya, and unlike the usual situation in the UK, former limestone quarries are seen as a positive resource. Without the twin constraints of

Table 1. Guidelines for the integration of environmental planning.

Level	Integration of Environmental policies and procedures.	EAP or Management/ Technical used.
National	Environmental policy included in national action plan.	Environmental profiles. International Assistance Agency Country Programming.
Regional	Economic, environmental development	Integrated regional development planning. Land use planning. Environmental Master Plans.
Sectoral	Sectoral review linked with other economic sectors.	Sector environmental guidelines. Sector review strategy.
Project	Environmental review of project activities. EIA procedures.	EIA. Environmental guidelines.

high human population density and limited land, and with searing Saharan winds, the development of a quarry in Libya is not generally contentious. Furthermore, at the termination of its industrial life it provides the community with a positive asset. This has been the way in Libya since before the Romans.

It is suggested that best practice guidance currently operating in the UK should be modified and carefully tailored to the Libyan system. It is not necessary to remove the present system, (and indeed for the social and historical reasons noted this would not work), but the relevant procedures could be taken, modified and integrated into the Libyan structures. Part of this would be provision of legislative and structural support, and a system of training and awareness raising for key personnel at all levels. A key factor will be the degree of "ownership" of the ideas and systems by the Libyan operators. If the systems appear to be imposed or even copied from the West, then they are far less likely to succeed.

The development of restoration practices that are fully embedded in the context of Libyan traditions will be a significant demonstration of the potential of the approach. This could be through the adoption of valley-side restoration for domestic use. Techniques to restore the remaining faces of former quarries to more environmentally valuable end-use could provide examples of how this approach might operate more widely.

ACKNOWLEDGEMENTS

We thank the Libyan Government for the generous provision of a postgraduate research grant, Sheffield Hallam University for use of facilities, and the many individuals and organisations who provided information. John Gunn is thanked for his critical assessment of the research.

REFERENCES

Anon. (1985) *The Libyan Mining and Quarrying Procedures Handbook.* Tripoli: Ministry of Industry Publications.

Anon. (1998) *Directory of Mining & Quarrying.* Nottingham and London: British Geological Survey Publications.

Barrow, C. (1997) *Industrial Relations – Law & Legilislation.* Cambridge: Cavendish.

Bradshaw, A.D. & Chadwick, M.J. (1980) *The Restoration of Land.* London: Blackwell.

Brundtland, G. (1987) *Our Common Future: The World Commission on Environment and Development.* Oxford: Oxford University Press.

Canter, L.W. (2000) *Environmental Impact Assessment.* New York, USA: McGraw-Hill.

Canter, L.W. (1996) *Environmental Impact Analysis.* New York, USA: McGraw-Hill.

DoE. (1989) MPG 7. *The Reclamation of Mineral Workings.* Cardiff: Welsh Office.

DoE. (1989) *Report on research and development.* London: HMSO.

DoE. (1994) PPG 9 – *Planning Policy Guidance: Nature conservation.* London: HMSO.

Elbah, S. (2001) An evaluation of the Environmental Impact Assessment within the planning process in Libya and in the UK in relation to cement manufacture. Sheffield: Unpublished MPhil Dissertation, Sheffield Hallam University .

English Nature, Quarry Products Association, and Silica & Moulding Sands Association (1999) *Biodiversity and minerals – extracting the benefits for wildlife.* Entec UK Ltd.

Gilbert, O.L. & Anderson, P. (1998) *Habitat Creation and Repair.* Oxford: Oxford University Press.

Gilpin, A. (1995) *Dictionary of environment & sustainable development.* London: Wiley.

Glasson, J., Therivel, R. & Chadwick A. (1994) *Introduction to environmental impact assessment: Principles and procedures, process, practice and prospects.* London: UCL Press.

Gunn, J., Bailey, D. & Handley, J. (1997) *The reclamation of limestone quarry using landform replication.* London: D.E.T.R.

Jalal, K.F. (1993) *Sustainable development, environment and poverty nexus.* Oxford: Oxford University Press.

Jordan, W.R, Gilpin M.E., & Aber, J.D. (Eds.). (1987) *Restoration Ecology: A synthetic approach to ecological research.* Cambridge: Cambridge University Press.

Jorissen, J. & Coenen, H. (1992) *The EEC Directive on EIA and its implementation in the EC member states in Colombo.* University of Lisbon Publications, Lisbon.

Land Use Consultants/DoE. (1992) *Amenity Reclamation of Mineral Workings.* London: HMSO.

Land Use Consultants and Wardell Armstrong (1996) *Reclamation of Damaged Land for Nature Conservation After-uses.* London: HMSO.

Lee, N. & George, C. (Eds.) (2000) *Environmental Assessment in Countries in Transition.* Occasional Paper 58. Manchester: Manchester University, School of Planning and Landscape.

Libyan Industrial Research Centre Report (1992) Ministry of Industry Annual Publications, Tripoli. (Translated from Arabic.)

Libyan Industrial Research Centre Report (1998) Ministry of Industry Annual Publications, Tripoli. (Translated from Arabic.)

Lohani, B.N. (1997) *Sustainable Development in Asia.* Manila, Philippines: World Bank Publications.

Moffatt, A. & McNeil, J. (1994) *Reclaiming disturbed land for Forestry.* Forestry Commission Bulletin 110. London.

Munn, J. (1979) *Community Leisure.* Birmingham: University of Birmingham, INLOGOV.

Ofori, S.C. (1991) *African Studies for Environmental Impact Assessment.* Trier, Germany: University of Trier Publications.

Wathern, P. (1992) *Environmental Impact Assessment: theory and practice.* London: Routledge.

Weston, J. (1997) *Planning and environmental impact assessment in practice.* London: Wesley Longman.

Wood, C. (1995) *Environmental Impact Assessment: a comparative review.* London: Longman.

Innovative Techniques

Land Reclamation – Moore, Fox & Elliott (eds)
© 2003 Swets & Zeitlinger, Lisse, ISBN 90 5809 562 2

Remediation technologies in UK: pitfalls and opportunities, selection and application

S. Ruzicka & E. Gustavsen
Remediation Technology Group, Waterman Environmental, London, UK

ABSTRACT: Despite the plethora of innovative remediation technologies arriving at the UK market the number of effective and reliable remedial solutions remains small. The introduction of novel techniques is being hampered by the lack of trial data, unproven reliability, licensing difficulties and not least by the pessimism or conservatism of key stakeholders. An overview of the techniques currently available in the UK is presented followed by a review of procedures associated with the decision making on the appropriate risk management solutions for contaminated land. A two stage assessment framework is proposed for the purpose of assessing the suitability of different remediation solutions, with simple examples from a small remediation project.

1 INTRODUCTION

Pragmatic and efficient management of risks arising from historically contaminated sites is a key prerequisite for successful redevelopment of brownfield sites. The risk management framework comprises identification and assessment of potential hazards (e.g. all pollutant linkages), estimation of the risks posed by these hazards and evaluation of their significance. Where the identified risks are deemed unacceptable suitable risk management solutions are developed for their mitigation. The general objective for these solutions, also known as remedial measures, is risk reduction through breaking the pollutant linkage by either source reduction, pathway management, control of the receptor exposure, or by a combination of these controls.

Careful selection of remediation methods is critical if resources are not to be wasted and environmental risks are to be minimised. However, despite the growing number of remedial solutions available many risk management decisions are still largely intuitive and impulsive, driven by the experience, or lack of it, of key stakeholders (land owner, consultant, regulator).

This paper presents an overview of key process-based technologies currently used in the UK and attempts to summarise the procedures involved in the decision making process for contaminated land management.

2 REMEDIATION TECHNOLOGIES IN THE UK

Although the range of novel and sustainable remediation technologies is steadily growing, only a limited number have been tested on the commercial scale, have a proven track record for the necessary licences to operate. Remediation technologies that are currently commercially available in the UK are summarised in Table 1 and discussed in the sections below.

2.1 *Biological methods*

Biological treatment technologies utilise the inherent ability and tendency of micro-organisms to break down hydrocarbon based compounds. In some environments the degradation is proceeding naturally at the optimum rate, while in others the process may be impeded by lack of oxygen or nutrients. Biodegradation can also be slow at high levels of free phase hydrocarbons or in the presence of high concentrations of heavy metals, some of which can be toxic to the micro-organisms.

For sites where the physical and geochemical properties of the aquifer and the nature of the contamination allow for efficient natural degradation of the contaminants, *monitored natural attenuation* (MNA) may be the preferred remedial option. Natural attenuation can be demonstrated by trends of lower down-gradient contaminant concentrations, changes in the chemical composition of the contaminants, i.e. the presence of known by-products down-stream, and by culturing indigenous microbes that are known to degrade the contaminants in question. A major drawback for MNA is the need for long term monitoring which can be costly and act as a deterrent in property transactions.

Where oxygen is the limiting factor, in-situ *bioventing* can stimulate the natural in-situ biodegradation of aerobically degradable compounds in the soil by providing oxygen to indigenous micro-organisms. The

Table 1. Remediation technologies available in the UK.

Biological	
In-situ	Monitored natural attenuation (MNA)
	Bioventing
	Enhanced bioremediation
Ex-situ	Landfarming
	Windrow turning
	Bio-piles
Chemical	
In-situ	Soil flushing
	Chemical oxidation
	Oxygen release compounds (ORC)
Ex-situ	Solvent extraction
Physical	
In-situ	Air sparging
	Soil vapour extraction
	Dual phase extraction (DPE)
Ex-situ	Soil washing
	Excavation and disposal
Thermal	
Ex-situ	Thermal desorption
	Incineration
Containment	
In-situ	Stabilisation
	Impermeable barriers
Ex-situ	Stabilisation
	Capping

technology uses direct low-flow air injection to provide just enough oxygen to sustain microbial activity. In addition to degradation of adsorbed residuals, volatile compounds are biodegraded as vapours move slowly past the microbes in the soil. The remediation process may take several years followed by a lengthy monitoring programme, which may be unfeasible when selling or developing a property, although it may be the preferred option for hot-spot remediation.

Enhanced bioremediation is an in-situ process in which the activity of naturally occurring or inoculated microbes is stimulated by circulating water-based solutions through contaminated soils to enhance in-situ biological degradation of organic contaminants, or immobilisation of inorganic contaminants. Nutrients, oxygen or other additives may be used to enhance bioremediation and contaminant desorption from sub-surface materials. Enhanced bioremediation can be especially effective for remediating low-level residual contamination in conjunction with source removal, although the process and following monitoring programme may be lengthy.

Ex-situ technologies offer more controlled processes and are more readily validated. For sites where there is sufficient space, hydrocarbon contaminated soil can be excavated and placed in *landfarms*, *windrows* or engineered *bio-piles*. Landfarms and windrows involve

placing the soil in piles, tilling the soil for mixing and oxygenation, and manually adding nutrients and moisture. Bio-piles have built-in aeration and irrigation systems and utilise bulking agents. Conditions can be fully optimised during ex-situ remediation, reducing the required treatment times, and ensuring that there is no need for long term post-remediation monitoring.

2.2 Chemical methods

Addition of chemical compounds, either in-situ or ex-situ, can break down the contaminants through oxidation or reduction, dissolve contaminants for extraction, or slowly release oxygen to enhance biodegradation. Chemical methods can be more costly compared to biological technologies on a cost per day basis, however significantly reduced treatment times lead to speedier re-development.

In-situ *soil flushing* is the extraction of contaminants from the soil with water or an organic solvent mixture. The flushing fluid is injected up-gradient of the plume and is extracted, with the dissolved contaminants, down-gradient and treated above ground before it is recovered and recycled. The method is effective for both inorganics and organics, although other technologies may be more cost effective for organic contaminants. Metals can also potentially be extracted. The technology should be used only where the flushing fluid and contaminants can be contained and recaptured. Above-ground separation and treatment costs for recovered fluids can be significant.

Chemical oxidation is a highly effective treatment method in which hazardous contaminants are converted to non-hazardous or less toxic compounds that are more stable, less mobile, and/or inert. The chemical oxidants can cause the rapid and complete chemical destruction of many toxic organic chemicals, while in other cases it is an aid to subsequent bio-remediation. It is essential that chemical oxidation is carried out under carefully controlled conditions as many of the reactions are exothermic and can cause violent eruptions within the ground if not controlled properly.

Injection of *oxygen release compounds* (ORC) is often used as a polishing process to target residual contamination following in-situ bioremediation. A slurry is injected, which contains chemicals that slowly release oxygen to enhance natural attenuation. The slurry can continue to release oxygen for several months without the need for further injections.

Ex-situ *chemical extraction* separates hazardous contaminants from soils, sludges, and sediments using an extracting chemical, thereby reducing the volume of the hazardous waste that must be treated. Physical separation steps are often used before chemical extraction to grade the soil into coarse and fine fractions, as the fines normally contain most of the contamination. Traces of solvent may remain within the treated soil matrix, so the toxicity of the solvent is an important consideration. The treated media are usually returned to the site if the chemical geotechnical properties are acceptable.

2.3 Physical methods

Physical technologies involve the use of non-chemical and non-biological methods of stripping the contaminants from the soil or groundwater for subsequent treatment or disposal, without altering the composition or nature of the contaminants.

In-situ methods utilise either injection of air, *air sparging*, or application of a vacuum, *soil vapour extraction* (SVE) or *dual phase extraction* (DPE), to volatilise the contaminants in the unsaturated zone or groundwater. Air sparging flushes the contaminants from the groundwater into the unsaturated zone where a vapour extraction system is usually installed to remove the vapour phase contamination. The oxygen added to the contaminated ground water and unsaturated zone soils can also enhance biodegradation. SVE and DPE both extract the volatile contaminants from the unsaturated zone through vertical extraction vents. Horizontal wells can also be used if warranted by the contaminant zone or access restrictions, e.g. underneath buildings. Geo-membrane covers are often placed over the soil surface to prevent short-circuiting and to increase the radius of influence of the wells. DPE also "slurps" groundwater for ex-situ treatment. By drawing down the water table around the wells, more of the adsorbed hydrocarbons in the unsaturated zone are exposed and accessible to vapour extraction. Use of SVE/DPE combined with bioremediation, air sparging or bioventing can shorten the cleanup time at a site.

Soil washing is an ex-situ soil separation processes in which soils are scrubbed with hot or cold water in a jet stream to remove contaminants such as organics, heavy metals and other inorganic contaminants. The process removes contaminants from soils by either dissolving or suspending them in the wash solution or by concentrating them into a smaller volume of soil through particle size or gravity separation, as most organic and inorganic contaminants tend to bind to the finer soil particles. For complex contaminant mixtures sequential washing, using different wash formulations and/or different soil to wash fluid ratios, may be required. Soil washing can usually be completed within relatively short timescales and the majority of materials, with exception of the fines, reused on site.

2.4 Thermal methods

Thermal desorption is a physical separation process in which wastes are heated at specifically designed temperatures and resident times to volatilise water and selected organic contaminants. Particulates and contaminants are then removed by conventional particulate removal equipment, such as wet scrubbers or fabric filters. Thermal Desorption can treat a range of contaminants, although soils with a high moisture content may require pre-treatment dewatering.

During *incineration*, high temperatures (870 to 1,200°C) are used to volatilise and combust halogenated and other refractory organics in hazardous wastes. Often auxiliary fuels are employed to initiate and sustain combustion. Over 99.99% of hazardous waste are removed from the soil and off-gases and combustion residuals are subsequently treated to remove particulates and neutralize and remove acid gases. Incineration can be costly and only a limited number of incinerators have permits to burn contaminants such as PCBs and dioxins.

2.5 Containment

In-situ *solidification/stabilization* is designed to reduce the mobility of contaminants in soil through both physical and chemical means. Grout or engineered clays (e-clays) are injected using auger systems either vertically or horizontally. Leachability testing is performed to measure the immobilization of contaminants. Utilisation of this technology is highly dependent on the physical properties of soil and is typically used on contaminants for which other in-situ techniques are unsuitable. Ex-situ solidification/stabilization typically requires off-site disposal of the resultant materials.

Physical barriers, or cut-off walls are used to contain or divert contaminated ground water, or provide a barrier for the ground water treatment system such as permeable reactive barriers. The cut-off walls consist of vertically excavated trenches that are filled with slurry, typically made from soil, bentonite and water. Certain contaminants may degrade the slurry wall components and reduce the long-term effectiveness.

2.6 Regulation

The use of the above methods within the UK market is being regulated by the Environment Agency through licensing process. Table 2 summarises mobile plant licences currently issued to service providers. For technologies without a current licence, the application process may take up to three months.

3 DECISION MAKING FRAMEWORK FOR RISK MANAGEMENT OF CONTAMINATED LAND

All risk management decisions involve three key stages: selection of appropriate solution, performance

Table 2. Mobile plant licences, Stevenson (2002).

Technology	No. of UK licences
Ex-situ stabilisation	19
Ex-situ bioremediation	18
Soil vapour extraction	13
Air sparging	13
In-situ bioremediation	11
Soil washing	11
In-situ stabilisation	7
Soil screening	2
Pump and treat	1
Thermal desorption	1

Table 3. General objectives for risk management in the UK.

Objectives	Key drivers
Protection of Human Health and the environment (controlled waters, ecological systems, property)	Environmental Protection Act 1990
Facilitate redevelopment (satisfy regeneration criteria and render land suitable for the intended use)	Town and Country Planning Act 1981
Economic investment (improve asset value, limit potential future liabilities)	Divestment or acquisition of property, business and/ or reputational risk
Repair of unsuccessful or insufficient remedial works	Any of the above

monitoring and validation and reassessment. In this paper we will primarily focus on the selection of risk management solutions.

3.1 General objectives

The implementation of risk management solutions in the UK is primarily driven by a combination of regulatory requirements and business needs. The most common objectives are summarised in Table 3.

Regardless of which of the above general objectives is driving the remediation, the chosen strategy must incorporate site specific objectives. In this process the solutions available for achieving the general goal are scrutinised and considered in the light of additional factors such as stakeholder satisfaction, cost effectiveness and sustainability (Bardos et al. 2002).

Examples of relevant stakeholders, in addition to those directly involved in the risk management (e.g. landowner, consultant, regulator), can be site workers and users, financial bodies, neighbours, campaigners and technical peers in the scientific community. This stakeholder diversity necessitates careful planning and early consultations when considering remedial solution with potentially significant impact.

Cost benefit considerations involve the assessment of the relative gain from each solution in terms of monetary value, balancing, among others, benefits achieved, impacts incurred and resources used up. Cost benefit analyses are notoriously difficult due to the variety of factors involved and the many non-generic issues which are likely to be specific to a particular site. In the UK, the Environment Agency issued a framework for comparison of remedial alternatives utilising multicriteria analysis (Postle et al. 1999, Hardisty & Ozdemiroglu 1999).

The necessity for sustainable development has to be considered in every risk management solution, and indeed, key principles are included in the above cost benefit framework. Here it should be noted that many impacts, both positive and negative in terms of sustainable development, cannot be easily defined in monetary terms. The impacts are socio-economic as well as environmental. The primary environmental considerations include the energy efficiency of the solution, renewability of the materials used, use of landfill resources and impact on soil and landscape value. Socio-economic issues may include loss of revenue to site owner, impact on local businesses, impact on the local community (infrastructure, employment, leisure) or public perception of the risk management approach.

The selection of appropriate solution must consider all of the above issues albeit in the knowledge that their individual significance will vary from site to site. Some of the more common issues are discussed below, complemented with real life examples, loosely based on a remediation of a former petrol station carried out in 1999.

3.1.1 Information requirements

The type and quality of information necessary for an effective and efficient risk management is generally given by the complexity of the site and the requirements of the risk assessment. As a general rule, staged environmental assessment comprising preliminary investigation (desk top, reconnaissance), detailed site investigation and risk assessment would be carried out. Often additional investigations, particularly aimed at the properties of pathways and the behaviour of potential contaminants in time and space are also undertaken (e.g. groundwater investigation, monitoring of groundwater, ground gas etc.).

It is essential that the data available is reviewed and gap analysis is undertaken to identify any further requirement that may be specific to a particular remedial solution.

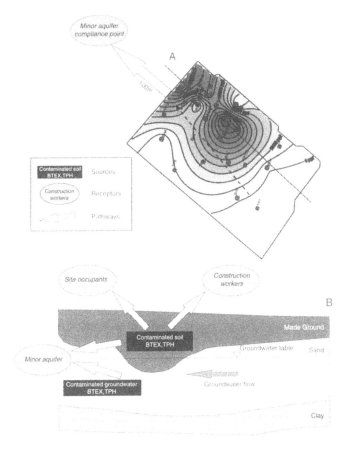

Figure 1. Simple conceptual model showing groundwater contamination contours (total petroleum hydrocarbons) and a cross section through the site.

A critical element for efficient decision making is an accurate conceptual model of the site. General guideline for the preparation of the model are given by Fookes (1997), ASTM (2000), McMahon et al. (2001), Nathanail et al. (2002).

A simple conceptual model for our case study site is shown in Figure 1.

Based on the conceptual model, critical pollutant linkages are assessed and the risks to each receptor are quantified during a site-specific risk assessment. Where a linkage is deemed to pose unacceptable risk to any associated receptor, the primary risk management objective is to reduce the risk to an acceptable level. For remedial solutions involving source reduction site specific target levels (SSTLs) are often calculated for each source (e.g. soil, groundwater), to ensure that particular receptor is adequately protected. The SSTLs then often act as performance requirements for the remedial works carried out.

3.1.2 Site specific objectives

For our case study example, specific objectives were formulated on the basis of the risk assessment and discussions with the client and the regulator:

1. to reduce the soil and groundwater contamination within the site to the levels below risk-based site specific target levels;
2. to minimise the impact of the remedial works on the proposed development of the site;
3. to minimise the impact of the remedial works on the ambient environment;
4. to provide a sustainable and economically viable solution; and
5. to render the site with minimal future liability with respect to ground contamination.

3.1.3 Review of available solutions

The purpose of the primary screening was to identify technologies which were technically suitable for the treatment of the subject site. In general, a single technology can rarely remediate an entire site. Consequently, two or more treatment technologies are usually combined at a single site to form what is known as a treatment train or treatment combination. For our case study site, where soil and groundwater was contaminated from a single source, one or two technologies should be sufficient. The following treatments and treatment combination which were available in the UK at the time were found to be suitable for the geological and hydrogeological conditions of the site:

– Monitored Natural Attenuation
– Excavation and Disposal combined with Pump & Treat

– Enhanced Natural Attenuation using the introduction of Oxygen Release Compound (ORC)
– Enhanced Natural Attenuation combined with Pump & Treat
– Low Vacuum Soil Vapour Extraction combined with Pump & Treat
– High Vacuum Dual Phase Extraction
– Air Sparging and Soil Vapour Extraction
– Bioremediation in situ
– Bioremediation ex situ (Bio-piles, Landfarming).

It is apparent from Table 4 that two technologies or technology combinations were found to be suitable, and three potentially suitable, for the remediation of the site. At this stage appropriate service providers were

Table 4. Suitability of remedial methodologies for achieving remedial objectives.

Technology	Code	Suitability for achieving objectives*					Comments
		1	2	3	4	5	
Monitored Natural Attenuation	MNA	✗	✓	✓	✓	✓	There is no evidence of contaminant plume shrinking; SSTLs are unlikely to be achieved within a reasonable timescale.
Enhanced Natural Attenuation (application of ORC)	ENA	?	✓	✓	✓	✓	Introduction of oxygen into the contaminated zone could accelerate natural degradation to acceptable levels, however, the performance of this scheme is difficult to control the treatment once introduced.
Enhanced Natural Attenuation combined with Pump & Treat	ENA/PT	✓	✓	✓	✓	✓	The employment of Pump & Treat would accelerate the process and will enable to control the treatment whilst operating.
Low Vacuum Soil Vapour Extraction combined with Pump & Treat	SVE/PT	?	?	✓	✓	✓	Achieving SSTLs for petroleum hydrocarbons could be compromised by low volatility of diesel range hydrocarbons. The requirement for abstraction wells and ducting would cause a delay to the development or necessitate alteration of the proposed foundation design.
High Vacuum Dual Phase Extraction	DPE	?	?	✓	✓	✓	The requirement for abstraction wells and ducting would cause a delay to the development or necessitate alteration of the proposed foundation design.
Air Sparging and Soil Vapour Extraction	AS/SVE	✗	?	✓	✓	✓	The effectiveness would be compromised by the insufficient thickness of the aquifer (2 to 1.5 m). In addition, the requirement for abstraction wells and ducting would cause a delay to the development or necessitate alteration of the proposed foundation design.
Bioremediation in-situ	BIS	✓	✓	✓	✓	✓	Is likely to be effective, however, there is insufficient evidence from pilot studies to guarantee the achievement of SSTLs.
Bioremediation ex-situ (Biopiling)	BES	✓	✗	✓	✓	✓	Not feasible as there is insufficient storage space within the site for biopiling.
Excavation and Disposal combined with Pump & Treat	ED/PT	✓	✓	✓	✗	✓	Not a sustainable solution. Unnecessary use of landfill space. Total cost of this approach was estimated around five fold the amount of the next most expensive scheme.

* for individual objective refer to section 3.1.2; ✓ – suitable; ✗ – unsuitable; ? – suitability uncertain.

348

contacted on a pre-tender basis in order to obtain information regarding technical feasibility, potential restrictions, timescales and costs. This information along with site specific issues and client's requirements were used for the selection of the most suitable methodology.

3.1.4 Assessment of suitable solutions

The purpose of this assessment is to evaluate the risks associated with the employment of the proposed technologies at the subject site and rank these according to their suitability. Generally, risks associated with site remediation can be divided into four categories:

– Operational (technical feasibility, result uncertainty)
– Environmental (impact on ambient environment, including properties and public)
– Commercial (cost implications, time-scales, necessary statutory consents)
– Legal (regulatory compliance, e.g. mobile plant licence, waste management licensing regulations, duty of care, health and safety, off-site contaminant migration).

Table 5 presents a breakdown of risks posed by the employment of treatment technologies or technology combination identified in the previous section for specific factors within these categories. The assessment is semi-quantitative whereby each factor is allocated a suitability ranking from 0.1 to 1.0, with 0.1 denoting the lowest suitability (highest risk) and 1.0

denoting highest suitability (none or negligible risk). This way, technical feasibility, for example, would be rated 0.9 for enhance natural attenuation, comprising drilling treatment wells and provision of oxygen source, but only 0.7 for dual-phase vacuum extraction as the installation of treatment wells, ducting, vacuum pump, scrubber unit etc. is far more elaborate. In the case study the rating was done qualitatively based on experience, however, once sufficient data is obtained on each technology, semi-quantitative scores can be derived for a particular solution. Each risk factor is given a weighting from 1, 2 or 3, corresponding to low, medium or high importance, respectively, which are allocated following discussions with key stakeholders, in this case the statutory authorities, residents representative and the client. The resulting average rank score for each individual treatment technology is calculated as a weighted average of all risk quotients and expressed in percentage of maximum possible score. The most suitable technology is then indicated by the highest score.

3.1.5 Evaluation and selection of preferred solution

Following the above assessment enhanced natural attenuation combined with pump and treat system was selected as the appropriate solution for managing ground and groundwater contamination risks in the subject site.

Table 5. Risk assessment for the employment of individual treatment technologies.

| Risk Factor | Weighting | Suitability ranking | | | | |
		ENA	ENA/PT	SVE/PT	DPE	BIS
Operational						
Technical feasibility	1	0.9	0.9	0.7	0.8	0.8
Result uncertainty	2	0.5	0.8	0.7	0.9	0.7
Environmental						
Environmental impact	1	0.9	0.8	0.6	0.6	0.8
Commercial						
Capital cost	2	0.8	0.7	0.5	0.5	0.7
Operation/maintenance cost	1	0.9	0.8	0.6	0.6	0.9
Monitoring cost	1	0.9	0.7	0.8	0.8	0.6
Treatment duration	2	0.2	0.6	0.8	0.8	0.6
Statutory consents, negotiations	1	0.2	0.8	0.8	0.8	0.5
Effect on development programme	2	0.9	0.9	0.2	0.2	0.8
Legal						
Regulatory compliance	1	1.0	0.8	0.8	0.8	1.0
Health liability	2	1.0	0.8	0.8	0.8	0.5
Off site migration	2	0.3	0.7	0.7	0.4	0.3
Average rank score*		68%	77%	65%	64%	66%

* ranking coefficient 1 corresponds to 100%.

349

4 CONCLUSIONS

Despite the plethora of innovative remediation technologies arriving at the UK market the number of effective and reliable remedial solutions is growing slowly. The introduction of novel techniques is being hampered by the lack of trial data, unproven reliability, licensing difficulties and not least by the pessimism or conservatism of key stakeholders.

This naturally limits the number of solutions available for formulation of efficient, effective and flexible remediation strategies. A careful analysis is needed at this stage in order to incorporate a variety of factors including stakeholder satisfaction, sustainability and cost-effectiveness.

The selection of the appropriate remedial solution should be carried out in a staged assessment of available techniques. The two stage system proposed in this paper comprises technology screening, based on: technical feasibility, availability in the UK and their suitability to achieve all site specific objectives, followed by a suitability ranking which judges the potentially suitable solution on their operational, environmental, commercial and legal risks.

The basis for this decision making framework is currently fully qualitative. Much more information is needed before more objective assessment can be made, perhaps drawing from an industry-wide information sources and databases.

As the variety of remediation technologies available in the UK market increases, the need for an integrated approach in the decision making on remedial solutions becomes an essential prerequisite for successful risk management of contaminated land. Bardos et al. (2002) hint that key issues are likely to be built into a model procedure for evaluation and selection remedial measures (DEFRA *in preparation*). Let us wait and see.

REFERENCES

ASTM (2000). Standard Guide for Developing Conceptual Site Models, American Society for Testing and Materials. Report No. E1689–95.

Bardos, P., Nathanail, J. & Pope, B. (2002). General principles for remedial approach selection. *Land Contamination & Reclamation* 10(3): 137–160.

Department for Environment, Food and Rural Affairs (DEFRA) (in preparation). Model procedures for the management of contaminated land. CLR11, DEFRA and the Environment Agency, in preparation.

Fookes, P. G. (1997). Geology for Engineers: the Geological Model, Prediction and Performance. *Quarterly Journal of Engineering Geology* 30: 293–424.

Hardisty, P. E. & Ozdemiroglu, E. (1999). Costs and benefits associated with remediation of contaminated groundwater: A review of the issues. Environment Agency, Technical Report P278, Swindon.

McMahon, A., Carey, M., Heathcote, J. & Erskine, A. (2001). Guide to Good Practice for the Development of Conceptual Models and the Selection and Application of Mathematical Models of Contaminant Transport Processes in the Subsurface. National Groundwater & Contaminated Land Centre report NC/99/38/2, Environment Agency (UK).

Nathanail, J., Bardos, P. & Nathanail, P. (2002). Contaminated Land Management: Ready Reference, Land Quality Press and EPP Publications.

Postle, M., Fenn, T. Grosso, A. & Steeds. J. (1999). Costs-benefit analysis for remediation of land contamination. Environment Agency, Technical Report P316, Swindon.

Stevenson, H (2002) Environment Agency Mobile Plant Licence List. Environment Agency Website: www.environment-agency.gov.uk (accessed 26.11.02).

Land Reclamation – Moore, Fox & Elliott (eds)
© *2003 Swets & Zeitlinger, Lisse, ISBN 90 5809 562 2*

An appraisal of remedial trials for styrene contaminated ground and groundwater at a former chemical manufacturing facility in South Wales

P.E. Russell & B. Ellis
CELTIC Technologies Ltd, Cardiff, Wales, UK

ABSTRACT: Site ground investigations undertaken at a former chemical manufacturing facility in South Wales have identified the ground and groundwater surrounding the facility to be grossly contaminated with aromatic hydrocarbons including benzene, ethyl benzene, toluene and styrene. The contaminants occupy an area of approximately 15,000 m^2 in the unsaturated zone, 40,000 m^2 in groundwater and some 300 to 700 tonnes of non-aqueous phase liquid located principally in a smear zone associated with the shallow water table. In order to reduce environmental risks a number of remediation trials have been carried out with both laboratory and pilot field studies. The principal aim of the remediation trials has been to assess the best techniques for removing the volatile hydrocarbons without impacting the local environment. The physico-chemical characteristics of the contaminants influence the choice and application of remediation technique. The high toxicity of benzene and styrene and the particular physical properties of styrene also require careful consideration in applying risk-based remediation without causing detrimental impacts on air quality or controlled waters. The results of the feasibility trials are presented and conclusions are drawn as to the success and merits of the remedial techniques undertaken.

1 INTRODUCTION

1.1 *Background*

The study area formerly comprised a chemical manufacturing facility that was used, predominantly, for the production of styrene and polystyrene. Ground contamination assessment undertaken over a number of years, identified the area of the former facility to be grossly contaminated. This contamination having occurred as a result of 37 years of operation during which a series of product spillages and tank leaks had occurred.

1.2 *Extent of the problem*

The contaminants are predominantly volatile organic carbons and consist of benzene, ethyl benzene, toluene, styrene and styrene monomer. The composition of the cocktail varies across the area, almost pure styrene and ethyl benzene product in some locations, ranging up to 50% benzene in others. A particular factor associated with this contamination is the additional presence of styrene polymer thought to have resulted from the polymerisation of styrene by air (oxygen) or other influences within the ground itself.

The contaminants occupy an area of approximately 15,000 m^2 in the unsaturated zone, 40,000 m^2 in groundwater and some 300 to 700 tonnes of non aqueous phase liquid (NAPL) located principally in a smear zone associated with the shallow water table, 2 to 3 metres below the ground surface.

The presence of the contamination is significant due to several key factors. The high toxicity of the compounds of concern, particularly benzene; the evidence that a significant mass of contamination has migrated in groundwater towards other land owners; the potential influence of such high masses of contamination on the natural attenuation capacity of the underlying aquifer; uncertainties associated with potential vapours and odours, particularly given the shallow groundwater level and possible (unpredictable) impact of ambient atmospheric conditions and/or precipitation and restrictions to future development.

As a first step to reducing environmental risks a number of feasibility trials have been carried out both in the laboratory and with pilot field studies. The principal aim of the feasibility trials has been to assess the best techniques for removing the volatile hydrocarbons without impacting the local environment. The feasibility trials were additionally conducted to evaluate the technical and financial viability of on-site treatment of the contaminants of concern, this information then being used to make a full cost benefit analysis against other potential remedial options. Other options considered include excavation to landfill, soil washing, containment systems and low temperature thermal desorption.

The laboratory and field trials that have been undertaken and are described include: column studies with soils in the laboratory to assess the potential for stripping using air or nitrogen; pilot scale testing including the installation of remediation trenches and abstraction boreholes with methods assessed including passive and aggressive extraction systems such as buoyancy and band skimmers, total fluids pumping, vacuum enhanced recovery and a pilot scale treatment study evaluating ex-situ bioremediation.

2 ENVIRONMENTAL SETTING

2.1 Geology

Investigations within the study area have indicated that ground conditions consist of fill materials, slag deposits, overlying geologically recent deposits of sand (4 to 6 metres thick) overlying a laminated clay layer of thicknesses varying between 1.5 to 2.5 metres, which itself overlies deeper sand of 5 to 10 metres thick. The deeper sand unit overlies a sequence of Boulder Clay. The shallow sand deposits comprise principally windblown fine to medium sands with very little silt, clay or organic content. The windblown sands are overlain by fill material and hard core in places up to 0.5 metres thick. The laminated clay under the windblown sand has been ascertained as being continuous across the study area.

2.2 Hydrogeology

Within the study area, the groundwater table appears flat, with groundwater flowing at shallow depth, generally in three directions. Groundwater flow patterns may however be influenced by underground services, such as abandoned water supply and fire hydrant service lines. Groundwater is present at approximately 1 metre below the surface, but fluctuates between 0.5 and 1.2 metres seasonally. During periods of heavy rainfall the ground within the area is known to flood, with groundwater levels approaching or close to the ground surface.

Grain size analysis on the windblown sand has identified that these deposits comprise 96 to 100% sand, with 0 to 4% clay and silt. Hydraulic testing undertaken in the windblown sands has indicated that the permeability of the sands are in the order of 2×10^{-4}

metres/second and that the effective porosity is approximately 15 to 20%.

The laminated clay acts as an aquiclude or aquitard and separates the two sand aquifers. The groundwater heads measured in the deeper sand aquifer is on average 1 to 1.5 metres below the heads measured in the windblown sand. This gives a potential for downward vertical head gradients. In addition to this it is known that piling through the laminated clay and historical boreholes completed through the laminated clay have resulted in "cross aquifer" groundwater flow and contamination.

2.3 Contaminant distribution

Investigations have indicated that the ground and groundwater contamination is a mixture of aromatic chemicals associated with the historical processes and storage within the study area. The nature of NAPL detected in the area varies and because of groundwater fluctuations in groundwater levels, much of this free product is associated with a smear zone present up to approximately 0.3 metres below the surface and approximately 1 metre in thickness. Concentrations of hydrocarbons vary within the unsaturated zone, ranging from 2% by weight, in the upper 0.3 metres below ground level, rising to 7% at depths approximately 2 metres below ground level. The residual contamination is present to the base of the windblown sands at 4 to 6 metres below the surface. In addition the residual contamination, 7% by weight, will have a significant bearing on the total porosity of the shallow sand aquifer, estimated to be 15 to 20%, and subsequent groundwater flow mechanisms in the study area.

It is estimated that based on an average of 2% hydrocarbons and assumptions on porosity and residual retention, between 300 and 700 tonnes of hydrocarbons may be present in the unsaturated zone. An additional 500 to 1000 tonnes is estimated to be present as "free product."

In the saturated zone below the capillary fringe of the study area, it is difficult to make any estimates of mass distribution, but again based on average concentrations measured in groundwater it is estimated that an additional 500 tonnes of hydrocarbons could be present. In addition to this a groundwater plume has been delineated which extends in three main vectors with further contamination of the underlying lower sand aquifer. Styrene is identified in groundwater up to 150 metres from the source area, whereas benzene contamination is present up 600 metres down hydraulic gradient.

2.4 Environmental liabilities

Based upon the contamination found in the study area it was clear that a number of issues are present associated with environmental liabilities, potential redevelopment and any remediation undertaken. These issues include: nuisance – the volatile nature of the contaminants present within the study area poses an ongoing problem of potential odours, for instance the odour threshold of styrene is 0.03 ppm in air; the potential impact on human health of future users of the site, including contact via inhalation, ingestion or skin contact, of vapours and dust borne contamination; migration of the contamination via groundwater, surface run-off or drainage conduits to third party properties; risks and liabilities associated with the potential impacts and restrictions to future development of the study area, including potential effects on building materials.

Based on the factors given above, but primarily because of the environmental liabilities and restrictions to development, a remedial strategy was instigated which would require hydrocarbon source control by containment, removal or treatment, in addition to any risk-based monitoring.

The physico-chemical characteristics of the contaminants offer some advantages in the choice and application of remediation technique. The high toxicity of benzene and styrene and the particular physical properties of styrene also require careful consideration in applying risk-based remediation without causing detrimental impacts on air quality or controlled waters. Therefore in order to assess the optimal remedial technique a number of laboratory and field feasibility studies have been undertaken.

The results of these studies would then be used to make a full appraisal of potential remedial technologies through cost benefit analysis from a better knowledge of technical feasibility.

3 LABORATORY TESTING

3.1 Introduction and objectives

A series of laboratory scale tests were initiated to evaluate the potential for air stripping styrene and other volatile organic carbon contaminants from soil and water matrices. The principal aim was to assess the properties of styrene, for the other compounds, benzene, ethyl benzene and toluene, more is known regarding their physio-chemical properties.

Styrene (also known as vinyl benzene, phenylethylene, ethenylbenzene or cinnamene) is an oily volatile liquid, which is flammable, toxic and potentially carcinogenic. The compound polymerises readily at temperatures above 65°C and on exposure to light and peroxide catalysts in the environment, to form styrene polymers. Despite low aqueous solubility, styrene will disperse quickly within the water environment, having little affinity to soil and sediments. In air styrene is reported to evaporate quickly and degrade rapidly. In contrast, styrene polymers, though lower in toxicity, will persist for much longer periods in the environment.

A styrene plant and styrene storage operated in the study area for a number of years and site investigation has identified ground and groundwater contamination with styrene. Additionally it has been observed in the study area that a degree of styrene polymerisation has taken place in the ground to produce gel-like matter. This polymerisation is thought to be caused by the presence of oxygen in air/water but might additionally be influenced by the underground services.

Based on developing an understanding of the properties of styrene laboratory tests were undertaken to: evaluate the effectiveness of stripping soil contaminated with mixed volatile organic carbons, including styrene; evaluate the effectiveness of stripping groundwater similarly contaminated; compare air stripping with nitrogen gas given the possibility that styrene monomer could polymerise in an air-rich environment.

3.2 Methods

Tests were undertaken on composite samples taken from trial pits and boreholes installed in the study area. Laboratory tests involved the sparging of soil and groundwater at a predetermined flow rate for approximately 48 hours. Sub-samples of each matrix were used before and after to determine the reductions of styrene, while serial on-line activated carbon traps were used to collect contaminants volatilised within the exhaust gases at set time periods. Traps were replaced after each sampling period. The data was used to estimate the amount of mass removed and relate this to gas delivery. In view of the potential for polymerisation a second set of tests were undertaken using nitrogen to sparge the matrix. For soil matrices, tests involved passing air vertically through a column of soil, water testing involved sparging samples in Glass Dreshel bottles. Because of the nature of styrene monomer, all apparatus was constructed from glass and kept away from light.

3.3 Results

The results for the air stripping of soil are presented in Table 1, below.

Table 1. VOC analysis for air stripping of soils.

Compound	Conc. mg/kg*		Loss	
	Zero time	45 hrs	mg/kg	%
Benzene	235.5	0	235.5	100
Toluene	285	40.25	244.8	86
Ethyl Benzene	6590	3255	3335	50
Styrene	6060	3810	2250	37

* Concentration taken as mean of samples at beginning of test and after 45 hours of air stripping.

The results indicated that air stripping was successful in the removal of the key compounds and that after 45 hours concentrations were significantly reduced. For styrene however the carbon traps contained a total mass of 5 mg over the test (styrene sorbed onto the carbon from the vapour phase) compared to a total mass lost of 450 mg. This difference could be due to the effects of styrene polymerisation. Unfortunately a method of measuring the concentrations of styrene polymer could not be accurately developed by the laboratory undertaking the analysis. Table 2 presents the results for air stripping of groundwater.

As with the stripping of the soil samples, the stripping of the groundwater demonstrated significant reduction in mass. However, based on the carbon traps

Table 2. VOC analysis for air stripping of groundwater.

Compound	Conc. mg/l*		Loss	
	Zero time	45 hrs	mg	%
Benzene	8.1	ND	1.22	100
Toluene	3.3	ND	0.5	100
Ethyl Benzene	19.8	0.0033	2.95	100
Styrene	41.3	0.0044	6.2	99.9

* Concentration taken as mean of samples at beginning of test and after 45 hours of air stripping.

Table 3. VOC analysis for nitrogen stripping of soils.

Compound	Conc. mg/kg*		Loss	
	Zero time	21 hrs	mg/kg	%
Benzene	235.5	10.7	224.8	96
Toluene	285	160	125	45
Ethyl Benzene	6590	4030	2560	39
Styrene	6060	4180	1880	31

* Concentration taken as mean of samples at beginning of test and after 21 hours of nitrogen stripping.

Table 4. VOC analysis for nitrogen stripping of groundwater.

Compound	Conc. mg/l*		Loss	
	Zero time	20 hrs	mg	%
Benzene	8.1	ND	1.22	100
Toluene	3.34	ND	0.5	100
Ethyl Benzene	19.8	0.014	2.95	100
Styrene	41.3	0.0076	6.2	99.8

* Concentration taken as mean of samples at beginning of test and after 20 hours of nitrogen stripping.

only 0.01 mg was retained compared to the loss of 6.2 mg. No physical evidence of polymerisation was seen in the samples, although there was evidence of this is the carbon filter traps.

Tables 3 and 4 present the results of stripping using nitrogen on soil and groundwater.

As with the sparging with air the carbon traps could only account for a small percentage of the styrene or the other volatile carbons lost during the nitrogen stripping.

3.4 Conclusions of stripping tests

Based upon the laboratory testing it appears that sparging with air or nitrogen could be an effective method for the remediation of VOCs and styrene both from soil and from groundwater matrices. Results suggested complete removal of volatiles from groundwater within a few hours and around 37% removal from soil in 41 hours. Extrapolating the data it would indicate that 10 to 20 days of stripping would be necessary to reduce soil concentrations to near zero. Based on observational indicators air sparging did not appear to cause styrene polymerisation as results were similar for air and nitrogen. However, mass calculations indicated discrepancies between VOCs removed and those collected on carbon traps. These discrepancies could be attributed to natural biochemical degradation, polymerisation in the case of styrene or laboratory analysis error. Polymer quantification could not be undertaken as a reliable method of its analysis could not be developed by the laboratory.

In general it was considered that the laboratory testing indicated that field scale stripping with air or nitrogen could be a successful remedial technique for removing contaminant mass from both soil and groundwater at least within the range of concentrations tested.

4 FIELD SCALE TESTING

4.1 Introduction and objectives

Laboratory studies are inevitably limited with respect to the potential applicability in the field, particularly full scale commercial systems. Factors include variability of in-situ air permeability within the ground, which may be impacted where residual is present, polymer is present, or where pore space is saturated. In the study area it is known that the water table is close to the ground surface, that styrene polmer is present and that there is a large residual mass. For this reason it was identified that stripping would only be successful as an integrated approach following initial Non Aqueous Phase Liquid (NAPL) removal. Field trials were conducted to assess the optimum NAPL removal approach.

In addition to assisting in any sparging techniques NAPL removal will also increase the potential for natural bio-chemical degradation. Whilst ongoing groundwater monitoring in the study area will form part of the assessment of this potential ex-situ bioremediation trials were undertaken during the field scale testing programme.

4.2   Scope of works

The field scale testing programme aimed to determine the most suitable product recovery and pump methodology and assess the efficiency of product recovery from shallow trenches or boreholes. Additionally the biodegradation of styrene and other VOCs in soil was evaluated by creating a biopile with the material excavated from the trenches.

4.3   Methods, product recovery

The techniques undertaken in boreholes and within recovery trenches included both passive and active systems such as skimming, total fluids pumping and groundwater pumping with skimming. There was examination of recovery rates and effectiveness, limitations with regards health and safety and practicality and cost implications for full scale design.

Four collection trenches were constructed in the study area based on the understanding of free phase thickness across the area. Each trench was approximately 3 metres deep by 1 metre by 6 metres in length. The trenches were backfilled with 75 mm aggregate to approximately 0.5 metres below the surface. Within the gravel aggregate a 600 mm diameter sump was installed and this was designed to allow the entry of LNAPL. Also installed within the trench was a 50 mm diameter vapour extraction point. The aggregate was covered with an impermeable membrane and had material compacted above it to prevent the release of vapours to the atmosphere.

The boreholes used for the trials were drilled to the laminated clay layer (approximately 5 metres) and installed with 100 mm internal diameter HDPE casing. The casing was slotted from 0.5 metres below the ground surface to the base of the borehole.

A variety of systems were trialled to examine the best method of product removal. Two types of skimmers were tested belt skimmers and float skimmers. Belt skimmers are based on an electrically powered engine rotating a suspended belt into a layer of LNAPL. The belt becomes coated in the product and this is then periodically scraped off. This technique has advantages for skimming product with a high viscosity, such as styrene. Floating vacuum skimmers used in the trials were based on central or peripheral positioned floats (tri-floating skimmers and central

buoyancy skimmers). The skimmers being driven by a pneumatic vacuum pump.

Total fluid pumping was undertaken using a pneumatic pump with air flow controlled via a pressure regulator. Abstracted fluids include free phase and groundwater. The intake of the pump is positioned however such that the amount of groundwater pumped is minimised.

Groundwater pumping with product skimming was undertaken by placing a pneumatic pump below the product layer and pumping groundwater. The skimmer is placed at the product interface. The abstracted groundwater is then treated in a separator.

Soil vacuum extraction was additionally trialled in the soil vapour extraction pipe installed in the trenches. The vacuum extraction was trialled as a means of establishing whether creating a pressure head gradient into the well and inducing vapour phase mass removal would significantly increase recovery rates.

4.4   Methods, ex-situ bioremediation

A biopile was created with the material excavated from the trenches installed on site as part of the product recovery works. The treatment area of 20 by 12 metres was bunded to contain any possible leachates. This consisted of a layer of clean washed sand with raised edges over a basal lining material. An organic fertiliser was used to condition the material while inorganic nutrient supplement was added to aid microbial degradation in the pile. The biopile was covered to control vapour emissions and odours and prevent water ingress. A vapour extraction system was used to remove any vapours that were generated.

Operational monitoring of the biopile included pH, temperature, moisture and in-situ atmospheric gases as well as VOCs in the extracted vapours.

4.5   Results, product recovery

The belt skimming trials produced an average of 45 litres of recovered product per hour. However, the system had the disadvantage of being high maintenance, the belt showed rapid deterioration due to contact with the hydrocarbon and scraping of the product off the fleece. On the health and safety aspect the belt skimmer is considered to produce a higher level of VOC emissions to air compared to the other techniques trialled.

Free product was effectively removed using the tri-float skimmers. Once the set up had been optimised the skimmer achieved near waterless product removal, with rates of product removal up to 220 litres per hour. The central buoyancy skimmers had a lower rate of success compared to the tri-float skimmers, with rates of 100 litres of product recovered per hour. In the boreholes the free product drained too quickly using this technique causing discontinuous flow to the well and replacement of the pore space surrounding the borehole with water.

Total fluid pumping from the trenches resulted in an average removal of 127 litres of product per hour and 150 litres of groundwater. Within the boreholes used for the trial an average of 67 litres were removed per hour with 185 litres of groundwater. This indicated that the trenches offered advantages to boreholes in encouraging product flow and improving recovery by increasing the area of influence.

Concurrent groundwater pumping with product skimming was trialled with the tri-float set up only. The results indicated that 170 litres of product were removed each hour at a pumping rate of 1000 litres a hour. On the basis of the results the NAPL recovery rates were not improved by inducing a cone of drawdown on the groundwater table, however the increased area of influence created compared to a passive skimming system may mean that fewer boreholes or trenches need to be installed during full remedial works across the area for remediation.

Soil vacuum extraction was tested as a means of both enhancing product recovery and vapour phase mass removal. Vacuum extraction was found however to be ineffective at encouraging free phase product into the well. No increase in product thickness was noticed after 120 hours of continuous operation. Vapour phase monitoring in the extracted exhaust flow was seen to decrease from 430 ppm to 30 ppm within the first 30 minutes of testing.

4.6 Results, ex-situ bioremediation

The data obtained were used to assess the potential for degradation of styrene, benzene, ethyl benzene and toluene from the soils in the study area. Figure 1 and Figure 2 show the two main mechanisms by which styrene is considered to aerobically degrade.

Degradation via these pathways are considered to result in half lives for styrene of a few days in surface waters compared to several months in low oxygen environments. It is these processes that the ex-situ bioremediation will be utilising to reduce contaminant mass within the biopile.

The results of the bioremediation trials are presented in Table 5.

The results demonstrated that a total of 690 kg of contaminant mass had been removed during the 5 weeks of the testing. Vapour monitoring indicated that an average concentration of 20 ppm was removed via VOC vapours abstracted from the biopile.

4.7 Conclusions of product recovery trials

The pilot study resulted in the removal of 2700 kg of product from all of the differing techniques over a

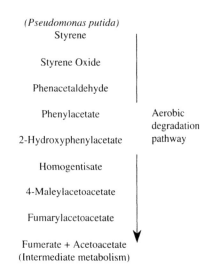

Figure 1. Aereobic degradation pathway with the microbial organism *Pseudomonas putida*.

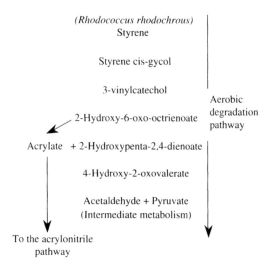

Figure 2. Aereobic degradation pathway with the microbial organism *Rhodococcus rhodochrous*.

Table 5. VOC analysis for bioremediation of soils.

Compound	Conc. mg/kg*		Loss	
	Zero time	5 wks	kg	Half life
Benzene	501	0.14	45.07	3.5
Toluene	263	0.21	23.65	4
Ethyl Benzene	4401	7.2	395.28	3.5
Styrene	2525	2.9	226.71	3.5

*Concentration taken as mean of samples at beginning of test and after 5 weeks. Half life measured approximately in days.

total operational period of 32 hours. This represents a high rate of product removal and indicates high recoverability potential. Based on an average recoverability of 80 litres per hour this would result in 340 tonnes of product over a 6 month period, assuming continuous operation.

The trials also indicated that trenches offer preferential recovery rates compared to boreholes. Float skimmers were considered to provide the best overall option with the ability to recover 200 litres per hour with negligible groundwater removal. Dewatering combined with product skimming will increase the radius of influence and this may result in less trenches over the area for remediation. However, the mass removal rate did not appear to be better than skimming alone.

Soil vacuum extraction was unsuccessful in increasing the rates of product recovery and only small quantities of volatile vapours were collected during this testing. However, volatile vapour emissions during the operation of the feasibility trials were problematic during the excavation of the trenches. This will be ongoing problem during a full scale remedial programme.

4.8 *Conclusions, ex-situ bioremediation*

The bioremediation system proved successful in reducing contaminant mass. However, potentially significant vapours were generated during the construction of the biopile and this would be a factor to consider during any full scale remediation of the study area.

5 CONCLUSIONS

The study area comprises an area of ground contamination of 15,000 m^2 and groundwater contamination of 40,000 m^2 in area. An estimated 300 to 700 tonnes in the unsaturated zone and 500 to 1000 tonnes of free phase is estimated to be present in the ground. The key contaminants of concern are benzene, ethyl benzene, toluene and styrene.

The environmental setting of the study area and subsequent environmental liability of the source-pathway-receptor linkages has resulted in a requirement for active remediation. In addition redevelopment drivers mean that active remediation is required in the study area.

Prior to undertaking full scale remediation a number of laboratory tests and field trials have been conducted to examine the most technically viable remedial

approach and techniques for addressing the contamination in the study area.

Laboratory testing has demonstrated that the contaminants of concern could be successfully stripped from both soil and groundwater matrices using both air and nitrogen. Given the reduction of the success of air or nitrogen sparging in the field due to the thickness of free phase present, the presence of any styrene in polymer form (having a gel like consistency) and the potential for significant reduction in air filled porosity due to residual product, it was concluded that, for the optimum approach the LNAPL product should be removed prior to sparging.

In order to evaluate the best technique for the removal of the LNAPL a series of passive and active field trials were conducted testing the effectiveness of product recovery from trenches and boreholes. The tests included product skimming, total fluids pumping and groundwater pumping with skimming. In addition to this a bioremediation trial was undertaken.

The product recovery trials resulted in the removal of some 2700 kgs of product from the techniques undertaken over a total operational period of 32 hours. Based on average rates it was crudely estimated that 340 tonnes of product could be removed over a 6 month period. The optimum approach was seen to be product skimming using tri-floating peripherally placed skimmers. However, it was considered that by combining skimming with groundwater pumping design savings could be made but these may be off-set by groundwater treatment costs.

The ex-situ bioremediation trial identified that aerobic biodegradation of the contaminants of concern was an important mechanism whereby contaminant mass could be reduced over a relatively short period of time. The results indicated that ex-situ treatment would be successful and suggests that in-situ bioremediation (natural attenuation) could be enhanced by inducing the right bio-chemical conditions.

The feasibility testing has acted as an important means of developing a robust remedial strategy for the study area and this will enable the detailed design of a full scale treatment system that is likely to be both technically feasible and can be evaluated financially against other remedial options.

The particular physio-chemical properties of styrene had led to uncertainties with regard to the likely success of differing remedial methods but the trials have indicated that significant mass reduction of styrene is possible via conventional techniques, such as air sparging, skimming and bioremediation.

Land Reclamation – Moore, Fox & Elliott (eds)
© 2003 Swets & Zeitlinger, Lisse, ISBN 90 5809 562 2

The development of a new low cost leachate treatment system for closed landfill site management

Steve Simmons
S³ Environmental Solutions, Builth Wells, Powys

Stuart Mollard
Babtie Group, Cardiff

ABSTRACT: In 1997, following a period of prolonged rainfall, Powys County Council were required to undertake emergency measures at their Nantmel former municipal waste landfill site to address uncontrolled leachate migration. These measures involved pumping and collecting leachate, before tankering it 25 miles to the nearest water treatment works. The cost to the Council exceeded £300,000 over a three-month period.

To provide long-term leachate management at the site, the Council sought proposals from specialist leachate treatment companies for the design, commissioning and management of a treatment plant which would allow discharge of treated leachate of the adjacent watercourse. As an alternative to the multi-million pound proposals that they received. Powys decided to build their own plant, using readily available construction materials and based on simple chemical principles. The subsequent plant was developed and constructed over a four-month period for a total cost of less then £50,000. The plant has now been operational for over three years, over which time it has continually performed and exceeded discharge consent requirements.

This paper will describe the background to and operational principles behind the Nantmel plant, including reference to various validatory studies which have been undertaken since its commissioning. It will outline how the design has been developed and applied to a further three sites in Powys, each with its own unique leachate characteristics, together with its application at a selection of other landfill sites in the UK.

1 BACKGROUND

Powys is one of the largest, most sparsely populated, Counties in England and Wales. With a population of only around 135,000 spread over 5000 square kilometers, collecting and disposing of household and commercial refuse was never going to be an easy task for the three small District Councils, who, prior to 1996, preformed the roles of Waste Collection, Disposal and Regulation Authorities. To put the task in context, Powys extends for over 122 miles from north to south, the same distance as Birmingham to Southampton. It encompasses parts of two National Parks and is famed for its remote mountain and hill landscapes, reservoirs, lakes and forests. Its County town, Llandrindod Wells has a population of around 5000, not much bigger than some English villages. The only industries to speak of are farming, forestry, tourism and a small number of manufacturing companies who brave the isolation and large transport distances in getting raw materials and products in and out of the County. The

communities in Powys are small and have to be largely self-reliant. Llanwrtyd Wells, for example, is Britain's smallest town with a population of only 600.

This picture of small self-reliant communities, isolated from one another, is an important factor in understanding the problems the County now faces in managing waste. Prior to 1994, there were a large number of small municipal waste landfills throughout the County. Powys' public register of closed landfills identified over 80 former household and commercial landfill sites that were operated by the former Districts of Breconshire, Radnorshire and Montgomeryshire, and prior to 1974, by a large number of Urban and Rural District Councils. With few exceptions, these sites were rudimentary, often being little more than a bog that was in-filled, sometimes covered with topsoil and then left. Many of the 80+ sites present very little risk to the environment or nearby communities. Some however are the cause of serious concern. In 1996, the newly formed Unitary Authority, Powys County Council, inherited responsibility for all the closed landfills from the predecessor District Authorities. At the same time, Waste Regulation passed from the local Environmental Health Departments to the newly formed Environment Agency. As many of Powys' Rivers have SSSI status and are renowned for their salmon, trout, otter and crayfish populations, any threat of water pollution was always going to be a high priority for the Environment Agency pollution officers. Naturally, attention fell on some of the worst offending former landfill sites.

2 NANTMEL TIP

One of the sites identified by the Agency as a cause for concern was a former municipal landfill, known as Nantmel Tip. It is located midway between Llandrindod Wells and Rhayader (Figure 1).

The tip was developed on low-lying marshy ground bounded by three watercourses. The land was too poorly drained to be of significant value for farming, so in 1960, around 29 acres were purchased for a new municipal refuse disposal site. Tipping commenced as straight landraising directly on top of the boggy ground. By the late 1980s, around 19 acres of the site had been landfilled, at its deepest to around

15 meters depth. Throughout the tipping period, leachate had always been a major headache. The site has artesian groundwater beneath it, and any disturbance of the marsh caused large springs to develop. As the site progressed, the quantities of leachate entering the three streams led to serious and escalating pollution problems (Figure 2).

By 1989, the leachate problems were serious enough for the District Council who operated the site to decide to close it prematurely and to commission consultants to devise a scheme to intercept leachate allowing its safe disposal. Three large cut-off walls and drainage pipes were installed to prevent lateral migration into the streams. The drains linked into a large leachate lagoon from where the leachate could be removed by tanker for disposal at Rhayader Sewage Treatment Works (STW). This system worked well for a few years; however by 1994, Welsh Water advised that Rhayader STW would no longer be able to accept leachate as this would be classed as industrial waste and would require them to obtain a Waste Management Licence. The nearest site that would be able to deal with leachate would be Hereford STW, some 50 miles away in England. The District Council had no real alternative but to devise a scheme to treat leachate on-site.

Again, consultants were engaged to oversee the development of a new facility. The leachate arisings were comparatively weak by now, being some 80–100 mg/l ammonium as N, Biological Oxygen Demand (BOD) around 80 mg/l, but with high levels of iron and other metals. The consultants devised a system based on two small vertical flow reed beds. Unfortunately, although the quality of the leachate was well characterised, the records used to determine the quantities of leachate were unreliable. The only volumetric data on arisings was the number of tanker loads delivered to the STW. What was not appreciated at this time was that the Council's Sewerage Gang,

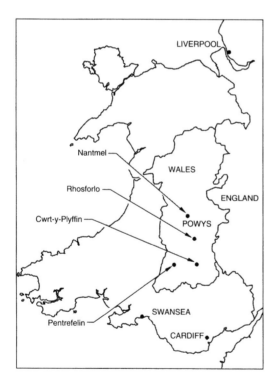

Figure 1.   Map of Wales showing location of Nantmel Tip, plus other sites referred to later in text.

Figure 2.   Leachate outbreak and surface ponding adjacent to stream.

who managed the site on a day-to-day basis, had developed their own informal means of reducing tanker costs. They had installed a network of spray irrigation pipes to re-circulate leachate across the tip and the surrounding marsh. This was effective in reducing disposal costs, but meant the new plant was designed to cope with only fraction of the actual arisings, some 10–15 m^3 per day. In fact the true levels of arisings were probably approaching 100 m^3 per day! Consequently the new reed beds were overloaded and failed to ensure effective treatment.

By 1997 the Environment Agency had lost patience with the situation at the site and threatened prosecution. By now Powys County Council had the problem and had no alternative but to re-commence haulage of the leachate to Hereford. This proved to be monumentally costly and gave significant impetus to the search for a robust solution. Cardiff University were commissioned to design a much larger reed bed capable of dealing with an additional 50 m^3 per day, as an interim measure. At the same time, investigations were started to obtain reliable data on leachate production. As work progressed, it became clear that not only did the site generate more leachate than had been previously assumed, but that in wet weather up to 400 m^3 per day was being produced. In addition, even the new reed beds were not working. The high levels of iron were causing pore spaces in the sand fill to clog, reducing leachate infiltration and leading to surface ponding and the creation of strongly anaerobic conditions which prevent the oxidation of ammonium to nitrate. Consequently, after discussions with the Environment Agency, the system of reed beds was mothballed.

The winter of 1997/98 proved to be disastrous. Following a period of prolonged heavy rainfall, tankering costs escalated and in January 1998 alone over £80,000 was spent on leachate disposal. As there was only limited storage capacity on-site, and with leachate volumes rising to around 600 m^3 per day during the Christmas period, tanker operations were now a seven-day a week affair, with up to twenty tanker loads per day. That year disposal costs exceeded £330,000.

The Council assembled a specialist team to tackle the problem. Their first step was to find ways of reducing the volumes of leachate that the site produced. After careful study, including development of full water balance models, it became clear that water was entering the site from a number of routes, including old land drains serving the surrounding agricultural land, groundwater ingress and inadequate capping. Even Council highways drains were found to have been connected directly into the soakaways in the site! A few days with an excavator and some new French drains resolved the problems arising from the highway and field drains. This alone had a dramatic impact on the volumes of leachate being produced, cutting average baseflow production levels by over 50% to around 45 m^3 per day and a peak of 250 m^3 per day in wet weather. Nothing could be done to control the groundwater ingress as the artesian springs were being fed from water contained in a layer of coarse pebbles and cobbles and the volumes were too great to attempt to lower pressures by controlled pumping. Having controlled the volumes, the second step was to provide greater on-site storage capacity. This would allow the leachate produced in wet weather to be stored, and then fed into a treatment plant in dry conditions. A simple automatic leachate balancing system was developed where leachate was pumped to a new 500 m^3 balancing lagoon during high level conditions and then siphoned back when levels dropped. The advantage of this arrangement was that the treatment plant could be constructed to accommodate a smaller volume at any one time. This significantly reduced the size of plant that would be required, and hence cost.

The Council's team then sought proposals from specialist leachate treatment companies for the design, commissioning and management of a treatment plant that would allow discharge of treated leachate to the adjacent watercourse. Having reviewed the proposals that were received, which exceeded £1,000,000 capital cost, Powys decided instead to build their own plant, using readily available construction materials and based on simple principles. In part this decision was based on cost, but also the Council was reluctant to construct a plant that would require a high level of technical expertise to operate.

The subsequent plant was developed and constructed over a four-month period for a total cost of less than £50,000. The plant has now been operational for over three years, over which time it has continually performed well, achieving treatment standards well below discharge consent requirements.

3 THE NANTMEL SYSTEM

The leachate produced by Nantmel Tip is characteristic of weak leachates, with moderate Ammonium levels (50–80 mg/l), low BOD (20–30 mg/l) and relatively high dissolved iron (10–30 mg/l). From the experience derived from the two reed bed systems at the site, it was clear that dissolved iron (Fe^{2+} ions) rapidly oxidize with dissolved oxygen through semi-catalytic and biological processes in the fine pore spaces in the sand beds. This forms insoluble iron hydroxide compounds which have a characteristic sol form (loosely bound flocculated solid). This rapidly fills the pore spaces and prevents further leachate infiltration. Once the pore spaces are blocked, remaining underlying levels of the reed bed rapidly turn anaerobic. When anaerobic conditions are present, the oxidation of ammonium is inhibited and the bed turns "sour".

The first stage of the treatment process was therefore a high rate vertical filter where leachate is cascaded down through a nitrifying media with a high surface area to volume ratio. This allows a thinly distributed leachate film to form which is in contact with a large volume of air. These conditions are ideal for the oxidation of Fe^{2+} ions to insoluble Fe^{3+} which rapidly deposits as a film on the media. Providing that the rate of application is high enough, the resulting iron film will regularly slough off, preventing blockage of the filter tower. The leachate is retuned to the main lagoon where finer iron sediments are precipitated. The lagoon is bright orange and has a heavy loading of fine iron precipitates. It was found by experimentation that the precipitation process worked best when the pH was around 8–8.5. Nantmel's natural pH was around 6.5–7.0. Therefore, the pH is corrected using Magnesium Carbonate granules through which the leachate is passed at a low flow rate. Magnesium Carbonate naturally buffers the pH to around 8.0.

Whilst the leachate is passing down through the filter tower, a second benefit is gained. The raw leachate contains dissolved ammonia gas. When this is in contact with the air in slightly alkaline conditions, ammonia gas will volatilize to be lost to the atmosphere, and hence reduce the total loading of ammonia in the leachate.

The partially treated leachate stays in the lagoon for around 3 days during which time it is recirculated through the filter towers on a regular basis. This achieves progressively higher removal rates and allows the next stage of treatment to initiate, the oxidation of ammonia through biological oxidation. Nitrifying bacteria are present in the tower and lagoon and providing that there is adequate dissolved oxygen, will start to convert ammonium ions into nitrate ions. The addition of Magnesium Carbonate raises the alkalinity of the leachate to create ideal conditions for the nitrification process. By the time the leachate has been re-circulated through the tower and lagoon over several days, the levels of dissolved iron have dropped to less than 1.0 mg/l and the levels of ammonium have dropped to around single figures. The next stage can then concentrate on lowering the ammonium levels and residual BOD. This is achieved by passing the leachate through a roughing filter constructed from an old cattle drinking trough filled with course Magnesian limestone (30–50 mm) then through two sequential gravel filter beds (Figure 3).

These are simple tanks filled with 5–10 mm gravel which the leachate stands in for around 15 minutes, and is then pumped out. Lowering the levels rapidly pulls fresh air down through the bed preventing

Figure 3. Nantmel gravel filter beds.

anoxic or anaerobic conditions from developing. In this environment, conditions are ideal for nitrifying bacteria, which form a thick film around the gravel. By now, the treated leachate is crystal clear (free from suspended solids) and the ammonium levels have dropped to around 1–2 mg/l after only 20 minutes in contact with the gravel. The leachate is now ideal for polishing treatment in the existing large reed bed. Leachate is distributed using trickle bars. Rather than a vertical flow system, the beds have been adapted to operate as a pump and fill system. This has the benefit of ensuring adequate oxygen levels throughout the bed. After passage through the reed bed, the leachate is now <0.1 mg/l ammonium and levels of BOD, iron and other metals are below detection or their relevant target concentrations. The only parameter to cause concern is nitrate which has been produced by the oxidation of ammonium ions. Nitrate is then reduced by passing the leachate through two large grassed beds constructed with bunded clay walls, a clay base and collector drains. The soil in the beds is a fine silty clay. After passing through these wetland areas, the levels of nitrate are reduced by around 50%, either through uptake by the growing grasses or adsorption on the clay particles in the oil matrix.

Figure 4 illustrates the fate of ammonium passing through the system (expressed as total ammoniacal-N). The starting concentration in raw leachate during the tests was 15 mg/l. This fell through volatilization and bio-oxidation to around 2 mg/l passing through the filter towers then fell to below detection (<0.1 mg/l) at the discharge point. The nitrate levels (expressed as NO3 – N) started at around 2 mg/l rising through the system as ammonium was oxidized to each a peak of around 31 mg/l in the reed beds then falling to around 12 mg/l after passing through the grass irrigation areas. The above results were obtained by the University

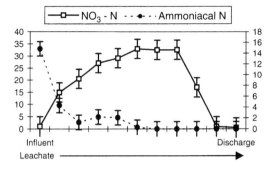

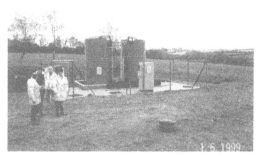

Figure 4. Variation in Ammoniacal Nitrogen (right-hand y axis, mg/l) and Nitrate (left-hand y axis, mg/l) at various stages in the treatment process from influent (left hand side) to discharge.

Figure 5. Nitrifying towers at Cwrt-y-Plyffin Landfill

College of Wales (Aberystwyth) who have undertaken various evaluation studies of the system. They have produced a scale model system at their laboratories in Aberystwyth and have managed to reproduce the results experienced in the full-scale plant. Further work is now underway using a 1/5th scale portable plant, which will be tested on a range of operational and closed sites throughout the County during 2003/04.

Figure 6. Rhosforlo plant, comprising compact system of storage tank, nitrifying tower and gravel filter.

4 OTHER EXPERIENCE

Using the same general principles of treatment that were developed at Nantmel, similar systems have now been developed on four further closed landfill sites in Powys (see Figure 1 for locations). These are:

- Cwrt-y-Plyffin Landfill (Figure 5) which closed in 1992/3 where the leachate is still relatively young with high levels of ammonium (around 250 mg/l), high BOD (around 100 mg/l) and moderate levels of dissolved iron;
- Pentrefelin Landfill which was closed in the mid 1980s and is characterised by a weak leachate but very high in dissolved metals;
- Rhosforlo Landfill (Figure 6) which was closed in the late 1980s and is characterised by a weak leachate, but is high in metals, particularly Cadmium; and
- Carreghofa Landfill which was closed in 1989 and is characterised by a strong leachate high in ammonium, BOD COD and metals.

The first three plants are fully operational and performing to consistently high levels of efficiency. Cwrt-y-Plyffin has proved to be particularly effective, reducing ammonium from around 250 mg/l to consistently less than 3 mg/l, and mainly less than 0.1 mg/l.

Figure 7. Activated sludge plant at closed landfill in southern England, proposed to be converted to "Powys type system" using existing tanks.

The plant at Carreghofa is still being commissioned, but is expected to be operational during early 2003. None of the plants has cost more than £50,000 to construct and all are characterised by low revenue costs, which comprise mainly energy costs (typically around £2500 per annum) analysis costs (typically around £1000 per annum), consumables (around £1000 per annum) and spares and maintenance (typically around £2000 per annum). All the plants are operated by manual staff who visit each of the sites at least weekly to undertake inspections and minor adjustments etc.

The experience and knowledge developed in the commissioning and operation of these plants is also now being applied to treat leachates produced at old landfill sites at a variety of locations in England and Wales (Figure 7).

5 SUMMARY AND CONCLUSIONS

Based on simple principles and using readily available materials, Powys County Council developed a treatment plant at their Nantmel Tip site specifically designed to treat the relatively weak leachate present at the site. Through a largely iterative process, the plant was fine tuned to achieve a high degree of efficiency and consistency in treating the leachate to levels sufficient to allow discharge to the adjacent high quality watercourse. Independent review and verification of the plant's performance has been provided by Environmental Consultants and Research Bodies. The system developed has also now proved to be transferable and, in appropriate situations, has been demonstrated to provide a relatively simple, low cost and sustainable approach to leachate treatment at closed landfill sites.

Land Reclamation – Moore, Fox & Elliott (eds)
© *2003 Swets & Zeitlinger, Lisse, ISBN 90 5809 562 2*

Remediation in practice – expect the unexpected

D.L. Avalle & R.P. Ashby
REC Ltd, Manchester, United Kingdom

ABSTRACT: Statistically speaking, we base contamination assessments on a far lower level of confidence than is commonly adopted by design engineers. Recent regulatory guidance in the UK provides a prescribed path that can limit developer's risk and ease the discharge of environmental planning conditions. Case studies are presented of recent remediation projects in the UK, where the clean-up has been completed or the process continues. The authors attempt to draw lessons from the projects and put forward challenging suggestions for consultants, contractors and developers.

1 INTRODUCTION

Resource & Environmental Consultants (REC) Ltd specialise in contaminated land risk assessment and remediation. The authors have drawn from project experience over the last four years to present case studies which contrast the actual outcome of a remedial program with the original perception from a site investigation.

The case studies also review how the application of recent Environment Agency guidance has, or could have, influenced the projects. All case study sites are in England. Details about the sites remain confidential and only selected information has been used.

2 CASE STUDY 1

2.1 *Background*

An infilled sand and gravel quarry, used as a distribution depot for heating fuels, was earmarked for a retail development. A limited trial pit study by others identified that some of the infill was gasworks waste, contaminated by cyanide.

A Source-Pathway-Target review, along the lines set out in Environment Agency Guidance P66 (2000), would have identified that the principal migration pathways, taking into account the site setting and proposed use, were groundwater and vapour migration.

However, the investigation by others focused only upon health implications of fill in the upper 3 m and did not assess migration risks.

A remedial scheme to remove impacted soil from site was enacted and then reviewed by REC Ltd when it became clear that the original budget was insufficient to achieve the aims of the program.

Following the review, REC Ltd installed a network of groundwater and free phase wells. The Environment Agency Model P13 (Quint et al 1996) and United States ASTM model RBCA (1995) was applied to assess groundwater and vapour phase migration risk.

The risk assessment demonstrated that the existing soil decontamination threshold, applied during the earlier phase of remediation, could be safely increased by 800%.

The findings of the assessment were accepted by the Environment Agency. The risk assessment prevented the unnecessary export of tens of thousands of tonnes of soil that were impacted, but not significantly contaminated, taking into account the site setting and proposed land use.

Further remedial work was confined to the removal of free phase diesel from the aquifer by a combination of vacuum extraction and skimming.

2.2 *Key factors*

- More thorough site investigation design, utilising a "conceptual model" and following protocols such as P66 Environment Agency (2000) would have identified the potential sources of contamination and also indicated the potentially significant receptors, allowing a more appropriate site investigation.
- Selection of an analytical suite taking into account former land use, following protocols such as CLR8 (2002b), would have identified more widespread contamination than was initially evident. The

analytical suite originally applied was based upon ICRCL Guidance Note 59/83 (1987) and was not completely applicable to the recent land use.

- The additional site investigation costs were repaid ten times over during remediation.
- Delays in site preparation works could have been avoided by having a better understanding of the site characteristics.

3 CASE STUDY 2

3.1 Background

The site of a former heavy engineering and chemical works had been partially redeveloped with a small supermarket and car park in the early 1990s. In 1999 it was proposed to redevelop the remaining 90% of the site for retail outlets and offices.

A significant thickness of cohesive glacial drift, interspersed with variable sand lenses, overlies a major aquifer at this site. There are, however, no sensitive groundwater abstractions in the vicinity. In general, the site had between 1 m and 3 m of made ground, overlying natural sand or clay.

The previous owner operated from one corner of the property, which was raised above the general site level. All former structures had been removed and the main area was fairly level and largely concrete-paved.

3.2 Issues identified pre-construction

The desk study and the two phases of intrusive investigation, carried out by REC Ltd at the pre-acquisition stage, produced little evidence of any significant ground contamination. In the raised part of the site, some boreholes were not able to penetrate buried concrete features and access to a number of locations was restricted by its use at the time as a sales yard.

3.3 Aspects arising during construction

With a desk study and two phases of site investigation previously completed, it was considered by the project team that sufficient work had been done to facilitate the proposed development. However as the programme of earthworks progressed three issues arose.

3.4 Fly-tipped waste

When work commenced to reduce the level of the raised area (the former sales yard), waste was uncovered between two retaining walls below a tarmac surface.

A subsequent review of aerial photographs showed that the fill had been placed between 1997 and the start of the site investigations in 1999. Further enquiries confirmed the earlier desk study information, that no

waste management licenses had been registered at this location.

3.5 PCB contamination

Due to commercial confidentiality, the Council only became involved at the application stage, after completion of the investigations and acquisition of the site. In view of the past site use, the Council Officers considered that poly-chlorinated biphenyls (PCBs) might have been associated with electricity transformer oil on the site and made testing for PCBs a requirement to facilitate the discharge of a planning condition.

As noted by the Environment Agency in CLR8 (2002b), PCBs can be found at any former industrial site at which there have been electricity sub stations. However PCBs are particularly immobile and previous experience has identified contamination only in drain sediment and soils in the immediate vicinity of the transformers.

The historical maps had not identified transformers on the site itself; therefore there was no indication where to target any site investigation. To meet the conditions imposed by the Council it was agreed to test three soil samples at random for PCBs. The testing identified significant concentrations of PCBs in two of the three samples.

Agreement on the scope of the validation exercise with the local authority had taken some time and earthworks had progressed across the site before the PCBs were identified.

A further phase of testing showed widespread distribution of PCBs possibly as a result of the historical use of impacted fill or "mixing" due to slab and foundation demolition and site regrading.

Risk assessments, showing that potential pollutant linkages would be broken by site cover, satisfied the Council that the proposed development could proceed. However a change of land use in the future may trigger further remedial work.

3.6 Hydrocarbon contamination

In the area of the former engineering works, large concrete bases were encountered, probably former gantry crane column footings, some as large as 2 m to 3 m cubes.

These were in the area where investigation boreholes did not penetrate obstructions some 3 m below the then existing site levels. Adjacent to some of these bases, when holes were excavated to establish their size in advance of breaking out, hydrocarbon contaminated soil and groundwater was encountered.

This triggered a groundwater and soil vapour risk assessment using the Environment Agency's P20 model (Marsland & Carey 2000) and the ASTM risk model RBCA (1995), which indicated that the levels of soil and groundwater contamination were not significant, taking into account the site setting and proposed land use.

However, it was decided to incorporate a gas-proof membrane in the floor slab because construction was moving ahead of regulatory approval.

The 6-week programme of this study was simultaneous with construction activities, as the contractor could not stop due to contractual obligations.

3.7 Warning signs

Although the Planning Condition was eventually discharged, a number of lessons can be extracted from this case study:

- The conclusions drawn from a desk study rely upon the quality of the information available. Following the standard lines of enquiry does not remove developer's risk. Redevelopment budgets should reflect an unidentified risk wherever possible.
- It is unwise to assume that the regulators will be satisfied with the scope of site investigations that are carried out in advance of the imposition and enactment of Planning Conditions. Try to consult as early as is possible, given commercial constraints. In this project the local authority took several months to respond and, in other recent cases, some Environment Agency offices are delaying the consultation process by two to three months.
- Where development schemes are unclear, the potential implication from reducing levels should be considered when assessing the significance of field observations.

4 CASE STUDY 3

4.1 Background

The site of a proposed retail outlet did not have a particularly chequered history. Whilst there had been development on the site for more than 150 years, it had mainly been terrace houses and a club, although there had been various potentially contaminative uses of nearby properties over the decades.

When REC Ltd became involved, the developers and their consultants had commissioned two stages of site investigation. The site had a covering of between 2 m and 4 m of made ground, which exhibited variable concentrations of heavy metals, and, in particular, some quite elevated levels of arsenic.

The solution adopted for site remediation required the removal off-site of a substantial amount of made ground, which exceeded the developer's allocated budget for development of the property.

4.2 Previous investigations

The results of the two previous phases of intrusive work were presented to REC Ltd. A review indicated that the data had been assessed with reference to the Soil Guidance Values (SGVs) issued by the Environment Agency 2002 (2002a).

Based upon the initial findings, a remedial scheme to remove an irregular shaped volume of made ground was proposed. The Council had approved this strategy.

However, the assessment had relied upon direct comparison of the soil data with the relevant SGV, rather than the statistical approach proposed by the Environment Agency in CLR7 (2002a).

4.3 Risk assessment and consultations

Our first action was to put all the data through the CLR7 statistical model (2002a), which highlighted the value of this approach. Although some results are well above the SGV, they all fell inside a normal distribution and the 95-percentile concentration fell within the SGV.

It was also established that there were no concerns about protecting water resources because the site overlies a minor aquifer with no nearby abstractions or sensitive water courses.

The next step was to establish the Council's position with respect to a major change in remediation strategy. After submitting the CLR7 analysis and a revised risk assessment, the Council indicated that they would be quite amenable to an alternative remediation plan, providing that a detailed method statement was submitted, supported by sufficient risk assessment and a programme of validation.

As it turned out, the revised remediation plan involved only the removal for off-site disposal of materials arising from the excavation of footing trenches. In addition, a clean capping, 600 mm thick, was specified for landscaped areas to provide a barrier to protect site users and neighbours and to provide a growth medium for vegetation.

4.4 Key factors

- The application of statistical methods proposed by the Environment Agency in CLR7 (2002a) can significantly influence the scope of a remedial project. However the application of the statistical techniques becomes more effective with increased sample numbers requiring a willingness to increase investigation costs.
- Clients frequently employ architects or structural engineers at an early stage, who then rely on site investigation contractors for site investigation design, implementation, chemical testing and data

interpretation. Intervention by risk assessment specialists at an early stage can ensure that remedial expenditure is minimised and targeted.

5 THE CHALLENGE

These case studies, along with other recent projects, have focused the authors' attention on the issue of the risk burden and how the various protagonists involved in the remediation of contaminated land share it.

It is the perception of the authors that there is sometimes a divergence of views between developers, engineering designers, contractors and environmental consultants, when it comes to the allocation of acceptable risk.

Although developers work in a particularly high risk environment, they anticipate a higher degree of certainty from their consultants than the reality of risk assessment should provide. Table 1 illustrates risk perceptions that the authors suggest might apply to a wide array of contaminated land projects.

The table highlights the challenge that we face in trying to satisfy government targets for the regeneration of derelict land, as well as maximising developer and contractor profits and maintaining professional standards.

The authors believe that the Environment Agency's recent series of Contaminated Land Reports offer a path to achieve all the above. However, consultants require a willingness from developers to undertake the following:

- Draw environmental specialists and risk assessors into projects at an early stage;
- Appreciate that investment in site investigation will often be re-paid many times over during remediation and development;

Table 1. Risk perceptions.

	Preferred level of certainty	Risk levels in normal business
Developers	Very high (100%?)	High
Design engineers	High (95%+)	Very low
Contractors	High	Moderate
Consultants	Moderate to high	Low (High?)

- Allow sufficient time for regulatory consultation to agree the findings of risk assessments that no longer rely on generic guidelines.

REFERENCES

Marsland, P.A. & Carey, M.A. 2000. *Methodology for the Derivation of Remedial Targets for Soil and Groundwater to Protect Water Resources.* Environment Agency R&D Publication P20.

Quint, M., Alexander, J., Curtis, S. & Irving, P. 1996. *Methodology to Determine the Degree of Soils Clean-up Required to Protect Water Resources.* Environment Agency R&D Publication P13.

ICRCL 1987. *Guidance on the Assessment and Redevelopment of Contaminated Land.* Interdepartmental Committee on the Reclamation of Contaminated Land ICRCL 59/83, 2nd ed.

ASTM 1995. *Standard Guide for Risk-Based Corrective Action Applied at Petroleum Release Sites.* American Society for Testing and Materials E 1739-95.

Environment Agency 2000. *Guidance for the Safe Development of Housing on Land Affected by Contamination.* Environment Agency R&D Report P66.

Environment Agency 2002a. *Assessment of Risks to Human Health from Land Contamination: An Overview of the Development of Soil Guideline Values and Related Research.* Environment Agency R&D Publication CLR7.

Environment Agency 2002b. *Priority Contaminants Report.* Environment Agency R&D Publication CLR8.

Land Reclamation – Moore, Fox & Elliott (eds)
© *2003 Swets & Zeitlinger, Lisse, ISBN 90 5809 562 2*

Look before you leap: the use of geoenvironmental data models for preliminary site appraisal

E. Hough, H. Kessler, M. Lelliott, S.J. Price, H.J. Reeves & D. McC Bridge
British Geological Survey, Keyworth, Nottingham, UK

ABSTRACT: In the urban environment, site investigation studies provide a wealth of information about the ground conditions of the shallow sub-surface. However, from the developers perspective, there is generally little incentive to integrate this information beyond the boundaries of the development site. By taking a more holistic view and combining knowledge of the near-surface geology with information on former landuse and groundwater regime across a wider area, it is possible to predict geological scenarios that may better inform ground investigation and reclamation strategies.

As part of its urban research programme, the British Geological Survey is currently integrating its data holdings across 75 km^2 of central Manchester and Salford. The aim is to develop a fully attributed 3D model of the shallow sub-surface that will provide information on the thickness, composition and geotechnical properties of the Quaternary superficial "drift" deposits and any artificial cover. It will also provide, at lower resolution, information on groundwater vulnerability, soil geochemistry and the potential of the ground to support Sustainable Urban Drainage Systems (SUDS). The model is based around a nucleus of 6500 boreholes, and is being developed with a range of 3D visualization and processing software.

This approach provides a means of identifying potential problems and opportunities at the desk study stage in any proposed development and, if implemented over a wider area, it could assist in designing site investigation strategies and reduce costs by ensuring a more focused approach to site appraisal.

1 INTRODUCTION

The role of geoenvironmental information is becoming increasingly important as legislative changes have forced developers, planning authorities and regulators to consider more fully the implications and impact on the environment of large-scale development initiatives. To comply with the principles of sustainable development, developers are increasingly required to demonstrate that proposals are based on the best possible scientific information and analysis of risk. Nowhere is this more relevant than in the context of urban regeneration.

The case for using geoenvironmental information to underpin preliminary site appraisal and for developing regional strategies has been made elsewhere (e.g. Bobrowsky 2002 and references therein; Culshaw and Ellison 2002; Ellison et al. 1998, 2002; McKirdy et al. 1998; Thompson 1998). In the UK, studies commissioned by the Department of the Environment in the 1980s and 1990s paved the way and promoted the use of applied geological maps to identify the principal geological factors which should be taken into account in planning for development (e.g. Forster et al. 1995).

Since this work was completed, advances in the use of Geographical Information Systems (GIS) and modeling packages have meant that there is now far greater opportunity to develop geoenvironmental products that take greater account of the third dimension. Because the information is captured and manipulated digitally, the outputs can be tailored to user needs, and more readily updated. This has clear advantages over hard copy maps and reports, which are difficult to update and provide limited flexibility in terms of usage.

The purpose of this paper is to demonstrate the potential that the new technology offers, and to illustrate, using examples from a major conurbation, the role that the 3D geological model is playing in deriving bespoke thematic products.

The area chosen for the study covers 75 km^2 of central Manchester and Salford (Figure 1). It is a predominantly urbanized area with a long history of intense, largely unrestrained industrialization, founded on coal-mining, chemical manufacture and the textile industry (including the bleaching and dyeing of cotton). These activities have left a legacy of contaminated land and groundwater pollution in what is one of the most densely populated areas of the UK.

The area includes Trafford Park, the largest industrial estate in Europe, Manchester city centre, still undergoing redevelopment following the 1996 terrorist attack, and east Manchester, an industrially-depressed area destined for urban renewal aided by £2bn of public and private investment over the next 15 years (Carroll 2000). Smaller areas of intense redevelopment include the former Bradford Colliery and gasworks, redeveloped as the focal site of the 2002 Commonwealth Games, and Salford Quays, formerly the Manchester Ship Canal docklands, but now home to the Lowry Centre and the Imperial War Museum North.

1.1 Background to the study area

Geologically, the Manchester and Salford region straddles the southern part of the Carboniferous South Lancashire Coalfield and the northern part of the Permo-Triassic Cheshire Basin. The coalfield was extensively worked up until the late 1970s from numerous collieries within the northern and eastern parts of the study area including Patricroft, Agecroft and Bradford. To the south and west, the Carboniferous Coal Measures are overlain by Permo-Triassic rocks of the Sherwood Sandstone Group, which is the second most important aquifer in the UK. Quaternary

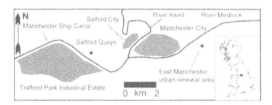

superficial deposits laid down during the Devensian glaciation mantle most of the area, locally reaching thicknesses in excess of 40 m. The deposits include glacial till (pebbly and sandy clay), glaciolacustrine deposits (laminated clays and sands) and glaciofluvial outwash (sands and gravels). Post-glacial deposits, associated with the proto-Irwell include alluvium, river terrace gravels, and peat (Figure 2). Extensive areas of made ground are present, and include colliery spoil tips, material dug during the construction of the Manchester Ship Canal and general inert and biodegradable fill. Many of the watercourses in south Lancashire have been culverted and their valleys infilled, as for example, at Crofts Bank and along much of the course of the lower Medlock and its tributaries.

1.2 Issues

Some of the geoenvironmental factors likely to influence the cost of developing a site in Manchester or Salford are known from anecdotal and published accounts. Difficult ground conditions are a material consideration throughout much of the region, because of the heterogeneous nature of the superficial deposits and the significant thickness of man-made deposits in certain areas. Damage caused to housing and roads in parts of Salford, as a result of piping or collapse of glaciofluvial sands, is well documented (Harrison & Petch 1985), as are the subsidence effects caused by undermining. There are also issues of contamination and groundwater protection. On a regional scale, uncertainty about the shallow groundwater regime, and the role played by the Manchester Ship Canal on aquifer recharge are issues of strategic concern.

- The present study is currently addressing a number of these issues, although some are still at an early stage of development; these are summarized below along with the main uses for the dataset. Topographic basemaps: various scales and vintages (backdrop for thematic mapping; historic and present-day landuse; potentially contaminating past landuse; hydrology, including springs and watercourses)

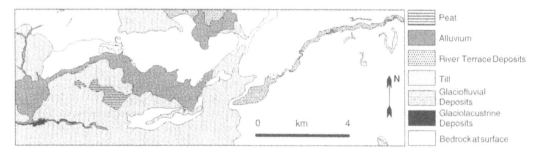

Figure 2. Geological map showing bedrock and superficial deposits present at surface within Manchester and Salford.

- Digital terrain model (DTM) (basis for 3D data modeling)
- Orthorectified aerial photographs (current land-use; surface sealing; visualization)
- Geological maps: various scales and vintages (regional illustrative overview of geology; details of local geology; distribution of aquifers; surface lithology; historical details about, e.g., pits and exposures)
- Borehole database and extracted downhole lithological information (3D geological characterization including rockhead)
- Baseline geochemical data (characterization of soils and geological units; contaminated land study)
- Geotechnical properties database (ground stability and strength assessment)
- Water levels data (water balance models; indication of near-surface water; ground engineering; groundwater vulnerability models)
- Seismic events database (history of seismic activity).

1.3 Building the 3D model

Software currently under development by Dr Hans Georg Sobisch of the University of Cologne is being used to construct a model of the superficial and artificial deposits. Eventually, the aim will be to extend the model to include the underlying solid geology.

Primary and derivative datasets essential to the model are listed in the previous section. Of these, the borehole geology database is the most fundamental, providing downhole information for some 6500 borehole sites, as well as other factual information including groundwater strikes and geotechnical test data.

The three dimensional configuration of the geological units in the sub-surface is built up from serial cross-sections, drawn interactively using mapface and downhole data (Figures 3a, b). Correlated surfaces are then gridded, and stacked to produce the final geological model. Accurate borehole correlation is critical to the final model, and care must be taken to ensure that each lithostratigraphical sub-unit is correctly attributed. This invariably involves some degree of subjective analysis to discriminate between deposits that are lithologically similar but may have been deposited in very different environmental settings (e.g. fluviatile, glacigenic, anthropogenic). Such deposits could reasonably be expected to exhibit different geotechnical or hydrogeological characteristics. Eight surfaces describing the subsurface alluvial and glacial geology have been identified and modeled in the Manchester and Salford area.

Figure 3c is an extract of the 3D model covering the western end of the project area. For clarity, only one lithostratigraphical unit (glaciolacustrine clay) is shown. The entire model, when fully attributed, will provide a greater level of understanding of the

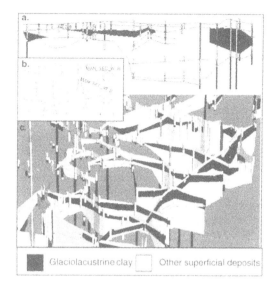

Figure 3. (a) A cross-section showing the distribution of glaciolacustrine clay proved by boreholes. (b) A plan view of the cross-sections that prove the distribution of glaciolacustrine clay in the south-western part (Trafford Park) part of Manchester. (c) Interrogation of the network of cross sections gives a representation of the 3D distribution of glaciolacustrine clay.

shallow sub-surface than is currently available for an urban centre in the UK.

1.3.1 Limitations of the model

The glacial deposits are represented by a range of lithofacies that were deposited in different environmental settings. Many of the deposits are laterally impersistent and their geometry is unpredictable. In such areas, borehole correlation becomes uncertain and the approach has been to combine genetically-related sequences into domains rather than to try and map out individual lithofacies.

The validity of the model also depends on accurate borehole-to-borehole correlation. Lithological descriptions, in themselves, do not necessarily provide an adequate basis for correlation as many geological units, such as river terraces, glacial ice-contact deposits and intra-till sand bodies are described in similar terms, but do not necessarily share the same depositional or hydrogeological characteristics. An understanding of the geological evolution of the area is, therefore, important, to avoid creating spurious linkages.

There are practical difficulties in recreating the ground surface. The Digital Terrain Model derived from Ordnance Survey 5 m contour data does not necessarily reflect the detailed variations observed in boreholes that have been levelled during the site investigation process. This a particular problem with thin, near-surface deposits such as made ground, where discrepancies between the DTM height and that of the levelled borehole (2 to 3 m) may be of the same order of magnitude as the deposit being modeled. Where possible these anomalies have been dealt with by re-hanging the boreholes to the modeled surface.

A further limitation of the model relates to its effective resolution. At a borehole density of between 1 and 257 data points per square kilometre, coverage in some parts of the study area is quite poor. This has a bearing on the applicability of the model at different scales of usage. It is important that data are processed in a way that ensures important relationships are not obscured at site or regional scale, and also that data are not over-interpreted beyond their intended useful range.

2 USE OF THE 3D MODEL FOR THEMATIC MAPPING

The potential of the model to deliver information relevant to a range of applications is illustrated by reference to three issues:

– Ground conditions
– Groundwater protection, and
– Sustainable Urban Drainage Systems.

2.1 Prediction of ground conditions

2.1.1 Natural superficial deposits
The superficial drift deposits of the area cover a spectrum of engineering soil types, some of which may not offer good foundation conditions. Coarse (0.06–60 mm) soils, represented by glaciofluvial sands, tend to be loose- to medium-dense and well graded with practically no binder. As noted earlier, these deposits are responsible for ground movements in parts of Salford. Excavations in these deposits may encounter problems from running sand and cut face instability may occur when the excavation is below the water table.

Similarly, soft fine-grained (less than 0.06 mm) soils (alluvium and glaciolacustrine clay) may pose problems due to their low strength giving a generally low bearing capacity.

By assigning geotechnical properties to particular sub-units of the model, very specific information can be displayed about the nature of the sub-unit, and its likely geotechnical performance. Plasticity is an important engineering characteristic of fine-grained soils and is commonly measured during routine ground investigations. It gives an empirical understanding of engineering soils behaviour, and their susceptibility to deformation and shrink-swell. In the example

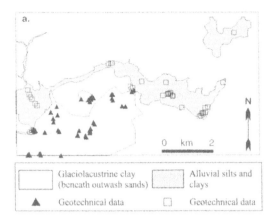

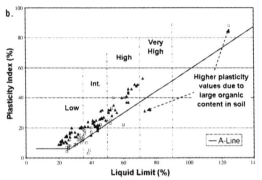

Figure 4. (a) Distribution of subsurface glaciolacustrine clay and surface alluvium in the south-western part of Manchester (Urmston), along with data points giving information on the geotechnical characteristics of the clays. The glaciolacustrine clay unit is almost entirely concealed by a thin veneer of glaciofluvial outwash sands which are shown (as Glaciofluvial Deposits) in Figure 2. (b) Plasticity chart of the alluvial silts and clays of the Irwell floodplain and the glaciolacustrine clays. The alluvial silt and clay displays low- to medium-plasticity; higher plasticity samples have a high organic (peat) content. The glaciolacustrine clay displays a low- to high-plasticity; higher plasticity samples again have a high organic (peat) content. In normal ground conditions, low plasticity values generally indicated a low shrink-swell potential and high plasticity values generally indicated a high shrink-swell potential.

(Figures 4a, b), plastic and liquid limit test results (British Standards 1990) are compared for the alluvial silts and clays of the River Irwell floodplain and the glaciolacustrine deposits that subcrop to the south of the Irwell. The alluvial deposits display low- to medium-plasticity, except for one or two samples with a high organic (peat) content, which fall into the high plasticity category. The glaciolacustrine clays cover similar fields, but generally have a higher plasticity. The implication is that, under normal ground conditions, the alluvial silts and clays have a lower shrink-swell potential than their glaciolacustrine counterparts.

By extending this work, it is hoped to characterize the foundation conditions across the whole of the urban area, and building on earlier studies (e.g. Paul & Little 1991), produce a predictive model that links the geotechnical performance of the glaciogenic deposits to their mode of deposition. This may shed light on the continuing debate as to whether the Cheshire Plain glacigenic sequences were deposited by more than one ice advance.

2.1.2 Assessing the subsurface morphology of made ground

An awareness of the presence, extent and composition of made ground is important, particularly as in areas of long historical development, like Manchester, made ground will be present beneath much of the urban area. The deposits are notoriously difficult to model because of their patchy distribution, but in areas of high borehole density, they can be delineated with some certainty. Made ground in excess of 0.5 m is recorded in 48% of the boreholes used within the study area. One of the largest backfilling operations involved the wholesale infilling of part of the River Irwell, which took place when the Manchester Ship Canal was cut in the late 19th century. The 3D model (Figure 5) depicts the morphology of the original river basin in the Salford Quays area prior to infilling.

2.2 The protection and management of groundwater in an urban setting

The European Water Framework Directive is due to be implemented in several key stages, commencing in 2003. The precise definitions used in the framework are currently ambiguous. However, one likely consequence is that surface and groundwater safeguards will become more stringent and cover all subsurface water within the saturation zone, regardless of potential yield of the host water-bearing unit, or potability. There will also be a requirement to ensure that the quality of existing groundwater bodies, classified as "good status," is maintained, and those that are of "poor status" will have to be improved.

In the UK, aquifer vulnerability maps, (e.g., Environment Agency 1996) provide basic information on the sensitivity of an aquifer to pollution. However, they take little account of the role of superficial deposits in determining recharge and run-off potential. Increased interest in the use and management of urban ground water for public and industrial supply and river augmentation has meant that drift characterization (thickness, lithology, grain size, porosity and hydraulic conductivity) is essential for a meaningful appraisal of aquifer sensitivity (Berg 2002).

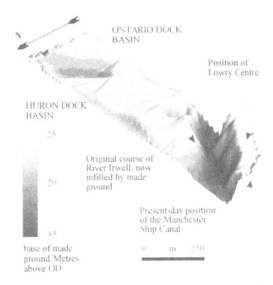

Figure 5. 3D representation of the base of made ground infill at Salford Quays. The base of the made ground has been modeled by geostatistial triangulation of provings from boreholes. The base of made ground is flatter in the east and drops considerably in the west, where the former channel of the River Irwell has been infilled during construction of the Manchester Ship Canal.

The use of domain maps in this context is now well established, particularly at catchment scale or larger (McMillan et al. 2000). The domain map constructed from the 3D geological model (Figure 6) provides a qualitative guide to aquifer sensitivity. The domains are based primarily on lithological criteria, associated with an estimate of the relative proportions of sand to clay in the sub-surface. It shows that the main areas of potential recharge occur along the River Irwell and beneath the Trafford Park Industrial Estate, where sandstone crops out or is overlain by permeable material (i.e., sandstone is in continuity with the surface). The potential for a pollution incident contaminating the groundwater in these areas must be a consideration.

Substantial infiltration to the aquifer through the "sand on clay-dominated", "clay-dominated" and "alluvium on clay-dominated" domains is unlikely as they contain clay-rich layers over 5 m thick. Perched water bodies are likely within the "sand" and "alluvium on till domains", which may be susceptible to contamination, depending on local groundwater flows and overlying seals to the groundwater body.

One aim of the project will be to refine the domain model with surface sealing data, and information on potential sources of pollution and their pathways. This will help delineate recharge areas more accurately and possibly restrict adverse land-use practices within sensitive areas.

373

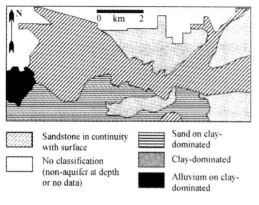

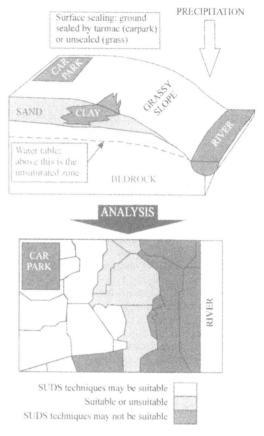

Figure 6. Extract of a map showing hydrogeological domains for Manchester and Salford. Recharge of the aquifer will occur where sandstone outcrops or is overlain by permeable material. Substantial infiltration through the sand on clay-dominated, clay-dominated and alluvium on clay-dominated domains to the aquifer is unlikely as they contain clay-rich layers over 5 m in thickness. Perched water bodies are likely within the sand and alluvium on clay-dominated domains which, depending on local groundwater flows and overlying seals to the groundwater body, may be susceptible to contamination.

2.3 Sustainable Urban Drainage Systems (SUDS)

SUDS are an alternative approach to conventional drainage systems, which replicate, as far as possible, the natural drainage and deal with runoff where it occurs. The successful implementation of SUDS techniques, including swales, balancing ponds and porous pavements can save money, reduce pollution and alleviate flood risk (CIRIA 2001). The system design and choice of devices depends on local factors but essentially relies on attenuation, treatment and infiltration techniques to deal with surface run-off.

The applicability of SUDS techniques to a particular geological situation can be assessed by reference to the 3D geological model. Information critical to the assessment includes the topographic slope angle, the transmissivity of the near-surface deposits, and the thickness of the unsaturated zone. Slope information can be calculated from the DTM; estimates of transmissivity for different lithologies are published in the hydrogeological literature (e.g. Allen et al. 1997; McMillan et al. 2000); an indication of the thickness of the unsaturated zone (or the depth to water table) may be derived from careful screening of first water strike as recorded in borehole logs. By combining this information as a simple tri-category map, areas more suited to infiltration techniques can be identified. Additional constraints (potential for contamination, surface sealing) can be incorporated to make the model more robust. In the example

Figure 7. Schematic representation of a SUDS suitability model. SUDS techniques are unlikely to be applicable near the river due to the effect of the slope and the close proximity of the water table to the ground surface. The sealing effect of the carpark, and clay-rich superficial deposits also make the success of SUDS techniques less likely. SUDS techniques in this setting are more likely to be successful where topographical gradients are low and the ground is unsealed and underlain by porus deposits such as sand-rich superficial deposits.

(Figure 7), the susceptibility polygons are based on present day land-use, rather than on a conventional rectangular grid. This approach allows areas of similar surface sealing (e.g. sealed: predominantly tarmac or unsealed: mostly grass) to be grouped. The preliminary results from Manchester using a simple weighting system indicates that SUDS techniques may be successful in up to 37% of the ground analyzed.

3 CONCLUSION

This paper illustrates some of the potential uses and benefits of utilizing readily available information relating to the ground in the preliminary stages of site development. The opportunity exists to provide a more relevant input of geoenvironmental data into the planning decision making process in the Manchester and Salford area, and to extend this methodology to other urban areas.

Geological and topographic data along with downhole information extracted from site investigation reports can be collated in a GIS and displayed in 3D to delineate areas that may be susceptible to a wide range of geohazards. These include ground instability and aquifer vulnerability, factors that may place additional financial and/or time costs early on in the site development process if not identified at an earlier stage. This study will provide a valuable set of background data to developers, local consultants and local planning authorities, outlining the general subsurface conditions that may be expected, so that these may be considered and, if necessary, mitigated during the planning stage of site development.

The models developed would benefit from the integration of additional datasets and this must be considered as a main area of future work in Manchester and Salford. These include the input of surface sealing data into the SUDS model, and information concerning past and present landuse and the geochemistry of the near-surface soils and fluids into the groundwater vulnerability models. A programme of ground truthing in order to adjust and refine the algorithms used (for example, defining the local transmissivity of the superficial deposits, and the properties of different surface sealing media in the urban area) would also provide a level of confidence to the historic third party data that has been used.

This paper is published with the permission of the Executive Director of the British Geological Survey (NERC).

REFERENCES

Allen, D.J., Brewerton, L.J., Coleby, L.M., Gibbs, B.R., Lewis, M.A., MacDonald, A.M., Wagstaff, S.J. & Williams, A.T. 1997. The physical properties of major aquifers in England and Wales. *British Geological Survey Technical Report* WD/97/34. *Environment Agency R&D Publication* 8, 312 pp.

Berg, R.C. 2002. Geoenvironmental mapping for groundwater protection in Illinois, USA. In P.T. Bobrowsky (ed), Geoenvironmental mapping: methods, theory and practice. Lisse: Swets and Zeitlinger B.V.

British Standards. 1990. Methods of test for soils for civil engineering purposes. *British Standards Institution*, BS 1377.

Bobrowsky, P.T. (ed). 2002. Geoenvironmental mapping: methods and practice. Lisse: Swets and Zeitlinger B.V.

Carroll, N. A sporting chance. In Manchester Focus. November 2000. Ashford: The MJ.

CIRIA. 2001. Sustainable urban drainage systems – best practice manual. Construction Industry Research and Information Association publication C523. London: CIRIA.

Culshaw, M.G. & Ellison, R.A. 2002. Geological maps: their importance in a user-driven digital age. In J.L. Van Roony & C.A. Jermy (eds), Proceedings of the 9th International Association for Engineering Geology and the Environment Congress, Durban, 16–20 September 2002. Keynote Lectures and Extended Abstracts Volume, 25–51. Pretoria: South African Institute of Engineering and Environmental Geologists [ISBN 0-620-28560-5]. Also pages 67–92 on CD Rom [ISBN 0-620-28559-1].

Ellison, R.A., Arrick, A., Strange, P.J. & Hennessey, C. 1998. Earth Science Information in support of major development initiatives. *British Geological Survey Technical Report* WA/97/84. 56pp.

Ellison, R.A., McMillan, A.A. & Lott, G.K. 2002. Ground characterization of the urban environment: a guide to best practice. *British Geological Survey Internal Report* IR/02/044. 40pp.

Environment Agency. 1996. Groundwater vulnerability of Derbyshire and north Staffordshire Sheet 17. Solihull: Environment Agency.

Forster, A., Stewart, M., Lawrence, D.J.D., Arrick, A., Cheney, C.S., Ward, R.S., Appleton, J.D., Highley, D.E., Macdonald, A.M. & Roberts, P.D. 1995. A geological background for planning and development in Wigan. A. Forster, A. Arrick, M.G. Culshaw & M. Johnston (eds). *British Geological Survey Technical Report* WN/95/3.

Harrison, C. & Petch, J.R. 1985. Ground movements in parts of Salford and Bury, Greater Manchester – aspects of urban geology. In R.H. Johnson (ed), The Geomorphology of north-west England. Manchester: Manchester University Press.

McKirdy, A.P., Thompson, A. & Poole, J. 1998. Dissemination of information on the earth sciences to planners and other decision-makers. In M.R. Bennett & P. Doyle (eds), Issues in Environmental Geology: a British Perspective. Oxford: The Geological Society of London.

McMillan, A.A., Heathcote, J.A., Klinck, B.A., Shepley, M.G., Jackson, C.P. & Degnan, P.J. 2000. Hydrogeological characterization of the onshore Quaternary sediments at Sellafield using the concept of domains. *Quarterly Journal of Engineering Geology and Hydrogeology* 33: 301–323.

Paul, M.A. & Little, J.A. 1991. Geotechnical properties of glacial deposits in lowland Britain. In J. Ehlers, P.L. Gibbard & J. Rose (eds), Glacial deposits in Great Britain and Ireland. Rotterdam: Balkema.

Thompson, A. 1998. *Environmental geology in landuse planning: A guide to good practice.* East Grinstead: Symonds Travers Morgan for the DETR.

Chinese Case Studies

Land Reclamation – Moore, Fox & Elliott (eds)
© 2003 Swets & Zeitlinger, Lisse, ISBN 90 5809 562 2

Study of the control of land desertification and ecological restoration in QUSHUI County in the watershed of the Lasa river in China

Meichen Fu, Zhenqi Hu & Jing Mi
China University of Mining and Reclamation (Beijing Campus), Beijing

ABSTRACT: Many kinds of desert land scattered throughout QUSHUI County have affected the local economy, human life and species diversity. The ecological restoration and the control of land desertification are essential measures for accelerating the development of Tibet's green economy, eliminating poverty and making the land-use sustainable. So far the desert land has reached 1.83×10^4 hm^2. The current situation and the causes of land desertification are introduced in this paper. Depending on the cause of land desertification, some earthworks, biological treatments and management measures to control desertification. Local sand-drifting is a function of atmospheric activity and local landform and the supply of loose sand. So improving the status of the vegetation of the earth's surface, closing lands for planting trees and grass, encouraging investors to control the land desertification, clarifying property rights are effective methods to control land desertification and ecological restoration.

1 INTRODUCTION

"Develop the west" is a large-scale, historical and systemic project in China. The construction of eco-environment is the base and the root of the project. Under the actions of climate change, unreasonable economic activities and the fragile ecology, the "YLN" region in Tibet has suffered land desertification or has the danger of potential desertification. So the ecological restoration and the control of land desertification is the essential measure that safeguards the acceleration of the development of Tibet's green economy, the elimination of poverty and sustainable land use (The subject group of "Studies on prevention and control of desertification (land degradation) in China" 1998). QUSHUI County of Lasa is the typical representative of the "YLN" region, and its experience and methods of desertification control can give a demonstration to the other regions in Tibet.

QUSHUI County lies in the great Fractured zone of the Brahmaputra between the Himalayas and the Nyainqentanghla, and its average altitude is 4669.5 m. Its annual average air temperature is 7.5°C. There is small difference between years, but there is larger difference between days and nights. Annual average rainfall is 444.8 mm, but there is large variation between years. Rainfall is mainly concentrated in May to September, which is 85% of the year-round rainfall. Evaporation is 2205.6 mm, sunshine hours are 3007.7 hr

per year; annual average wind speed is 2.1 m/s, and its maximum wind speed is up to 32.3 m/s. About 35 days per year experience 7–8 force strong winds.

QUSHUI County is of a semi-droughty climate of temperate zone, so there is no natural forest. The dominating vegetations of this area are frigid meadow; the second is frigid brushy meadow and brushy plants. The soil from the highest mountain to the sand beach is classified into (as) frigid frozen soil, alpine meadow soil, semi-alpine meadow soil, semi-alpine grassland soil, meadow soil, new deposited soil, moist soil and so on (Land administration bureau of Tibet Autonomous Region 1992).

2 CURRENT SITUATION AND CAUSES OF LAND DESERTIFICATION

2.1 *Current situation of land desertification in QUSHUI County*

According to the investigation of land desertification, the area of all kinds of land desertification in QUSHUI County is 1.83×10^4 ha. The area of shifting dunes is 4416.5 ha, accounting for 24.0% of the total desert area; the area of semi-fixed dunes is 3468.7 ha, accounting for 18.9% of the total desert area; the area of gravel land is 5617.1 ha, accounting for 30.7% of the total desert area.

From the distribution of different types of land desertification, we can see that the shifting dunes are mainly located on the river shoal, beach, terrace and alluvial fans and hillsides along the river area in river valley in Lasa. According to its vegetation, the land of semi-fixed dunes is classified as arbor semi-fixed dunes land, frutex semi-fixed dunes land and herbage semi-fixed dunes land. The arbor semi-fixed dunes land is mainly the under-age forest land, which was planted in recent years and mainly lies on floodplains. Frutex semi-fixed dunes land is mostly bosky sand land, which is mainly planted with sophora in gravel and located on terraces and alluvial fans along rivers of area and some relatively high river channel bars, but the area is relatively small. Herbage semi-fixed dunes land is mainly with fixed-sand grass and three assassinate grass, etc, distributed on terraces, tableland, hillside footslopes and alluvial fans. According to the investigation, the fixed dunes land can be classified in the same way with the semi-fixed dunes land. But there is no arbor fixed dunes land, and the frutex fixed dunes land only exist at the joint of Brahmaputra and Lasa River. The area of herbage fixed dunes land is relatively large.

2.2 Causes of land desertification

The causes of land desertification in Tibet are the results of natural factors and human activities. The main causes are as follows:

2.2.1 Many sandy sediments
The valley of the Brahmaputra and the lower reaches of Lasa river in QUSHUI County is large. It is usually 3–5 km wide, and the river way slopes gently (the average is 0.068%). The valley contains large spreads of mud and sand derived from upriver and the mountain slopes either side of the river. The sediments usually come out of the river in the low water season, and is drifted by wind. It becomes the primary cause of desertification. Furthermore, the topsoil of QUSHUI County, especially the surface of the cultivated soil and the natural soil, has high content of sand. Its texture is sandy soil or sand. So under the condition of excess cultivation and herding, it is easy to be drifted by wind and form desert land. So the county is in a danger of potential desertification with such large areas of sandy land and sandy sediments located at the broad valley of Brahmaputra and Lasa river.

2.2.2 Dry and windy climate
The annual average rainfall in QUSHUI County is only 444.8 mm, and its rainfall is mainly concentrated in May to September (more than 85% of the year-round rainfall). But its evaporation is 2205.6 mm per year, which is 5 times of the precipitation, so it belongs to the semi-arid climatic region, especially in

November to April when, the weather is dry and cold. QUSHUI County is located in an area of strong winds. Its annual average wind speed is 2.1 m/s; the fastest wind speed is up to 32.3 m/s. On average about 35 days have 7–8 force strong wind, with a maximum up to 65 days. Sand storms occur on average 6 days per year, reaching a maximum of 27 days. So, there is a strong driving force resulting in land desertification in QUSHUI County.

In recent years, with the warming of global weather, the climate of Tibet Plateau is becoming worse. Some glacier ice and firn is melting and the depth of the active layer in the permafrost zone is increasing. As a result soil layers becomes loose in summer which results in land sliding, aggravating the danger of potential land desertification in Tibet.

2.2.3 Sparse vegetation
Because of low temperature, long frost season and poor soil condition, few or no plants grow on the vast hillsides and weathered materials. Only some low and sparse shrub grows there. The ground cover is low, and it is easily eroded by wind, resulting in land desertification.

2.2.4 Excess and unreasonable land utilization
Because of poor natural economical condition and lagged mode of production, people lack financial and material resources, scientific and technological means to deal with the impact that the natural calamity brings to the fragile ecological environment. In some areas excessive land reclamation and cultivation, ignoring the effective natural limits of agriculture and animal husbandry. This is an important human factor in land desertification of QUSHUI County. Because of the increase of the population, the demand for the grain increases constantly. Restricted by weather condition, the cultivated land of QUSHUI County can only harvest one season in one year. In the spring and winter, the large area of cultivated land is bare without any vegetation. In addition this cultivated land is distributed in Brahmaputra and area along Lasa river mainly. The sand content of soil is high. With the wind-force, the farmland itself has potential of desertification, which results in a decline of land fertility and poor soil quality.

The whole county has 40290 hm^2 of degraded meadow. Overstocking and overgrazing cause meadows to degrade and serious desertification to occur.

Another important human factor aggravating the development of serious desertification in QUSHUI County is the excessive picking of firewood and plants causing serious destruction to natural vegetation and planted woods, etc. Because of the activities of people and domestic animals the natural vegetation is totally destroyed, the wind erosion of the ground is serious, and ditch sedimentation rapid. According to studies of this county in recent years, in total more than 5 million kilograms of bush is cut down (mainly wolf's fang stung etc.) and is used as firewood every year.

2.3 Developing tendency of land desertification

Land desertification is the result of the interaction of natural factors (abundant sand material source, arid windy weather, sparse and low vegetation etc.) and human factors (arbitrary cultivation, excess stocking, firewood gathering etc.). In 15–30 years in the future, the glacier and firn will melt gradually with the warming of global weather. The change of the human factor will become the key factor influencing desertification.

With the progress of society, the development of economy, and the increase of population, land desertification will be even more serious if essential preventative measure are not taken. The ecological environment of this county will fall into the vicious circle, and will be a great harm to the people's life and the production.

3 NECESSITY OF PREVENTION AND TREATMENT OF LAND DESERTIFICATION

3.1 Serious desertification endanger

The extensive distribution of many kinds of land desertification in QUSHUI County have already caused very significant danger and loss to the local economy and people's life and influenced the sustainable development of economy seriously.

3.1.1 The influence on agricultural production
Decrease of farmland: With gradual expansion of land desertification area, county's usable farmland and grassland has already and being reduced constantly.

Land quality is also reduced day by day. Because of desertification and wind action, the clay content and nutrition element of the earth's surface such as organic matter, nitrogen, phosphorus and potassium etc, are lost constantly. Thus, it causes land quality to be reduced day by day.

Land productivity drops. Emergence and development of the desertification not merely causes land quality to be reduced, the usable land area to be reduced, and land productivity to be lowered, it also can blow away seed, pull out seedlings, blow and reveal root systems. Aridity increases and deficient, groundwater resources also are detrimental to the growth of crops and herbs. Thus productivity levels of land drop.

3.1.2 Destruction to water conservancy projects
Water conservancy is the precondition for QUSHUI County to develop crop cultivation. But during the course of the desertification, sand blown by the wind, deposition of dust and moving of the sand dunes cause all sorts of harm and destruction to various kinds of water conservancy projects.

3.1.3 Endangerment of communications and transportation
Because of land desertification, railway and highway may be blocked with sand dune. Especially the emergence of the sandstorm in recent years has already influenced the normal operation of the air transportation of airport of Lasa.

3.1.4 Pollution on the environment
With the development of land desertification, a large amount of dust and sand material will be put into suspension by the wind and encroach on people's lives, pollute the air, sources of water and food, etc. This has directly endangered the health and survival of people and domestic animals. In addition, when dust and sand material settles and enters factory buildings, industrial and mining, it can damage the machinery and instruments, especially the precision and service life of the precision instruments. This can jeopardize the safety in production of the industrial and mining enterprises.

In a word, the danger of land desertification are many. The endangerment of desertification nearly involves the urban district of Lasa and all aspects of life of QUSHUI County. The problem of land desertification has already become an obstacle to the development of the local economy and the safeguarding of the ecological balance.

3.2 Enormous potential threats

Research indicates that land desertification not merely endangers the people's production and life, the protection and improvement of ecological environment; it but also forms a potential threat to environment and social economic development of Tibet.

The expansion of land desertification will cause following main consequences in ecological environment:

– The threat from land desertification will worsen day by day, while degradation in the environment such as ecology and soils of QUSHUI County, will become an important aspect of the environment degradation of this county;
– As desertification, increases the variety of vegetation and the height and coverage of vegetation will reduce. Thus losses of bio-diversity are the result of desertification;
– In the course of desertification, because of wind erosion and loss of soil the land resource will decline and productivity level fall;
– Desertification expansion can cause population, resource and economy imbalance, the weather to

worsen, the ecosystem to be more fragile, aggravate the emergence of the natural hazards and jeopardize social economic sustainable development greatly. Hence, the emergence and development of desertification has already posed serious potential threat to the improvement of ecological environment of this county.

3.3 *The necessity to prevent and treat land desertification*

Land desertification has already lead to serious impacts on the production and life of people. It is an enormous obstacle to the the development of economy and the improvement of the ecological environment of QUSHUI County at present. In addition people in some areas realize that it is one of the key problems in shaking off poverty in the county. Land desertification has posed enormous potential threat to the continuation and stability of the development of the social economy in the whole county. The control of the land desertification has already become a very essential and urgent task. And the control of land desertification not only is the important measure that renovates the territory, improves the ecological environment, ensures the farming and animal husbandry to develop with high yield, but also is a major issue that bring about an advance in its social economy, such that people reach a well-to-do level and benefiting descendants especially. The full realisation of the necessity, importance and urgency of the prevention and cure of desertification in this county is essential. Prevention and control of desertification in QUSHUI County as soon as possible is paramount.

4 PREVENTION AND CONTROL MEASURES OF LAND DESERTIFICATION

Prevention and control of land desertification should meet with the strategic aim of sustainable development, and take out the policy of "put prevention first, the comprehensive planning, integrated control, measures to suit local conditions, strengthen management, pay attention to benefit". There is also a need to take suitable measures to local conditions, set up defenses according to harm, combine prevention and control and development and utilization together, give top priority to what is the most important, be easy first and difficult later, implement by stages, and realize the best benefit of ecological economy progressively. Land desertification is mainly caused by the wind acting on a sandy surface. It depends on three conditions: wind speeds – greater than or equal to critical levels which can cause the wind erosion and sand blowing, sandy surfaces and sparse surface vegetation cover. The key to preventing and controlling

desertification lies in its forming conditions. In the QUSHUI district improving vegetation cover and weakening the wind directly acting on earth's surface are effective ways to fix sandy surfaces.

4.1 *Engineering measures*

Artificial sand hinders can be used on dunes. Wheat grass or highland barley can be used. The main sand hinder should be put at right angles to the main wind direction, and the secondary sand hinder should be interwoven parallel to the main wind direction. Changing slope angles between 5° and 25° with terraces, and shelter forest, also prevent soil erosion and land desertification.

4.2 *Biological measures*

With the implementation of development of the western regions of China, Tibet has begun to return the land to forestry, and grassland. This will be favorable to the ecological recovery of Tibet and the control of land desertification.

QUSHUI County's main measures to prevent and cure desertification are: directly broadcasting sand born plants in sand hinder panels before the rainy season; planting trees in shallow areas to construct a group of cubic arbor and a stretch of bush woods for cutting to surround the sand dunes and stop the movement. On the middle-under part of the sand dune facing the windward slope, planting fix-sand plants under the protection of the sand hinder, or dense lines of fix-sand plants without sand hindering protection, help to pin down the mobile dunes.

Through building of sand-fixation forest, new charcoal woods, high forest, economic forest, etc, are created. In addition the setting up of windbreak systems with a shape of strip, patch or network improves the ecological environment of farmland, reducing the wind speed, weakening the sand flow blown by the wind, and reducing soil erosion.

Based on the local conditions of QUSHUI County the selected species of plants are: white poplar, Tibet-Sichuan poplar, thin leaf red willow, long stamen willow, sea-buckthorn, sand catch Chinese scholar tree, weeping willow, wild peach, walnut, honey peach, and matrimony vine. The preferred herbs are high grasses, big flower high grasses, Sichuan wormwoods, narrow fruit grass, Tibet white fleabane, high mountain line chrysanthemum, sand-fix grass, and white grass.

4.3 *Management measures*

After planting and irrigating, the root system has not resumed, so it is difficult to absorb water. Watering effectively in time can improve the survival rate of nursery stock. After selecting and wiping the main bud, the stalwart sprout is left to promote stem growth. Through pruning and control of side shoots trunk quality can be improved. Other measures may include: strengthening propagation and education, strengthening management and law enforcement, strict supervision of the quality of the desertification control project, especially of the quality of re-vegetation with grassland and woodland, closing lands to plant trees (grass), putting some cultivated lands to recreation use, adjust the structure of animal husbandry, strengthening and raising the quality of farm animals, reducing the stocking on pasture, protection of the wildwood, returning the hillside fields above 25° to forestry (grass), pest control, especially mice and worms. In addition the adoption of many kinds of development, management styles, the setting up and amplification of the finance mechanisms makes for favorable industrial policy. By using the principle that "who plants trees and grass, who invest, who deal in, that someone is benefited", investment from each side of the society is encouraged and helps to control desertification. It is important to encourage the right people and operators to devote more efforts to investing in, accelerating the rate of ecological restoration.

4.4 Adopting many kinds of development, management styles to develop water conservancy projects

Regional precipitation is low and evaporation is high. Sunshine time is long. Hence the dry season is obvious and long. The local water resources can be fully utilized by building water conservancy projects to irrigate the woods and grasses and improve the survival rate of woods and grasses.

4.5 Accelerating the control of land desertification, by relying on science and technology

Demonstration projects of science and technology, adopting advanced desertification control technology and scientific and technical results, will help to popularize the successful experience of controlling desertification, and improving the ecological environment.

In a word a mass, social undertaking with very strong public welfare is needed to prevent and control desertification, and it is a major environmental protection and ecological restoration project.

The QUSHUI County of Lasa is only one of the typical representatives of land desertification areas of Tibet. Preventing and controlling land desertification has already became the main measure to develop the green economy of Tibet. Ecological environmental protection will play an exemplary role in the development of the western regions. We should seize the favorable opportunity of agricultural products and adopt comprehensive measures such as "conceding the land to forestry (grass), closing off hillsides for afforestation, giving relief to local residents in the form of grain, encouraging private actions" to develop the green economy of Tibet and the west of china.

ACKNOWLEDGEMENTS

*Supported by National Natural Science Foundation of China 49701010, 40071045, Ministry of Land and Resources and The Ministry of Education's "Trans-century Outstanding Young Scientist Program".

REFERENCES

The subject group of "Studies on prevention and control of desertification (the land degradation) in China". 1998. Studies on prevention and control of desertification (the land degradation) in China. Beijing: China Environmental Science Press.
Land administration bureau of Tibet Autonomous Region. 1992. The land utilization of Tibet Autonomous Region. Beijing: Science Press.

Land Reclamation – Moore, Fox & Elliott (eds)
© *2003 Swets & Zeitlinger, Lisse, ISBN 90 5809 562 2*

Perspectives on non-filling reclamation techniques for subsided land in Chinese coal mining areas

Zhenqi Hu & Changhua Liu
China University of Mining and Technology (Beijing campus), Beijing

ABSTRACT: In China, coal is the most important energy source and underground mines produce more than 96% of coal output. Coal mining has resulted a large amount of subsided lands, which has led to farmland losses and caused severe conflicts between farming and mining. So reclaiming subsided land is a very important task in China. Although the development of subsidence areas is a brand new field in China, some good technical approaches for developing subsidence areas have been presented and some great achievements have been made in recent years. Filling and non-filling methods for reclaiming subsided land are typical techniques in China. This paper reviews non-filling reclamation techniques for subsided land in Chinese coal mining area. Direct recondition, drainage and digging deep to fill shallow are typical non-filling reclamation techniques in China. The application conditions for each technique are introduced. Problems in technical process and reclaimed soils are revealed. The characteristics of reclaimed soils and their improvement should be the focus of future study in China.

1   INTRODUCTION

Coal is the main energy resource and the largest mining industry of mineral resource in China. More than 96% of coal outputs are from underground mining. With the excavation of coal from underground, severe land subsidence often occurs, which always causes huge losses of cultivatable lands. According to statistics, the area of subsiding and subsided land from coal mining is more than 400,000 ha with a increase of 22,000 ha each year (Hu 1996). Thus subsidence land reclamation has become the focus of research activities in China.

Land reclamation in China was started very late. Frequently particular emphasis has been laid on engineering practice, and little attention and study on reconstructed soil so that many reclaimed lands have low productivity with worse soil conditions and high salt contents. Usually, land reclamation techniques could be classified into two types: filling and non-filling. Filling method is the use of filling materials such as coal wastes, fly ash, etc for filling subsided lands. Non-filling methods are the one without filling materials, e.g., using digging deep to fill shallow. This paper will focuses on the non-filling techniques.

2   MAJOR NON-FILLING RECLAMATION TECHNIQUES FOR SUBSIDED LANDS IN CHINA

Three types of non-filling methods are usually used for reclaiming subsided land in China:

1. Direct reconditioning: If there is no water in some shallow subsidence areas and the subsurface water level is not very high, the method of direct reconditioning the subsided land can be used. Usually, leveling of the subsided land by bulldozers or manual work is often used. If the slope of the subsided land is large, terraces should be used. This sort of reclamation method is usually used in the mining area with low subsurface water level. Thus, it is widely used in northern China. However, there are still some subsidence lands in China that can use this method, for example, 13.6 ha. of subsidence prone areas in Feicheng, Shangdong province were directly graded.

2. Drainage: In some coal mines, the subsidence depth is not very deep and the depth of impounded water in the subsidence trough is shallow. For this case, establishment of a system of drains such as a

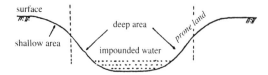

Figure 1. Profile of subsidence trough.

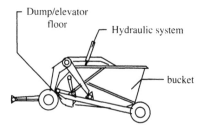

Figure 2. Towed scraper used in reclaiming subsided land in China.

trench can make the impounded water drain away and lower the subsurface water level so that the subsidence land is relatively raised. After the establishment of the system of drains, leveling the subsidence land is needed if the topography affects agricultural production very much. This technical approach is simple and cost-effective, which is mainly for reclaiming farmlands. But it requires a drainage trench which is deep and wide enough to drain impounded water away and lower the subsurface water to a available level for plant growth. In Pingdingshang coal mine, the drainage trench is 3 m wide and 2 m in depth.

3. Digging deep to fill shallow: This is a popular, simple and practical method for reclaiming the subsidence prone areas in China. It usually divides the subsidence prone area into two parts: deep and shallow (see Figure 1). The deep area is made deeper by an excavator such as Hydraulic Dredge Pump (HDP) (Hu 1994), bulldozer and towed scraper for digging a fish pond. The soil excavated from the deep area by excavators is used for filling the shallow area so that the shallow area can reach a desired elevation for crop growth. Obviously, this technical approach for developing the subsidence area is used to construct both lands and fish ponds. Therefore, the effectiveness of this reclamation method is very high, which has become the most encouraging treatment in China.

The Hydraulic Dredge Pump (HDP), in fact, is a set of machines for earthwork, which includes a high-pressure pump, hydraulic monitor, a slurry pump, two electric machines, some steel and plastic pipes, etc. It is widely used in excavating fish ponds, dredging rivers or irrigation ditches, building river banks, etc. The basic principle of this reclamation method is that: using the HDP machine, simulating the natural water erosion and turning the mechanical and electrical power into hydraulic power for digging, transporting and filling of soils. The operation procedures are: (1) Production of high-pressure and high-speed water by a high-pressure pump (usually the water speed is $50 \, m^3/h$). (2) Excavation of soils by use of hydraulic monitors with the high-pressure and high-speed water, which makes the soil become slurry. (3) Transportation of the slurry to the "shallow area". (4) Filling and settling: the slurry can be filled in the

designated "shallow area" by moving the transportation pipes, then letting the slurry settle down naturally. A drainage system is needed for quickly settling down the slurry. Generally, the filled lands need more than 5 months to settle down after the cessation of the filling work. (5) Leveling the reclaimed land by hand or bulldozers. This method has many advantages such as: the equipment is simple, the cost is low, the operation efficiency is high and the operation is convenient and not affected by weather.

A scraper towed by tractor or bulldozer is a new method to reclaim the subsided land introduced in recent years (see Figure 2) (Hu 2001). Usually the power source and bucket are small such as 58 horsepower and $3 \, m^3$. Usually the soil condition and shape of fishponds reclaimed by towed scraper is much better than that by HDP. Thus, this technique is becoming the popular method for reclaiming subsided land.

In short, reclamation techniques (1) and (2) have some limiting factors for use. Reclamation techniques (3) could be used in many situations. Thus, it has become the focus of research activities.

Non-filling reclamation has become main common technology in the restoration of subsided land due to coal mining for more than 10 years of history. Digging deep to fill shallow method has reclaimed more than $30,000 \, hm^2$ of subsided lands, and is reclaiming about $6000 \, hm^2$ per year. Thus, the ways of non-filling reclamation are mainly by hydraulic dredge pump, bulldozer and towed scraper.

3 RESEARCH FOCUS AND PROGRESS OF NON-FILLING RECLAMATION TECHNOLOGY IN CHINA

3.1 Technical process

Many reclamation projects were fulfilled by HDP method during 1983–1996. Thus the principles and technical processes of the HDP method was studied deeply (Hu 1996, 1994). As soil conditions reclaimed by HDP were poor, therefore, a new reclamation model with topsoil removal and a new technical process were developed (Hu 1994, 1996).

The 2nd mine of Liuqiao applied towed scraper to reclaim subsided land (Hu 2000). The deep area was divided into several sections that could be determined by the number of mechanics and the size of land range based on the reclamation design. Machines were used to dig earth and refill it at the same time. Fertile soil was then refilled to the top of reclaimed land with 30 cm of average thickness in the shallow area after having refilled up to a certain height. Grading and raking continued to cultivation, eventually application of fertilizer and planting was carried out. The technical process of reclaiming subsided lands by towed scraper was also developed (Hu 2000).

3.2 Physical and chemical properties of reclaimed soil

On the basis of sampling and chemical examination of soil reclaimed by hydraulic dredge pump that have been cultivated for different times, the physicochemical properties of reclaimed soils have been analyzed systemically. The temporal variability of physicochemical property of soil has also been studied.

The soils reclaimed by hydraulic dredge pump in the Xuzhou mining area in different periods and layers of land have been monitored and analyzed. The spatial-temporal variation of chemical characteristics of reclaimed soil was revealed (Chen & Dong 2000). The results indicated that the content of organic matter and nutrients tend to be lower in the surface layer, the content of salt and pH value tended to be lower in all layers during the early stage of reclamation. Over time the content of organic matter and nutrients keep increasing in the surface layer and decreasing in the ground layer, while the content of salt and pH values keep decreasing in all layers, and the levels of these chemical characteristics tend to be steady. The contents of organic matter and nutrients in the surface layer of reclaimed soils were close to those of the normal farmland soil after 13 years. The total nitrogen and organic matter of reclaimed soils were the most sensitive with time. Salt content in reclaimed soils was also sensitive with time. The results of study on the temporal and spatial variation of the physical properties of the reconstructed soils showed that in comparison with normal farmland soil, the soil texture tended to be clayey in the topsoil and sandy in the subsoil (Chen & Dong 2000). The soil bulk density tended to be higher in the topsoil and lower in the subsoil. Meanwhile, the content of soil water-stable granular structure tended to be lower. As time went on, the reclaimed soil showed a decreasing tendency in soil bulk density in the topsoil and an increasing one in the subsoil whereas the content of water-stable granular structure increased. The 0.5~3 mm granular structure

formed quicker. After 13 years of cultivation on the reclaimed land the soil bulk density and granular structure reached basically that of the normal farmland soil.

Similar studies were also conducted in Henan and Anhui Provinces. It was found that the physical and chemical characteristics of the newly reclaimed soil by HDP method were obviously poor compared with the undamaged farmland. The soil profile of the arable layer was wholly destroyed because the soil was the mixture of all the original soil layers. The newly reclaimed soil had high clay content and no distinct horizontal layers. Its characteristics of porosity, bulk density, moisture content, infiltration capability were severely damaged: the moisture content was very high; the effective capillary porosity were very little; the bulk density in the upper layers of 0–20 cm was relatively high; the cracks were found everywhere on the surface; the infiltration rate was much slower. And the newly reclaimed soil had much lower contents of organic matter and N, P, K elements because of soil layers' hydraulic mixing and nutrient loss with the drainage. All these above affect crop growth significantly.

After several years of cultivation, the physical and chemical characteristics of the reclaimed soil by HDP method were improved year after year, especially on the upper layer of 0–20 cm. The moisture content, infiltration capability, porosity and bulk density became relatively suitable for planting. And because of the cultivation and fertilization, the contents of organic matter and N, P, K elements in the reclaimed soil were gradually enhanced. But compared with the high quality farmland, the difference was obvious: the content of organic matter, the content of total N, P, K and the content of rapidly available nutrients of N, P, K were still very low. The changes were even smaller in the layer below 20 cm. Whilst the elements of P and K can be improved in the short term, the contents of organic matter and element of N can not be enhanced rapidly.

However, the studies on the new soil formed by the non-filling reclamation technology were very few, which directly influenced the promotion of the productivity of reclaimed soil and the innovation of reclamation procedure. Another problem was that the soil characteristics that had been studied usually came from fewer samples with few parameters and inadequate analysis. There was much less study on the characters of soil reconstructed by the other non-filling reclamation methods. Temporal and spatial variation of the soil properties and compaction problems of reconstructed soils need to be researched deeply. Soil improvements for reclaimed lands have also not been studied in depth. Recently, use of fly ash and sewage slurry for improving soil quality is under study in China.

387

4 CONCLUSIONS AND PROSPECTS

Non-filling reclamation techniques have become an important method for reclaiming subsided land. Digging deep to fill shallow is the popular non-filling reclamation technique in China compared to drainage and direct recondition methods. Although the development of subsidence areas is a brand new field in China, some good technical approaches for developing subsidence areas have been presented and some great achievements have been made in recent years. From the practice of the development, we recognize that the selection of the technical approaches developing subsidence areas should be based on the subsidence characteristics and the local social and economical conditions. Meanwhile, by analyzing and assessing comprehensively the progress of non-filling reclamation research in China, it is concluded that the present research is in its early stages. Thus, we will have to pay more attention to the physical, chemical and biological characteristics of reclaimed soil and its improvement.

ACKNOWLEDGEMENTS

Supported by National Natural Science Foundation of China 49701010, 40071045, Ministry of Land and Resources and The Ministry of Education's "Trans-century Outstanding Young Scientist Program."

REFERENCES

Hu, Zhenqi. He, R. & Wei, Z. 2001. New land reclamation technology for subsidence ground. *Coal Science and Technology* Vol. 29 (1):17–19 (in Chinese).

Hu, Zhenqi. 1996. *Administration and Reclamation of Subsidence Land Resource due to Coal Mining.* Beijing: Coal Industry Press.

Hu, Zhenqi. 1994. The technique of reclaiming subsidence areas by use of a hydraulic dredge pump in Chinese coal mines, *International Journal of Surface Mining, Reclamation and Environment* 8(4):137–140.

Chen, L. & Deng, K. 2000. Spatial-temporal variation of chemical properties of soil reclaimed by hydraulic dredge pump in mining areas. *Journal of China University of Mining & Technology* 29(3):262–265.

Author index